ELECTRICAL MACHINES

Jaydeep Chakravorty
Head
Department of Electrical Engineering
Indus University
Ahmedabad

Universities Press

ELECTRICAL MACHINES

UNIVERSITIES PRESS (INDIA) PRIVATE LIMITED

Registered office
3-6-747/1/A & 3-6-754/1, Himayatnagar, Hyderabad 500 029, Telangana, India
info@universitiespress.com; www.universitiespress.com

Distributed by
Orient Blackswan Private limited

Registered office
3-6-752 Himayatnagar, Hyderabad 500 029, Telangana, India

Other offices
Bengaluru, Bhopal, Chennai, Guwahati, Hyderabad, Jaipur, Kolkata, Lucknow, Mumbai, New Delhi, Noida, Patna, Vijayawada

ISBN: 978-93-86235-18-3

Cover and book design
© Universities Press (India) Private Ltd 2017

Typeset in Times New Roman PS 10.5/12.5 by
Cameo Corporate Services, Coimbatore, Tamil Nadu, India

Printed at Rajiv Book Binding House, Delhi

Published by
Universities Press (India) Private Limited
3-6-747/1/A & 3-6-754/1, Himayatnagar, Hyderabad 500 029, Telangana, India

Dedicated to my family

Dedicated to my family

Table of Contents

Preface

This book aims to provide a strong foundation in the field of electrical machines for B.E./ B.Tech. students who study this subject as part of their curriculum in electrical, electronics and industrial engineering courses. It will also be useful to candidates appearing for AMIE, IETE, GATE, UPSC and other professional entrance examinations.

The exposition in each chapter has been developed in a systematic manner with emphasis on clarity, building on the basic concepts, step by step. The book lays out several solved problems in the course of the discussion to drive home the theory and augments them with unsolved problems for practice at the end of each chapter. In addition, there are also multiple choice questions and short questions provided for the student's benefit.

Spread across 11 chapters, the book has been written in such a way that no prior knowledge of the topics is required to understand the subject. Chapter 1 deals with electromechanical energy conversion. It explains the behaviour of magnetic circuits as they appear in a transformer and in other electrical machines, serving as an intermediary stage in energy conversion process. Chapter 2 explains the theory behind DC generators, their construction, operation, design, parallel operations and applications. Chapter 3 describes the function of a DC motor in detail, focussing on its characteristics, starting methods, speed control and braking methods. Chapters 4, 5 and 6 explain the working of the electrical transformer in detail, with an exclusive chapter each, on single-phase transformer, three-phase transformer and special transformer respectively. Chapters 7 and 8 delve into the principles governing a three-phase induction motor and a single-phase induction motor, elucidating their construction, operation, speed control and starting methods. Chapter 9 expounds on the synchronous generator while Chapter 10 analyses the functions of a synchronous motor. Synchronous generator and synchronous motor have been discussed separately in two different chapters to enable students understand the difference between the two clearly. Chapter 11, the last chapter of the book, deals with special machines, machines that have some specific applications, and machines having some unique construction or operations.

While every effort has been made to make this work error-free, it is possible that inadvertent lapses may have crept in. Such flaws, if noticed, may please be pointed out to the author or the publisher.

Jaydeep Chakravorty

Acknowledgements

I am thankful to Mr Pradip Shrivastava, Hon'ble Chancellor, Baddi University, and Prof. (Dr) Shakti Kumar, Hon'ble Vice-chancellor, Baddi University, for their support and encouragement in writing this book. I am obliged to Prof. (Dr) Sandeep Chakravorty, Dean–Academics, Baddi University, for providing the necessary technical support. I am indebted to Prof. (Dr) Vinay Bhatia, Dean–SW, for his constant encouragement. I thank Dr Ajay Goel (HOD–CSE), for providing IT support to this endeavour as and when required. I thank Dr Ishbir Singh, Head, Department of Mechanical Engineering, for helping me with the manuscript.

I am grateful to my parents Mr P Chakravorty and Ms J Chakravorty for their constant encouragement while I was writing this book. I also thank my brother Mr S Chakravorty and his wife Ms S Chakravorty for their constant encouragement. Last but not the least, I thank my wife Ms S Chakravorty for helping me in editing and also in typing this book.

Jaydeep Chakravorty

CHAPTER 1

ELECTROMECHANICAL ENERGY CONVERSION

1.1 INTRODUCTION

Electromechanical devices convert energy from electrical to mechanical form and vice-versa. These devices can be classified into the following three categories

i) Transducers
ii) Devices that can produce force, for example, solenoids, relays etc.
iii) Energy conversion equipments, for example, generators, motor, etc.

Electromechanical energy conversion theory helps in representing electromagnetic force or torque in terms of such parameters as electric current or displacement. Electromechanical conversion takes place through electrical or magnetic field or through suitable conversion devices.

1.2 EXPRESSION FOR INDUCED E.M.F.

An alternating e.m.f. can be generated in two ways,

i) When a coil is rotated in a stationary magnetic field.
ii) When a stationary coil is placed in a rotating magnetic field.

The magnitude of e.m.f. generated will depend on the number of turns in the coil, strength of the magnetic field and the speed of rotation of the coil or the magnetic field.

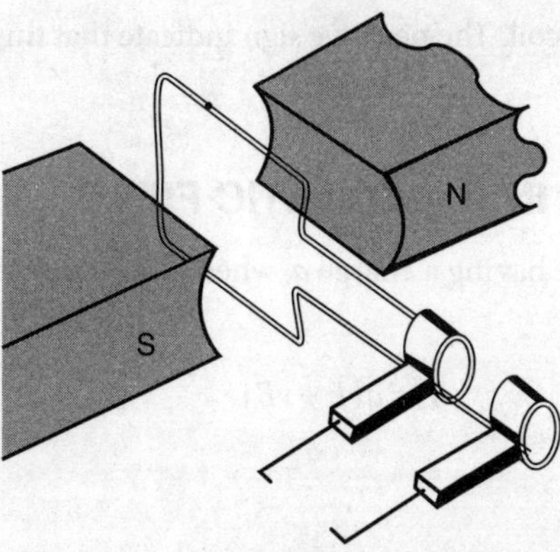

FIG.1.1: Coil rotating in a magnetic field

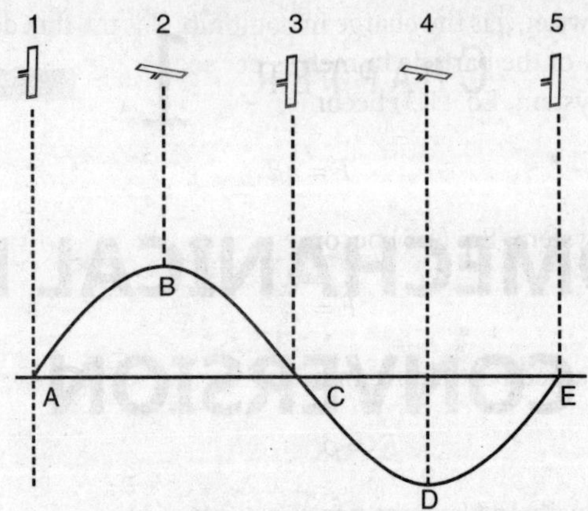

FIG.1.2: Waveform of the induced e.m.f.

Let us consider a multi-turn coil of length l metres, width b metres and having T turns rotated at an angular speed $\omega = 2\pi n$ rad / sec inside a magnetic field as shown in Fig.1.1. Here, n is the speed of the coil in r.p.s. (revolutions per second).

Let,

B be the flux density in the air gap.

$N = 60 \times n$ be the speed of the rotation of the coil in r.p.m. (revolutions per minute)

$\vartheta = \pi bn$ be the tangential velocity of the coil.

At a particular instant of time (t), the coil will make an angle of θ radian with respect to the horizontal. Here, $\theta = \omega t$ radian, is the angular displacement. Then, the e.m.f. induced on one side of the coil will be $Blv \sin\theta$. So, the e.m.f. induced in a single turn coil will be $2Blv\sin\theta = 2Bl\pi bn\sin\theta$.

Hence, the total e.m.f. induced in a multi turn coil will be

$$E = T \times 2Bl\pi bn \sin\theta = E_m \sin\theta \quad (1.1)$$

Where,

$E_m = 2TBl\pi bn$ is the maximum value of induced e.m.f.

The direction of induced e.m.f. is obtained by "Flemings left-hand rule". If the coil is having T turns, then the induced e.m.f. is also obtained by the expression,

$$e = -T\frac{d\varphi}{dt} \quad (1.2)$$

Where, φ is the flux linkage in the coil. The negative sign indicate that this e.m.f. will oppose the cause of its production.

1.3 FORCE AND TORQUE IN A MAGNETIC FIELD

The force acting on a moving particle having a charge q, when placed in a magnetic field, is given by the Lorentz's force law. Mathematically,

$$F = q(E + vB) \quad (1.3)$$

Where, F is the force in Newton, q is the charge in coulomb, B is the flux density in Tesla, E is the volts per meter and v is the velocity of the particle in metres per sec.

For a pure electrical field system, Eq. (1.3) becomes

$$F = qE \tag{1.4}$$

For a pure magnetic field system, Eq. (1.3) becomes

$$F = qvB \tag{1.5}$$

When large numbers of charged particle are in motion, then Eq. (1.3) becomes,

$$F_v = \rho\left(E + vB\right) \tag{1.6}$$

Where, F_v is the current density in Newtons per cubic metre,

The product ρv is called the current density and is denoted by J, that is,

$$J = \rho v \text{ amp/sq metre} \tag{1.7}$$

So, for a pure magnetic field system

$$F_v = JB \tag{1.8}$$

1.4 ENERGY BALANCE CONCEPT

The law of principle of conservation of energy states that, *energy cannot be created or destroyed but it is possible to transform energy from one form to the some other form.*

In this transformation of energy from one form to another, there will be always some loss (energy loss) due to practical devices. Thus, we can say that it is not possible to transform the entire energy from one form to the other, some energy will always be lost in the transformation process. Some part of the energy also gets stored in a medium like magnetic field during the conversion process. This part of energy that gets stored is called storage energy.

Based on these principles of transformation and storage of energy, an energy balance equation now can be written as follows,

For motoring action:

Energy input = Mechanical output + Energy loss + Total energy stored.

For generating action:

Mechanical input = Electrical output + Energy loss + Total energy stored.

This equation will have a positive value for motoring action and a negative value for generating action. The analysis of electromechanical devices can be done based on such energy balance equation.

1.5 LAWS OF ELECTROMAGNETISM

There are four laws of electromagnetism:

i) Faraday's law.
ii) Lenz's law

iii) The Biot–Savart law
(iv) Ampere's law

1.5.1 Faraday's Laws of Electromagnetic Induction

Faraday's First Law: When the magnetic flux linking with the coil changes, an e.m.f. is induced in the coil.

Faraday's Second Law: The magnitude of induced e.m.f. is the rate of change of flux linkage in it. Mathematically,

$$\text{Induced e.m.f.}(e) = N\frac{\varphi_2 - \varphi_1}{t} \tag{1.9}$$

$$\text{or, induced e.m.f.}(e) = -N\frac{d\varphi}{dt} \tag{1.10}$$

Where,
φ_1 is the initial value of the flux.
φ_2 is the new value of the flux after time t.
N is the number of turns in the coil.

1.5.2 Lenz's Law

Lenz's law states that the direction of induced e.m.f. in the conductor will always oppose the cause of its production. The negative sign in Eq. (1.10) signifies this effect.

1.5.3 The Biot–Savart Law

According to Biot–Savart law, any current element projects into space a magnetic field, the magnetic flux density of which, dB, is directly proportional to the length of the element dl, the current I, the sine of the angle θ between direction of the current and the vector joining a given point of the field and the current element and is inversely proportional to the square of the distance of the given point from the current element, r.

$$dB = k\frac{Idl\sin\theta}{r^2} \tag{1.11}$$

where, k is a constant.

1.5.4 Ampere's Law

Ampere's law states that for any closed loop, the sum of the length elements times the magnetic field in the direction of the length element is equal to the permeability times the electric current enclosed in the loop.

The formula for this law is

$$\oint \vec{B}d\vec{l} = \mu_0 I$$

Where,
μ_0 is the relative permeability of the medium
B is the magnetic field
l is the infinitesimal length of conductor
I is the current flowing in the loop

1.6 INDUCED E.M.F.

When a conductor is placed in a varying magnetic field, an e.m.f. is induced in the coil. This induced e.m.f. is classified into two categories.

Dynamically induced e.m.f.: When the conductor is in motion and the field is stationary an e.m.f. is induced in the coil. This e.m.f. is called the dynamically induced e.m.f.

Statically induced e.m.f.: When the conductor is stationary and the field is changing, an e.m.f. is induced in the coil. This e.m.f. is called the statically induced e.m.f. This statically induced e.m.f. is further classified as follows,

(a) Self-induced e.m.f.
(b) Mutually induced e.m.f.

1.6.1 Self-induced e.m.f.

Self inductance is a phenomenon by which a change in the current flowing through an inductive coil changes, produces an induced e.m.f. in the coil. Mathematically, it can be written as

$$\text{Induced e.m.f.} = -L\frac{di}{dt} \tag{1.12}$$

Where, L is the self inductance of the coil.

1.6.2 Mutually induced e.m.f.

It is the phenomenon by which a change of current in one coil causes an induced e.m.f. in another coil placed near to the first coil. The coil in which current is changed is called primary coil and the coil in which e.m.f. is induced is called secondary coil. Mathematically, the mutually induced e.m.f. in the secondary coil due to the change of current (I_p) in the primary coil is given by the expression

$$\text{Mutually induced e.m.f.} = -M\frac{dI_P}{dt} \tag{1.13}$$

Where, M is the mutual inductance of the two coils.

1.7 ENERGY STORED IN A MAGNETIC FIELD

Consider a coil with N turns wound on a magnetic material. This magnetic material will act as a core. When the coil is supplied with a voltage of v volts, current i will flow through it. Let R be the resistance of the coil. Then from KVL we can write,

$$v = iR + e \tag{1.14}$$

Where, iR is the voltage drop across the resistance and e is the voltage drop across the coil.

The power equation will be

$$P = vi = i^2R + ei \tag{1.15}$$

Let a DC voltage be applied to the coil during the period $t = 0$ to t = T. During this period, let the current be changed to I. Then the input energy will be,

$$W_i = \int_0^T P dt = \int_0^T \left(i^2R + ei\right)dt = \int_0^T i^2R\,dt + \int_0^T ei\,dt \tag{1.16}$$

Thus, the input energy has two parts:

$\int_0^T i^2 R dt$ is the energy lost due to the loss in resistance which is present in the coil.

$\int_0^T ei dt$ is the energy stored in the magnetic field.

Let,

$$W_f = \int_0^T ei dt \tag{1.17}$$

Again, from Faraday's law, we can write,

$$e = \frac{d(N\varphi)}{dt} \tag{1.18}$$

Substituting, we get

$$W_f = \int_0^T \frac{d(N\varphi)}{dt} idt = \int_0^{\varphi_m} Nid\varphi = \int_0^{\varphi_m} Fd\varphi \tag{1.19}$$

Where, $F = Ni$ = mmf.

1.8 SINGLY EXCITED LINEAR ACTUATOR

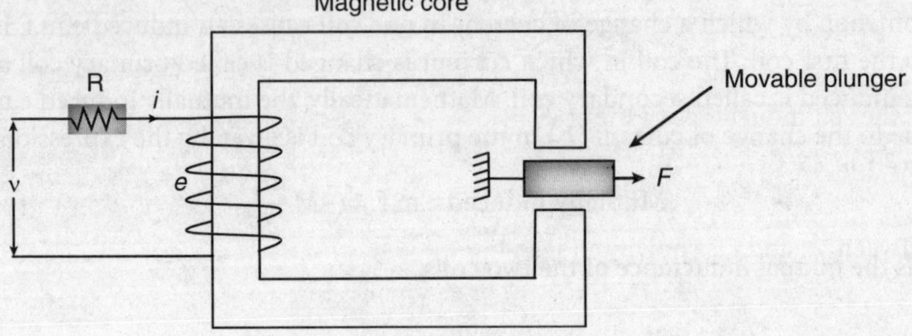

Fig.1.3: Linear actuator

Let us consider a linear actuator as shown in Fig.1.3. A coil having resistance R is wound on a magnetic core. When this coil is excited with a voltage v, current i will flow through the coil. Let us assume that after some time dt, the movable plunger is moved by a distance dx under the action of force F.

Then, the mechanical work done will be,

$$dw_m = Fdx \tag{1.20}$$

The electrical energy that has been transferred into the magnetic field and converted into mechanical work will be,

$$dw_e = dw_f + dw_m = vidt - Ri^2 dt \tag{1.21}$$

where,

dw_m is the differential mechanical energy output.

dw_f is the differential change in magnetic storage energy.
dw_e is the differential electrical energy input.

From Faraday's law, we can write,

$$e = \frac{d\lambda}{dt} = v - Ri \tag{1.22}$$

where, λ is the flux linkage.

Again,

$$dw_f = dw_e - dw_m = eidt - Fdx$$

$$= id\lambda - Fdx \tag{1.23}$$

From the above equation, we can conclude that the energy stored in the magnetic field is a function of the flux linkage and the position of the plunger.

When the flux linkage is increased from 0 to λ, the energy stored in the field will be,

$$w_f = \int_0^\lambda id\lambda \tag{1.24}$$

For a linear system,

$$w_f = \frac{1}{2}\frac{\lambda^2}{L} \tag{1.25}$$

The force acting on the plunger will be,

$$F = \frac{\delta w_f}{\delta x} = \frac{1}{2}\left(\frac{\lambda}{L}\right)^2 \frac{dL}{dx} = \frac{1}{2}i^2\frac{dL}{dx} \tag{1.26}$$

Where, L is the self inductance of the coil.

$$L = \frac{\lambda}{i} \tag{1.27}$$

The plot between λ and i is shown in Fig.1.4

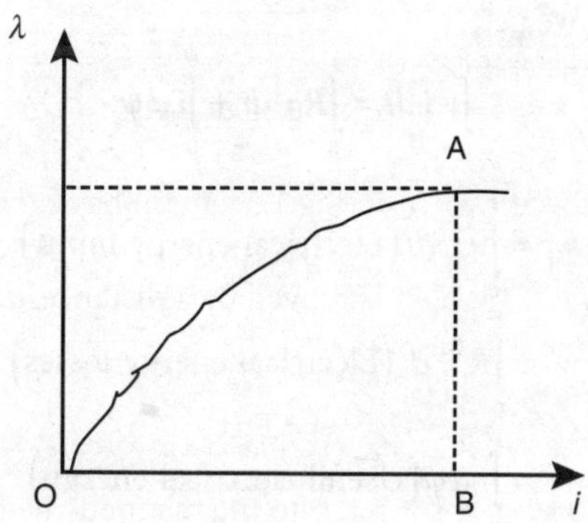

Fig.1.4: Plot of flux linkage and current

The area OABO is called the co-energy and is denoted by w'_f. Co-energy has no physical significance but it is important in obtaining the magnetic force.

1.9 SINGLE EXCITED ROTATING SYSTEM

Consider a single excited rotating system as shown in Fig.1.5

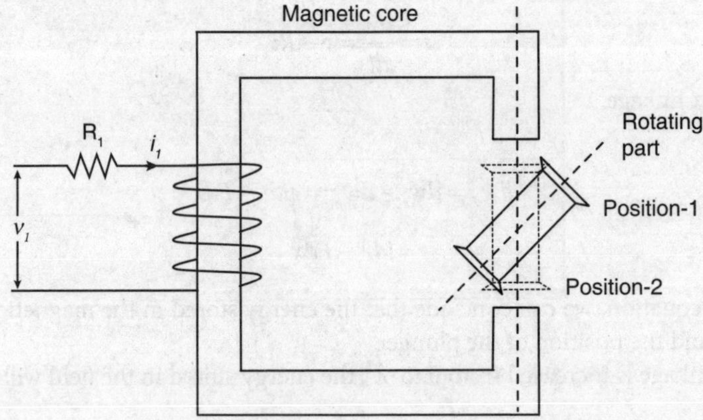

Fig.1.5: Single excited rotating system

A coil having resistance R_1 is wound on a magnetic core. When this coil is excited with a voltage v_1 current i_1 flows through the coil. Let us assume that all the flux produced by the coil will pass through the core, that is, there is no leakage flux.

The movable part will experience torque and will try to position itself in such a way that minimum reluctance is given to the magnetic flux.

From KVL, we can write,

$$v_1 = R_1 i_1 + \frac{d\psi}{dt} \tag{1.28}$$

where, ψ is the flux linkage

Multiplying the equation by i_1 on both sides we get,

$$v_1 i_1 = R_1 i_1^2 + i_1 \frac{d\psi}{dt} \tag{1.29}$$

Integrating with respect to dt, we get,

$$\int_0^t v_1 i_1 dt = \int_0^t R_1 i_1^2 dt + \int_0^\psi i_1 d\psi \tag{1.30}$$

Let,

$$w_e = \int_0^t v_1 i_1 \, dt \left(\text{Electrical energy input}\right) \tag{1.31}$$

$$w_l = \int_0^t R_1 i_1^2 \, dt \left(\text{Electrical energy losses}\right) \tag{1.32}$$

$$w_f = \int_0^\psi i_1 d\psi \left(\text{Useful electrical energy}\right) \tag{1.33}$$

Then we can write,

$$w_e = w_l + w_f \tag{1.34}$$

Again,

$$w_f = \int_0^\psi i_1 d\psi = \int_0^\psi \frac{\psi}{L} d\psi = \frac{\psi^2}{2L} = \frac{(Li_1)^2}{2L} = \frac{1}{2} L i_1^2 \left(\because \psi = L i_1\right) \tag{1.35}$$

The plot of flux linkage (ψ) versus current i_1 is shown in Fig.1.6

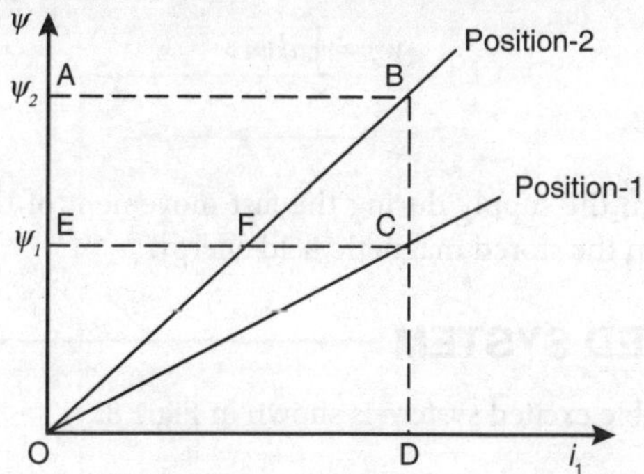

FIG.1.6: Plot of flux linkage versus current

Area OCDO and OBAO is the stored energy when the movable rotor is in Position-1 and Position-2 respectively. Let us assume that the rotor is initially positioned at Position-1 (operating point C in the Fig.1.6). At this position, let the current through the circuit be i_1. Let us assume that the rotor is moving very slowly towards Position-2. Since the movement of the rotor is very slow, the current through the circuit will remain constant (i_1). When the rotor reaches Position-2, the flux linkage changes from ψ_1 to ψ_2.

Then,

$$w_f = \text{Area}\, OFBAEO - \text{Area}\, OCFEO$$

$$w_e = \text{Area}\, CBAEFC$$

$$w_l = \text{Area}\, OCBFO$$

Then we can write,

$$w_e = w_l + w_f \tag{1.36}$$

The electromagnetic torque will be,

$$\tau = \frac{i_1^2}{2}\frac{dL}{d\theta} \tag{1.37}$$

Now, if the rotor is moved very fast from Position-1 to Position-2, the flux linkage cannot change so fast. So at this instant, the operating point moves horizontally from C to F. The modified plot is shown in Fig.1.7

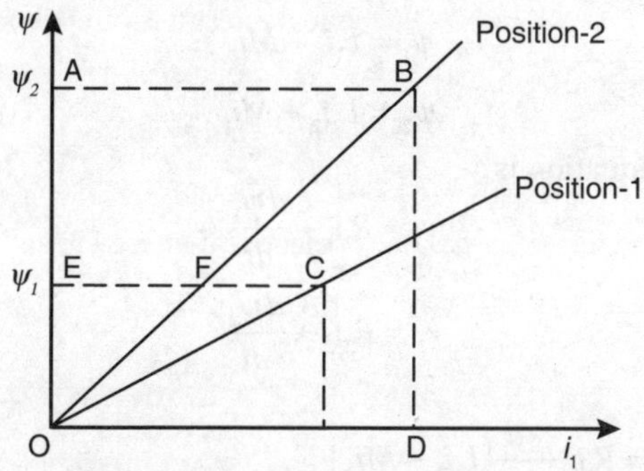

FIG.1.7: Plot of modified flux linkage versus current

In due course of time the flux linkage ψ and current i_1 will reach the steady state value (Position-B). Again,

$$w_e = \int_{\psi_1}^{\psi_2} i\,d\psi \tag{1.38}$$

But at this position, $\psi_1 = \psi_2$

So, $w_e = 0$

So, no energy is taken from the supply during the fast movement of the rotor and the mechanical energy output is obtained from the stored magnetic field energy.

1.10 DOUBLE EXCITED SYSTEM

A schematic diagram of a double excited system is shown in Fig.1.8.

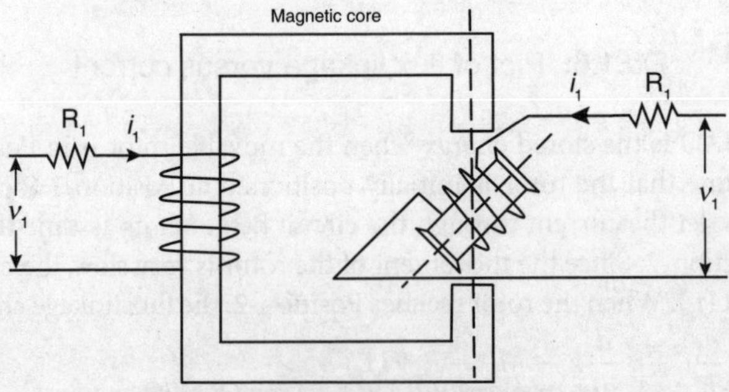

Fig.1.8: Double excited system

The movable part is called the rotor and the stationary part is called the stator. Let us assume that the stator and the rotor are having N_s and N_r number of turns respectively. Most of the electromechanical energy conversion devices have multiple excited magnetic field system, for example, DC machine, synchronous machines etc.

As shown in the circuit (Fig.1.8), the stator is fed from a voltage source of v_1 because of which current i_1 flows through it. Let R_1 be the resistance of the stator winding. Similarly, the rotor is fed from a voltage source of v_2 because of which current i_2 is flowing through it. Let R_2 be the resistance of the rotor winding. Let the self inductance of the stator and the rotor be L_1 and L_2 respectively and M be the mutual inductance between them.

The flux linkage equations for the two windings are

$$\psi_1 = L_1 i_1 + M i_2 \tag{1.39}$$

$$\psi_2 = L_2 i_2 + M i_1 \tag{1.40}$$

The instantaneous voltage equation is

$$v_1 = R_1 i_1 + \frac{d\psi_1}{dt} \tag{1.41}$$

$$v_2 = R_2 i_2 + \frac{d\psi_2}{dt} \tag{1.42}$$

Substituting we get,

$$v_1 = R_1 i_1 + \frac{d}{dt}\left[L_1 i_1 + M i_2\right]$$

$$= R_1 i_1 + \frac{d}{dt}(L_1 i_1) + \frac{d}{dt}(M i_2)$$

$$= R_1 i_1 + L_1 \frac{d}{dt} i_1 + i_1 \frac{d}{dt} L_1 + M \frac{d}{dt} i_2 + i_2 \frac{d}{dt} M \tag{1.43}$$

$$v_2 = R_2 i_2 + \frac{d}{dt}[L_2 i_2 + M i_1]$$

$$= R_2 i_2 + \frac{d}{dt}(L_2 i_2) + \frac{d}{dt}(M i_1)$$

$$= R_2 i_2 + L_2 \frac{d}{dt} i_2 + i_2 \frac{d}{dt} L_2 + M \frac{d}{dt} i_1 + i_1 \frac{d}{dt} M \tag{1.44}$$

Multiplying Eq. 1.43 by i_1 we get,

$$v_1 i_1 = R_1 i_1^2 + L_1 i_1 \frac{d}{dt} i_1 + i_1^2 \frac{d}{dt} L_1 + M i_1 \frac{d}{dt} i_2 + i_1 i_2 \frac{d}{dt} M \tag{1.45}$$

Multiplying Eq. 1.44 by i_2 we get,

$$v_2 i_2 = R_2 i_2^2 + L_2 i_2 \frac{d}{dt} i_2 + i_2^2 \frac{d}{dt} L_2 + M i_2 \frac{d}{dt} i_1 + i_1 i_2 \frac{d}{dt} M \tag{1.46}$$

Integrating Eq. (1.45) and Eq. (1.46) with respect to dt and adding we get,

$$\int (v_1 i_1 + v_2 i_2) dt = \int (R_1 i_1^2 + R_2 i_2^2) dt +$$
$$\int \left(L_1 i_1 \frac{d}{dt} i_1 + i_1^2 \frac{d}{dt} L_1 + M i_1 \frac{d}{dt} i_2 + i_1 i_2 \frac{d}{dt} M + L_2 i_2 \frac{d}{dt} i_2 + i_2^2 \frac{d}{dt} L_2 + M i_2 \frac{d}{dt} i_1 + i_1 i_2 \frac{d}{dt} M \right) dt$$

or
$$\int (v_1 i_1 + v_2 i_2) dt - \int (R_1 i_1^2 + R_2 i_2^2) dt =$$
$$\int \left(L_1 i_1 \frac{d}{dt} i_1 + i_1^2 \frac{d}{dt} L_1 + M i_1 \frac{d}{dt} i_2 + i_1 i_2 \frac{d}{dt} M + L_2 i_2 \frac{d}{dt} i_2 + i_2^2 \frac{d}{dt} L_2 + M i_2 \frac{d}{dt} i_1 + i_1 i_2 \frac{d}{dt} M \right) dt \tag{1.47}$$

The difference,

$\int (v_1 i_1 + v_2 i_2) dt - \int (R_1 i_1^2 + R_2 i_2^2) dt$ is called the useful electrical energy input.

$\int \left(L_1 i_1 \frac{d}{dt} i_1 + i_1^2 \frac{d}{dt} L_1 + M i_1 \frac{d}{dt} i_2 + i_1 i_2 \frac{d}{dt} M + L_2 i_2 \frac{d}{dt} i_2 + i_2^2 \frac{d}{dt} L_2 + M i_2 \frac{d}{dt} i_1 + i_1 i_2 \frac{d}{dt} M \right) dt$ is the combination of energy to field storage in the electrical system and electrical to mechanical energy.

To determine the mechanical torque produced, let us assume that the rotor has been displaced by a small angle $d\theta$. Then, the differential electrical energy input during this movement is given by,

$$dw_e = i_1 d(L_1 i_1 + M i_2) + i_2 d(L_2 i_2 + M i_1)$$

$$= i_1^2 dL_1 + L_1 i_1 di_1 + i_1 M di_2 + i_1 i_2 dM + i_2^2 dL_2 + L_2 i_2 di_2 + i_2 M di_1 + i_1 i_2 dM \tag{1.48}$$

The differential magnetic energy stored during this movement is given by,

$$dw_f = \frac{1}{2} i_1^2 dL_1 + L_1 i_1 di_1 + M i_1 di_2 + \frac{1}{2} i_2^2 dL_2 + L_2 i_2 di_2 + M i_2 di_1 + i_1 i_2 dM \tag{1.49}$$

The differential mechanical work done, is given by,

$$dw_l = T d\theta$$

We know that,

$$dw_l = dw_e - dw_f$$

Substituting we get,

$$Td\theta = \left[i_1^2 dL_1 + L_1 i_1 di_1 + i_1 M di_2 + i_1 i_1 dM + i_2^2 dL_2 + L_2 i_2 di_2 + i_2 M di_1 + i_1 i_2 dM\right]$$
$$-\left[\frac{1}{2} i_1^2 dL_1 + L_1 i_1 di_1 + M i_1 di_2 + \frac{1}{2} i_2^2 dL_2 + L_2 i_2 di_2 + M i_2 di_1 + i_1 i_2 dM\right]$$

$$or\, T = \frac{1}{2} i_1^2 \frac{dL_1}{d\theta} + \frac{1}{2} i_2^2 \frac{dL_2}{d\theta} + i_1 i_1 \frac{dM}{d\theta} \tag{1.50}$$

1.11 DYNAMIC EQUATION OF ELECTROMECHANICAL SYSTEMS

Let us consider a simple transducer system, as shown in Fig.1.9. When the push-button switch is closed, the solenoid moves towards the right. The spring action (k) brings the solenoid back to its original position when pressure on the push-button switch is released.

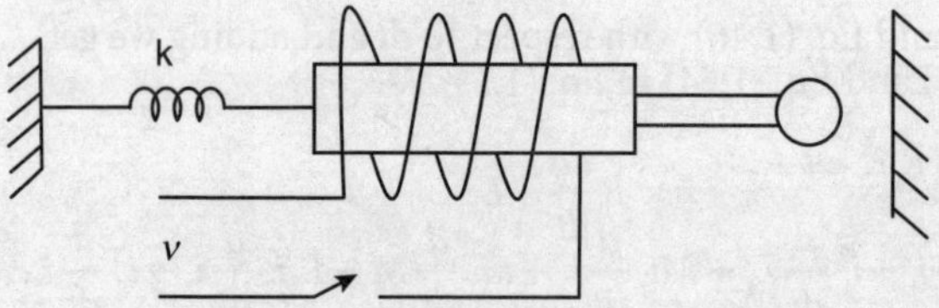

Fig.1.9: Transducer system

This electromechanical system can be divided into three parts

- An electrical circuit. (Fig.1.10 (a))
- A mechanical circuit (Fig.1.10 (b))
- Transducer (Fig.1.10 (c))

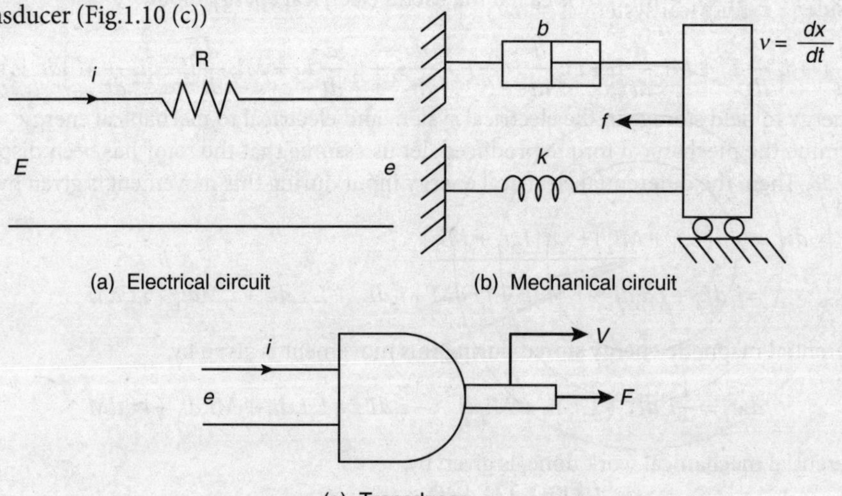

(a) Electrical circuit

(b) Mechanical circuit

(c) Transducer

Fig.1.10:

From the electrical circuit we can write,

$$E = Ri + e = R\frac{dq}{dt} + e \tag{1.51}$$

Where, q is the charge and e is the voltage that appears across the transducer.

From the mechanical circuit, using constitutive relations for the spring and the dashpot, we get,

$$m\frac{d^2x}{dt^2} = -f - kx - b\frac{dx}{dt} \tag{1.52}$$

where, f is the force on the transducer.

From the transducer circuit we get,

$$e = \frac{d\lambda}{dt} = \frac{d}{dt}(Li) = L\frac{d^2q}{dt^2} + L'\frac{dq}{dt}\frac{dx}{dt} \tag{1.53}$$

And

$$f = \frac{1}{2}L'\left(\frac{dq}{dt}\right)^2 \tag{1.54}$$

Substituting Eq. (1.53) and Eq. (1.54) in Eq. (1.51) and Eq. (1.52), we get,

$$E = R\frac{dq}{dt} + L\frac{d^2q}{dt^2} + L'\frac{dq}{dt}\frac{dx}{dt} \tag{1.55}$$

and

$$m\frac{d^2x}{dt^2} + b\frac{dx}{dt} + kx = \frac{1}{2}L'\left(\frac{dq}{dt}\right)^2 \tag{1.56}$$

This is the direct method of solving an electromechanical system.

1.12 THE TORQUE EQUATION OF A MACHINE WITH CYLINDRICAL AIR GAP

Let us consider a cylindrical system as show in Fig.1.11.

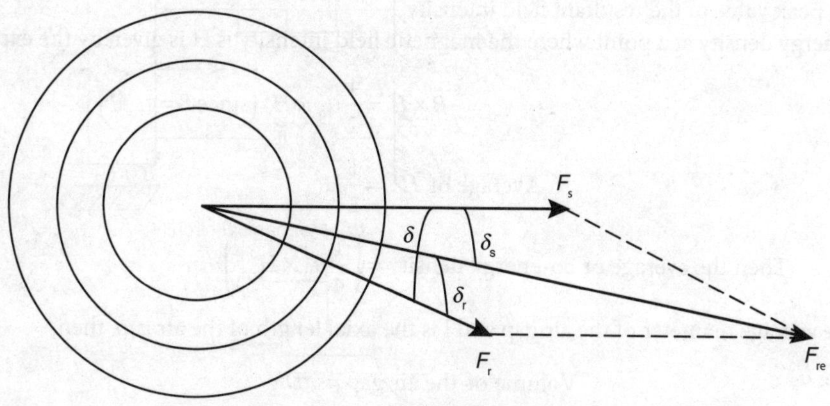

Fig.1.11: Cylindrical system

This system can be considered of having two windings, the stator winding and the rotor winding. The stator is generally the stationary part of the system and the rotor is the rotating part of the system. In AC machines both the stator and the rotor windings carry currents and thus produce their own magnetic fields. Torque is produced due to the tendency of these two magnetic fields to line up in the same direction, which in turn makes the rotor to rotate. Voltage is produced due to the relative motion between the two windings.

To derive the torque equations, the following assumptions are made

- The air gap in the round rotor machine is uniform
- The m.m.f. produced by the stator and the rotor are sinusoidal in nature.
- The tangent component of the magnetic field in the air gap is negligible compared to radial component.
- The radial length *g* of the air gap is very small compared with the radius of the rotor and the stator.
- The reluctance of the iron path in negligible.
- The flux produced by both the windings crosses the air gap and links with the other windings.

In the Fig.1.11, F_r is the axis of the rotor field and F_s is the axis of the stator field and the angle between then is δ. Therefore, the resultant of the fields will be,

$$F_{res}^2 = F_r^2 + F_s^2 + 2F_rF_s\cos\delta \tag{1.57}$$

The m.m.f. across the air gap is,

$$F_{ag} = H \times g \tag{1.58}$$

where,
H is the field intensity along the air gap.
g is the radial length of the air gap.
Since the reluctance of the iron path is neglected, we can write,

$$F_{ag} = F_{res}$$

$$= H_{res} \times g \tag{1.59}$$

where,
H_{res} is the peak value of the resultant field intensity.
The co-energy density at a point where the magnetic field intensity is H is given by the expression,

$$\frac{1}{2}B \times H = \frac{1}{2}\mu_0 \times H^2 \left(\text{since}\, B = \mu_0 H\right) \tag{1.60}$$

$$\text{Average of } H^2 = \frac{1}{2}H_{res}^2 \tag{1.61}$$

$$\text{Then the average of co-energy density} = \left(\frac{1}{4}\mu_0 \times H_{res}^2\right) \tag{1.62}$$

If D is the average diameter of the air gap and l is the axial length of the air gap, then,

$$\text{Volume of the air gap} = \pi Dlg \tag{1.63}$$

$$\text{Total co-energy of the field } (W_f') = (\text{Average of co-energy density}) \times (\text{Volume of the air gap})$$

$$= \left(\frac{1}{4}\mu_0 \times H_{res}^2\right) \times (\pi Dlg)$$

$$= \frac{1}{4}\left(\frac{F_{res}}{g}\right)^2 \pi\mu_0 Dlg$$

$$= \frac{\pi\mu_0 Dl}{4g} F_{res}^2$$

$$= \frac{\pi\mu_0 Dl}{4g}\left(F_r^2 + F_s^2 + 2F_r F_s \cos\delta\right) \tag{1.64}$$

The torque equation is obtained by the partial differentiation of W_f' with respect to δ.

$$Torque(T) = \frac{\partial W_f'}{\partial \delta} = -\frac{\mu_0 \pi Dl}{2g} F_r F_s \sin\delta \tag{1.65}$$

This torque equation is for machines having two poles. The generalised torque equation for machines having *P* poles will be,

$$Torque(T) = -\frac{P}{2} \times \frac{\mu_0 \pi Dl}{2g} F_r F_s \sin\delta \tag{1.66}$$

1.13 FORCE AND CO-ENERGY

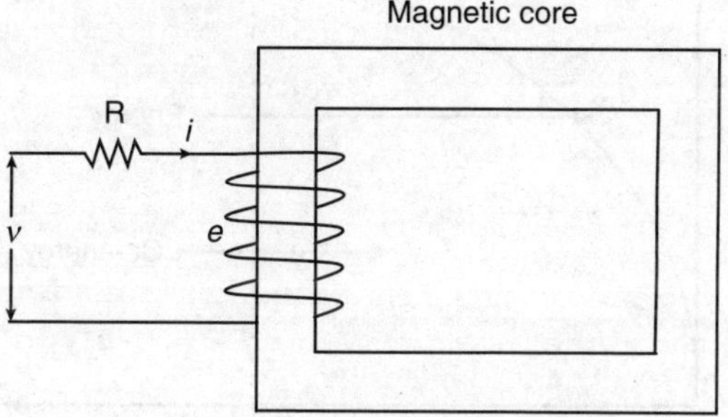

Fig.1.12: Magnetic circuit

Co-energy is the dual of energy. It is a non-physical quantity useful for theoretical analysis of systems storing and transforming energy. Co-energy is expressed in the same units as energy and is especially useful for calculation of magnetic forces and torque in rotating machines. The co-energy is zero for systems incapable of storing energy.

Let us consider a magnetic circuit that is excited by a coil carrying a current *i*, as shown in Fig.1.12. The resulting magnetising curve is shown in Fig.1.13.

The electrical energy supplied to the loss less coil of Fig.1.12, as the current increases from 0 to I_0, must be equal to the energy stored in the magnetic field of the ferromagnetic core, which is given by the expression,

$$W_f = \int_0^{\lambda_0} i\,d\lambda \tag{1.67}$$

From Eq. (1.67), it is clear that the stored magnetic energy is given by the area to the left of the magnetising curve and below $\lambda = \lambda_0$.

Integrating Eq. (1.67) by parts, we get,

$$W_f = i\lambda - \int \lambda di \tag{1.68}$$

The term $\int \lambda di$ of Eq. (1.68), is called the co-energy, as shown in Fig.1.14.

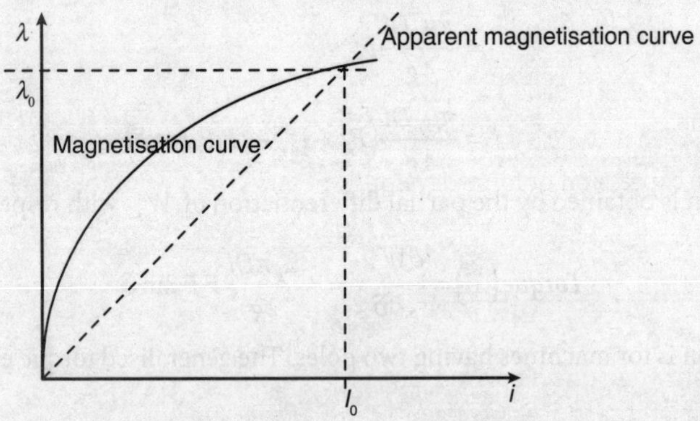

Fig.1.13: Resulting magnetising curve

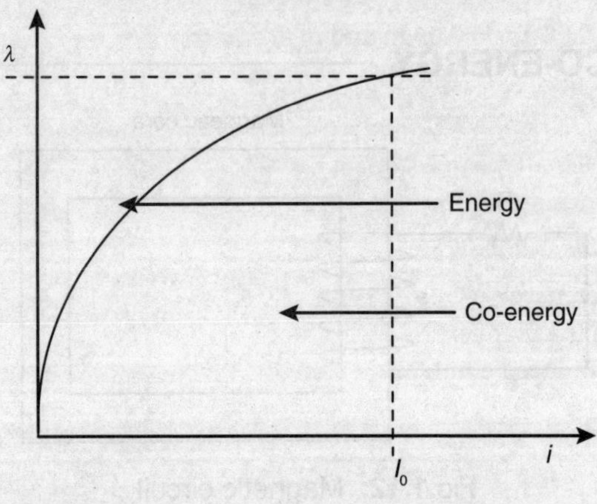

Fig.1.14: Co-energy plot

Let,

$$W'_f = \int \lambda di$$

Then we can write

$$i\lambda = W_f + W'_f$$

Applying total differentiation we get,

$$id\lambda + \lambda di = dW_f + dW'_f$$

Or

$$id\lambda - dW_f = dW'_f - \lambda di$$

Let,

$$F_d dx = id\lambda - dW_f$$

Then,

$$F_d dx = \frac{\delta W'_f}{\delta x}dx + \frac{\delta W'_f}{\delta i}di - \lambda di$$

$$F_d = \frac{\delta W'_f}{\delta x} + \left(\frac{\delta W'_f}{\delta i} - \lambda\right)\frac{di}{dx}$$

$$or\, F_d = \frac{\delta W'_f}{\delta x} \tag{1.69}$$

Eq. (1.69), gives the expression of force developed in terms of magnetic co-energy.

SOLVED PROBLEMS

Ex.1.1 A rotor with only one turn is placed in a uniform magnetic field of 0.08 T. The coil sides has a radius of 0.06 m and carries a current of 15 A. If the length of the rotor is 0.5 m, obtain the torque acting on the rotor.

Solution

Given, I=15 A, B = 0.08 T, r = 0.06 m and l = 0.5 m.

$$T = -2IBrl\sin\propto = -2\times15\times0.08\times0.06\times0.5\times\sin\propto = -0.072\sin\propto$$

Ex.1.2 Consider a single excited linear system as shown in Fig.E1.2
Where, N = 1500 turns, g = 1 mm, d = 0.2 m, x = 0.07 m, l = 0.2 m and i = 15 A. Obtain an expression for the following.

(i) Magnetic stored energy
(ii) Force acting on the plunger
(iii) Expression for induced e.m.f.

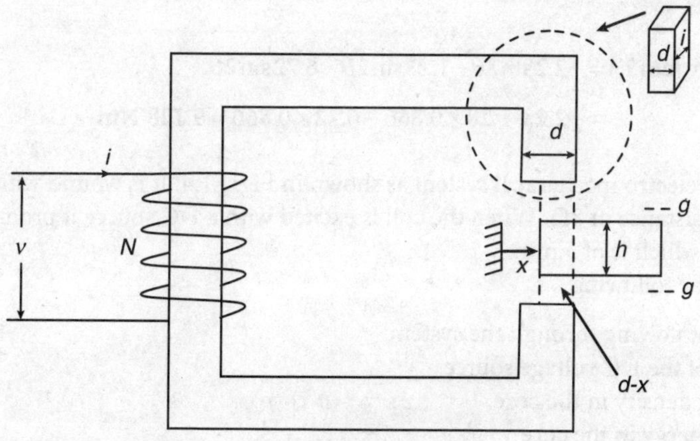

FIG. E1.2:

Solution

$$\text{Inductance}(L)=\frac{(\mu_0 N^2 l)\times(d-x)}{2g}=\frac{(4\pi\times10^{-7}\times1500^2\times0.2)\times(0.2-0.07)}{2\times1\times10^{-3}}=36.75H$$

$$\text{Magnetic stored energy}(W_{me})=\frac{1}{2}Li^2=\frac{1}{2}\times36.75\times15^2=4134.38J$$

$$\text{Force acting on the plunger}(F)=-\frac{\mu_0 l(Ni)^2}{4g}=-\frac{4\pi\times10^{-7}\times0.2\times(1500\times15)^2}{4\times1\times10^{-3}}=31.81\times10^3 N$$

$$\text{Induced e.m.f. }(e)=L\frac{di}{dt}+i\frac{dL}{dx}\frac{dx}{dt}$$

$$=L\frac{di}{dt}-i\frac{\mu_0 N^2 l}{2g}v$$

If current i is DC:

$$\text{Induced e.m.f.}(e)=-i_{dc}\frac{\mu_0 N^2 l}{2g}v$$

If current $I=I_m\sin\omega t$

$$\text{Induced e.m.f.}(e)=\frac{I_m\mu_0 N^2 l}{2g}\left[(d-x)\omega\cos\omega t-v\sin\omega t\right]$$

Ex.1.3 Consider the double excited system as shown in Fig.1.8. If $L_1=20+3\cos3\theta$, $L_2=17+2\cos2\theta$, $M=6\cos2\theta$, I_1 = 0.7 A and I_2 = 0.8 A. Obtain an expression for torque and find the value of torque for $\theta=30^0$

Solution

$$\text{Torque }(T)=\frac{1}{2}I_1^2\frac{dL_1}{d\theta}+\frac{1}{2}I_2^2\frac{dL_2}{d\theta}+I_1I_2\frac{dM}{d\theta}$$

$$=(0.5)(0.7)^2\frac{d}{d\theta}[20+3\cos3\theta]+(0.5)(0.8)^2\frac{d}{d\theta}[17+2\cos2\theta]+(0.7)(0.8)\frac{d}{d\theta}[6\cos2\theta]$$

$$=-2.2\sin3\theta-1.28\sin2\theta-6.72\sin2\theta\text{ Nm}$$

For $\theta=30^0$

$$\text{Torque }(T)=-2.2\sin3\theta-1.28\sin2\theta-6.72\sin2\theta$$

$$=-2.2-1.28\times0.866-6.72\times0.866=9.128\text{ Nm}$$

Ex.1.4 Consider a electro mechanical system as shown in Fig. E1.4. It is wound with a coil of 300 turns having a resistance of 8Ω. When the coil is excited with a DC source it produces a flux of 2 T in the air gap which is of 5 mm.
Calculate the following

(i) Current flowing through the system
(ii) Value of the DC voltage source
(iii) Energy density in the core
(iv) Field energy in the core
(v) Energy density in the air gap

(vi) Field energy in the air gap
(vii) Total stored

Given, magnetic intensity of the core (H_c) = 800 AT/m.

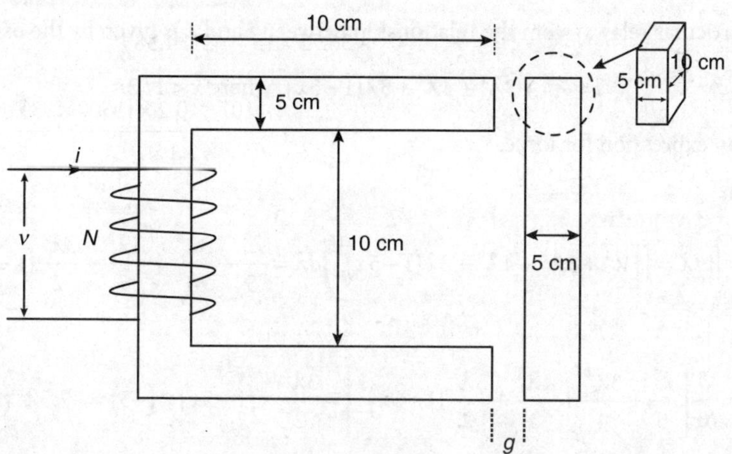

FIG. E1.4:

Solution

$$\text{Flux path length} = (10+5)+(10+5)+(10+5)+(10+5) = 60 \text{ cm}$$

$$\text{Magnetic field intensity in the air gap } (H_{air}) = \frac{B}{\mu_0} = \frac{2}{4\pi\times 10^{-7}} = 1591549 \text{ AT/m}$$

$$\text{Total } mmf = H_c l_c + H_{air} l_g = 800\times 0.6 + 1591549\times 2\times 0.005 = 16395 \text{ AT}$$

Again,

$$Ni = 16395 \text{ AT}$$

$$i = \frac{16395}{300} = 54.65 \text{ A}$$

$$\text{value of DC voltage source} = iR = 54.65\times 8 = 437 \text{ V}$$

$$\text{Energy density in the core } (W_{core}) = \frac{1}{2}BH_c = 0.5\times 2\times 800 = 800 \text{ J/m}^3$$

$$\text{Volume of the core} (V_c) = 2(0.05\times 0.1\times 0.2) + 2(0.05\times 0.1\times 0.1) = 0.003 \text{ m}^3$$

$$\text{Field energy in the core } (W_f) = W_{core}\times V_c = 800\times 0.003 = 2.4 \text{ J}$$

$$\text{Energy density in the air gap } (W_{eair}) = \frac{B^2}{2\mu_0} = \frac{2^2}{2\times 4\pi\times 10^{-7}} = 1591549 \text{ AT/m}$$

$$\text{Volume of the air gap} (V_{gap}) = 2(0.05\times 0.1\times 0.005) = 5\times 10^{-5} \text{ m}^3$$

$$\text{Field energy in the air gap}\left(W_g\right) = W_{eair} \times V_{gap} = 79.58 \text{ J}$$

$$\text{Total stored energy} = W_T = W_f + W_g = 2.4 + 79.58 = 81.98 \text{ J}$$

Ex.1.5 For a particular relay system the relationship between i and λ is given by the expression

$$i = \lambda^4 + 3\lambda^3 + 4\lambda^2 + 3\lambda(1-5x)^2 \text{ where } x < 1/2$$

Derive an expression for force.

Solution

$$W = \int_0^\lambda i d\lambda = \int_0^\lambda \left[\lambda^4 + 3\lambda^3 + 4\lambda^2 + 3\lambda(1-5x)^2\right] d\lambda = \frac{\lambda^5}{5} + \frac{3\lambda^4}{4} + \frac{4\lambda^3}{3} + \frac{3\lambda^2}{2}(1-5x)^2$$

Again,

$$\frac{\delta W}{\delta x} = \frac{\delta}{\delta x}\left[\frac{\lambda^5}{5} + \frac{3\lambda^4}{4} + \frac{4\lambda^3}{3} + \frac{3\lambda^2}{2}(1-5x)^2\right] = \frac{3\lambda^2}{2} \times (1-5x) \times (-5) = -7.5\lambda^2(1-5x)$$

Then,

$$\text{Force}\left(F\right) = -\frac{\delta W}{\delta x} = 7.5\lambda^2(1-5x)N$$

Ex.1.6 Consider two inductive coils with self inductance of L_1=4 H and L_2 = 2 H and the mutual inductance between them is M=1 H. Find the expression for current flowing through each coil when they are

(i) Connected in series
(ii) Connected in parallel

Across a source of 20 cos 314*t* volts.

Solution

Let the induced e.m.f. the first and the second coils be e_1 ande$_2$ respectively. Then,

(i) For series connection

$$\text{For the first coil: } e_1 = 4\frac{di}{dt} + 1\frac{di}{dt}$$

$$\text{For the second coil: } e_2 = 2\frac{di}{dt} + 1\frac{di}{dt}$$

Where, *i* is the current flowing through the series circuit.
Again, in case of a series circuit,

$$V = e_1 + e_2$$

Where, *V* is the supply voltage.
Then,

$$e_1 + e_2 = 20\cos 314t$$

$$\text{or } 4\frac{di}{dt} + 1\frac{di}{dt} + 2\frac{di}{dt} + 1\frac{di}{dt} = 20\cos 314t$$

$$\text{or } 8\frac{di}{dt} = 20\cos 314t$$

$$di = \frac{20}{8}\cos 314t\,dt$$

Integrating we get,

$$i = \frac{20}{8}\frac{\sin 314t}{314}\text{ A}$$

(ii) When the coils are connected in parallel

$$\text{For the first coil: } e_1 = 4\frac{di_1}{dt} + 1\frac{di_2}{dt} = 20\cos 314t$$

$$\text{For the second coil: } e_2 = 2\frac{di_2}{dt} + 1\frac{di_1}{dt} = 20\cos 314t$$

where, i_1 and i_2 are the currents flowing through Coil 1 and Coil 2 respectively. Solving the above two equations we get,

$$\frac{di_1}{dt} = 2.86\cos 314t$$

Integrating we get,

$$i_1 = \frac{2.86}{314}\sin 314t\text{ A}$$

and

$$\frac{di_2}{dt} = 8.57\cos 314t$$

Integrating we get,

$$i_2 = \frac{8.57}{314}\sin 314t\text{ A}$$

Ex.1.7 The expression for self and mutual inductance in Henry's two coupled coils is given by the equations

$$L_1 = 2 + \frac{1}{3x}$$

$$L_2 = 3 + \frac{1}{2x}$$

$$M = \frac{1}{2x}$$

Calculate the following:

(a) Mechanical work done when x is changed from 0.5 to 1.5
(b) Energy supplied by each electrical source
(c) Change in field energy

Where, x is the displacement in meters. Assume I_1 = 15 A and I_2 = 2 A.

Solution

$$W_{fld} = \frac{1}{2}L_1 I_2^2 + \frac{1}{2}L_2 I_2^2 + M I_1 I_2$$

$$= \frac{1}{2}\times\left(2 + \frac{1}{3x}\right)\times 15^2 + \frac{1}{2}\times\left(3 + \frac{1}{2x}\right)\times 2^2 + \frac{1}{2x}\times 2\times 15$$

$$= 231 + \frac{53.4}{x}$$

$$f_e = \frac{\partial W_{fld}}{\partial x} = -\frac{53.4}{x^2}$$

$$W_{mec} = \int_{0.5}^{1.5} f_e dx = \int_{0.5}^{1.5} -\frac{53.4}{x^2} dx = -71.2 \text{ Watt-sec}$$

$$\psi_1 = L_1 I_1 + M I_2$$

$$= \left(2 + \frac{1}{3x}\right) \times 15 + \frac{1}{2x} \times 2 = 7.5 + \frac{6}{x}$$

$$\psi_2 = L_2 I_1 + M I_1$$

$$= \left(3 + \frac{1}{2x}\right) \times 2 + \frac{1}{2x} \times 15 = 6 + \frac{8.5}{x}$$

$$W_{elect} = I_1 \psi_1 \Big|_{0.5}^{1.5} + I_2 \psi_2 \Big|_{0.5}^{1.5}$$

$$= \left[15 \times \left(7.5 + \frac{6}{1.5}\right) - 15 \times \left(7.5 + \frac{6}{0.5}\right)\right] + \left[2 \times \left(6 + \frac{8.5}{1.5}\right) - 2 \times \left(6 + \frac{8.5}{1.5}\right)\right]$$

$$= -193.67 \text{ Watt-sec}$$

$$\text{Change in } W_{fld} = \left(231 + \frac{53.4}{1.5}\right) - \left(231 + \frac{53.4}{0.5}\right) = -71.34 \text{ Watt-sec}$$

Ex.1.8 Consider the system shown in Fig. E1.8

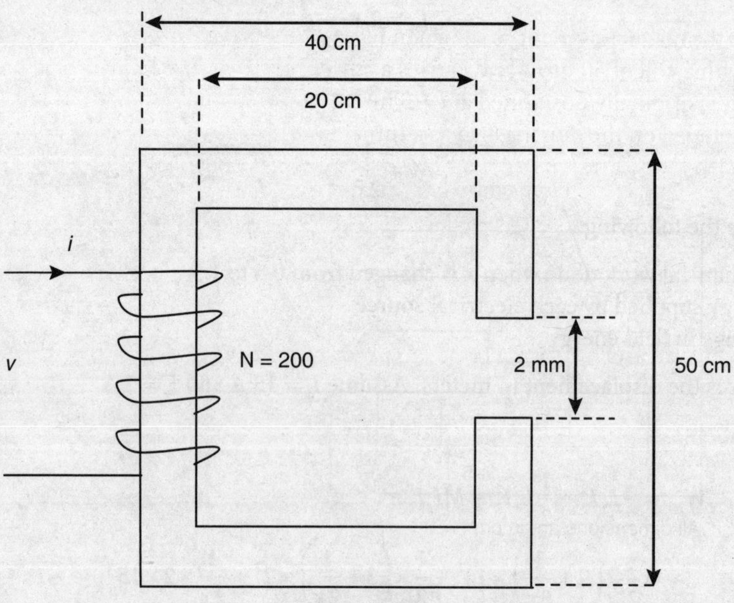

Fig. E1.8:

The system has a cross sectional area of 100 cm² and depth of 10 cm. If the coil current is 12 A calculate the m.m.f. across the core and also across the air gap. Assume $\mu_c = 2000\mu_0$.

Solution

$$\text{Mean length } (l) = 2\times 40 + 2\times 30 - 0.2 = 139.8 \text{ cm}$$

$$\text{Coil m.m.f.} = NI = 200\times 12 = 2400 \text{ AT}$$

$$\text{Flux } (\varphi) = \frac{m.m.f.}{\dfrac{l}{A\times\mu_c}+\dfrac{g}{A\times\mu_o}} = \frac{2400}{\dfrac{1.398}{0.01\times 2000\mu_0}+\dfrac{0.002}{0.01\times\mu_o}} \frac{2400}{55.62+159.2} = 11.174 \text{ m Wb}$$

$$\text{Core m.m.f.} = \frac{11.174}{55.62} = 621.6 \text{ AT}$$

$$\text{Air gap m.m.f.} = \frac{11.174}{159.2} = 1778.4 \text{ AT}$$

Ex.1.9 The air gap length of a ring shaped electromagnet is of 5 mm. The cross sectional area is of 15 cm². Calculate the m.m.f. required to produce a flux density of 0.5 Wb/m². Take the mean length of the core to be 50 cm and the relative permeability to be 400.

Solution

$$\text{Flux } (\varphi) = B\times A = 0.5\times 15\times 10^{-4} = 7.5\times 10^{-4} \text{ Wb}$$

$$\text{Reluctance } (S) = \left[\frac{50\times 10^{-2}}{4\times\pi\times 10^{-7}\times 400\times 15\times 10^{-4}} + \frac{5\times 10^{-3}}{4\times\pi\times 10^{-7}\times 15\times 10^{-4}}\right] = 3.316\times 10^{6} \text{ AT/Wb}$$

$$\text{m.m.f.} = \varphi\times S = 7.5\times 10^{-4}\times 3.316\times 10^{6} = 2487 \text{ AT}$$

Ex.1.10 Consider the magnetic circuit as shown in Fig.E1.10. It has a depth (d) of 5 cm. The fixed arm is wound with a coil of 50 turns and carries a current of 10 A. The moving arm has negligible air gap and it is originally positioned at l = 5 mm. Determine the coil flux and inductance and the force developed on the moving arm as a function of x

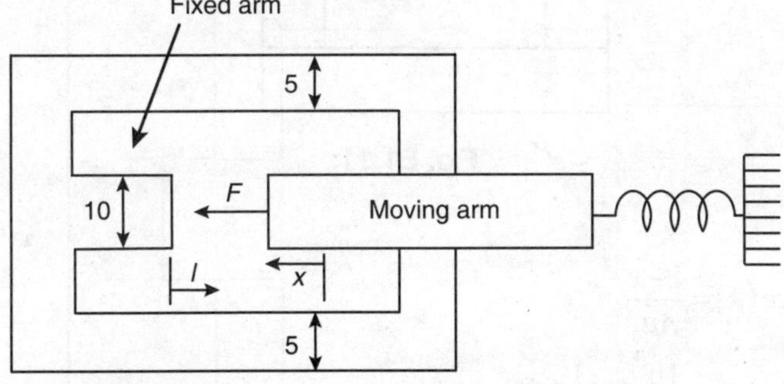

Fig. E1.10:

Solution

$$\lambda = \mu_0 N^2 i\left(\frac{wd}{l-x} + 0.52(w+d) + 0.308(l-x)\right)$$

$$= 4\pi\times10^{-7}\times50^2\times10\left(\frac{0.1\times0.05}{0.005-x} + 0.52(0.1+0.05) + 0.308(0.005-x)\right)$$

$$= \frac{1.571\times10^{-4}}{0.005-x} + 2.5\times10^{-3} - x9.68\times10^{-3}\ \text{Wb}-t$$

$$L = \frac{\lambda}{i} = \frac{\left[\dfrac{1.571\times10^{-4}}{0.005-x} + 2.5\times10^{-3} - x9.68\times10^{-3}\right]}{10} = \frac{1.571\times10^{-5}}{0.005-x} + 2.5\times10^{-4} - x9.68\times10^{-4}\ H$$

$$F = \frac{N^2 i^2 \mu_0}{2}\times\left(\frac{wd}{(l-x)^2 - 0.308}\right) = \frac{50^2\times10^2\times4\pi\times10^{-7}}{2}\times\left(\frac{0.1\times0.05}{(0.005-x)^2 - 0.308}\right)$$

$$= \frac{0.001}{(0.005-x)^2} - 0.062\ \text{N}$$

Ex.1.11 Consider the magnetic circuit shown in Fig.E1.11. The fixed arm in wound with a coil of N_1 turns with self-inductance of L_1 and carrying a current of i_1. Similarly, the moving arm is wound with a coil of N_2 turns with self inductance of L_2 and carrying current i_2. The mutual inductance between the coils is M. Find an expression for the force acting on the moving arm. Neglect the effect of leakage flux and also the effect of air gap fringing. Take the area of the air gap to be A.

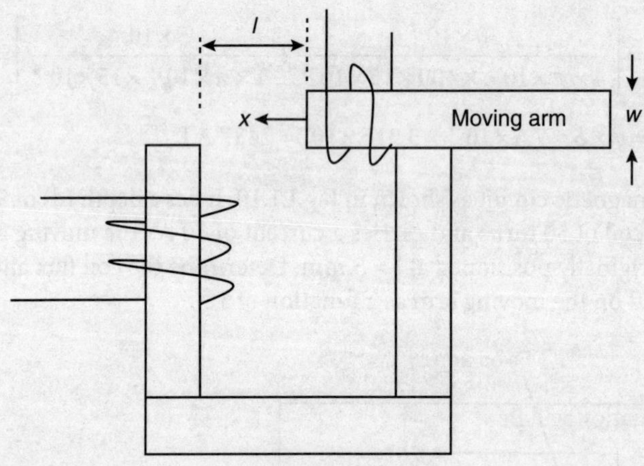

FIG. E1.11:

Solution

$$\text{Reluctance}\,(\text{R}) = \frac{l-x}{A\mu_0}$$

$$\text{Force}\,(F) = \frac{1}{2}L_1 i_1^2 + \frac{1}{2}L_2 i_2^2 + M i_1 i_2$$

$$= \frac{1}{2}\left(\frac{\mu_0 N_1^2 A}{l-x}\right) i_1^2 + \frac{1}{2}\left(\frac{\mu_0 N_2^2 A}{l-x}\right) i_2^2 + \left(\frac{\mu_0 N_1 N_2 A}{(l-x)^2}\right) i_1 i_2$$

Ex.1.12 An iron ring has a cross sectional area of 3 cm^2 and a mean diameter of 25 cm. An air gap of 0.4 mm has been cut on the ring. The ring is wound with a coil of 200 turns with a current of 2 A. Find the relative permeability of iron if the total magnetic flux is 0.24 mWb.

Solution

$$B = \frac{\varphi}{A} = \frac{0.24\times 10^{-3}}{3\times 10^{-4}} = 0.8 \text{ Wb/m}^2$$

$$\text{AT of the ring} = \frac{Bl}{\mu_0\mu_r} = \frac{0.8\times 0.25}{4\pi\times 10^{-7}\times \mu_r} = \frac{1.59\times 10^5}{\mu_r}$$

$$\text{AT of air gap} = \frac{Bl}{\mu_0} = \frac{0.8\times 0.4\times 10^{-3}}{4\pi\times 10^{-7}} = 255$$

$$\text{Total AT} = 200\times 2 = 400$$

$$\text{Then, } \frac{1.59\times 10^5}{\mu_r} + 255 = 400$$

$$\text{or } \mu_r = 1096$$

Ex.1.13 For a double excited system of Fig.1.8, find the value of torque under the following conditions:

$L_1 = 12 + 2\cos 5\theta$

$L_2 = 21 + 5\cos 9\theta$

$M = 10\cos 3\theta$

$I_1 = 1$ A

$I_2 = 2$ A

$\theta = 60^0$

Solution

The torque equation will be

$$\text{Torque} = (0.5)(1)^2\frac{d}{d\theta}[12+2\cos 5\theta] + (0.5)(2)^2\frac{d}{d\theta}[21+5\cos 9\theta] + (1)(2)\frac{d}{d\theta}[10\cos 3\theta]$$

$$= -5\sin 5\theta - 90\sin 9\theta - 6\sin 3\theta$$

Now for and angle of 60^0, the torque will be,

$$\text{Torque} = -5\times(-0.866) - 90\times 0 - 6\times 0 = 4.33 \text{ Nm}$$

Ex.1.14 Obtain an expression for the force on the system where the relationship between i and λ is expressed as,

$$i = 7\lambda^5 + 5\lambda^4 + 8\lambda^3 + +2\lambda^2 + 6\lambda(1-5x)$$

Solution

$$W = \int_0^\lambda \left[7\lambda^5 + 5\lambda^4 + 8\lambda^3 + +2\lambda^2 + 6\lambda(1-5x)\right] d\lambda$$

$$= \frac{7\lambda^6}{6} + \frac{5\lambda^5}{5} + \frac{8\lambda^4}{4} + \frac{2\lambda^3}{3} + \frac{6\lambda^2}{2}(1-5x)$$

So,

$$\text{Force} = -\frac{\delta W}{\delta x} = -\frac{\delta}{\delta x}\left[\frac{7\lambda^6}{6} + \frac{5\lambda^5}{5} + \frac{8\lambda^4}{4} + \frac{2\lambda^3}{3} + \frac{6\lambda^2}{2}(1-5x)\right] = -\frac{6\lambda^2}{2} \times (1-5x) \times (-5)\text{N}$$

Ex.1.15 Find an expression for current that will flow through the two inductive coils having the self inductance of 6 H and 3 H and mutual inductance of 2 H, with a source of 5 cos 2πft. Consider both the conditions (when connected in series and when connected in parallel)

Solution

If the coils are connected in series, the current through both the coils will be same; let it be i. Then,

$$e_1 = 6\frac{di}{dt} + 2\frac{di}{dt} \text{ for the first coil}$$

$$e_2 = 3\frac{di}{dt} + 2\frac{di}{dt} \text{ for the second coil}$$

From Kirchoff's law

$$V = e_1 + e_2$$

$$\text{or } 5\cos 2\pi ft = e_1 + e_2$$

$$\text{or } 5\cos 2\pi ft = 6\frac{di}{dt} + 2\frac{di}{dt} + 3\frac{di}{dt} + 2\frac{di}{dt}$$

$$\text{or } 5\cos 2\pi ft = 13\frac{di}{dt}$$

Integrating, we get,

$$i = \frac{5}{13}\frac{\sin 2\pi ft}{2\pi f} \text{ A}$$

When the coils are connected in parallel, the voltage drop across each coil will be same and will be equal to the supply voltage. Let the current through the first coil be i_1 and the current through the second coil be i_2. So,

$$5\cos 2\pi ft = 6\frac{di_1}{dt} + 2\frac{di_2}{dt} \text{ for the first coil}$$

$$5\cos 2\pi ft = 3\frac{di_2}{dt} + 2\frac{di_1}{dt} \text{ for the second coil}$$

Solving the above two equations we get,

$$\frac{di_1}{dt} = \frac{1}{14}5\cos 2\pi ft$$

Integrating we get,

$$i_1 = \frac{5}{14}\frac{\sin 2\pi ft}{2\pi f}$$

$$\frac{di_2}{dt} = \frac{2}{14}5\cos 2\pi ft$$

Integrating we get,

$$i_2 = \frac{10}{14}\frac{\sin 2\pi ft}{2\pi f}$$

Ex.1.16 A ring type electromagnet with cross sectional area of 20 cm² has an air gap of 6 mm. If the mean length of the core is taken to be 70 cm and the relative permeability assumed as 800, calculate the reluctance of the coil.

Solution

$$\text{Reluctance }(S) = \left[\frac{70\times10^{-2}}{4\times\pi\times10^{-7}\times800\times20\times10^{-4}} + \frac{6\times10^{-3}}{4\times\pi\times10^{-7}\times20\times10^{-4}}\right]$$

$$[348151.44 + 2387324.146]$$

$$= 2.735\times10^{6}\ \text{AT/Wb}$$

Ex.1.17 What will be the inductance of a coil having 600 turns and producing a flux of 1 mW at a current of 25 A?

Solution

$$\text{Inductance} = \frac{600\times1\times10^{-3}}{25} = 0.024\ \text{H}$$

Ex.1.18 The inductance of a coil of a magnetic circuit, having single turn stator and rotor when placed in a non uniform air gap varies as follows,

$$L = 0.05 + 0.04\cos 2\theta + 0.003\sin 4\theta$$

Obtain the torque for a current of 4 A.

Solution

$$T = \frac{4^2}{2}\frac{d}{d\theta}[0.05 + 0.04\cos 2\theta + 0.003\sin 4\theta]$$

$$= \frac{4^2}{2}[-0.08\sin 2\theta + 0.012\cos 4\theta]\ \text{Nm}$$

Ex.1.19 Calculate the force acting on a 100 cm long coil with 200 turns carrying a current of 3 A, when placed in a magnetic field of 0.1 Wb/m^2

Solution

$$F = 2 \times 200 \times 1 \times 3 \times 0.1 = 120 \text{ N}$$

Ex.1.20 What will be the mutual inductance between two coils having 120 and 250 turns respectively, wound on a magnetic circuit having 4 m length and cross sectional area of 400 cm^2. Assume μ_r to be 2000.

Solution

$$M = \frac{120 \times 250 \times 4 \times \pi \times 10^{-7} \times 2000 \times 0.04}{4} = 0.754 \text{ H}$$

EXERCISES

1. What is co-energy?
2. What are the dimensions of co-energy?
3. What are actuators?
4. List the losses that take place during electro-mechanical energy conversion.
5. How can electromechanical devices be classified?
6. State the law of conservation of energy.
7. What are the different parts of an electromechanical system?
8. State the following laws:
   (i) Faraday's Law.
   (ii) Lenz's Law
   (iii) The Biot–Savart law
   (iv) Ampere's law

CHAPTER 2

DC GENERATOR

2.1 INTRODUCTION

A DC generator converts mechanical energy into electrical energy (DC). The working of a DC generator is based on the principle that when a conductor cuts a magnetic field, an e.m.f. is induced in the conductor. A dynamically induced e.m.f. will be produced in the conductor when it will cut the magnetic flux as per the laws of electromagnetic induction. If the circuit is closed, this induced e.m.f. will cause a current to flow through circuit.

A DC machine is an alternating current machine. The e.m.f. generated in the conductor is alternating in nature. With the help of the commutator segments this alternating e.m.f. is converted into a direct one.

2.2 CONSTRUCTION OF DC MACHINE

A DC generator has mainly two parts, namely, the stator and the rotor. Stator is the stationary part and rotor is the rotating part of the generator. They are made up of ferromagnetic material. The winding which is present on the stator is called the field winding and the winding which is present on the rotor is called the armature winding. The cross-sectional view of a DC generator is shown in Fig.2.1.

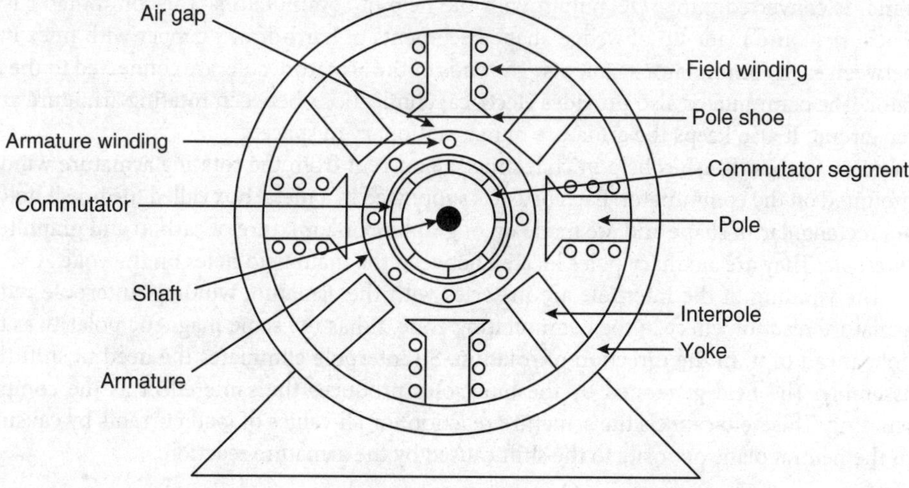

FIG: 2.1: Cross-sectional view of a DC generator

Some other parts of a DC machine are as follows:

i) *Poles:* For s small-sized DC machine the poles are made up of permanent magnet. But for a bigger rated machine, electromagnets are used for poles. Poles are made up of laminated material (thin sheets of steel), that are insulated from each other and riveted together. This lamination reduces eddy current losses. The poles are generally welded or bolted to the yoke.

   Each pole core is wound with one or more field winding. For the production of alternating north and south poles in the direction of rotation, these field coils are connected in series with one another. Since these poles project inwards they are called salient poles.

   The portion of the pole which is near to the armature is slightly stretched and has a curved surface. This curved surface is called the pole shoe and it has three main functions:

   - It holds the field winding.
   - Since it is stretched, the cross-sectional area of the magnetic circuit increases, thus reducing its reluctance.
   - It also reduces the flux bunching effect of armature reaction.

ii) *Armature:* Armature is the rotating part of a DC machine. It consist of a laminated cylinder or drum mounted on a shaft. Laminated armature core reduces eddy current loss. The armature core is separated from the field poles by an air gap, which allows the armature to rotate freely. The air gap is kept as small as possible so that the magnetic reluctance is reduced. The armature core has grooves or slots on its surface to hold the armature winding. Due to the slots of the armature core, a pulsating flux is produced on the stator pole surface. This pulsating flux can be reduced by laminating the pole shoes. Insulated conductors are placed on the armature slots, which are wedged and bands of steel wires are fastened round the core to prevent the conductors from flying out during the rotation of the armature. There are two types of windings used in the armature – the wave winding and the lap winding.

iii) *Yoke:* It is the outer frame of the machine and it carries the magnetic flux produced by the poles. Yoke is generally made up of cast steel or fabricated rolled steel. In some DC machines, yoke is also laminated to reduce eddy current loss. Field poles are bolted to the yoke.

iv) *Commutator:* The armature winding of a DC machine generates alternating e.m.f. This alternating e.m.f. is converted into a DC output with the help of commutators. The commutator is a cylindrical structure made up of wedge-shaped segments of hard-drawn copper with mica insulation between each commutator segments. The ends of the armature coils are connected to the commutator. The commutator also provides electrical connections between rotating armature and external circuit. It also keeps the armature m.m.f. stationary in space.

v) *Brushes:* Carbon brushes help in collecting the current from the rotating armature winding. It is mounted on the commutator. Each brush is supported in a metal box called the brush holder. They are rectangular in shape and are made up of graphite or a mixture of carbon and graphite.

vi) *Interpole:* They are auxiliary poles located between the main field poles on the yoke.

   The winding of the interpole are in series with the armature winding. Interpole reduces the armature reaction effect in the commutating zone. It has the same magnetic polarity as the main pole ahead of it, in the direction of rotation. So interpole eliminates the need to shift the brush assembly. The field generated by the interpoles produces the same effect as the compensating winding. This field cancels the armature reaction for all values of load currents by causing a shift in the neutral plane opposite to the shift caused by the armature reaction.

2.3 WORKING PRINCIPLE OF DC GENERATOR

A DC generator works on the principle of Faraday's law of electromagnetic induction. That is, when a conductor cuts a magnetic field, an e.m.f. is induced in the conductor. The magnitude of induced e.m.f. depends on the speed of rotation of the conductor, the number of conductors connected in series and the flux in the magnetic field.

Let us consider a single-turn conductor in a uniform magnetic field. When the conductor is in the position shown in Fig.2.2 (a), the plane of the coil is parallel to the direction of the magnetic flux. Under this condition the flux linkage is maximum, but the rate of change of linking flux is zero. So at this position, the induced e.m.f. is zero.

As the coil rotates, the coil sides begin to cut the magnetic flux at a gradually increasing rate. Hence, the magnitude of induced e.m.f. also gradually increases and becomes maximum when the coil rotates by an angle of 90^0, as shown in Fig.2.2.(b). Clockwise rotation of the coil has been assumed here. The direction of induced e.m.f. in the coil can be obtained by Fleming`s Right Hand Rule.

As the coil further rotates from position 90^0 to 180^0, the rate of change of flux linkage gradually decreases. Thus, the induced e.m.f. also gradually decreases and at the position 180^0 it becomes zero. (Fig.2.2(c)).

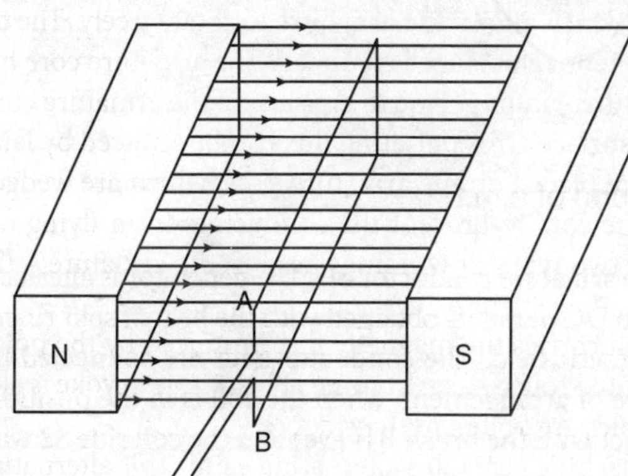

Fig. 2.2(a): Initial position of coil

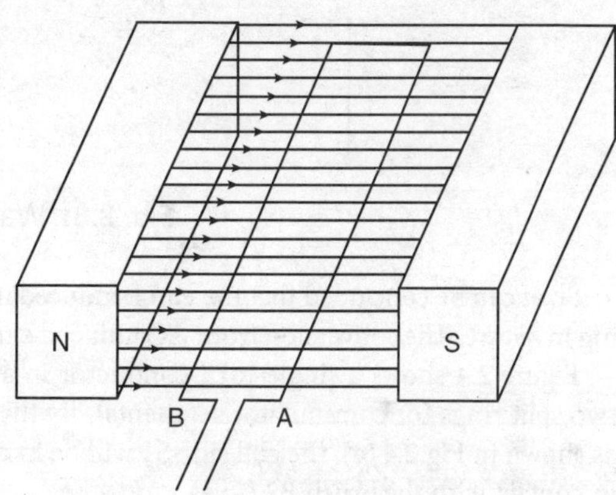

Fig. 2.2(b): Coil rotated by 90°

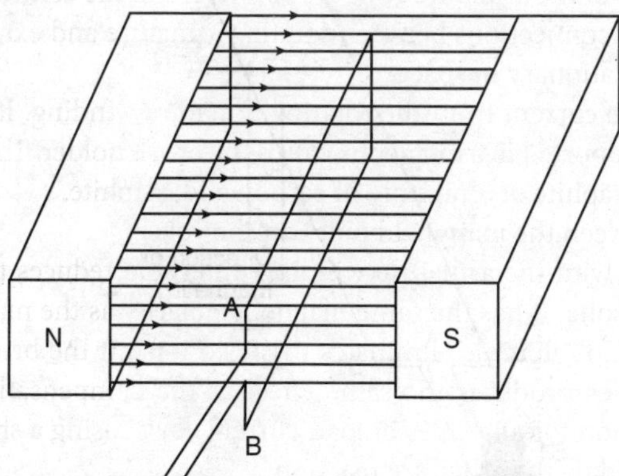

Fig. 2.2(c): Coil rotated by 180°

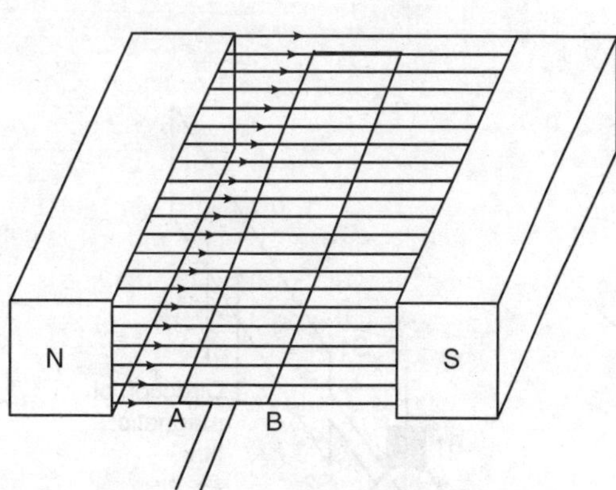

Fig. 2.2(d): Coil rotated by 270°

Again, as the coil rotates from the position 180⁰ to 270⁰, Fig.2.2 (d), the induced e.m.f. starts increasing from 0, but in the opposite direction. As the coil again moves from 270⁰ to 360⁰, the induced e.m.f. starts decreasing from its maximum value and reaches zero again. The magnitude of induced e.m.f. in the coil with respect to its position is shown in Fig.2.3.

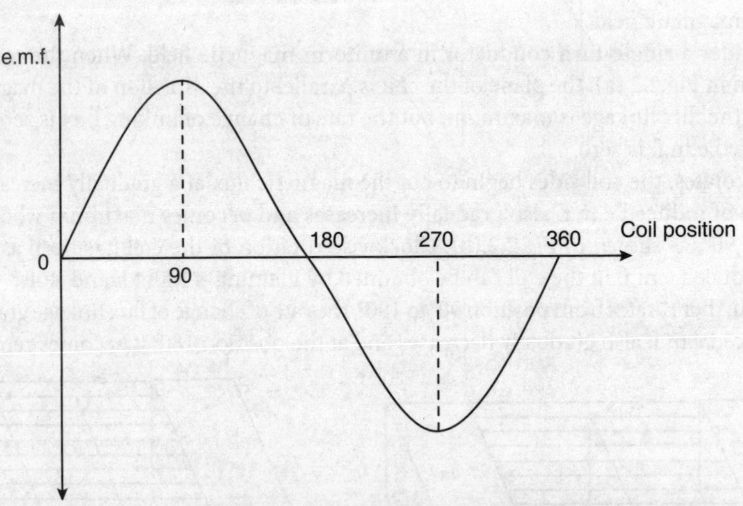

Fig. 2.3: Wave form of e.m.f.

So, it can be concluded that the e.m.f. induced in the armature conductor of a DC generator is alternating in nature. The conversion from AC induced e.m.f. to DC output is obtained with the help of split rings.

Figure 2.4 shows a single-turn conductor in a magnetic field. The conductor ends are connected to two split rings (or commutator segments). By this type of arrangement, when the coil is in the position as shown in Fig.2.4 (a), the coil side S1 will be in contact with the brush B1(+ve) and the coil side S2 will in contact with the brush B2 (–ve).

After 90⁰ rotation, when the coil is in position shown in Fig.2.4 (b), the coil is short-circuited and so the external load current is zero.

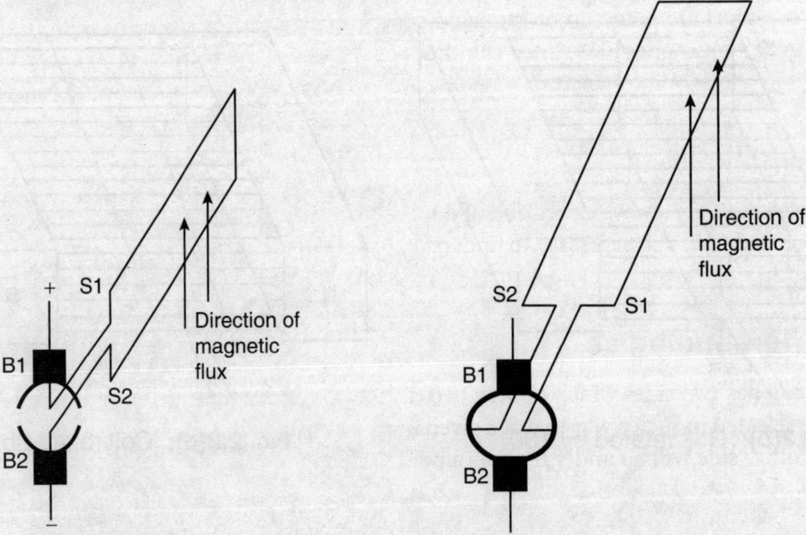

Fig. 2.4 (a): **Fig. 2.4 (b):** Slip rings are short-circuited

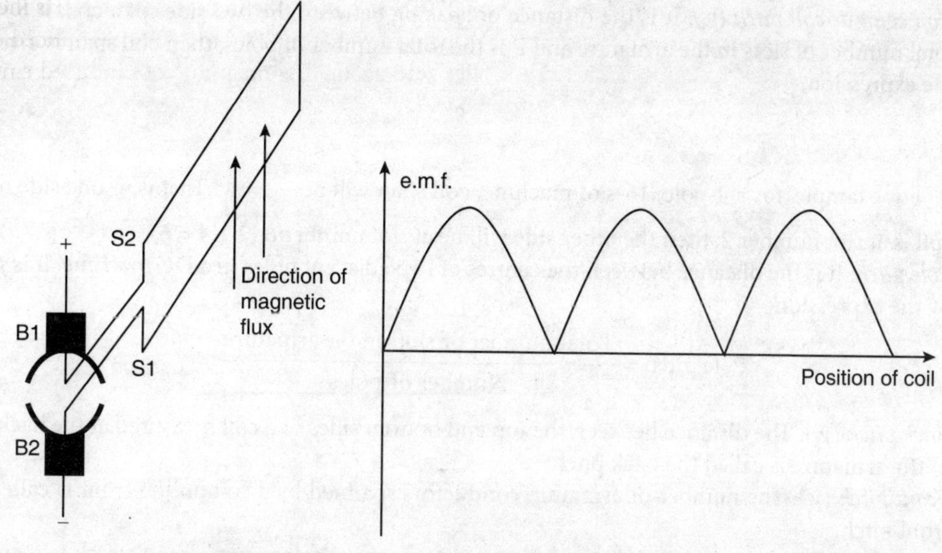

Fig. 2.4 (c):

Fig. 2.4 (d): Wave form of induced e.m.f.

When the direction of the current reverses, the position of the coil also reverses and as a result, Sides S1 and S2 will be in contact with the negative and positive brush respectively. So, the direction of the load current in the external circuit remains the same.

Thus the alternating e.m.f. that is produced in the armature is now rectified with the help of split rings. This rectified output is not a steady direct current but it is pulsating in nature, as shown in Fig.2.4 (d). This output can be made practically constant by having large number of coils and commutator segments.

2.4 ARMATURE WINDING

Armature is the most important part of a machine. This is because armature is the place where e.m.f. is produced (in the case of a generator) and torque is developed to turn the rotor (in case of a motor).

Armature windings are made up of insulated copper conductors. The junctions of the consecutive coils are terminated on copper bars called commutator segments. There are two basic methods of making the end connections of the armature.

i) Simplex lap winding
ii) Simplex wave winding

Armature windings are always closed and of double layer type. Before studying about these two winding connections in detail, it is necessary to understand some of the terminologies associated with armature coils.

2.4.1 Coil Terminologies

i) *Coil:* A coil has two sides which occupy two distinct positions or slots in the armature. These two slots are selected in such a way that at a given time, one side of the coil will be under the north pole and the other side will be under the south pole.

ii) *Coil span or coil pitch* (y_s): It is the distance or spacing between the two sides of a coil. If S is the total number of slots in the armature and P is the total number of poles, then coil span is given by the expression,

$$y_s = \frac{S}{P} \tag{2.1}$$

For example, for a 4-pole, 16-slot machine, coil span will be $\frac{16}{4} = 4$. That is, if one side of the coil is at slot number 2, then the other side will be at slot number 6 (2 + 4 = 6).

iii) *Pole pitch*: It is the distance between the centres of two adjacent poles in a DC machine. It is given by the expression,

$$\text{Pole pitch} = \frac{\text{Total number of slots in the armature}}{\text{Number of poles}} \tag{2.2}$$

iv) *Back pitch* (y_b): The distance between the top and bottom sides of a coil measured at the back-end of the armature is called the back pitch.

v) *Front pitch* (y_f): The number of armature conductors spanned by a coil on the front is called the front pitch.

vi) *Commutator pitch* (y_c): It is the distance between two commutator segments in which two ends of same armature coil are connected.

vii) *Resultant pitch*(y_r): The distance between the beginning of one coil and the beginning of the next coil is called resultant pitch.

viii) *Full pitch*: If the coil span becomes equal to pole pitch, it is called a full pitch coil. The phase difference between the e.m.f. in the two coil sides will be zero.

ix) *Fractional pitch*: If the coil span is less than the pole pitch, it is called the fractional pitch coil. The phase difference between the e.m.f. in the two coil sides will not be zero in this case.

2.5 LAP WINDING DESIGN

In this section, we will be designing a lap winding for a DC machine having 4 poles and 16 slots. The commutator pitch (y_c), for lap winding is ±1 and the coil span $y_s \approx$ pole pitch.

The 16 slots of the armature of a DC machine are shown in Fig.2.5. Note that in actual case, these slots are present at the outer periphery of the armature drum and thus they are arranged in circular formation. But to make the design simpler these 16 slots are shown horizontally.

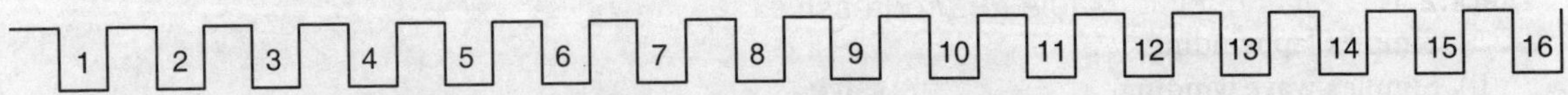

Fig. 2.5: Representation of slots

Similarly, the 16 commutator segments are also shown horizontally instead of in a circular formation for making the design simple and easy to understand. The 16 commutator segments are shown in Fig.2.6.

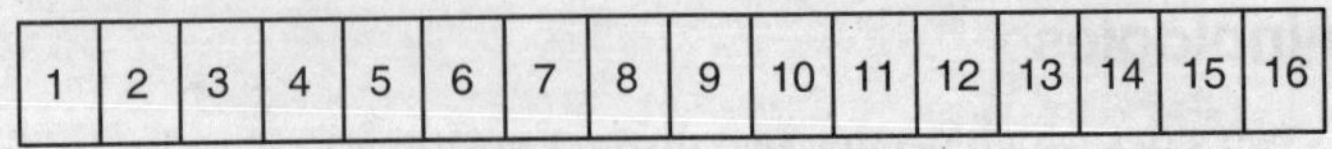

Fig. 2.6: Representation of commutator segments

For a 4-pole, 16-slot DC machine, the coil span is,

$$\text{Coil span}(y_s) = \frac{16}{4} = 4$$

The lap winding can be designed in two ways:

i) *Progressive lap winding*: In this case, $y_b > y_f$ (that is, $y_b = y_f + 2$). This winding progresses in the clockwise direction as seen from the commutator end and is designed to be from left to right. In the case of progressive lap commutator pitch

$$y_c = +1.$$

ii) *Retrogressive lap winding*: In this case, $y_b < y_f$ (that is, $y_b = y_f - 2$). This winding progresses in the anticlockwise direction as seen from the commutator end and its design is from right to left. In the case of retrogressive lap commutator pitch $y_c = -1$.

In this design of the armature winding, the upper side of the coil has been represented by a firm line and the lower side of the coil has been represented by a dotted line. For the proper design we have to properly number each coil side. The complete numbering of the coils for a 16 slot armature is shown in Fig.2.7. In the figure, the upper side of the coils has been numbered as 1,2,3,...while the lower side of the coils has been identified as 1′, 2′, 3′,.....

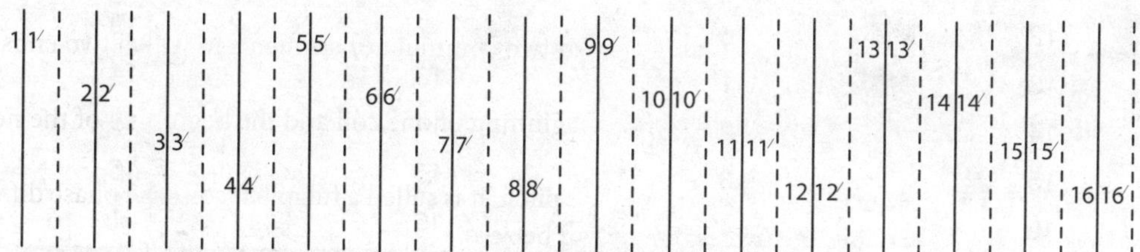

Fig. 2.7: Representation of coil sides

Now, to connect these coils as per our requirement a proper coil connection table is required. The coil connection table will be different for progressive and retrogressive lap winding.

2.5.1 Progressive Lap Winding Design

The detailed coil connection table for a progressive lap winding design is shown in Table.2.1. In the case of progressive lap winding commutator pitch, $y_c = +1$ and coil span$(y_s) = \frac{16}{4} = 4$

Table.2.1: Coil connection table for progressive lap winding

Progressive lap winding			
Coil details		Commutator segment details	
Upper side coil number (C_U)	Lower side coil number (C_L) $C_L = C_U$ + Coil span	Upper side of the coil (S_U)	Lower side of the coil (S_L) $S_L = S_U + y_c$
1	5′	1	2
2	6′	2	3
3	7′	3	4
4	8′	4	5

(continued)

Table 2.1: (continued)

Progressive lap winding			
Coil details		**Commutator segment details**	
Upper side coil number (C_U)	**Lower side coil number (C_L) $C_L = C_U$ + Coil span**	**Upper side of the coil (S_U)**	**Lower side of the coil (S_L) $S_L = S_U + y_c$**
5	9′	5	6
6	10′	6	7
7	11′	7	8
8	12′	8	9
9	13′	9	10
10	14′	10	11
11	15′	11	12
12	16′	12	13
13	1′	13	14
14	2′	14	15
15	3′	15	16
16	4′	16	1

The first entry of Table 2.1 indicates that, one of the terminals of the upper side of Coil 1 will be connected with the one of the terminals of the lower side of Coil 5′.

The other terminal of upper side Coil 1 will be connected with Commutator Segment 1. Similarly, the other terminal of lower side Coil 5′ will be connected with Commutator Segment 2, as indicated in the coil connection diagram shown in Fig.2.8

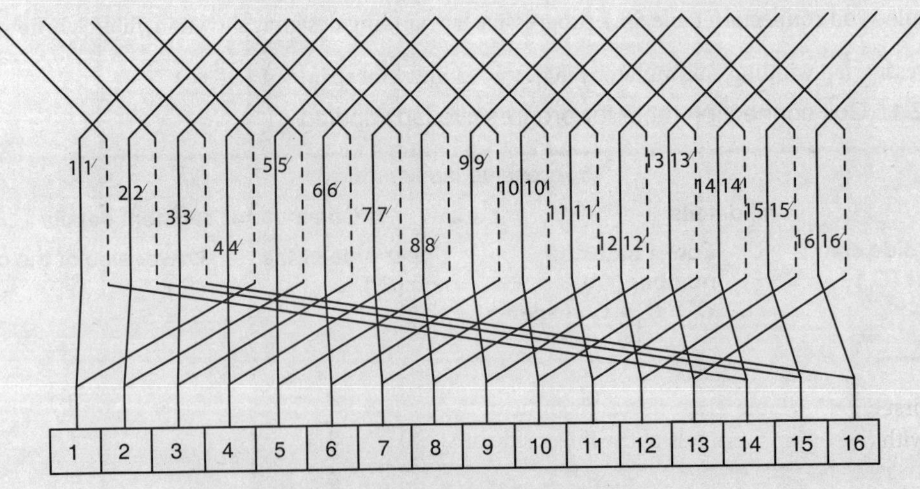

Fig. 2.8: Progressive lap winding

Similarly, the second entry of Table 2.1 indicates that one of the terminals of the upper side of Coil 2 will be connected with the one of the terminals of the lower side of Coil 6′. The other terminal of upper side Coil 2 will be connected with Commutator Segment 2. Similarly, the other terminal of lower side Coil 6′ will be connected with the Commutator Segment 3, as indicated in the coil connection diagram shown in Fig.2.8.

Proceeding on similar lines, the complete coil connection can be made.

2.5.2 Retrogressive Lap Winding Design

The detailed coil connection table for a retrogressive lap winding design is shown in Table.2.2. In the case of retrogressive lap winding commutator pitch $y_c = -1$ and

$$\text{coil span}(y_s) = \frac{16}{4} = 4$$

TABLE.2.2: Coil connection table for retrogressive lap winding

Retrogressive lap winding			
Coil details		**Commutator segment details**	
Upper side coil number (C_U)	**Lower side coil number (C_L) $C_L = C_U$ + coil span**	**Upper side of the coil (S_U)**	**Lower side of the coil (S_L) $S_L = S_U - y_c$**
1	5′	1	16
2	6′	2	1
3	7′	3	2
4	8′	4	3
5	9′	5	4
6	10′	6	5
7	11′	7	6
8	12′	8	7
9	13′	9	8
10	14′	10	9
11	15′	11	10
12	16′	12	11
13	1′	13	12
14	2′	14	13
15	3′	15	14
16	4′	16	15

The first entry of Table 2.2 indicates that one of the terminals of the upper side of Coil 1 will be connected with one of the terminals of the lower side of Coil 5′.

The other terminal of upper side Coil 1 will be connected with Commutator Segment 1. Similarly, the other terminal of lower side Coil 5′ will be connected with Commutator Segment 16, as indicated in the coil connection diagram shown in Fig.2.9.

Similarly, the second entry of Table 2.2 indicates that one of the terminals of the upper side of Coil 2 will be connected with the one of the terminals of the lower side of Coil 6′. The other terminal of upper side Coil 2 will be connected with Commutator Segment 2. Similarly, the other terminal of lower side Coil 6′ will be connected with Commutator Segment 1, as indicated in the coil connection diagram shown in Fig.2.9.

Proceeding on similar lines, the complete coil connection can be made.

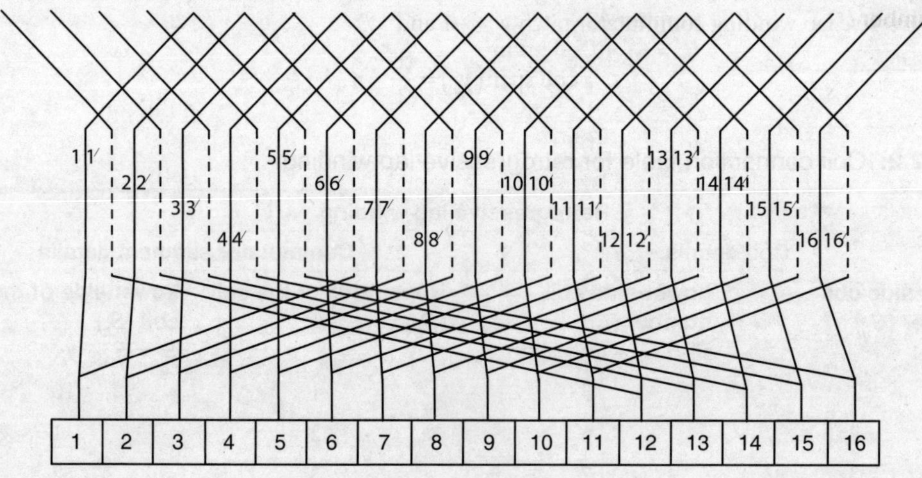

Fig. 2.9: Retrogressive lap winding

2.6 WAVE WINDING

For the wave winding design, let us consider the same DC machine which we had taken for lap winding design. That is a DC machine with 16 slots and 4 poles. For wave winding design, the commutator pitch is given by the expression,

$$y_c = \frac{2(S \pm 1)}{P} \tag{2.3}$$

Like lap winding, wave winding can also be designed in two ways:

i) *Progressive wave winding*: In this case, the commutator pitch is given by,

$$y_c = \frac{2(S+1)}{P}$$

ii) *Retrogressive wave winding*: In this case, the commutator pitch is given by,

$$y_c = \frac{2(S-1)}{P}$$

2.6.1 Progressive Wave Winding

For this design, $S = 16$, $P = 4$ and

$$y_c = \frac{2(S+1)}{P} = \frac{2(16+1)}{4} = 8.5$$

For proper design, y_c must be a whole number; so in this case, the value of S is taken as 17 instead of 16 so that y_c becomes 9. This extra coil is called the dummy coil and is used to make the armature dynamically balanced. The coil connection table is shown in Table.2.3.

Table 2.3: Coil connection table for progressive wave winding

Progressive wave winding			
Coil details		**Commutator segment details**	
Upper side coil number (C_U)	**Lower side coil number (C_L) $C_L = C_U$ + Coil span**	**Upper side of the coil (S_U)**	**Lower side of the coil (S_L) $S_L = S_U + y_c$**
1	5′	1	10
2	6′	2	11
3	7′	3	12
4	8′	4	13
5	9′	5	14
6	10′	6	15
7	11′	7	16
8	12′	8	1
9	13′	9	2
10	14′	10	3
11	15′	11	4
12	16′	12	5
13	1′	13	6
14	2′	14	7
15	3′	15	8
16	4′	16	9

The first entry of Table 2.3 indicates that one of the terminals of the upper side of Coil 1 will be connected with the one of the terminals of the lower side of Coil 5′.

The other terminal of upper side Coil 1 will be connected with Commutator Segment 1. Similarly, the other terminal of lower side Coil 5′ will be connected with Commutator Segment 10, as indicated in the coil connection diagram shown in Fig. 2.10.

Similarly, the second entry of Table 2.3 indicates that one of the terminals of the upper side of Coil 2 will be connected with one of the terminals of the lower side of Coil 6′. The other terminal of upper side Coil 2 will be connected with the Commutator Segment 2. Similarly, the other terminal of lower side Coil 6′ will be connected with the Commutator Segment 11, as indicated in the coil connection diagram shown in Fig.2.10.

Proceeding in a similar way, the complete coil connection can be made.

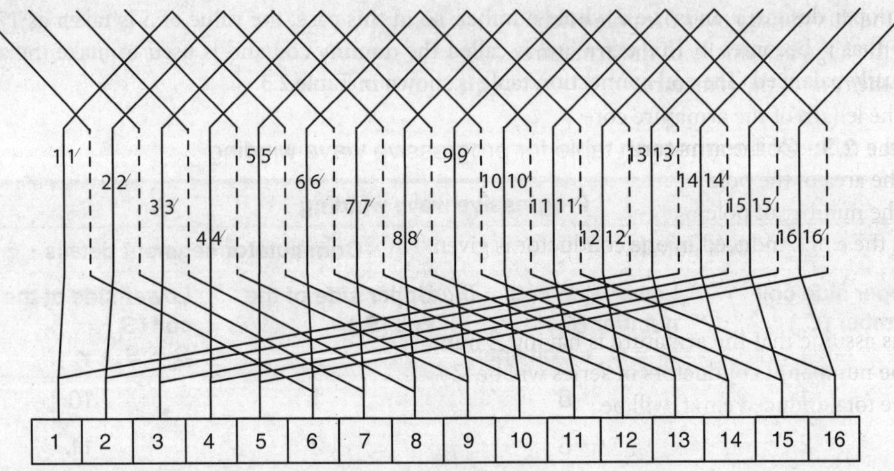

FIG. 2.10: Progressive wave winding

The coil connection for retrogressive wave winding is left for the reader to design.

2.7 DUMMY COIL

As seen in Section 2.6, the commutator pitch for a wave winding is given by the expression,

$$y_c = \frac{2(S \pm 1)}{P}$$

and y_c it must be an integer. However, sometimes the value of P and S may not give an integer value of y_c, as seen in the winding design in Section 2.6.1. Under such conditions, dummy coils are needed. These coils are used to make the armature dynamically balanced. They do not contribute to the induced e.m.f. (for DC generator) or production of torque (for DC motor)

2.8 EQUALISER RING

A lap winding has many parallel paths. The e.m.f. induced in each parallel path may not be same. This may be because of non-uniformity of air gap, wear and tear of brushes etc. This unbalanced e.m.f. will circulate a considerable amount of current in the armature conductors. Since the armature resistance is low, the magnitude of this circulating current will be high. This, in turn, will produce heat and cause commutation difficulties. To overcome this problem, the same potential points in different parallel paths are connected together by a copper ring. Each ring is insulated from each other. These rings are called equaliser rings. They keep the circulating current inside a small shorted section.

2.9 E.M.F. EQUATION OF A DC GENERATOR

As discussed earlier, when a conductor cuts across a magnetic field an e.m.f. is induced in the conductor. The induced e.m.f. generated in a DC generator depends on the following factors:

i) Magnetic flux
ii) Speed of the machine

Let,

φ be the magnetic flux per pole.

B be the flux density over the pole pitch.
ω be the angular velocity of the armature.
N be the rpm of the armature.
l be the length of the armature core
r be the radius of the armature
a is the area of the pole
P is the number of pole
Then the e.m.f. induced in one conductor is given by the expression.

$$e = Bl\omega r \text{ volts} \tag{2.4}$$

Let us assume that the armature is having Z number of conductors and A number of parallel paths. Then, the number of conductors in series will be Z/A.

So the total induced e.m.f. will be

$$E = Bl\omega r\frac{Z}{A} \tag{2.5}$$

The flux density, B, is given by the expression,

$$B = \frac{\varphi}{a} \tag{2.6}$$

Again,

$$\text{Area}(a) = \frac{2\pi rl}{P} \tag{2.7}$$

Substituting Eq. (2.6) and Eq. (2.7) in Eq. (2.5), we get,

$$E = \frac{\varphi Pl\omega r}{2\pi rl} \times \frac{Z}{A} = \frac{\varphi PZ\omega}{2\pi A} \tag{2.8}$$

Again, substituting

$$\omega = \frac{2\pi N}{60} \text{ in Eq. } (2.8), \text{ we get,}$$

$$E = \frac{\varphi ZNP}{60A} \tag{2.9}$$

Let,

$$K = \frac{ZP}{60A} = \text{Constant}$$

Then, Eq. (2.9) becomes,

$$E = K\varphi N \tag{2.10}$$

Similarly, Eq. (2.8) becomes,

$$E = K\varphi\omega \tag{2.11}$$

2.10 ARMATURE REACTION

The DC generator has two kinds of magnetic fluxes:

i) The stator flux, also called the main flux, which is because of the current flowing through the stator conductors.
ii) The armature flux, which is because of the current flowing in the armature.

The armature magnetic field de-magnetises or weakens the main flux and also cross-magnetises or distorts it; thus, the overall effective flux in DC generator decreases. This mutual action of armature flux on the main field flux is known as armature reaction. Figure 2.11 shows the stator flux only, that is, when the current through the armature conductor is zero.

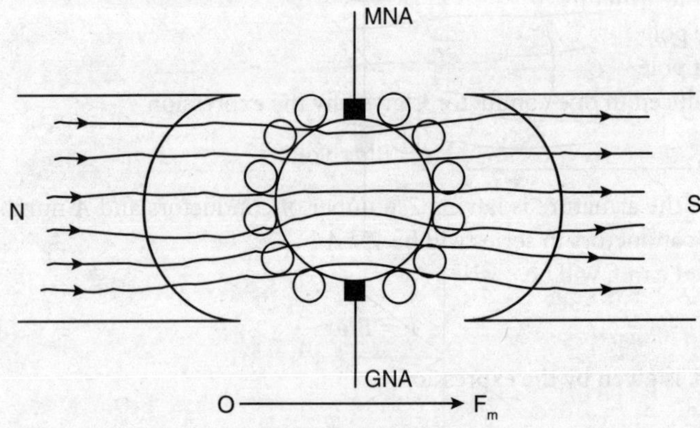

Fig. 2.11: Stator flux

From Fig. 2.11, we can say that the flux across the air gap is uniform. Under this condition, the magnetic neutral axis (MNA) coincides with the geometrical neutral axis (GNA). Magnetic neutral axis may be defined as the axis along which no emf is produced in the armature conductors because they move parallel to the lines of flux. MNA can also be defined as the axis which is perpendicular to the flux passing through the armature. Vector OF_m represents both magnitude and direction of the m.m.f producing the main flux. Note that OF_m is perpendicular to GNA.

Figure 2.11 (a) shows the field (or flux) set up by the armature conductors alone when carrying current, the field coils being unexcited. The current direction is downwards in conductors under N-pole and upwards in those under S-pole. The direction of magnetic lines of force can be found by the cork screw rule. It is clear that armature flux is directed downward parallel to the brush axis. The m.m.f. producing the armature flux is represented in magnitude and direction by the vector OF_A.

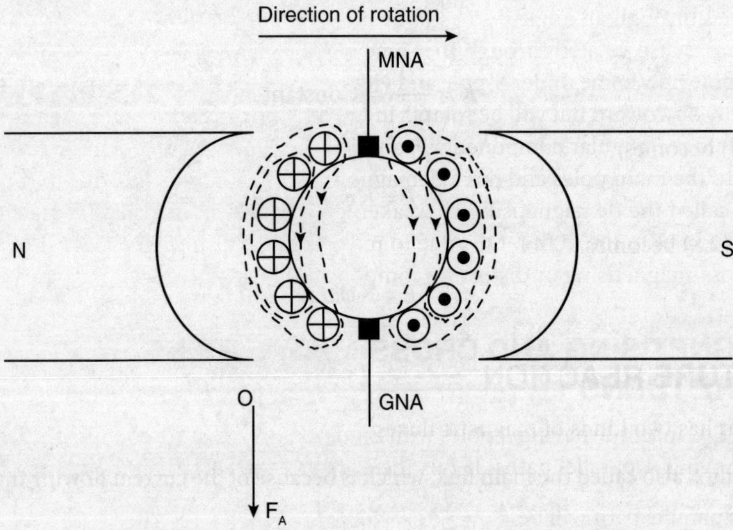

Fig. 2.11 (a): Flux by armature conductor

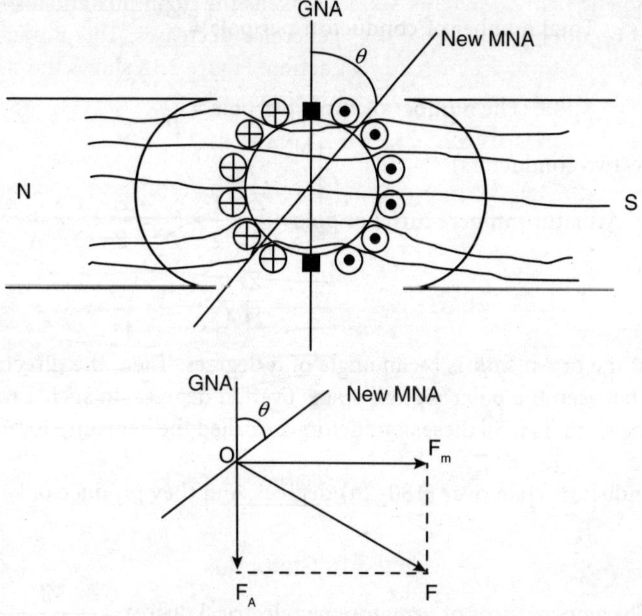

FIG. 2.11 (b): Resultant m.m.f

Under actual load conditions, the two m.m.f exist simultaneously in the generator as shown in Fig. 2.11(b). It is seen that the flux through the armature is no longer uniform and symmetrical about the pole axis; rather, it is distorted. The flux is seen to be crowded at the trailing pole tips but weakened or thinned out at the leading pole tips (the pole tip which is first met during rotation by armature conductors is known as the leading pole tip and the other, as trailing pole tip).

The resultant m.m.f. OF is the vector sum of OF_m and OF_A, as shown in Fig.2.11 (b). Since MNA is always perpendicular to the resultant m.m.f., the MNA is shifted through an angle θ. Note that MNA is shifted in the direction of rotation of the generator.

In order to achieve spark-less commutation, the brushes must lie along the MNA. Consequently, the brushes are shifted through an angle θ so as to lie along the new MNA. Due to brush shift, the m.m.f. F_A of the armature is also rotated through the same angle θ. Thus, some of the conductors which were earlier under N-pole now come under S-pole and vice-versa. The result is that armature m.m.f. F_A will no longer be vertically downward but will be rotated in the direction of rotation through an angle θ. Now, F_A can be resolved into rectangular components F_c and F_d. The component F_d is in direct opposition to the m.m.f. OF_m due to the main poles and has a demagnetising effect on the flux due to the main poles. For this reason, it is called the demagnetising or weakening component of armature reaction. The component F_c is at right angles to the m.m.f. OF_m due to main poles. It distorts the main field. For this reason, it is called the cross-magnetising or distorting component of armature reaction.

2.11 DEMAGNETISING AND CROSS-MAGNETISING AMPERE TURNS

Let us consider a DC machine having P poles with Z number of armature conductors. Let us assume that the armature is having A parallel paths. If I_a is the armature current, then the current flowing through the armature per parallel path will be $I_{ap} = \frac{I_a}{A}$.

$$\text{Total number of conductors per pole} = \frac{Z}{P}$$

$$\text{The number of turns per pole} = \frac{Z}{P} \times \frac{1}{2}$$

(Since one turn has two conductors)

$$\text{Armature ampere turn per pole} = \frac{Z}{P} \times \frac{1}{2} \times I_{ap} = \frac{Z}{P} \times \frac{1}{2} \times \frac{I_a}{A}$$

$$= \frac{ZI_a}{2PA} \tag{2.12}$$

Let us assume that the brush shift is by an angle of α degrees. Then, the direction of current in the group of conductors between the polar regions spans over 2α degrees, in such a way that it produces a flux which opposes the main flux. So these conductors are called the demagnetisation armature conductors.

The remaining conductors span over (180–2α) degrees, and they produce only cross-magnetisation effect.

From Eq. (2.12),

$$\text{The ampere turns of armature per electrical degree} = \frac{ZI_a}{2PA \times 180}$$

[Since the ampere turn spans over one pole pitch (=180^0 electrical)]

$$\text{Demagnetising ampere turn per pole} = \frac{ZI_a}{2PA \times 180} \times 2\alpha = \frac{ZI_a\alpha}{180PA} \tag{2.13}$$

Similarly,

$$\text{Cross-magnetising ampere turns per pole} = \frac{180 - 2\alpha}{180} \times \frac{ZI_a}{PA}$$

$$= \left(1 - \frac{2\alpha}{180}\right)\frac{ZI_a}{PA} \tag{2.14}$$

2.12 COMMUTATION IN DC GENERATOR

Commutation process involves conversion of the generated AC current into DC current and transfer of this current from the rotating armature to an external circuit with the help of brushes. The reversal of current in the armature coil by means of brush and commutator is called the commutation process. A brush will have to make contact with two adjacent commutator segments during this process. Also during this process, the direction of current will change from one direction to the other. The time taken for this process is called commutation period.

In order to explain the complete commutation process, let us first consider an armature coil segment shown in Fig. 2.12. For the armature segment shown in Fig.2.12 (a), current *I* from Conductor B and also current *I* from Conductor C will pass through the brush into the load. So, the load current will be 2*I*. As the armature rotates, the position of the brush starts changing. In Fig.2.12, the direction of rotation of the armature is assumed from left to right.

For the brush position shown in Fig.2.12 (b), the Commutator Segments 2 and 3 are shorted. The current *I* from Conductor A and from Conductor C will pass through the brush into the load. The total current coming to the load will still be 2*I*.

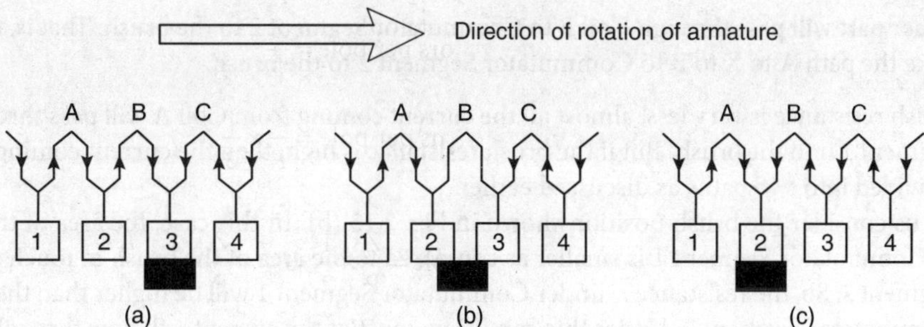

Fig. 2.12: Commutation process

When the brush shifts to Commutator Segment 2, the direction of current in Conductor B changes (Fig.2.12 (c)). This change of direction of current in Coil B takes place during the short-circuit period, that is, when Commutator Segment 2 and Commutator Segment 3 are shorted. If the reversal of current attains its maximum value during the short-circuit period, then the commutation is said to be an ideal commutation. However, if the current in Coil B does not attain its maximum value in the reversed direction during this period, then the difference between the currents of Coil B and Coil C will flow through the commutator into the brush in the form of sparks. These sparks will heat up the commutator surface and thus may damage it. So, for satisfactory commutation, it is necessary that the current in the coil undergoing commutation reverses completely during the commutation period.

2.12.1 Methods to Improve Commutation

There are four methods of making the current attain its full value in the reverse direction at the end of the commutation period so that no spark is produced and the commutator and the brushes are not damaged.

Use of high-resistance brush

In this method, the commutation is improved by increasing the contact resistance between the commutator segments and the brushes. In the coil arrangement of Fig.2.13 (a), two commutator segments are shown. The direction of rotation has been assumed to be from left to right.

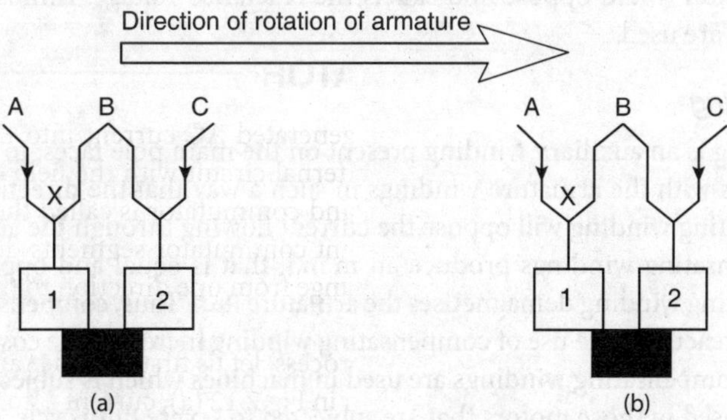

Fig. 2.13: Improvement of commutation

As the brush moves from Commutator Segment 1 to Commutator Segment 2 and touches both the commutator segments, the current from Coil A will get divided into two parts at the junction X.

- One part will pass through Commutator Segment 1 to the brush. That is, the current will take the path from A to X to the brush.

- The other part will pass through Coil B to Commutator Segment 2 to the brush. That is, the current will take the path A to X to B to Commutator Segment 2 to the brush.

If the brush resistance is very less, almost all the current coming from Coil A will pass through Commutator Segment 1 into the brush. But if the brush resistance is high, then the current coming from Coil A will get divided into two paths as discussed earlier.

Now, let us consider the brush position shown in Fig. 2.13 (b). In this case, the area of the brush in touch with Commutator Segment 1 is smaller as compared to the area of the brush in touch with Commutator Segment 2. So, the resistance r_1 under Commutator Segment 1 will be higher than the resistance r_2 under Commutator Segment 2. Under this condition, most of the current will flow through Commutator Segment 2 into the brush. Thus, this high contact resistance has a tendency to force the current in the short-circuit coil to change accordingly, as per the commutation process. The use of carbon and graphite for the brush can increase the contact resistance.

Brush shifting

If the brushes are shifted from the geometric neutral plane to a new magnetic neutral plane, it is possible to get sparkles commutation. By shifting the brushes they are made to lie on the fringe of the next pole. An e.m.f. will be induced in the coil while it undergoes commutation due to the effect of the immediate next pole and the induced e.m.f. will cancel the reactance voltage. The brushes are shifted in the direction of rotation in case of a DC generator. In the case of a DC motor they are shifted backward with respect to the rotation of the machine. The main disadvantage of this method is that the amount of brush shift depends on the load, or in other words, it depends on the armature current, which keeps changing as the load changes.

Interpoles

Interpoles are auxiliary poles situated between the main field poles. The windings of the interpoles are connected in series with the armature windings. So, the armature flux in commutating zone is neutralised by the interpole flux and the tendency of the shift of magnetic neutral axis is eliminated. For a DC generator, the polarity of the interpole must be the same as that of the next field pole in the direction of rotation. However, for an AC generator, the polarity of the interpole must be opposite to that of the next field pole in the direction of rotation. If the interpole is of correct polarity, the induced e.m.f. in the coil undergoing commutation would oppose and cancel the reactance voltage. Almost in all DC machines above 1 HP, interpoles are used.

Compensating winding

Compensating winding is an auxiliary winding present on the main pole faces, in slots. These windings are connected in series with the armature windings in such a way that the direction of current flowing through the compensating winding will oppose the current flowing through the adjacent armature conductors. Thus, compensating windings produce an m.m.f. that is equal and opposite to the armature m.m.f. This compensating winding demagnetises the armature flux. Thus, compensating winding cancels the effect of armature reaction. The use of compensating winding increases the cost as well as the copper loss in the machine. Compensating windings are used in machines which is subjected to heavy overload or plugging. It is also used in those motors that are subjected to severe duty cycle, such as rolling mills.

2.13 EQUIVALENT CIRCUIT OF A DC MACHINE

For the purpose of analysis, the armature and the field winding of a DC machine is denoted by a symbol as shown in Fig. 2.14, where R_a and R_f indicates armature resistance and field resistance respectively. The two black square blocks in the armature circuit indicate the two brushes. Note that the direction of the flow of current in the armature circuit indicates whether it is a DC generator or a DC motor. In case of DC generator, current flows out from the armature circuit and in the case of DC motor, current flows into the armature circuit. Now depending upon how these armature and field windings are connected with each other, we get different types of DC machines, as discussed in the next section.

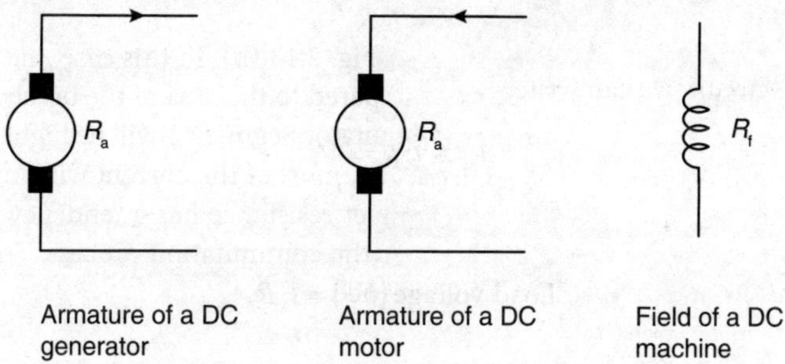

FIG. 2.14: Symbol of DC machine

2.14 CLASSIFICATION OF DC GENERATOR

DC machine can be broadly classified into two categories, namely,

i) Separately excited DC machine
ii) Self-excited DC machine

If the poles of a DC machine are made up of permanent magnet, then such machines are called permanent magnet DC machine.

2.14.1 Separately Excited DC Generator

As the name suggests, in this type of machine, the field of the machine is excited by an independent external source. This source can be a battery, rectifier circuit or a DC generator. The circuit of a separately excited DC generator is shown in Fig.2.15.

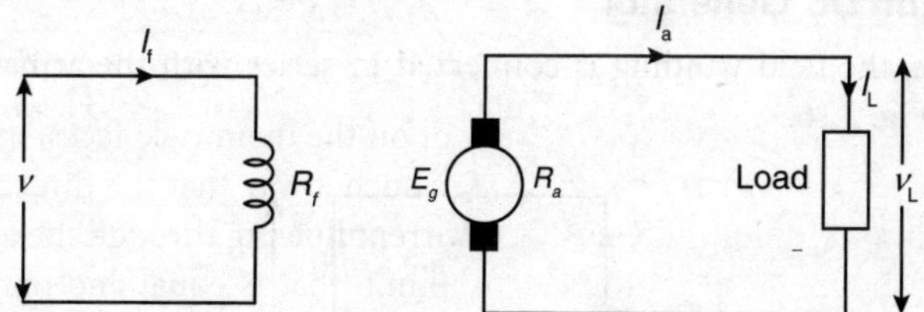

FIG. 2.15: Separately excited DC generator

Let,

V be the field excitation voltage
R_f be the field resistance.
I_f be the field current
E_g be the generated e.m.f. in the armature
R_a be the armature resistance

I_a be the armature current
I_L be the load current
V_L be the load voltage
Then from Fig.2.15, we can write,

$$V = I_f R_f \tag{2.15}$$

From the armature circuit, we can write,

$$E_g = I_a R_a + V_L \tag{2.16}$$

Again,

$$\text{Load voltage } (V_L) = I_L R_L \tag{2.17}$$

And $I_a = I_L$
From Eq. 2.11, we can also write,

$$E_g = K\varphi\omega$$

Where,
φ is the magnetic flux per pole.
ω is the angular velocity of the armature.

2.14.2 Self-excited DC Generator

When the field winding of a DC generator is excited by the current from the armature of the generator, it is called the self-excited DC generator. Depending upon how the field winding is connected with the armature winding, a self-excited DC generator is further classified as,

i) Series-wound DC generator.
ii) Shunt-wound DC generator.
iii) Compound-wound DC generator.

2.14.2.1 Series-wound DC Generator

In this type of machine, the field winding is connected in series with the armature winding of a DC generator, as shown in Fig. 2.16.

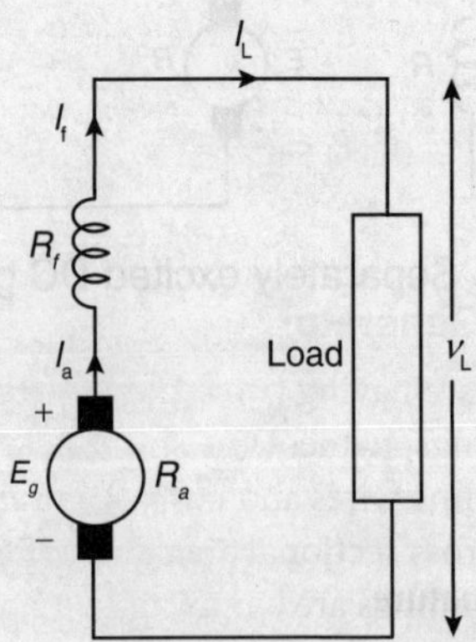

Fig. 2.16: Series DC generator

In this type of machine, since the field winding carry the full armature current, the field winding has less number of turns of wire of large cross sectional area. From the machine circuit shown in Fig.2.16, we can write,

$$I_a = I_f = I_L \tag{2.18}$$

$$E_g = V_L + I_a\left(R_a + R_f\right) \tag{2.19}$$

2.14.2.2 Shunt-wound DC Generator

In this type of machine, the field winding is connected in parallel with the armature winding as shown in Fig.2.17. In this machine, the voltage across the field winding is same as the output voltage of the generator. The field winding of such a type of machine is made up of a large number of turns of fine wire, so that it carries very less field current. This is because, the effective power of the generator is proportional to the current delivered at the load.

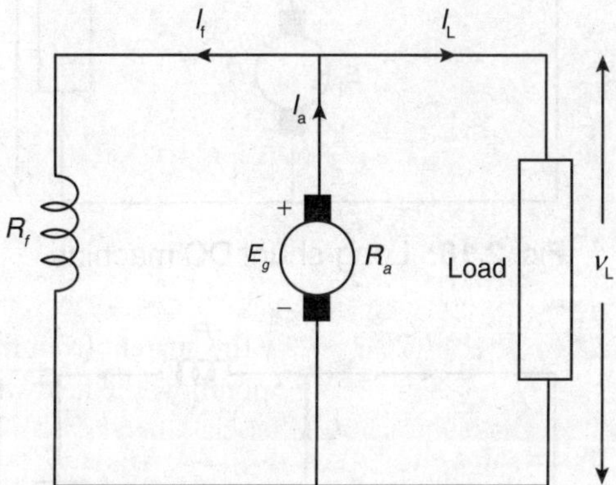

Fig. 2.17: Shunt DC generator

From the motor circuit shown in Fig.2.17, we can write,

$$I_a = I_f + I_L \tag{2.20}$$

$$E_g = V_L + I_a R_a \tag{2.21}$$

$$I_f = \frac{V_L}{R_f} \tag{2.22}$$

2.14.2.3 Compound-wound DC Generator

In this type of DC machine, the poles are having two separate windings. That is for a compound wound DC machine there are two field windings, namely, the shunt field winding and the series field winding. The shunt field winding is made up of fine wires and has large number of turns. Whereas, the series field winding is made up of wires of large cross sectional area and has only few turns.

Depending upon how these two windings are connected with the armature of the DC machine, they are classified as,

i) Long-shunt compound DC machine.
ii) Short-shunt compound DC machine.

If the shunt field winding is connected in parallel with the armature winding and the series field winding, such kind of machine is called long-shunt compound DC machine, as shown in Fig.2.18.

If the shunt field winding is connected in parallel with the armature alone, then such type of machine is called short-shunt compound DC machine, as shown in Fig.2.19.

In the circuit shown in Figs. 2.18 and 2.19,

R_{se} denotes the series field winding and R_{sh} denotes the shunt field winding

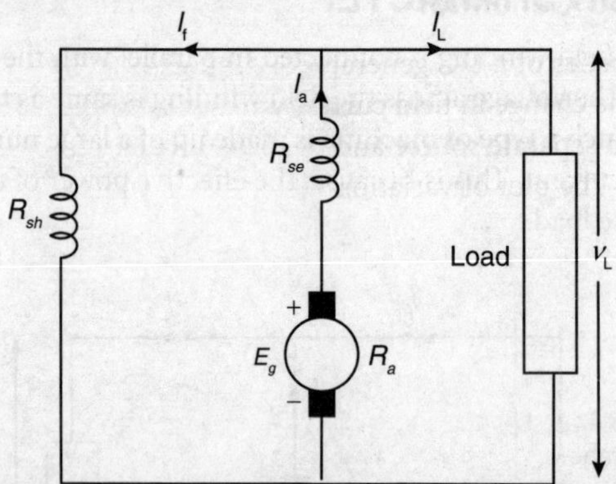

Fig. 2.18: Long-shunt DC machine

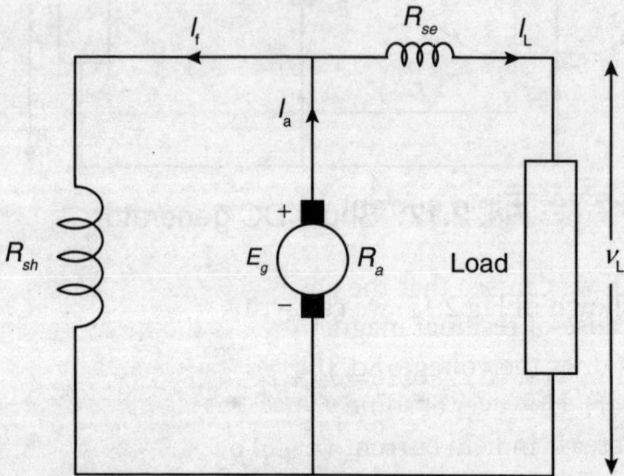

Fig. 2.19: Short-shunt DC machine

From the circuit of Fig. 2.18

$$I_a = I_L + I_f \tag{2.23}$$

$$I_f = \frac{V_L}{R_{sh}} \tag{2.24}$$

$$E_g = I_a R_a + I_a R_{se} + V_L \tag{2.25}$$

From the circuit of Fig. 2.19

$$I_a = I_L + I_f I_f = \frac{V_L}{R_{sh}}$$

$$I_f = \frac{V_L}{R_{sh}}$$

$$E_g = I_a R_a + I_L R_{se} + V_L \tag{2.26}$$

2.15 MAGNETISATION CHARACTERISTICS OF A DC GENERATOR

The magnetisation characteristics of a DC generator are represented by a plot of variation of generated voltage (at no load) against the change in field current at constant speed. In a separately excited DC generator, the field is excited by a separate source and the armature of the generator is rotated at a constant speed by some prime mover. The plot of variation of generated voltage with the change in field current is shown in Fig.2.20.

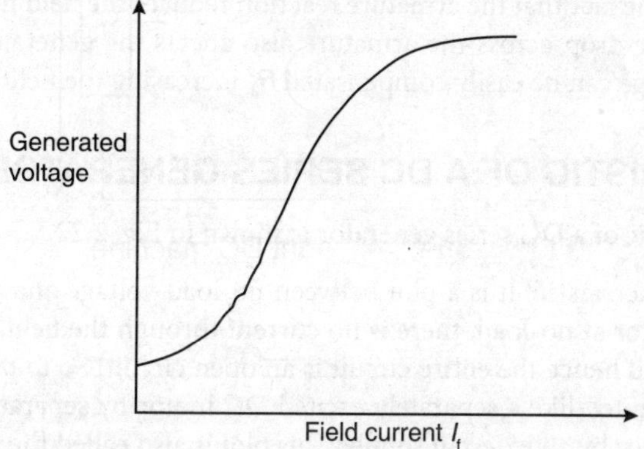

Fig. 2.20: Field current vs generated voltage plot

From the plot of Fig. 2.20, we can see that the curve does not start from zero, but from some value higher than zero. This is because of residual magnetism. As the field current is increased, the flux also proportionately increases, as does the voltage. So, the plot between the field current and the generated voltage is a straight line (linear). However at some value of field current, I_f, the magnetic circuit gets saturated. Hence any further increase in field current, I_f, will not increase the flux. Therefore, the generated voltage will now change very slightly with the change in field current.

This characteristic is called the magnetic characteristic or Open Circuit Characteristic (O.C.C.) of a separately excited DC generator.

2.16 CHARACTERISTIC OF A SEPARATELY EXCITED DC GENERATOR

We know that in a separately excited DC generator, the excitation current or the field current is independent of the load current. So, the plot between the field flux and the load current will be a straight line parallel to the x-axis. Since the voltage is proportional to the field flux, the plot between the voltage and the load current will also be a straight line parallel to x-axis. The plot is shown in Fig. 2.21(a).

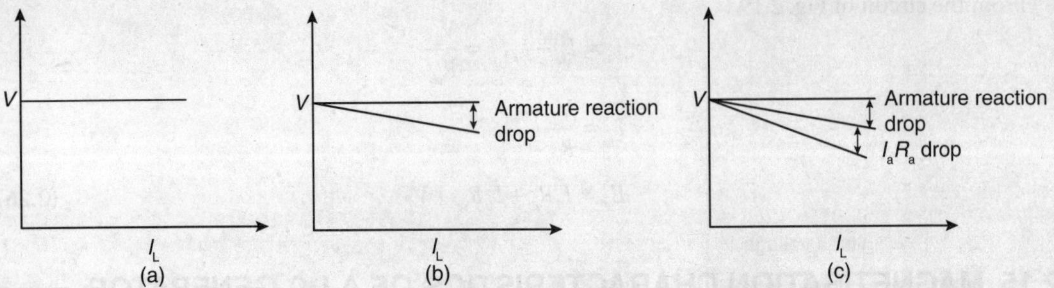

Fig. 2.21: Plot between voltage and load current

If the armature reaction is considered, then the graph will be slightly drooping in nature as shown in Fig. 2.21(b). Again, if I_aR_a drop is also taken into account, then the graph will further droop down shown in Fig.2.21(c).

Thus, from the plot it can be concluded that as the load current increases, the terminal voltage decreases. This is due to the fact that the armature reaction reduces the field flux, which in turn reduces the generated voltage. The drop across the armature also affects the generated voltage. However, this decrease in terminal voltage can be easily compensated by increasing the field current.

2.17 CHARACTERISTIC OF A DC SERIES GENERATOR

The complete characteristic of a DC series generator is shown in Fig. 2.22.

i) **Open-circuit characteristic**: It is a plot between no-load voltage and field excitation. However, for a DC series motor at no-load, there is no current through the field, since at no-load the load terminal is open and hence the entire circuit is an open circuit. So to plot this characteristic, DC series motor is operated like a separately excited DC motor by separating the field winding and exciting the generator by an external supply. This plot is also called the magnetic characteristic of a DC generator.

ii) **Internal characteristic: It is** the plot between the voltage generated in the armature and the load current. This plot can be obtained by subtracting the effect of armature reaction drop from the no-load voltage. Thus, the actual generated voltage (E_g) will be less than the no-load voltage (E_0) of a DC generator. Because of this armature reaction drop, the curve drops slightly from the open-circuit characteristic curve.

iii) **External characteristic**: It is the plot between terminal voltage (V) and the load current (I_L). Terminal voltage is obtained by subtracting the drop across the armature and the drop across the field from the generated e.m.f. it is given by the expression,

$$\text{Terminal voltage } V = E_g - I\left(R_a + R_f\right) \tag{2.27}$$

From the graph, we can say that the terminal voltage is less than the generated voltage. It is also seen from the graph that with increase in load current the terminal voltage of the machine increases up to a certain point and after that it starts to decrease due to excessive demagnetising effect of armature reaction (shown in the plot by a dotted line). In a DC series generator, there is an increase in both field current and armature current with increase in load. However, due to saturation effect, the magnetic field does not change and so, neither does the induced voltage. However, due to increased armature current, the armature reaction will increase thereby reducing the load voltage. As per Ohm`s law, with decrease in load voltage, the load current will also decrease proportionately. Due to this cumulative effect, there will be no significant change in load current in the dotted portion of external characteristics of series DC generator. That is why, series DC generator is called constant current DC generator.

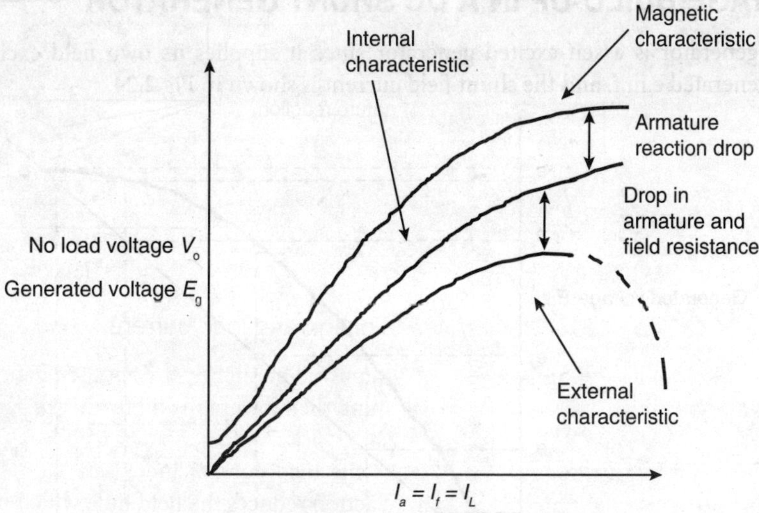

Fig. 2.22: Characteristic of DC series generator

2.18 CRITICAL LOAD RESISTANCE

Consider the characteristic shown in Fig.2.23. Let the line OA represent the resistance of the external load circuit. The point of intersection of this line with the external characteristic gives the current (*x*-axis) and the terminal voltage (y-axis) for that particular load resistance. Now, if the load resistance is chosen in such a way that the line is a tangent to the external characteristics (line OB), then that resistance is called the critical load resistance. This is the maximum load resistance which the generator will be able to excite. If the load resistance increases beyond this point, then it will not cut the external characteristic and hence the generator will not excite and there will be no current.

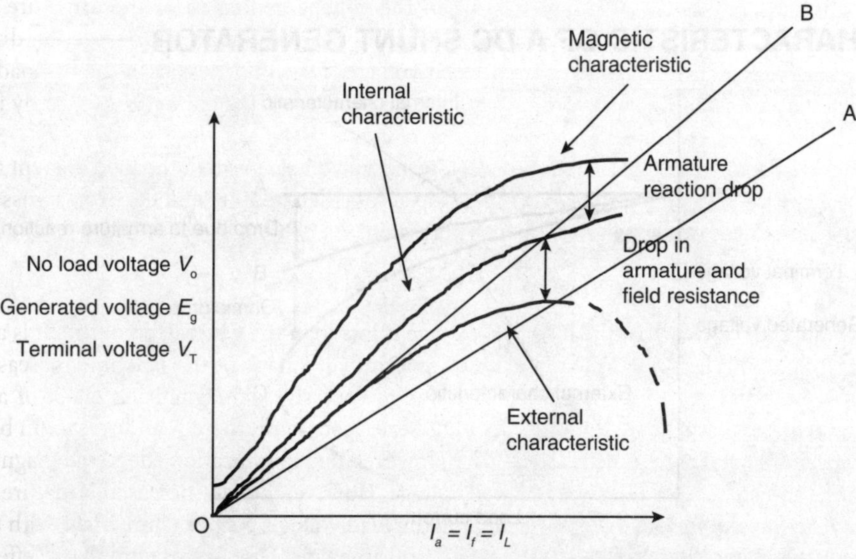

Fig. 2.23: Obtaining critical load resistance

2.19 VOLTAGE BUILD-UP IN A DC SHUNT GENERATOR

A DC shunt generator is a self-excited generator since it supplies its own field excitation. The plot between the generated e.m.f. and the shunt field current is shown in Fig.2.24.

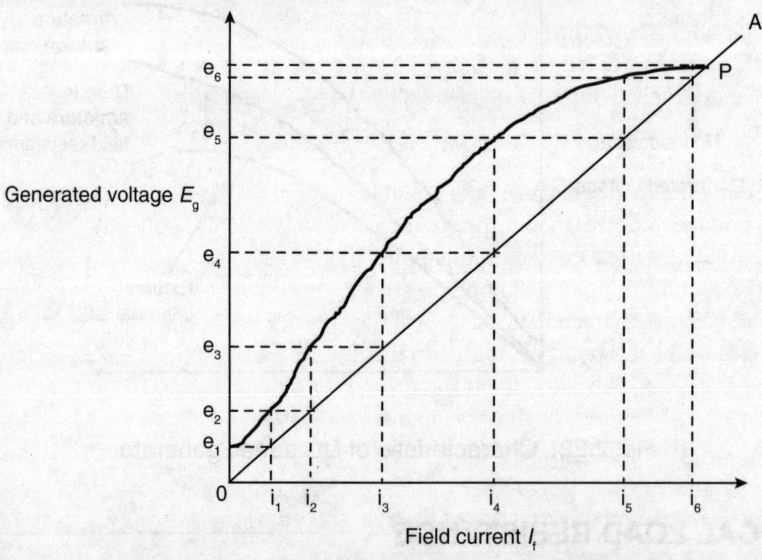

Fig. 2.24: Voltage build-up in DC shunt generator

In the plot, the line OA represents the shunt field resistance. Due to the residual flux in the field poles of the generator, a small voltage (O_{e1}) will be produced when the generator is activated. Due to this voltage, a current (O_{i1}) will flow through the field winding. At this field current, the generated e.m.f. will be O_{e2}. This, in turn, will give rise to a field current of O_{i2}, which will generate an e.m.f. of O_{e3}. This process of voltage build up will continue until it reaches the point *P*. It is the point where the line OA cuts the e.m.f. plot.

2.20. CHARACTERISTIC OF A DC SHUNT GENERATOR

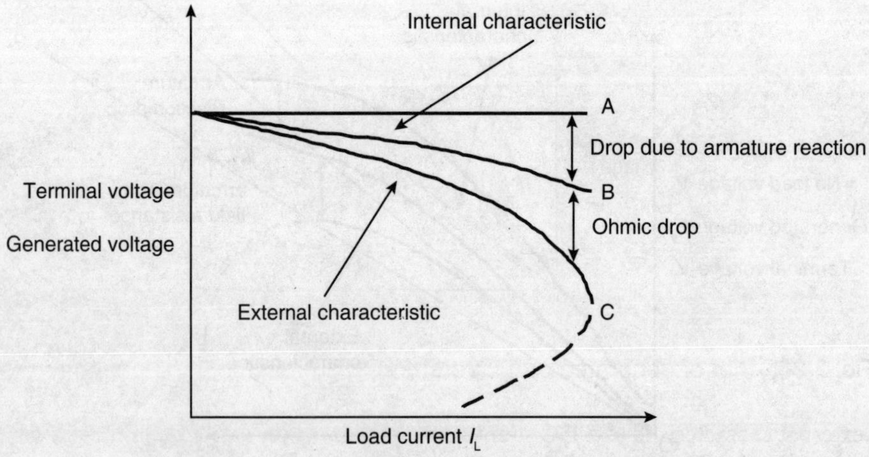

Fig. 2.25: Characteristic of DC shunt generator

i) **Internal Characteristic**: It is the plot between the generated voltage (E_g) and the load current (I_L). The generated voltage is generally affected by armature reaction during loaded condition. So, the generated voltage on loaded condition will be slightly less than the generated voltage at no-load condition. In the plot shown in Fig.2.25, Graph A shows the generated voltage at no-load condition and Graph B shows the generated voltage during loaded condition.

ii) **External Characteristic**: It is the plot between the terminal voltage and load current. Due to the armature resistance drop, this terminal voltage is slightly less than the generated voltage. Thus, we can write

$$V_T = E_g - I_a R_a \tag{2.28}$$

By decreasing the load resistance, it is possible to increase the load current, which will cause the terminal voltage to decrease. But this load current will increase only up to a certain point (Point C in Fig. 2.25). The load current at this point is called the full load current. Decreasing the load resistance beyond this point will not increase the load current but rather, will cause it to decrease. Hence, the external characteristic curve will turn back, as shown in Fig. 2.25 by a dotted line. This is because beyond the point of full load current, the demagnetising effect of armature reaction and the voltage drop will cause the load current to decrease instead of increasing. That is why the load resistance of the machine must be maintained properly. The point at which the machine gives maximum current output is called *breakdown point*.

2.21 CRITICAL FIELD RESISTANCE AND CRITICAL LOAD RESISTANCE FOR A DC SHUNT GENERATOR

The plot of generated e.m.f and the field current is shown in Fig.2.26. Let the line OA represent the field resistance. From the graph, when the field resistance is OA, the generated e.m.f. will be the value where the line OA cuts the curve. The field resistance represented by the line OB is called the critical field resistance. This is because, if the field resistance is increased beyond this point (line OC), the generator will not excite. This is borne out by the fact that in Fig. 2.26, the line OC does not intersect the magnetisation characteristic.

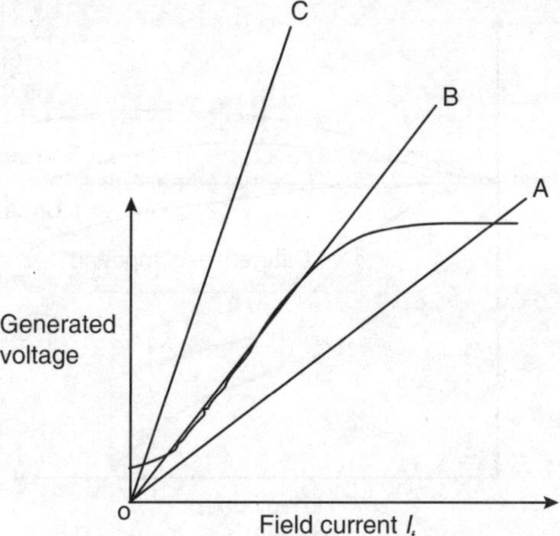

Fig. 2.26: Plot of generated e.m.f. versus field current for DC shunt generator

In the external characteristic of a DC shunt generator, shown in Fig. 2.27, let Line A represent the external load resistance. On decreasing the load resistance, we get the load resistance Line B. On further decreasing the load resistance we get the load Line C. The load Line C is called the critical load resistance, because if the load resistance is decreased further the generator will fail to build up its voltage.

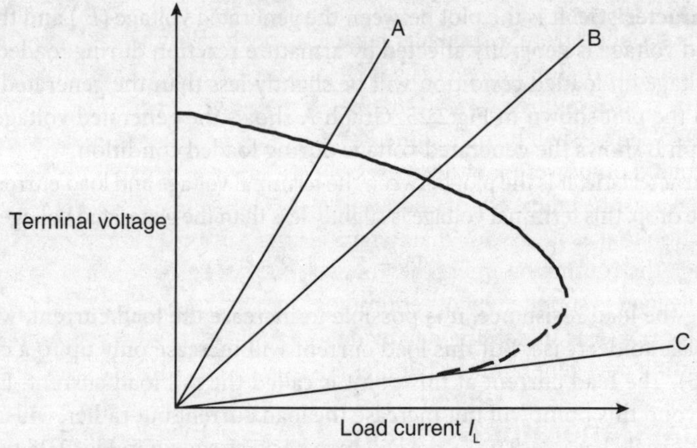

Fig. 2.27: External characteristic of DC shunt generator

2.22 CHARACTERISTIC OF A COMPOUND DC GENERATOR

The external characteristic of a compound generator is shown in Fig.2.28. The characteristic curves for a cumulative compound generator is a combination of those for series and shunt generator. At no-load, the shunt field winding provides a field flux and when a load is applied, the flux of series field winding adds up with the shunt field flux.

If the series turns are so adjusted that with increase in load current the terminal voltage also increases, then the generator is said to be over-compounded. If the series turns are so adjusted that with the increase in load current the terminal voltage remains constant, then the generator is considered as flat-compounded. If the series field winding has lesser number of turns, then the rated terminal voltage becomes less than the no-load voltage, and the generator becomes under-compounded.

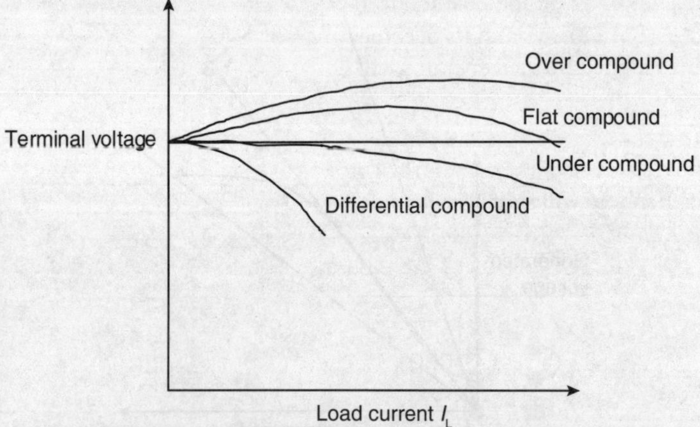

Fig. 2.28: Compound DC generator characteristic

2.23 PARALLEL OPERATION OF DC GENERATOR

In a power plant, power is usually supplied from several generators of small ratings connected in parallel rather than from one large generator. This arrangement has many advantages,

i) If the power is supplied from a single large generator, then in the case of breakdown, the whole plant will be shut down. But, if power is supplied from a number of small generators operating in parallel, then in case of failure of a few generators, the power plant will not shut down completely, and continuity of supply will be maintained by the other generators present there.

ii) If the load demand on power plant decreases, then it is possible to shut down one or more generators. This increases the efficiency of operation.

iii) Parallel operation makes it easy for maintenance and repair of generators. If generators are operated in parallel, the routine or emergency operations can be performed by isolating the affected generator while load is supplied by other units.

iv) The demand for power is increasing day by day. So, this increase in power demand can be easily met by connecting new generators in parallel

2.23.1 DC Shunt Generator Connected in Parallel

Consider a bus bar and two generators connected as shown in Fig. 2.29. Let us assume that generator, G1is already connected to the bus bar and now generator G2 needs to be connected in parallel with the existing generator G1. To connect the generator G2 with the bus the following procedure is followed.

- The prime mover of the generator to be connected to the bus (generator G2) is brought to its rated speed.
- Switch SG_3 is closed, which connects the field of the generator G2.
- The circuit breaker CBG2 is closed and the excitation of the generator G2 is changed so that the generated voltage of the generator becomes equal to the bus bar voltage. This can be easily checked with the help of a voltmeter.
- Once the generated voltage becomes equal to the bus bar voltage the switch SG_4 is closed thereby connecting the generator G2 with the bus. Now, both the generators G1 and G2, are connected in parallel.
- At this stage no load is given to the generator G2 and hence it is said to be "floating".
- To bring the generator G2, in the position to take the load, its field current is increased, so that its generated voltage (E) increases and its becomes greater than the bus bar voltage. Under this condition, the current supplied by the generator will be

$$I_2 = \frac{G_2 - G_1}{R_2}$$

Where, R_2 is the resistance of the armature of generator G2.
Now both the generators will share the load demand.

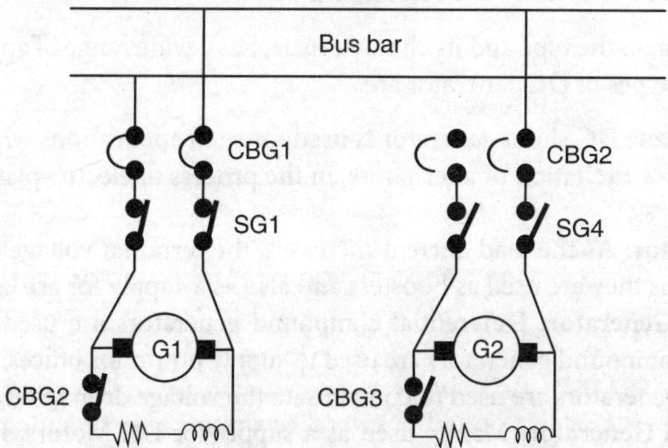

Fig. 2.29: Shunt generators in parallel

2.23.2 Load sharing

The load sharing between the generators connected in parallel with the bus bar can be done by adjusting the field excitation. If field excitation is changed, then the voltage of the generator will change and hence the current output of the generators.

2.23.3 Compound Generators in Parallel

To operate the compound generators in parallel, its series fields must be connected in parallel. This is done with the help of an equaliser bar as shown in Fig.2.30.

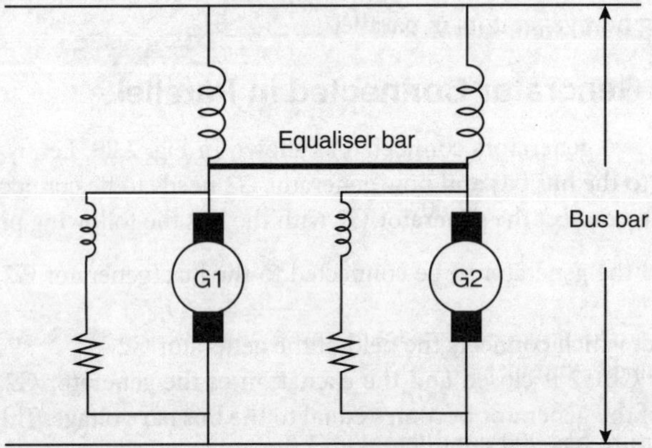

Fig. 2.30: Compound generators in parallel

Without the equaliser bar, if current supplied by the generator G1 increases by any means, it will cause the current of generator G2 to decrease, since the load supplied by the system is constant. This effect is cumulative and hence, a stage will reach when the generator G1 will take the entire load causing the generator G2 to act as a motor. Eventually, the machine G2 (now acting as a motor) will act as a short circuit, in parallel to generator G1. This will bring the entire system into stand still.

With the presence of the equaliser bar neither machine tends to take all the load and hence the system will not come to a standstill.

2.24 APPLICATION OF DC GENERATOR

DC generator, depending on the type and its characteristic, has a wide range of applications. Some of the applications of different types of DC generator are:

i) **DC shunt generator:** DC shunt generator is used for such applications where constant voltage is required, such as for excitation of alternators, in the process of electro-plating, for battery charging, etc.

ii) **DC series generator:** As the load current increases, the terminal voltage of the series generator also increases. Thus they are used as boosters and also as a supply for arc lamps.

iii) **DC Compound Generator:** Differential compound generators are used to supply dc welding machines. Level compound generators are used to supply power for offices, hostels and lodges etc. Over compound generators are used to compensate the voltage drop in feeders.

iv) **Separately Exited Generator:** Mainly used as a supply for DC Motors, it is also used for such applications where a wide range of voltage is required for the purposes of testing.

SOLVED PROBLEMS

Ex.2.1 A 16-pole armature has 990 conductors. It is rotating at 600 rpm. Calculate the induced e.m.f. on open circuit, when the DC generator is having,

(i) Lap winding
(ii) Wave winding

Take flux per pole to be 50×10^{-4} Wb

Solution

Given, $P = 16$, $N = 600$ rpm, $\varphi = 50 \times 10^{-4}$ Wb, $Z = 990$.

(i) For Lap winding
$A = P = 16$.

$$\text{Induced e.m.f.}(E) = \frac{\varphi ZNP}{60A} = \frac{50 \times 10^{-4} \times 990 \times 600 \times 16}{60 \times 16} = 49.5 \text{ V}$$

(ii) For wave winding
$A = 2$.

$$\text{Induced e.m.f.}(E) = \frac{\varphi ZNP}{60A} = \frac{50 \times 10^{-4} \times 990 \times 600 \times 16}{60 \times 2} = 396 \text{ V}$$

Ex.2.2 An 8-pole armature has 900 conductors and has wave winding. When it runs at 1500 rpm it generates an e.m.f. of 900 V. Calculate the flux per pole.

Solution

Given, $P = 8$, $N = 1500$ rpm, $Z = 900$, $E = 900$, $A = 2$.
We know that,

$$\text{Induced e.m.f.}(E) = \frac{\varphi ZNP}{60A} \qquad \text{(i)}$$

From Eq. (i), we can write,

$$\varphi = \frac{60AE}{ZNP} = \frac{60 \times 2 \times 900}{900 \times 1500 \times 8} = 0.01 \text{ Wb}$$

Ex.2.3 An 8-pole armature of a DC generator is rotating with a speed of 1000 r.p.m. and it is having 50 slots with 10 conductors per slot. Calculate the induced e.m.f. if the flux per pole is 30 mWb. Consider, (i) a wave winding generator and (ii) a lap winding generator.

Solution

Given, $N = 1000$ r.p.m., $Z = 50 \times 10 = 500$, $\varphi = 30$ mWb.
Induced e.m.f. is given by the expression,

$$E = \frac{\varphi ZNP}{60A} \qquad \text{(i)}$$

(i) For wave winding, $A = 2$, so Eq.(i) becomes,

$$E = \frac{30 \times 10^{-3} \times 500 \times 1000 \times 8}{60 \times 2} = 1000 \text{ V}$$

(ii) For lap winding, $A = P$, so Eq. (i) becomes,

$$E = \frac{30\times10^{-3}\times500\times1000\times8}{60\times8} = 250\text{ V}$$

Ex.2.4 A 4-pole DC generator generates 600V with the armature having 40 slots with 6 conductors per slot. If the flux per pole is 20mWb, calculate its speed of rotation for a (i) wave-wound armature and (ii) lap-wound armature.

Solution

Given, $P = 4$, $E_b = 600V$, $Z = 40\times6 = 240$, $\varphi = 20\text{mWb}$

(i) For wave winding (A = 2).
We know that

$$E = \frac{\varphi ZNP}{60A} \qquad \text{(i)}$$

From, Eq. (i),

$$N = \frac{60AE}{\varphi ZP} = \frac{60\times2\times600}{20\times10^{-3}\times240\times4} = 3750\text{ r.p.m.}$$

(ii) For lap winding, $A = P = 4$.

$$N = \frac{60AE}{\varphi ZP} = \frac{60\times4\times600}{20\times10^{-3}\times240\times4} = 7500\text{ r.p.m.}$$

Ex.2.5 A 4-pole, wave-wound DC generator rotates with a speed of 550 r.p.m. The armature is 30 cm long and 25 cm in diameter, with 200 coils of 12 turns each. Calculate the induced e.m.f., if the flux density in the air gap is 0.5 Wb/m² and each pole subtends an angle of 120⁰.

Solution

Given, $P = 4$, $l = 30$ cm = 0.3 m, $A = 2$ (wave winding), $N = 550$ r.p.m., diameter (d) = 25 cm = 0.25 m, $B = 0.5$ Wb/m², $z = 200\times12\times A = 200\times12\times2 = 4800$.

$$\text{Radius}\,(r) = \frac{d}{2} = \frac{0.25}{2} = 0.125\text{ m}$$

$$\text{Area of air gap}\,(a) = \frac{2\pi rl}{P}\times\frac{120}{180}$$

$$= \frac{2\times\pi\times0.125\times0.3}{4}\times\frac{120}{180} = 0.0393\text{ m}^2$$

Again,

$$\varphi = B\times a = 0.5\times0.0393 = 0.01965\text{ Wb}$$

Then,

$$E = \frac{\varphi ZNP}{60A} = \frac{0.01965\times4800\times550\times4}{60\times2} = 1729.2\text{ V}$$

Ex.2.6 A 4-pole lap wound DC generator rotates at 550 r.p.m. and generates a voltage of 900 V. The armature has 1200 conductors. The diameter of the pole shoe is 0.25 m and the length of the shoe is 0.3m. If the ratio of the pole arc to pole pitch is 0.9, find the flux per pole and flux density.

Solution

Given, $P = 4$, $N = 550$ r.p.m., $E = 900V$, $Z = 1200$, $d = 0.25$ m, $l = 0.3$ m, $A = P = 4$.

$$\text{Pole pitch} = \frac{\pi d}{P} = \frac{\pi \times 0.25}{4} = 0.1963 \text{ m}$$

Given,

$$\frac{\text{Pole arc}}{\text{Pole pitch}} = 0.9$$

$$\text{Pole arc} = 0.9 \times \text{Pole pitch} = 0.9 \times 0.1963 = 0.17667 \text{ m}$$

$$\text{Area of pole arc} = 0.17667 \times l = 0.17667 \times 0.3 = 0.053 \text{ m}^2.$$

Then,

$$\text{Flux per pole}(\varphi) = \frac{60AE}{ZNP} = \frac{60 \times 4 \times 900}{1200 \times 550 \times 4} = 81.818 \text{ mWb}$$

$$\text{Flux density}(B) = \frac{\varphi}{\text{Area}} = \frac{81.818 \times 10^{-3}}{0.053} = 1.544 \text{ Wb/m}^2$$

Ex.2.7 A DC generator is having 100 slots with 12 conductors per slot. When it runs at 900 r.p.m., it generates a no-load e.m.f. of 400 V. What will be its speed to generate a voltage of 300 V?

Solution

Case-1: When the speed of the generator (N_1) = 900 r.p.m., the generated voltage (E_1) = 400 V. For the generator, flux φ, Z, P and A are constant. So the e.m.f. equation becomes,

$$E = KN \qquad \text{(i)}$$

Where, K is a constant.
So for the Case-1, Eq. (*i*), becomes,

$$E_1 = KN_1 \qquad \text{(ii)}$$

For the generated voltage (E_2) = 300V, let the speed of the motor be N_2. Then Eq. (i) becomes,

$$E_2 = KN_2 \qquad \text{(iii)}$$

Dividing Eq. (iii) by Eq. (ii), we get,

$$\frac{E_2}{E_1} = \frac{N_2}{N_1}$$

$$or, N_2 = \frac{E_2}{E_1} \times N_1 = \frac{300}{400} \times 900 = 675 \text{ r.p.m.}$$

Ex.2.8 The armature of a 16-pole DC generator is having 900 conductors and is carrying a current of 382 A. Find the demagnetising ampere turns per pole and the cross-magnetisation ampere turns per pole. When the armature conductor is having,

(i) Wave winding
(ii) Lap winding

Under the following conditions:
(a) Brush shift is by 4^0 electrical
(b) Brush shift is by 2^0 mechanical
(c) Brushes are on GNP

Solution

Given, $P = 16$, $Z = 900$, $I_a = 382$ A.

(i) For wave winding (A=2):

(a) $\alpha = 4^0$ electrical

$$\text{Demagnetising ampere turns per pole} = \frac{ZI_a \propto}{180AP} = \frac{900 \times 382 \times 4}{180 \times 2 \times 16} = 239 \text{ AT / pole}$$

$$\text{Cross-magnetising ampere turns per pole} = \frac{ZI_a}{AP} \times \left(\frac{180 - 2\propto}{180}\right)$$

$$= \frac{900 \times 382}{2 \times 16} \times \left(\frac{180 - 2 \times 4}{180}\right)$$

$$= 10266 \text{ AT / pole}$$

(b) 2^0 mechanical

$$1^0 \text{ mechanical} = \frac{P}{2} \text{electrical degree}$$

$$so, 2^0 \text{ mechanical} = \frac{P}{2} \times 2 = \frac{16}{2} \times 2 = 16^0 \text{ electrical}$$

$$\text{Demagnetising ampere turns per pole} = \frac{ZI_a \propto}{180AP} = \frac{900 \times 382 \times 16}{180 \times 2 \times 16} = 955 \text{ AT / pole}$$

$$\text{Cross-magnetising ampere turns per pole} = \frac{ZI_a}{AP} \times \left(\frac{180 - 2\propto}{180}\right)$$

$$= \frac{900 \times 382}{2 \times 16} \times \left(\frac{180 - 2 \times 16}{180}\right)$$

$$= 8834 \text{ AT / pole}$$

(c) Brushes are on GNP

Since the brushes are on GNP, $\alpha = 0$. So,

$$\text{Demagnetising ampere turns per pole} = \frac{ZI_a \propto}{180AP} = 0 \text{ AT / pole}$$

$$\text{Cross-magnetising ampere turns per pole} = \frac{ZI_a}{AP} = \frac{900 \times 382}{2 \times 16} = 10744 \text{ AT / pole}$$

(ii) For lap winding (A = P = 16):

(a) $\alpha = 4^0$ electrical

$$\text{Demagnetising ampere turns per pole} = \frac{ZI_a \propto}{180AP} = \frac{900 \times 382 \times 4}{180 \times 16 \times 16} = 30 \text{ AT / pole}$$

$$\text{Cross-magnetising ampere turns per pole} = \frac{ZI_a}{AP} \times \left(\frac{180 - 2\propto}{180}\right)$$

$$= \frac{900 \times 382}{16 \times 16} \times \left(\frac{180 - 2 \times 4}{180}\right)$$

$$= 1283 \text{ AT / pole}$$

(b) 2^0 mechanical

$$1^0 \text{ mechanical} = \frac{P}{2} \text{ electrical degree}$$

$$so,\ 2^0 \text{ mechanical} = \frac{P}{2} \times 2 = \frac{16}{2} \times 2 = 16^0 \text{ electrical}$$

$$\text{Demagnetising ampere turns per pole} = \frac{ZI_a \propto}{180AP} = \frac{900 \times 382 \times 16}{180 \times 16 \times 16} = 119 \text{ AT / pole}$$

$$\text{Cross-magnetising ampere turns per pole} = \frac{ZI_a}{AP} \times \left(\frac{180 - 2\propto}{180}\right)$$

$$= \frac{900 \times 382}{16 \times 16} \times \left(\frac{180 - 2 \times 16}{180}\right)$$

$$= 1104 \text{ AT / pole}$$

Brushes are on GNP

Since the brushes are on GNP, $\alpha = 0$. So,

$$\text{Demagnetising ampere turns per pole} = \frac{ZI_a \propto}{180AP} = 0 \text{ AT / pole}$$

$$\text{Cross-magnetising ampere turns per pole} = \frac{ZI_a}{AP} = \frac{900 \times 382}{16 \times 16} = 1344 \text{ AT / pole}$$

Ex.2.9 The output terminal voltage of a DC shunt generator is 300V and the armature current and armature resistance is 120A and 0.2Ω respectively. Find the value of the induced e.m.f.

Solution

Given, V=300V, I_a = 120A and R_a = 0.2 Ω.

$$\text{The induced e.m.f.}\left(E_g\right) = V + I_a R_a = 300 + 120 \times 0.2 = 324 \text{ V}$$

Ex.2.10 The armature and the field resistance of a shunt generator is 0.2Ω and 80Ω respectively. The load connected across the output terminal draws 80 kW power at 300V. Calculate the following:

(i) Load current
(ii) Field current
(iii) Armature current
(iv) Generated e.m.f.

Solution

Given, R_a = 0.2Ω, R_f = 80Ω, P_L = 80kW, V_L = 300V.

$$\text{Load current } (I_L) = \frac{P}{V_L} = \frac{80 \times 10^3}{300} = 267\text{A}$$

$$\text{Field current } (I_f) = \frac{V_L}{R_f} = \frac{300}{80} = 3.75 \text{ A}$$

$$\text{Armature current } (I_a) = I_f + I_L = 3.75 + 267 = 270.75 \text{ A}$$

$$\text{Generated e.m.f.}\left(E_g\right) = V + I_a R_a = 300 + (270.75 \times 0.2) = 354.15\text{V}$$

Ex.2.11 If the load voltage and the load current of a DC shunt generator are 500V and 120A respectively, calculate the generated e.m.f. Take armature and field resistances as 0.3Ω and 90Ω respectively.

Solution

Given, V_L = 500V, I_L = 120A, R_a = 0.3Ω, R_f = 90Ω.

$$\text{Field current } (I_f) = \frac{V_L}{R_f} = \frac{500}{90} = 5.556 \text{ A}$$

$$\text{Armature current } \left(I_a\right) = I_L + I_f = 120 + 5.556 = 125.556 \text{ A}$$

$$\text{Generated e.m.f.}\left(E_g\right) = V_L + I_a R_a = 500 + 125.556 \times 0.3 = 537.667\text{V}$$

Ex.2.12 A shunt DC generator is connected across a lamp load having 100 lamps of 40W each. The terminal voltage is 230V. The armature and field resistance of the generator are 0.2Ω and 40Ω respectively. Calculate the following:

(i) The generated e.m.f.
(ii) Current in each armature conductor.

Assume the drop across the brush to be 0.8V per brush. Solve the problem for

(a) A wave-wound generator with 8 poles
(b) A lap-wound generator with 8 poles.

Solution

Given, V_L = 230V, R_a = 0.2Ω, R_f = 40Ω, Number of poles (P) = 8, Load = 100 lamps of 40W each.

$$\text{Field current } \left(I_f\right) = \frac{V_L}{R_f} = \frac{230}{40} = 5.75 \text{ A}$$

$$\text{Armature current } \left(I_a\right) = \frac{\text{Power}}{V_L} = \frac{100 \times 40}{230} = 17.39 \text{ A } \left(\text{Power factor is} = 1\right)$$

$$\text{Load current } \left(I_L\right) = I_a + I_f = 5.75 + 17.39 = 23.14 \text{ A}$$

(a) For a wave wound generator (A = 2)

$$\text{Current in each armature conductor} = \frac{17.39}{A} = \frac{17.39}{2} = 8.695 \text{ A}$$

$$\text{Generated e.m.f.}\left(E_g\right) = V_L + I_a R_a + \text{Brush voltage drop}$$

$$= 230 + (17.39 \times 0.2) + (2 \times 0.8) = 235.078\text{ V}$$

(b) For a wave lap generator ($A = P = 8$)

$$\text{Current in each armature conductor} = \frac{17.39}{A} = \frac{17.39}{8} = 2.174\text{ A}$$

$$\text{Generated e.m.f.}\left(E_g\right) - V_L + I_a R_a + \text{Brush voltage drop}$$

$$= 230 + (17.39 \times 0.2) + (2 \times 0.8) = 235.078\text{V}$$

Ex.2.13 The armature resistance, shunt field resistance and series field resistance of a DC compound generator are 0.2Ω, 40Ω and 0.09Ω respectively. The load connected across the generator draws a current of 100A at 230V. Calculate the generated e.m.f. when,

(i) It is a long shunt connection
(ii) It is a short shunt connection

Solution

Given, $V_L = 230\text{V}$, $I_L = 100\text{A}$, $R_{sh} = 40\Omega$, $R_{se} = 0.09\Omega$, $R_a = 0.2\Omega$.

(i) Long shunt connection

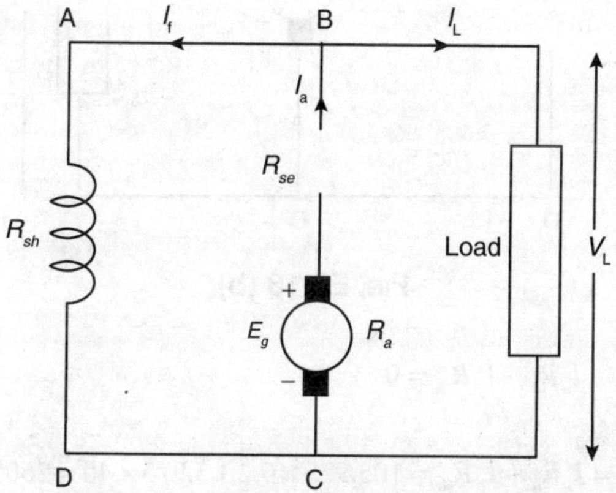

Fig. E2.13 (a):

From the long shunt connection shown in Fig. E2.13(a), we can write,

$$I_f = \frac{V_L}{R_{sh}} = \frac{230}{40} = 5.75\text{ A}$$

$$I_a = I_L + I_f = 5.75 + 100 = 105.75\text{ A}$$

Applying KVL in ABCDA

$$E_g - I_a R_a - I_a R_{se} - I_f R_{sh} = 0$$

$$or, E_g = I_a R_a + I_a R_{se} + V_L \because V_L = I_f R_{sh}$$

$$or, E_g = 105.75 \times 0.2 + 105.75 \times 0.09 + 230 = 260.67 \text{ V}$$

(ii) Short shunt connection

From the short shunt connection shown in Fig. E2.13(b), we can write,

$$V_{sh} = V_L + I_L R_{se} = 230 + 100 \times 0.09 = 239 \text{ V}$$

$$I_f = \frac{V_{sh}}{R_{sh}} = \frac{239}{40} = 5.975 \text{ A}$$

$$I_a = I_L + I_f = 100 + 5.975 = 105.975 \text{ A}$$

Applying KVL in ABCDA, we get,

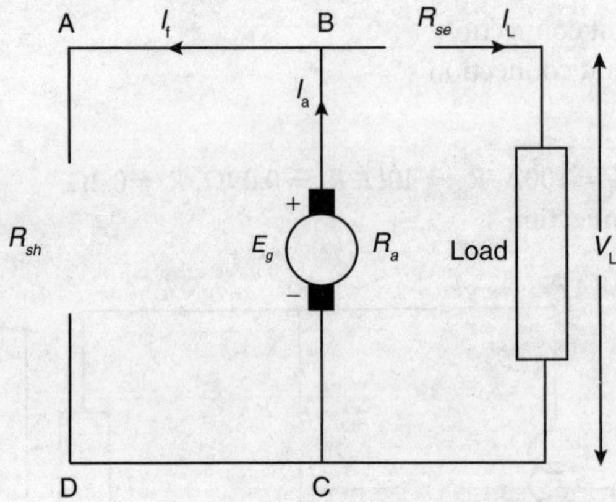

Fig. E2.13 (b):

$$E_g - I_a R_a - I_f R_{sh} = 0$$

$$or, E_g = I_a R_a + I_f R_{sh} = 105.975 \times 0.2 + 5.975 \times 40 = 260.195 \text{ V}$$

Ex.2.14 Consider a short-shunt DC generator with a load current of 90A at 200V. Calculate the generated e.m.f. if,

(i) The armature resistance is 0.2Ω, shunt field resistance is 150Ω, series field resistance is 0.1Ω,

(ii) Voltage drop per brush = 1V.

Solution

Given, $I_L = 90\text{A}$, $V_L = 200\text{V}$, $R_a = 0.2\Omega$, $R_{sh} = 150\Omega$, $R_{se} = 0.1\Omega$.

$$\text{Brush voltage } (V_b) = 1 \times 2 = 2\text{V (since there are two brushes)}$$

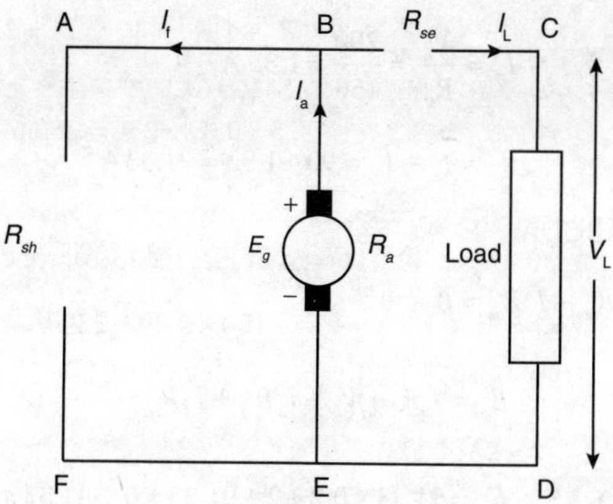

Fig. E2.14:

Applying KVL in ABCDEFA, we get,

$$I_L R_{se} + V_L - I_f R_{sh} = 0$$

$$or, (90 \times 0.1) + 200 - 150 I_f = 0$$

$$I_f = \frac{(90 \times 0.1) + 200}{150} = 1.39 \text{ A}$$

$$I_a = I_f + I_L = 1.39 + 90 = 91.39 A$$

Applying KVL in EBAFE, we get,

$$E_g - I_a R_a - V_b - I_f R_{sh} = 0$$

$$E_g = I_a R_a + V_b + I_f R_{sh} = 91.39 \times 0.2 + 2 + 1.39 \times 150 = 228.78 \text{ V}$$

Ex.2.15 Repeat Ex.2.14 for a long-shunt DC generator.

Solution

Given, $I_L = 90$A, $V_L = 200$V, $R_a = 0.2\Omega$, $R_{sh} = 150\Omega$, $R_{se} = 0.1\Omega$.

$$\text{Brush voltage } (V_b) = 1 \times 2 = 2\text{V (since there are two brushes)}$$

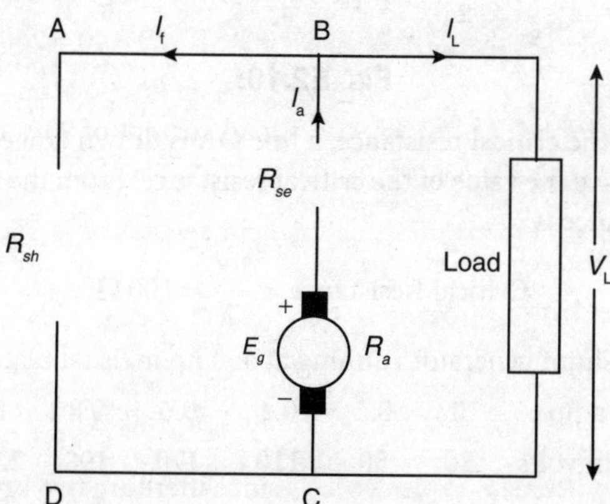

Fig. E2.15:

$$I_f = \frac{V_L}{R_{sh}} = \frac{200}{150} = 1.33\text{ A}$$

$$I_a = I_L + I_f = 90 + 1.33 = 91.33\text{ A}$$

Applying KVL in ABCDA

$$E_g - I_a R_a - V_b - I_a R_{se} - I_f R_{sh} = 0$$

$$E_g = I_a R_a + V_b + I_a R_{se} + I_f R_{sh}$$

$$E_g = 91.33 \times 0.2 + 2 + 91.33 \times 0.1 + 1.33 \times 150 = 228.89\text{ V}$$

Ex.2.16 The O.C.C of a DC shunt generator running at 1000 r.p.m. is tabulated as:

Field current (I_f) in amp	0	0.5	1	2	3	4	5	6	7
Induced e.m.f. (E_g) in volts	15	56	103	175	220	245	252	258	260

(i) Draw the O.C.C. from the given data.
(ii) Find the value of the critical resistance.

Solution

The O.C.C. plot is shown in Fig.E2.16.

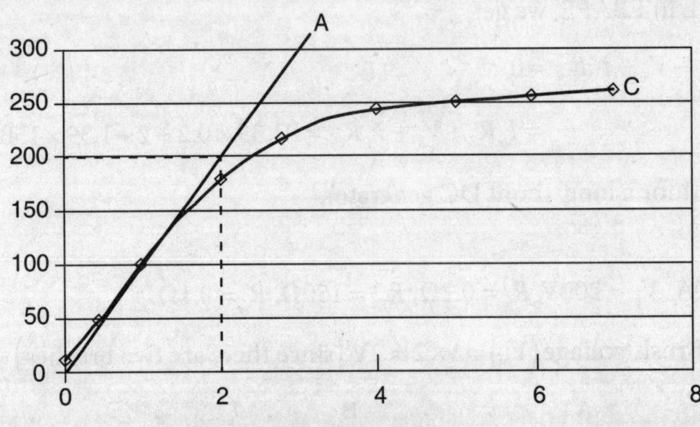

FIG. E2.16:

To find the value of the critical resistance, a line OA is drawn tangent to O.C.C. plot. The slope of the line OA will give the value of the critical resistance. From the plot of Fig.E.A.16, the value of the critical resistance is

$$\text{Critical Resistance} = \frac{200}{2} = 100\ \Omega$$

Ex.2.17 The O.C.C of a DC shunt generator running at 450 r.p.m. is tabulated as:

Field current (I_f) in amp	0	0.2	0.4	0.6	0.8	1	1.2	1.4	1.6	1.8
Induced e.m.f. (E_g) in volts	20	50	110	170	198	220	240	245	250	256

The field resistance of the DC machine is adjusted to 150 Ω. Calculate the following:

(i) Draw the O.C.C. from the given data.
(ii) Find the value of the critical resistance.
(iii) Find the no-load voltage
(iv) Find the extra resistance to be added in the field to reduce the no load voltage to 175V.

Solution

(i) The O.C.C plot is show in Fig. E2.17.

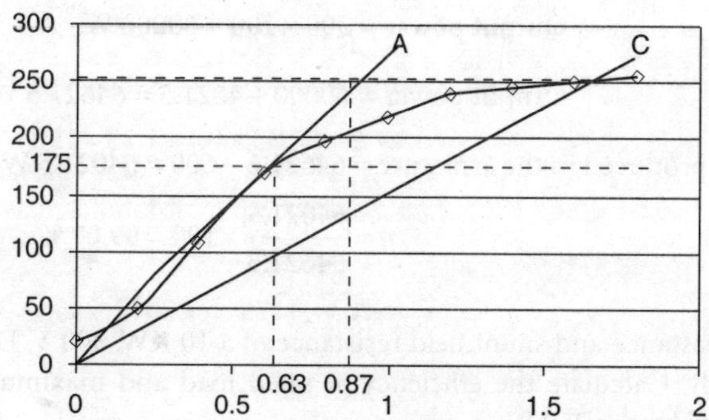

FIG. E2.17:

(ii) To find the value of the critical resistance a line OA is drawn tangent to O.C.C. plot. The slope of the line OA will give the value of the critical resistance. From the plot of Fig. E2.17, the value of the critical resistance is

$$R_{critical} = \frac{250}{0.87} = 287\ \Omega$$

(iii) To find the no-load voltage line OC is drawn, with a slope of 150 Ω. The intersection of the line OC with the O.C.C. plot, gives, Vno-load = 250V

(iv) From the O.C.C. plot

$$R_{inst} = \frac{175}{0.63} = 277.78\ \Omega$$

So the extra resistance required will be = 277.78 – 150 =127.76 Ω

Ex.2.18 The armature and the shunt field resistance of a shunt generator is 0.06 Ω and 60 Ω respectively. It has a terminal potential difference of 300V and delivers a current of 200A. Calculate the following, when the friction and iron losses equal to 600 W.

(i) Field current and armature current
(ii) Generated e.m.f.
(iii) Armature copper loss, shunt loss, stray loss and the total loss.
(iv) Output power, Input power and efficiency

Solution

$$\text{Field current}\left(I_{sh}\right) = \frac{300}{60} = 5\ \text{A}$$

$$\text{Armature current}\left(I_a\right) = 200 + 5 = 205\ \text{A}$$

$$\text{Generated } e.m.f. = 300 + I_a R_a = 300 + 205 \times 0.06 = 312.3 \text{ V}$$

$$\text{Armature copper loss} = I_a^2 R_a = 205^2 \times 0.06 = 2521.5 \text{ W}$$

$$\text{Shunt loss} = V \times I_{sh} = 300 \times 5 = 1500 \text{ W}$$

$$\text{Stray loss} = 600 \text{ W}$$

$$\text{Total loss} = 2521.5 + 1500 + 600 = 4621.5 \text{ W}$$

$$\text{Output power} = 300 \times 200 = 60000 \text{ W}$$

$$\text{Input power} = 60000 + 4621.5 = 64621.5 \text{ W}$$

$$\text{Electrical power produced in the armature} = 64621.6 - 600 = 64021.5 \text{ W}$$

$$\eta = \frac{64021.5}{64621.5} \times 100 = 99.07\ \%$$

Ex.2.19 The armature resistance and shunt field resistance of a 10 KW, 300 V, DC generator is 1Ω and 300Ω respectively. Calculate the efficiency at rated load and maximum efficiency. Take the no-load rotational loss as 250 W.

Solution

$$\text{Field current } (I_f) = \frac{300}{300} = 1\Omega$$

$$\text{Constant loss} = 300 \times 1 + 250 = 550 \text{ W}$$

$$\text{Load current } (I_L) = \frac{10 \times 10^3}{300} = 33.33 \text{ A}$$

$$\text{Armature current } (I_a) = I_f + I_L = 1 + 33.33 = 34.33 \text{ A}$$

$$\text{Armature copper loss} = I_a^2 R_a = 34.33^2 \times 1 = 1178.54 \text{ W}$$

$$\text{Input power} = 10 \times 10^3 + 550 + 1178.54 = 11728.54 \text{ W}$$

$$\text{Efficiency at rated load } (\eta_{rated}) = \frac{10 \times 10^3}{11728.54} \times 100 = 85.26\ \%$$

At maximum efficiency,

$$\text{Copper loss} = \text{Constant loss} = 550 \text{ W}$$

$$\text{Total loss} = 550 + 500 = 1100 \text{ W}$$

$$I_a^2 R_a = 550$$

$$so,\ I_a = \sqrt{\frac{550}{1}} = 23.45 \text{ A}$$

$$\text{Output current of the generatotr} = I_a - I_f = 22.45 \text{ A}$$

$$\text{Generator output} = 300 \times 22.45 = 6735 \text{ W}$$

$$\text{Generator input} = 6735 + 1100 = 7835 \text{ W}$$

$$\text{Maximum efficiency } (\eta_{max}) = \left(1 - \frac{1100}{7835}\right) \times 100 = 85.96\ \%$$

Ex.2.20 A 4-pole DC shunt generator with 492 armature conductors, supplies a current of 143 A and the shunt field current is 10 A. At full-load, the brushes are given an actual lead of 10^0. Calculate the demagnetising ampere turns per pole and also the number of extra shunt field turns required to neutralise the demagnetisation. Take,

(i) The armature winding to be wave-wound
(ii) The armature winding to be lap-wound.

Solution

Since the generator is shunt connected,

$$I_a = I_a + I_{sh} = 143 + 10 = 153\text{ A}$$

(i) For wave winding, A = 2.

$$\text{Current per conductor} = I_c = \frac{I_a}{A} = \frac{153}{2} = 76.5\text{ A}$$

$$\text{Demagnetising AT per pole} = (AT_d) = Z \times I_c \times \frac{\theta}{360} = 492 \times 76.5 \times \frac{10}{360} = 1045.5$$

$$\text{Shunt field turn required} = \frac{1045.5}{10} = 105\text{ turns}$$

(ii) For lap winding, A = P = 4.

$$\text{Current per conductor} = I_c = \frac{I_a}{A} = \frac{153}{4} = 38.25\text{ A}$$

$$\text{Demagnetising AT per pole} = (AT_d) = Z \times I_c \times \frac{\theta}{360} = 492 \times 38.25 \times \frac{10}{360} = 523$$

$$\text{Shunt field turn required} = \frac{523}{10} = 52\text{ turns}$$

Ex.2.21 A 4-pole, 440 V, 25 kW, DC generator has 846 conductors in the armature. Calculate the number of turns required on each interpole, if the flux density in the air gap under the interpoles is 0.5 Wb/m^2 at full-load and the radial gap length is 0.4 cm. Assume wave winding armature conductors and neglect the shunt winding. [UPTU-2005]

Solution

Since the shunt field is neglected.

$$\text{Armature current } (I_a) = \text{Load current } (I_L) = \frac{25 \times 10^3}{440} = 56.82\text{ A}$$

$$\text{Current per conductor} = I_c = \frac{I_a}{A}$$

$$\text{Armature cross AT per pole} = \frac{Z}{2P} \times I_a = \frac{Z}{2P} \times \frac{I_a}{A} = \frac{846}{2 \times 4} \times \frac{56.82}{2} = 3004\text{ AT}$$

$$\text{AT requited for commutating flux} = \frac{0.5}{4\times\pi\times10^{-7}}\times0.004 = 1592\text{ AT}$$

$$\text{Total interpole AT required} = 3004+1592 = 4596\text{ AT}$$

$$\text{Number of turns on each interpole} = \frac{4596}{56.82} = 81$$

Ex.2.22 A short shunt DC compound generator has armature resistance of 0.15 Ω, series field resistance of 0.1 Ω, shunt field resistance of 100 Ω. Calculate the e.m.f. generated if the brush drop is 2V per brush. Assume the generator supplies a load of 150 A at 230V.

Solution

$$\text{Series field drop}\left(V_{se}\right) = I_{se}R_{se} = 150\times0.1 = 15\text{ V}$$

$$\text{Shunt field drop}\left(V_{se}\right) = I_{se}R_{se} = 230+15 = 245\text{ V}$$

$$\text{Shunt field current}\left(I_{sh}\right) = \frac{245}{100} = 2.45\text{ A}$$

$$\text{Armature current}\left(I_{a}\right) = \frac{245}{100} = 150+2.45 = 152.45\text{ A}$$

$$\text{E.M.F.generated} = 230+150\times0.1+152.45\times0.15+2\times2 = 271.86\text{ V}$$

Ex.2.23 Consider two shunt generators, G1 and G2. The generator G1 is of 50 kW and G2 is of 100 kW and they are operated in parallel. The total load current is 250 A and the voltage ratio of both the machine is 500 V. Calculate the current generated by each generator and the terminal voltage if the voltage regulation of generator G1 and G2 is 6% and 4% respectively.

Solution

Generator G1:

$$I_1 = \frac{50\times10^3}{500} = 100\text{ A}$$

$$\text{Full load voltage drop} = 0.06\times500 = 30\text{ V}$$

$$\text{Voltage drop per ampere of current} = \frac{30}{100} = 0.3\text{ V / A}$$

Generator G2:

$$I_2 = \frac{100\times10^3}{500} = 200\text{ A}$$

$$\text{Full- load voltage drop} = 0.04\times500 = 20\text{ V}$$

$$\text{Voltage drop per ampere of current} = \frac{20}{200} = 0.1\text{ V / A}$$

Let the current supplied by the generator G1 and G2 be i_1 and i_2 respectively and the terminal voltage be V_t.

$$\text{Terminal volatge of generator G1}(V) = 500 - 0.3i_1 \quad \text{(i)}$$

$$\text{Terminal volatge of generator G2}(V) = 500 - 0.1i_2 \quad \text{(ii)}$$

From Eq. (i) and (ii)

$$i_2 = 3i_1 \quad \text{(iii)}$$

Again,

$$i_1 + i_2 = I_L = 250 \quad \text{(iv)}$$

Solving Eqs. (i), (ii), (iii) and (iv), we get,

$$i_1 = 62.5 \text{ A}$$

$$i_2 = 187.5 \text{ A}$$

$$V = 481.25 \text{ V}$$

Ex.2.24 Two shunt generators G1 and G2 connected in parallel. The e.m.f. of generators G1 and G2 are 250V and 260 V respectively. The field and the armature resistances of the generators are 100 Ω and 0.05 Ω respectively. Calculate the terminal voltages of the two generators and also the output power if the common load is 1500 A.

Solution

Let i_1 and i_2 be the current supplied by the generators G1 and G2 respectively, and let the terminal voltage be V_t. Then,

$$i_1 + i_2 = 1500 \quad \text{(i)}$$

For generator G1:

$$V_t = 250 - \left(i_1 + \frac{V_t}{100}\right) \times 0.05 \quad \text{(ii)}$$

For generator G2:

$$V_t = 260 - \left(i_2 + \frac{V_t}{100}\right) \times 0.05 \quad \text{(iii)}$$

Solving Eqs. (i), (ii) and (iii), we get,

$$i_1 = 650 \text{ A}$$

$$i_2 = 850 \text{ A}$$

$$V_t = 217.4 \text{ V}$$

$$\text{Power output of generator } G1 = V_t i_1 = 217.4 \times 650 = 141.31 \text{ kW}$$

$$\text{Power output of generator } G2 = V_t i_2 = 217.4 \times 850 = 184.79 \text{ kW}$$

Ex.2.25 A 250 V, 10 kW DC generator has 1400 turns on each pole. At shunt field current of 2 A, no-load voltage is 250 V (at rated speed). The same load voltage of 250 V can also be produced by a field current of 2.2 A at rated load. Load voltage is to be maintained constant without changing the field current, by adding a series field winding. Calculate the number of series field turns required for long shunt connection.

Solution

$$\text{At rated load, the mmf required} = 2.2 \times 1400 = 3080 \text{ AT}$$

$$\text{No load mmf} = 2 \times 1400 = 2800 \text{ AT}$$

$$\text{mmf required to be supplied by series field winding} = 3080 - 2800 = 280 \text{ AT}$$

$$\text{Series field current at full load} = \frac{10 \times 10^3}{250} = 40 \text{ A}$$

$$\text{Series field turns required} = \frac{280}{40} = 7 \text{ turns}$$

Ex.2.26 A DC shunt generator with 8 poles and field and armature resistance of 110Ω and 0.5Ω respectively supplies a load of 15 A at 220V. Calculate the generated e.m.f.

Solution

$$\text{Shunt field current } (I_{sh}) = \frac{220}{110} = 2 \text{ A}$$

$$\text{Armature current } (I_a) = 15 + 2 = 17 \text{ A}$$

$$\text{Generated e.m.f.} = 220 + 17 \times 0.5 = 228.5 \text{ V}$$

Ex.2.27 What will be the new generated e.m.f. if 1.2V per brush voltage drop is considered for the generator of Ex.2.26.

Solution

$$\text{Generated e.m.f.} = 228.5 + 1.2 \times 2 = 230.9\text{V}$$

Ex.2.28 A DC series generator with 8 poles and field and armature resistance of 0.4Ω and 0.5Ω respectively supplies a load of 15 A at 220V. Calculate the generated e.m.f. Take 1.2 V per brush voltage drop

Solution

$$\text{Generated e.m.f.} = 220 + 15 \times (0.4 + 0.5) + 2 \times 1.2 = 235.9 \text{ V}$$

Ex.2.29 Consider a generator with a 4-pole armature having 700 conductors and wave winding running at 2000 rpm and generating a voltage of 500 V. What is the flux per pole.

Solution

$$\varphi = \frac{60AE}{ZNP} = \frac{60 \times 2 \times 500}{700 \times 2000 \times 4} = 0.0107 \text{ Wb}$$

Ex.2.30 A 8-pole DC generator with the armature having 30 slots with 6 conductors per slot. If the flux per pole is 15 mWb, calculate its speed of rotation for a (i) wave wound armature and (ii) lap wound armature. Assume that it is generating a voltage of 500V.

Solution

For wave winding ($A = 2$), then

$$\text{Speed}(N) = \frac{60AE}{\varphi ZP} = \frac{60 \times 2 \times 500}{15 \times 10^{-3} \times 180 \times 8} = 2778 \text{ r.p.m.}$$

For lap winding ($A = P = 8$), then

$$N = \frac{60AE}{\varphi ZP} = \frac{60 \times 8 \times 500}{15 \times 10^{-3} \times 180 \times 8} = 11111 \text{ r.p.m.}$$

Ex.2.31 A lap-wound DC generator with 8 poles is rotating at 1000 r.p.m. The armature with 170 coils of 10 turns each is 20 cm long and 12 cm in diameter. The flux in the air gap is 0.02 Wb and each pole subtends an angle of 120°.

Solution

$$\text{Radius}(r) = \frac{0.12}{2} = 0.06 \text{ m}$$

$$\text{Area of air gap}(a) = \frac{2 \times \pi \times 0.06 \times 0.2}{8} \times \frac{120}{180} = 0.00628 \text{ m}^2$$

$$E = \frac{\varphi ZNP}{60A} = \frac{0.02 \times 170 \times 10 \times 1000 \times 8}{60 \times 8} = 566.67 \text{ V}$$

Ex.2.32 A 4-pole wave wound DC generator, generates a voltage of 1000V at a speed of 600 r.p.m.. The armature is having 910 conductors. The diameter of the pole shoe is 0.2m and the length of the shoe is 0.4m. If the ratio of the pole arc to pole pitch is 1, find flux density.

Solution

$$\text{Pole pitch} = \frac{\pi d}{P} = \frac{\pi \times 0.2}{4} = 0.1571 \text{ m}$$

Given,

$$\frac{\text{Pole arc}}{\text{Pole pitch}} = 1$$

$$\text{Pole arc} = 1 \times 0.1571 = 0.1571 \text{ m}$$

$$\text{Area of pole arc} = 0.1571 \times 0.4 = 0.0628 \text{ m}^2.$$

Then,

$$\varphi = \frac{60AE}{ZNP} = \frac{60 \times 2 \times 1000}{910 \times 600 \times 4} = 0.0549 \text{ Wb}$$

$$\text{Flux density}(B) = \frac{0.0549}{0.0628} = 0.8742 \text{ Wb/m}^2$$

Ex.2.33 The load connected across the output terminal of a shunt generator with the armature and the field resistance of 0.25Ω and 92Ω respectively, is 72 kW at 230V. Calculate the generated e.m.f.

Solution

$$\text{Load current}(I_L) = \frac{72 \times 10^3}{230} = 313.04 \text{ A}$$

$$\text{Field current}(I_f) = \frac{230}{92} = 2.5 \text{ A}$$

For shunt generator,

$$\text{Armature current}(I_a) = 2.5 + 313.04 = 315.54 \text{ A}$$

$$\text{Generated e.m.f.}\left(E_g\right) = 230 + (315.54 \times 0.25) = 308.885\ \text{V}$$

Ex.2.34 A 4-pole, wave wound, shunt DC generator is connected across a resistive load of 5kW. The load voltage is 300V. If the armature and field resistance of the generator are 0.25Ω and 80Ω respectively, calculate the generated e.m.f. Assume the drop across the brush to be 1V per brush.

Solution

Given, $V_L = 230\text{V}$, $R_a = 0.2\Omega$, $R_f = 40\Omega$, Number of poles (P) = 8, Load= 100 lamps of 40W each.

$$\text{Field current}\left(I_f\right) = \frac{300}{80} = 3.75\ \text{A}$$

$$\text{Armature current}\left(I_a\right) = \frac{5\times10^3}{300} = 16.67\ \text{A} \qquad \text{For shunt generator}$$

$$\text{Load current}\left(I_L\right) = I_a + I_f = 16.67 + 3.75 = 20.42\ \text{A}$$

$$\text{Generated e.m.f.}\left(E_g\right) = V_L + I_a R_a + \text{Brush voltage drop}$$

$$= 300 + (16.67 \times 0.25) + (2 \times 1) = 306.167\ \text{V}$$

EXERCISES

1. What is the principle of operation of a DC generator?
2. What is the function of commutator?
3. List and explain the different parts of a DC machine.
4. Explain why the poles of a DC machine are laminated.
5. What is the function of a pole shoe?
6. Explain why the air gap between the pole and the armature is kept small?
7. What are the two different types of armature winding?
8. What is the function of yoke?
9. What is the function of brush?
10. What material is the brush made up of?
11. What is the function of interpole?
12. Why is interpole used in a DC generator?
13. Explain the working of a DC generator.
14. Prove that the e.m.f. induced in the armature of a DC generator is alternating in nature.
15. What is the function of a slip ring?
16. Show that with the help of a slip ring, it is possible to convert an alternating e.m.f into a pulsating DC.
17. Explain the following terminologies:
    (i) Coil
    (ii) Coil pitch
    (iii) Pole pitch
    (iv) Back pitch
    (v) Front pitch
    (vi) Commutator pitch

(vii) Resultant pitch
(viii) Full pitch
(ix) Fractional pitch

18. Design a progressive lap winding of a DC machine having 8 poles and 24 slots.
19. Design a retrogressive winding of a DC machine having 8 poles and 24 slots.
20. Design a progressive wave of a DC machine having 8 poles and 24 slots.
21. What is the use of dummy coil in DC machine?
22. What is the function of equiliser ring?
23. Derive the e.m.f. equation of a DC generator.
24. On what factor does the e.m.f. equation of a DC generator depends?
25. Explain armature reaction with respect to DC generator.
26. What do you mean by demagnetising and cross-magnetisation ampere turns?
27. Explain the commutation process with respect to DC generator.
28. What are the different methods of improving commutation?
29. What is the use of compensating winding?
30. Obtain the equivalent circuit of a DC machine.
31. Give the different classification of a DC generator.
32. Obtain an expression for generated e.m.f. (Eg) for the following types of DC generator
    (i) Separately excited DC generator.
    (ii) Series wound DC generator.
    (iii) Shunt wound DC generator.
    (iv) Compound wound DC generator.
33. Draw and explain the magnetisation characteristics of a DC generator.
34. Draw and explain the open-circuit characteristic of a separately excited DC generator.
35. Explain the characteristics of a separately excited DC generator.
36. Explain the following characteristic with respect to a DC series generator
    (i) Open-circuit characteristic
    (ii) Internal characteristic
    (iii) External characteristic
37. Explain the following characteristic with respect to a DC shunt generator
    (i) Internal characteristic
    (ii) External characteristic
38. What do you mean by critical load resistance?
39. What will happen if the load resistance increases beyond critical load resistance?
40. What is voltage build up in a DC shunt generator?
41. Draw and explain the characteristic of compound DC generator.
42. What are the advantages of running many small DC generators in parallel?
43. What is the step-by-step procedure to connect DC shunt generator in parallel?
44. How to connect compound generators in parallel?
45. What is load sharing?
46. What are the applications of DC generator?
47. An 8-pole armature has 600 conductors. It is rotating at 1000 rpm. Calculate the induced e.m.f. on open circuit, when the DC generator is having,
    (i) Lap winding
    (ii) Wave winding

    Take flux per pole to be 50×10^{-4} Wb

48. A 16-pole armature has 1000 conductors and its windings are of wave winding. When it runs at 600 rpm it generates an e.m.f. of 450 V. Calculate the flux per pole.

49. An 8-pole armature of a DC generator is rotating with a speed of 1500 r.p.m. and it has 30 slots with 8 conductors per slot. Calculate the induced e.m.f. if the flux per pole is 20 mWb. Consider, (i) a wave winding generator and (ii) a lap winding generator.

50. An 8-pole DC generator generates 900V with the armature having 60 slots with 8 conductors per slot. If the flux per pole is 40mWb, calculate its speed of rotation for a (i) wave-wound armature and (ii) lap-wound armature.

51. An 8-pole, wave wound DC generator rotates with a speed of 690 r.p.m. The armature is 20 cm long and 25 cm in diameter, with 150 coils of 10 turns each. Calculate the induced e.m.f., if the flux density in the air gap is 0.8 Wb/m2 and each pole subtends an angle of 1200.

52. A 4-pole lap wound DC generator rotates at 900 r.p.m. and generates a voltage of 1200 V. The armature is having 1000 conductors. The diameter of the pole shoe is 0.2m and the length of the shoe is 0.4m. If the ratio of the pole arc to pole pitch is 0.8, find the flux per pole and flux density.

53. A DC generator is having 110 slots with 10 conductors per slot. When it runs at 700 r.p.m., it generates a no-load e.m.f. of 300 V. What will be its speed to generate a voltage of 200 V.

54. The armature of a 16-pole DC generator has 800 conductors and carries a current of 400 A. Find the demagnetising ampere turns per pole and the cross-magnetisation ampere turns per pole when the armature conductor is having, (i) wave winding and (ii) lap winding
Under the following conditions:
(a) Brush shift is by 8^0 electrical
(b) Brush shift is by 6^0 mechanical
(c) Brushes are on GNP

55. The output terminal voltage of a DC-shunt generator is 450V and the armature current and armature resistance is 100A and 0.5Ω respectively. Find the value of the induced e.m.f.

56. The armature and the field resistance of a shunt generator are 0.5Ω and 90Ω respectively. The load connected across the output terminal draws 95 kW power at 350V. Calculate the following:
(i) Load current
(ii) Field current
(iii) Armature current
(iv) Generated e.m.f.

57. If the load voltage and the load current of a DC shunt generator is 450V and 100A respectively, calculate the generated e.m.f. Take armature and field resistances as 0.6Ω and 85Ω respectively.

58. A shunt DC generator is connected across a lamp load having 98 lamps of 35W each. The terminal voltage is 200V. The armature and field resistance of the generator are 0.4Ω and 50Ω respectively. Calculate the following:
(i) The generated e.m.f.
(ii) Current in each armature conductor.
Assume the drop across the brush to be 0.8V per brush. Solve the problem for
(a) A wave-wound generator with 8 poles
(b) A lap-wound generator with 8 poles.

59. The armature resistance, shunt field resistance and series field resistance of a DC compound generator are 0.4Ω, 60Ω and 0.13Ω respectively. The load connected across the generator draws a current of 150A at 300V. Calculate the generated e.m.f. when,
(i) It is a long-shunt connection
(ii) It is a short-shunt connection

60. Consider a short-shunt DC generator with a load current of 70A at 230V. Calculate the generated e.m.f. if,the armature resistance is 0.25Ω, shunt field resistance is 200Ω, series field resistance is 0.15Ω and voltage drop per brush = 1.2V.

61. The O.C.C of a DC shunt generator running at 660 r.p.m. is tabulated as:

Field current (I_f) in amp	0	0.1	0.3	0.5	0.7	0.9	1.1	1.3	1.5	1.7
Induced e.m.f. (E_g) in volts	20	50	110	170	198	220	240	245	250	256

The field resistance of the DC machine is adjusted to 100 Ω. Calculate the following:

(a) Draw the O.C.C. from the given data.
(b) Find the value of the critical resistance.
(c) Find the no load voltage
(d) Find the extra resistance to be added in the field to reduce the no-load voltage to 200V.

62. A 30 kW, 300V separately excited DC machine is running at a constant speed of 2000 rpm. Its open circuit armature voltage is210V and its armature resistance is 0.07 ohm. Calculate the armature current, terminal power, electromagnetic power and torque when the terminal voltage is 130V

Chapter 3

DC MOTOR

3.1 INTRODUCTION

A motor is a machine which converts electrical energy into mechanical energy. A DC motor finds application in rolling mills, in traction system and overhead cranes. The principle of operation of DC motor is that when a current carrying conductor is kept in a magnetic field, a force is produced in that conductor. The construction of DC motor and DC generator is similar and hence it is not repeated again in this chapter.

3.2 OPERATING PRINCIPLE OF DC MOTOR

The operation of a DC motor is based on the principle that when a conductor carrying current is placed in a magnetic field, the conductor experiences a force.

These conductors are present in the armature and the current is fed into these conductors with the help of brushes and commutators. The magnetic field is provided by an electromagnet or permanent magnet. The of rotation of the armature conductors can be obtained by Fleming's left-hand rule. Thus in case of a motor, the function of the commutator is to convert a DC input into an AC output for the armature conductors, so that the group of conductors under the successive field poles carry currents in the opposite directions.

3.3 CONCEPT OF BACK E.M.F.

Whenever a current carrying conductor rotates in a magnetic field, an e.m.f. is induced in these conductors. In the case of a DC motor, this e.m.f. is called the back e.m.f. or the counter e.m.f. It is given by the expression

$$E_{\mathrm{b}} = \frac{\varphi ZNP}{60A} \tag{3.1}$$

Where,

φ is the magnetic flux per pole.

Z is the total number of armature conductors

N is the rpm of the armature.

P is the number of pole.

A is the number of parallel paths.

Note that Eq. (3.1) is the same e.m.f. equation as discussed in case of a DC generator.

3.4 ELECTROMAGNETIC TORQUE EQUATION OF A DC MOTOR

Let,

φ be the flux per pole.
P be the number of poles.
Z be the total number of armature conductors.
N be the speed of rotation of armature in r.p.m.
A be the number of parallel paths. For lap winding A=P and for wave winding A=2.
E_b be the back e.m.f.
R_a be the resistance of the armature.
I_a be the armature current.
V be the supply voltage.
T_e be the electromagnetic torque equation.
The voltage equation of a DC motor is

$$V = E_b + I_a R_a \tag{3.2}$$

Multiplying Eq. (3.2) by I_a on both sides, we get,

$$VI_a = E_b I_a + I_a^2 R_a \tag{3.3}$$

Equation (3.3) is called the power equation of a DC motor, where,
VI_a is the power supplied to the motor .
$E_b I_a$ is the power developed by the motor .
$I_a^2 R_a$ is the power loss in the armature conductor.
Again,

$$\text{Power developed} = T_e \omega \tag{3.4}$$

Where,
T_e is the electromagnetic torque.
ω is the speed of rotation in rps.
Electrical equivalent of mechanical power developed is given by,

$$E_b I_a \tag{3.5}$$

Equating Eq.(3.4) and Eq. (3.5), we get,

$$T_e \omega = E_b I_a$$

$$or, T_e = \frac{E_b I_a}{\omega} = \frac{E_b I_a}{\frac{2\pi N}{60}} = \frac{60 \times E_b I_a}{2\pi N} \tag{3.6}$$

Substituting the value of E_b from Eq.(3.1) to Eq. (3.6) we get,

$$T_e = \frac{60 \times \frac{\varphi ZNP}{60A} I_a}{2\pi N}$$

$$or, T_e = \frac{PZ\varphi I_a}{2\pi A} = 0.159\frac{PZ\varphi I_a}{A} \quad (3.7)$$

For a given machine $\frac{PZ}{2\pi A} = K = \text{Constant}$

So, Eq. (3.7), can be written as,

$$T_e = \frac{PZ\varphi I_a}{2\pi A} = K\varphi I_a \quad (3.8)$$

$$or, T_e \propto K\varphi I_a \quad (3.9)$$

The torque is uniform for a given flux per pole and armature current. In actual practice, the air path of the slots has large reluctance, due to which the flux density in the slots becomes very low. Hence, there is practically no force on the conductor. This force is due to the interaction of the main flux and the armature flux and it is excited on the teeth.

3.4.1 Condition for Maximum Power Output

From Eq. (3.5), mechanical power (P_m) is given by the expression,

$$P_m = E_b I_a \quad (3.10)$$

Again from Eq. (3.3), we can write,

$$E_b I_a = VI_a - I_a^2 R_a \quad (3.11)$$

Combining Eq.(3.10) and Eq. (3.11), we get,

$$P_m = VI_a - I_a^2 R_a \quad (3.12)$$

For maximum power output,

$$\frac{dP_m}{dI_a} = 0$$

Thus, differentiating Eq. (3.12) with respect to I_a, we get,

$$V - 2I_a R_a = 0$$

$$or, I_a R_a = \frac{V}{2} \quad (3.13)$$

From Eq. (3.2),

$$V = E_b + I_a R_a$$

$$or, E_b = V - I_a R_a$$

Substituting Eq. (3.13) we get,

$$E_b = V - \frac{V}{2}$$

$$\text{or, } E_b = \frac{V}{2} \tag{3.14}$$

Thus, the motor will develop maximum powerwhen the back e.m.f. (E_b) becomes equal to the applied voltage.

3.5 COMMUTATION ACTION IN DC MOTOR

In the case of a motor, the armature current flows in the opposite direction as compared to generated e.m.f. (or back e.m.f.).So in case of a motor, the current reversal is in the opposite direction as compared with the current reversal of generator. Then, the e.m.f. that is required for a better commutation will also be in the opposite direction to that of a DC generator. Thus, when the operation of the DC machine changes from generating action to motoring action, the direction of armature current reverses. As in the case of DC generator, interpole is also present in the DC motor. The polarity of the interpole must be same as the polarity of the main pole, immediately preceding the interpole.

3.6 ARMATURE REACTION IN DC MOTOR

The reason for armature reaction in a DC motor issame as explained in the case of a DC generator, but the effect of armature reaction in a DC motor is different than in a DC generator.

In a DC generator, the armature reaction effectreduces the generated e.m.f. but in the case of a DC motor, this armature reaction effect will increase the speed. If the magnetic lines of force of the armature is compared with the magnetic lines of force of the field, it is found that the main field increases under the leading pole and decreases under the trailing pole.

To overcome this effect, the brushes are moved backward with increase in motor loads to make the brush axis into Magnetic Neutral Plane (MNP). This backward movement of brush is carried out by the demagnetising and cross-magnetisation action.

In most of the DC shunt and compound motor, an extra winding called the stabilising winding is placed in series with the armature. This reduces the weakening of the main field flux because of armature reaction. Heavy-duty motors are generally provided with compensating winding to prevent armature reaction.

3.7 TYPES OF DC MOTOR

Depending upon how the field and armature windings are connected with each other, DCmotor can be categorised into following categories,

i) Permanent magnet DC motor
ii) Separately excited DC motor
iii) Series DC motor
iv) Shunt DC motor
v) Compound DC motor

3.7.1 Permanent Magnet DC Motor

In this type of DC motor, the field is made up of permanent magnet and hence, it requires no field winding.

3.7.2 Separately Excited DC Motor

As the name suggest, in this type of DC motor, both the field winding and the armature winding are fed from two different sources, as shown in Fig. 3.1.

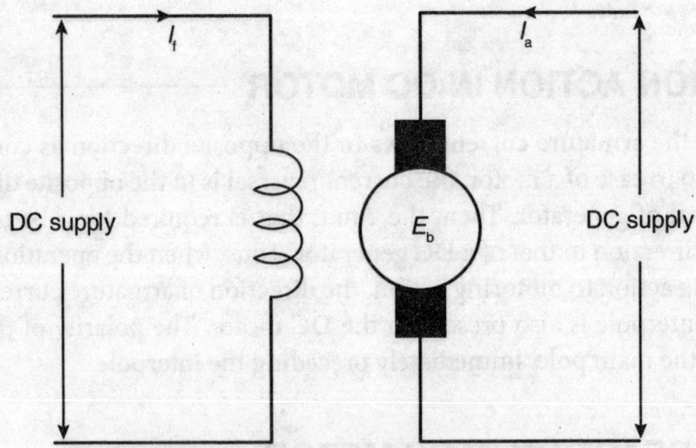

Fig. 3.1: Separately excited DC motor

We know that,

$$\text{Back e.m.f.}(E_b) = V - I_a R_a$$

$$\text{Power drawn} = VI_a$$

Where, V is the supply voltage.

$$\text{Mechanical power}(P_m) = VI_a - I_a^2 R_a$$

$$= I_a(V - I_a R_a)$$

$$= I_a E_b \qquad (3.15)$$

3.7.3 Series DC Motor

In this type of DC motor, the field winding is connected in series with the armature winding, as shown in Fig. 3.2.

From the circuit of Fig. 3.2, we can write,

$$I = I_f = I_a$$

Where, $I$is the supply current and I_f is the field current.

Applying KVL in the circuit of Fig. 3.2, we get,

$$V = E_b + I(R_a + R_f) \qquad (3.16)$$

Where, R_f is the resistance of the field winding

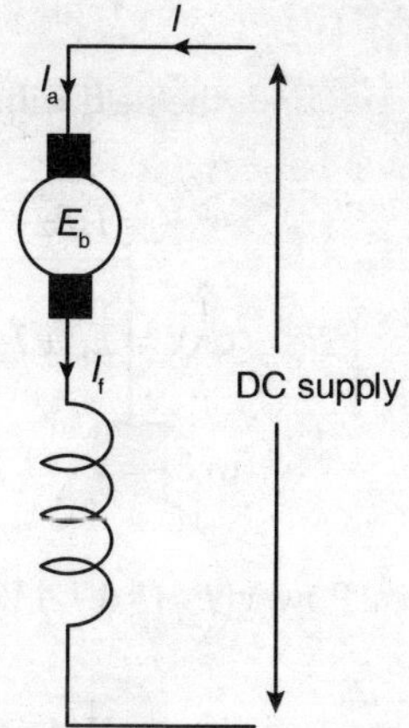

FIG. 3.2: Series DC motor

Series DC motor
Multiplying Eq. (3.16), by I, we get,

$$VI = IE_b + I^2(R_a + R_f) \tag{3.17}$$

Substituting from Eq. (3.15), we get,

$$VI = P_m + I^2(R_a + R_f)$$

$$or\, P_m = VI - I^2(R_a + R_f)$$

$$or\, P_m = I\left[V - I(R_a + R_f)\right]$$

Substituting from Eq. (3.16), we get,

$$P_m = IE_b$$

3.7.4 Shunt DC Motor

In this type of DC motor, the field and armature windings are connected in parallel with each other, as shown in Fig. 3.3.

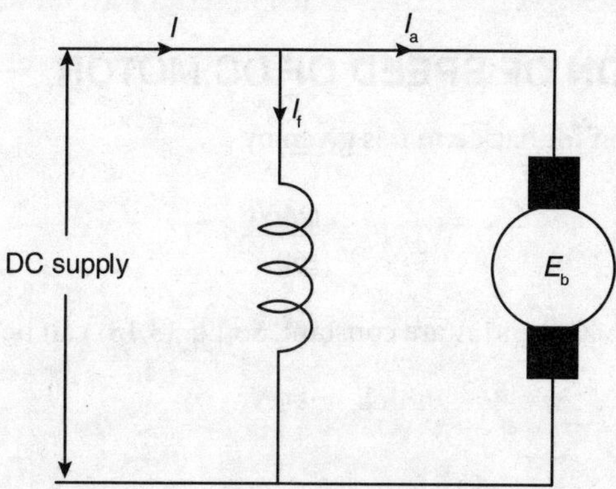

FIG. 3.3: Shunt DC motor

From the circuit of Fig. 3.3, we get,

$$I = I_a + I_f$$

$$V = I_f R_f$$

$$\text{or}\, V = E_b + I_a R_a$$

$$\text{or}\, E_b = V - I_a R_a$$

$$\text{Mechanical Power}\,(P_m) = VI - VI_f - I_a^2 R_a$$

$$= V(I - I_f) - I_a^2 R_a$$

$$= VI_a - I_a^2 R_a$$

$$= I_a(V - I_a R_a)$$

$$= E_b I_a$$

3.7.5 Compound DC Motor

A compound DC motor has both series and shunt field winding. It can be a short-shunt DC compound motor or a long-shunt DC compound motor. Again, depending on how these two windings are connected, they can be classified into cumulatively compound motor and differentially compound motor. The detailed functions of these motorshave already been discussed in Chapter2.

3.8 REVERSING THE DIRECTION OF ROTATION OF DC MOTOR

The direction of rotation of a DC motor can be reversed by changing the polarity of the field or by changing the polarity of the armature terminals. However, if the polarity of both the field and armature is changed simultaneously, then the direction of rotation of the DC motor will remain the same.

3.9 REPRESENTATION OF SPEED OF DC MOTOR

For a DC motor the expression for back e.m.f. is given by,

$$E_b = \frac{\varphi ZNP}{60A} \tag{3.18}$$

For a particular DC motor, Z, P and A are constant. So,Eq. (3.18) can be written as,

$$E_b = k\varphi N \tag{3.19}$$

Where,

$$k = \frac{ZP}{60A}$$

Again, the back e.m.f. can also be written as,

$$E_b = V - I_a R_a \tag{3.20}$$

Comparing Eq.(3.19) and Eq. (3.20), we can write,

$$k\varphi N = V - I_a R_a$$

$$or\, N = \frac{V - I_a R_a}{k\varphi} \tag{3.21}$$

From Eq. (3.21), we can say that the speed of a DC motor depends on the following factors:

i) Supply voltage
ii) Armature current
iii) Armature resistance
iv) Field flux.

3.10 CHARACTERISTICS OF DC MOTOR

DC motor characteristic is basically a graphical plot between two dependent quantities. Performance of the DC motor and its suitability for a particular application are determined from its characteristics. In this section, we will study three different characteristics of a DC motor.

i) Speed versus armature current characteristics.
ii) Torque versus armature currentcharacteristics.
iii) Speed versus torque characteristics.

3.10.1 Characteristics of a DC Shunt Motor

Speed versus armature current characteristics

From Eq. (3.21), the speed of a DC motor is given by the expression,

$$N = \frac{V - I_a R_a}{k\varphi} \tag{3.21}$$

Where,
V is the supply voltage.
I_a is the armature current
R_a is the armature resistance
φ is the flux per pole.

Let us assume that the supply voltage V is constant. If V is constant, then the field current I_f will also be constant. ($I_f = \frac{V}{R_f}$). Now, if I_f is constant then the flux φ, which is due to the current I_f will also be constant. So, the denominator of Eq. (3.21) will be constant and can write Eq. (3.21) as,

$$N \propto \left(V - I_a R_a\right) \tag{3.22}$$

From Eq. (3.22), we can say, as the armature current increases, the product ($I_a R_a$) increases and thus the difference (V–$I_a R_a$) decreases, thereby reducing the speed, as shown in Fig. 3.4.

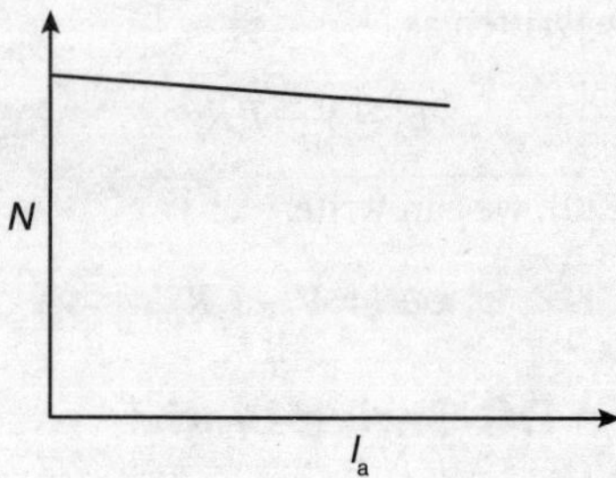

Fig. 3.4: Speed armature current characteristic

Since the voltage drop in the armature is very small at full load as compared with the applied voltage, the speed variation from no load to full load is also very small. This small variation can be made up by inserting a resistance in the shunt field, which reduces the flux. Thus DC shunt motor is used as a constant speed motor and finds application in lathes, line shafts, milling machines, fans, conveyors etc.

Torque versus armature current characteristics

The torque equation of a DC motor is given by the expression.

$$T = k\varphi I_a \tag{3.23}$$

Since the supply voltage is assumed to be constant, the flux φ, is also constant. So,Eq. (3.23) becomes,

$$T \propto I_a$$

Thus, the torque armature current characteristic is a straight line passing through the origin, as shown in Fig. 3.5.

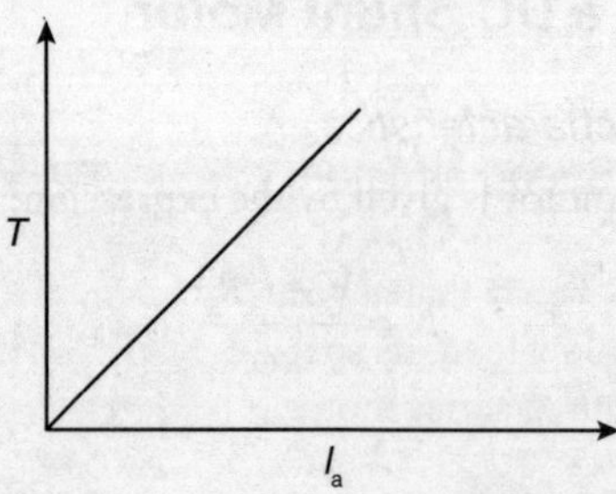

Fig. 3.5: Torque armature current characteristic

Speed versus torque characteristics

From the above two characteristics we can say, $Torque\left(T\right) \propto I_a$ and speed (N) is inversely proportional to I_a. From these two characteristics we can conclude that torque (T) is inversely proportional to armature current I_a. The characteristic is shown in Fig. 3.6.

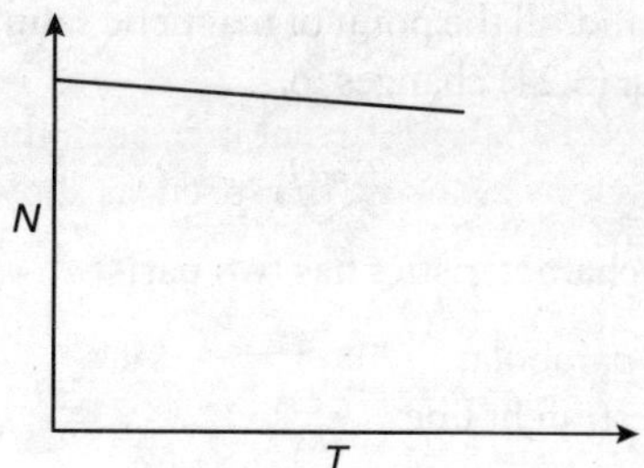

Fig. 3.6: Speed torque characteristic

3.10.2 Characteristics of a DC Series Motor

Speed versus armature current characteristics

From Eq. (3.21), we can say the motor speed is inversely proportional to the flux φ. Again, for a DC series motor, $I_a = I_f = I$. where, I_a is the armature current, I_f is the field current and I is the main supply current. So for a DC series motor, we can say, flux $\varphi \propto I_a$. If the supply voltage, V is constant then the speed N is inversely proportional to armature current I_a. So, for a DC series motor, we can say, speed N is inversely proportional to flux φ, under constant voltage V.

The plot of speed and armature current will be a rectangular hyperbola, untilthe magnetic saturation point. Up to the point of magnetic saturation, the speed decreases abruptly with increase in current. After saturation, the flux φ becomes constant and the speed current characteristic becomes a straight line. The speed decreases slightly due to the drop in the field and the armature winding. The plot is shown in Fig. 3.7.

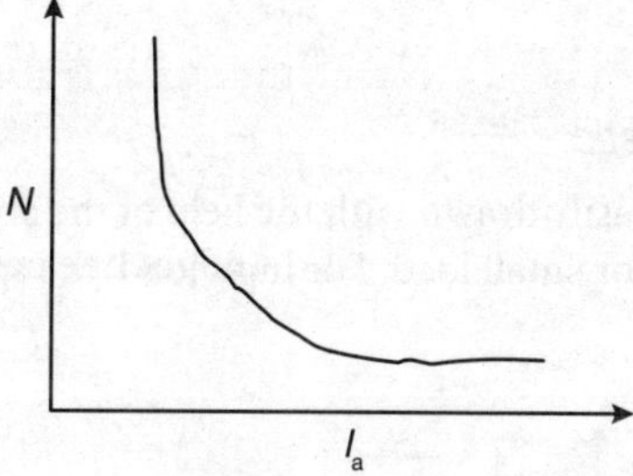

Fig. 3.7: Speed armature current characteristc

From the speed–current plot, we can say that at very low load the speed of the motor is very high. If a series motor is made to run at a very low load, its speed may exceed the normal speed of the motor and thus the motor may get damaged. Thus, a series motor is never started at no load.

Torque versus armature current characteristics

From Eq. (3.23), the torque equation of a DC motor is given by the expression.

$$T = k\varphi I_a$$

Again, the flux, $\varphi \propto I_a$ for a DC series motor, since, $I_a = I_f$. So, the torque given by Eq. (3.23) now becomes,

$$T \propto I_a^2 \tag{3.24}$$

The expression of Eq. (3.24), is valid till the point of magnetic saturation. When saturation is reached, the flux φ becomes constant and Eq. (3.24) changes to,

$$T \propto I_a \tag{3.25}$$

So, the torque armature current characteristics has two parts

- Before saturation: the curve is parabola.
- After saturation: the curve is a straight line.

The torque armature current plot is shown in Fig. 3.8.

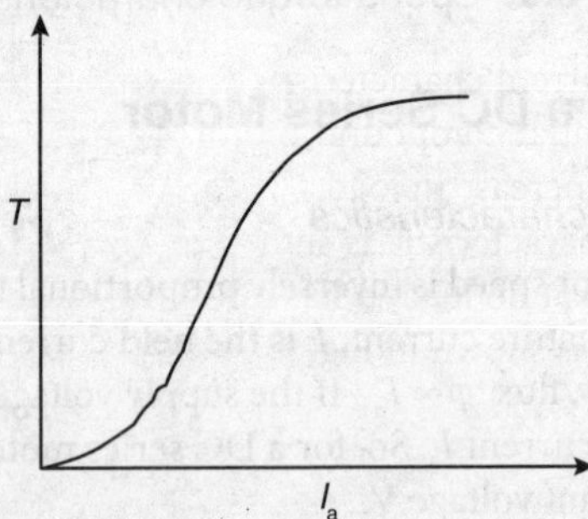

Fig. 3.8: Torque armature current characteristic

For a DC series motor, the starting torque is very high; it is proportional to the square of the armature current. So series motor is used for those applications, where high starting torque is required, as in the case of hoists, railways, trolleys etc.

Speed versus torque characteristic

Speed torque characteristic can be easily drawn with the help of the above two characteristics. Speed will fall rapidly with increase in torque for small load. For high load, the speed torque characteristic is linear. The plot is shown in Fig. 3. 9.

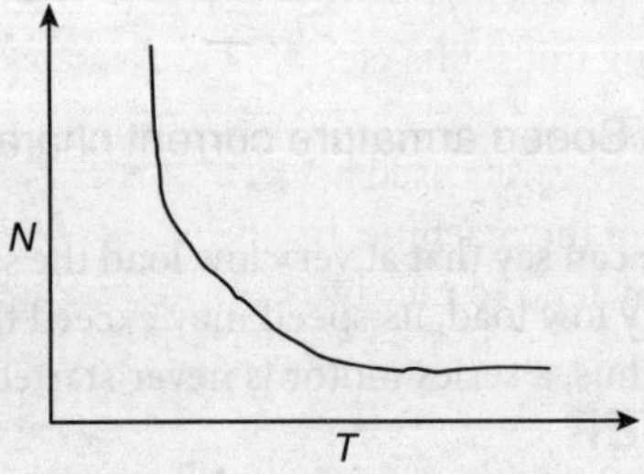

Fig. 3.9: Speed torque characteristic

3.10.3 Characteristic of a Compound DC Motor

The characteristic of a compound motor is a combination of shunt and series motor. Its characteristics are intermediate between the shunt and the series motor.

The speed armature current characteristic is shown in Fig. 3.10.

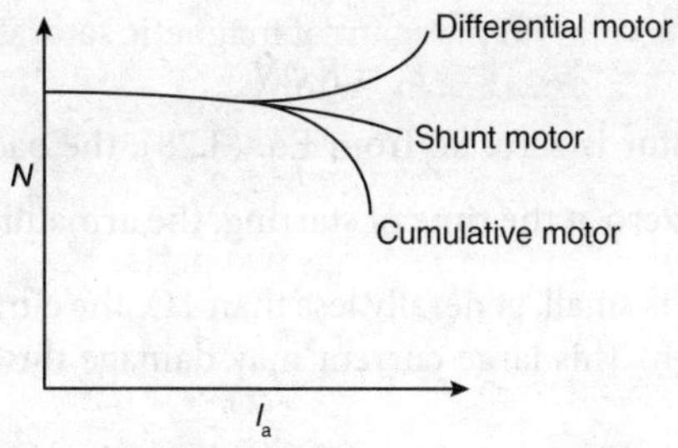

FIG. 3.10: Speed armature current characteristic

The shape of speed armature current characteristic of a cumulative motor is more drooping as compared to that of a shunt or differential compound motor. The speed of a differential compound motor remains almost constant with the increase in load.

The torque armature current characteristic is shown in Fig. 3.11. The torque of a compound motor increases with increase in load and at no load, it has a definite speed.

The speed torque characteristic is shown in Fig. 3.12.

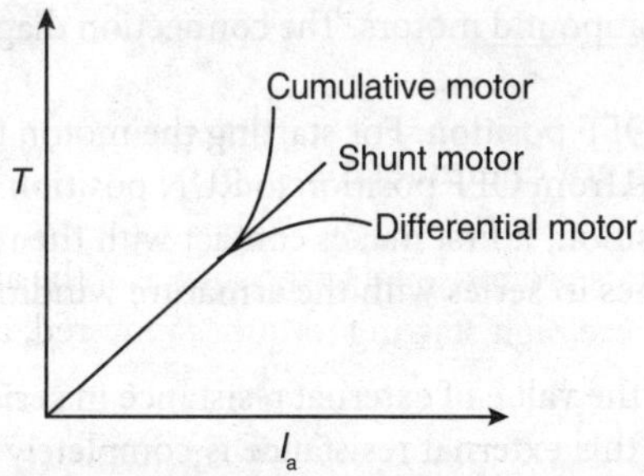

FIG. 3.11: Torque armature current characteristic

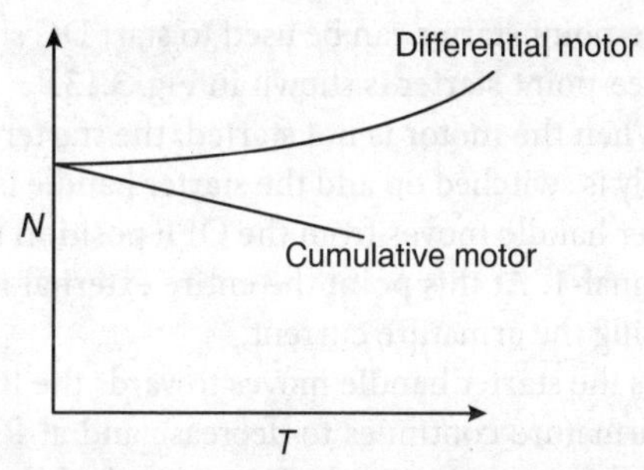

FIG. 3.12: Speed torque characteristic

It can be concluded that compound motor has high starting torque with a safe no-load speed. Cumulative compound motors are used for such applications where the motor is subjected to sudden application of heavy loads,such as rolling mills, lifts etc. They are also used for such applications where large starting torque is required.

Differential compound motors are rarely used because of its differential arrangement. It creates problems during overload and starting of the machine. In a heavy-load condition, the field may get weakened to a large extent and the motor may tend to race by taking a very large current.

3.11 DC MOTOR STARTER

A starter is a device that is used to start a DC motor from standstill. The e.m.f. equation of a DC motor is, $E_b = V - I_a R_a$. From this e.m.f. equation, the armature current is given by the expression,

$$I_a = \frac{V - E_b}{R_a} \tag{3.26}$$

Again, we know that the speed of a DC motor is given by the expression,

$$N = \frac{V - I_a R_a}{K\varphi} = \frac{E_b}{K\varphi} \tag{3.27}$$

From Eq. (3.27), we get,

$$E_b = K\varphi N \tag{3.28}$$

At start, the speed of the motor is zero. So from Eq. (3.28), the back e.m.f. will also be zero when starting. Since the back e.m.f. is zero at the time of starting, the armature current I_a will be, $I_a = \frac{V}{R_a}$.

Since the armature resistance is small, generally less than 1Ω, the current through the armature at the time of starting will be very high. This large current may damage the armature winding, brushes and commutator segments.

But once the motor has started and it picks up speed, the back e.m.f. increases and thus the armature current gradually decreases. It ultimately comes down to its rated value when the motor has reached its normal speed.

Due to the high inrush of current at the time of starting the motor, it becomes necessary to limit this current to a safe value, which is generally obtained by connecting an extra resistance in series with the armature winding. This extra resistance will limit the armature current to a safe value. When the motor picks up speed and the back e.m.f. is developed, this extra resistance is removed from the armature circuit. If this resistance is not removed, itwill causespeed loss in the motor there by reducing its efficiency.

3.11.1 Three-point Starter for DC Shunt Motor

Three-point starter can be used to start DC shunt or DC compound motors. The connection diagram of a three-point starter is shown in Fig. 3.13.

When the motor is not started, the starter handle is in OFF position. For starting the motor, the DC supply is switched on and the starter handle is slowly moved from OFF position to RUN position. As the starter handle moves from the OFF position to the ON position, it first makes contact with the resistive terminal-1. At this point the entire external resistance comes in series with the armature winding, thus limiting the armature current.

As the starter handle moves towards the RUN position, the value of external resistance in series with the armature continues to decrease and at RUN position, this external resistance is completely cut off from the armature circuit. The starter handle is held at RUN position with the help of an electromagnet, called the hold-on electromagnet. This electromagnet keeps the starter handle in position by attracting the soft iron core, which is present at the starter handle, towards itself.

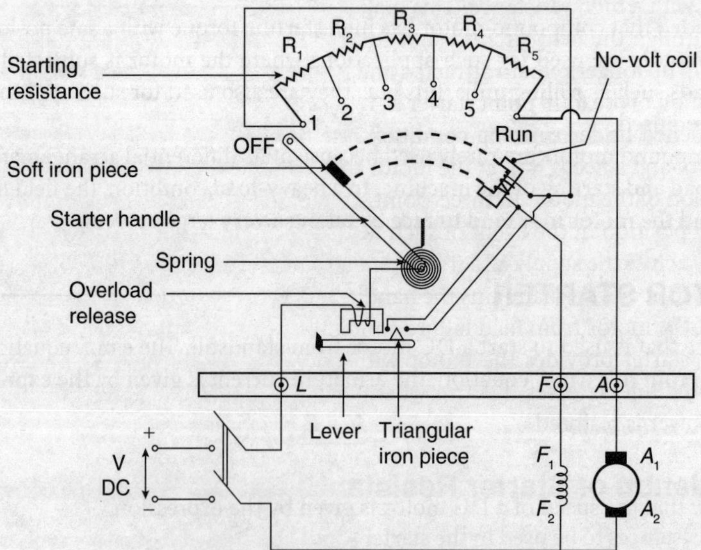

Fig. 3.13: Three-point starter

This electromagnet is energised by a no-voltage coil. When the supply fails or gets disconnected, the electromagnet gets demagnetised and the starter handle is pulled back to OFF position by the spring.

The starter is also provided with an overload protection. This overload coil carries the full armature current. If the load increases, the current drawn by the armature also increases. If this current exceeds due to overload, it pulls the triangular iron piece and short circuits the no-voltage coil. Short circuiting the no-voltage coil demagnetises the hold-on electromagnet and hence, the starter handle is pulled back to the OFF position and the motor stops.

3.11.2 Four-point Starter for DC Shunt Motor

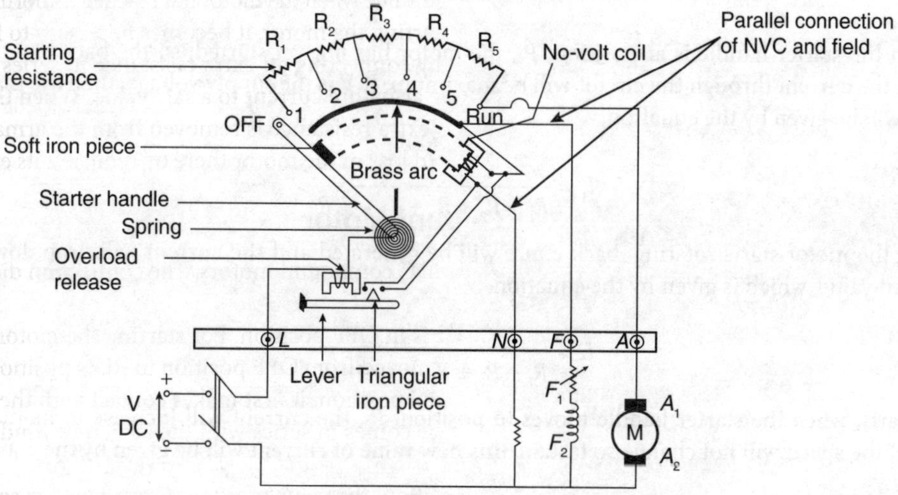

Fig. 3.14: Four-point starter

In a four-point starter, the drawbacks of three-point starter are eliminated. In the case of four-point starter, the holding coil circuit is not connected to the field circuit of the motor. Rather, an additional terminal is provided with a high resistance. This makes the system more reliable because if, by chance, the current flowing through the field circuit is not sufficiently high to magnetise the holding coil, then the soft iron keeper will no longer remain attracted and the starting arm will not lie at the final stud position.

The only limitation of the four-point starter is, it does not provide high-speed protection to the motor. If the field gets opened under running condition, the field current reduces to zero. But there is some residual flux present and since $N \propto 1/\varphi$, the motor tries to run at a dangerously high speed. This is called high-speeding action of the motor. In three-point starter as NVC is in series with the field, it releases the handle to the OFF position in the event of such field failure. However, in four-point starter the NVC is connected directly across the supply and its current is maintained irrespective of the current through the field winding. Hence, it always maintains the handle in the run position, as long as there is supply. Thus, it does not protect the motor from field failure conditions which result in high speeding of the motor.

The four-point starter provides the motor with no voltage protection. The holding magnet, being connected across the line, releases the arm when voltage drops below a specific value andprotects the motor when the power is restored.

3.11.3 Calculation of Starter Resistance for DC Shunt Motor

The value of the resistances to be used in the starter should be chosen is such a way that the current at the time of starting remains between its maximum and minimum value. Let I_1 and I_2 be the maximum and minimum value of currents respectively, for which the starter needs to be designed. Let the starter to be designed for a DC shunt motor havensections of resistances, as shown in Fig. 3.15.

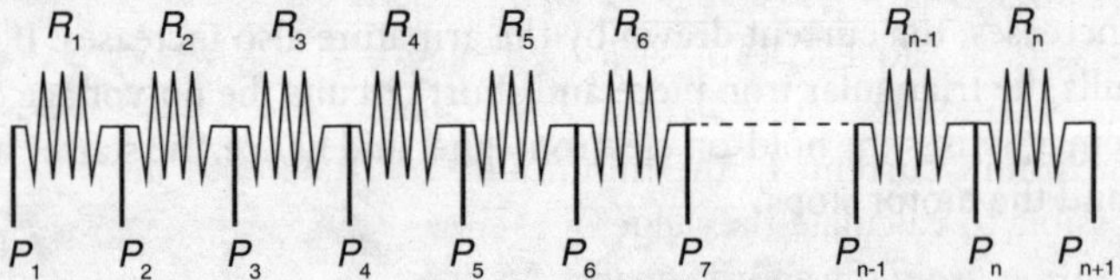

Fig. 3.15: Starter resistances

When the starter handle is at position, P_1, the motor has not yet started, so the back e.m.f. will be zero and the current through the circuit will be maximum. If V is the supply voltage, then the maximum current will be given by the equation,

$$I_1 = \frac{V}{R_1 + R_2 + R_3 + \cdots\cdots + R_n} \tag{3.29}$$

Once the motor starts rotating, back e.m.f. will be generated and the current will drop down to its minimum value, which is given by the equation,

$$I_1 = \frac{V - E_{b1}}{R_1 + R_2 + R_3 + \cdots\cdots + R_n} \tag{3.30}$$

Similarly, when the starter handle moves to position, P_2, the current will increase to its maximum value but the speed will not change so fast.So, this new value of current will be given by the equation,

$$I_1 = \frac{V - E_{b1}}{R_2 + R_3 + \cdots\cdots + R_n} \tag{3.31}$$

After the motor speed changes, the current will drop down to its minimum value, which is given by the equation,

$$I_2 = \frac{V - E_{b2}}{R_2 + R_3 + \cdots\cdots + R_n} \tag{3.32}$$

Again, when the starter handle moves to position, P_3, the current will increase to its maximum value but the speed will not change so fast.Hence,this new value of current will be given by the equation,

$$I_1 = \frac{V - E_{b2}}{R_3 + \cdots\cdots + R_n} \tag{3.33}$$

Dividing Eq.(3.31) by Eq. (3.30), we get,

$$\frac{I_1}{I_2} = \frac{R_1 + R_2 + R_3 + \cdots\cdots + R_n}{R_2 + R_3 + \cdots\cdots + R_n} \tag{3.34}$$

Again, dividing Eq.(3.33) by Eq. (3.32), we get,

$$\frac{I_1}{I_2} = \frac{R_2 + R_3 + \cdots\cdots + R_n}{R_3 + \cdots\cdots + R_n} \tag{3.35}$$

Combining Eqs.(3.34) and Eq. (3.35), we get,

$$\frac{I_1}{I_2} = \frac{R_1 + R_2 + R_3 + \cdots\cdots + R_n}{R_2 + R_3 + \cdots\cdots + R_n} = \frac{R_2 + R_3 + \cdots\cdots + R_n}{R_3 + \cdots\cdots + R_n} = k^n \tag{3.36}$$

If the value of maximum current, I_1, the minimum current, I_2 and the armature resistance, R_a is known, then it is possible to calculate the value of total resistance, $R_1 + R_2 + R_3 + \cdots\cdots + R_n$, k and the total number of sections, n, from the above equations.

3.12 SPEED CONTROL OF DC SHUNT MOTOR

From Eq. (3.21), the speed of a DC motor is represented by the equation,

$$N = \frac{V - I_a R_a}{k\varphi} \tag{3.37}$$

From the above equation, we can say that the speed of a DC motor can be controlled by,

i) Changing the flux, φ.
ii) Changing the armature resistance, R_a.
iii) Changing the supply voltage, V.

The different methods of speed control of DC shunt motor are as follows:

i) Shunt field control or flux control method.
ii) Armature resistance control or armature control method.
iii) Voltage control method.

3.12.1 Shunt Field Control or Flux Control Method

In this method the speed of a DC shunt motor is controlled by varying the field flux, φ. The circuit for this method is shown in Fig. 3.16.

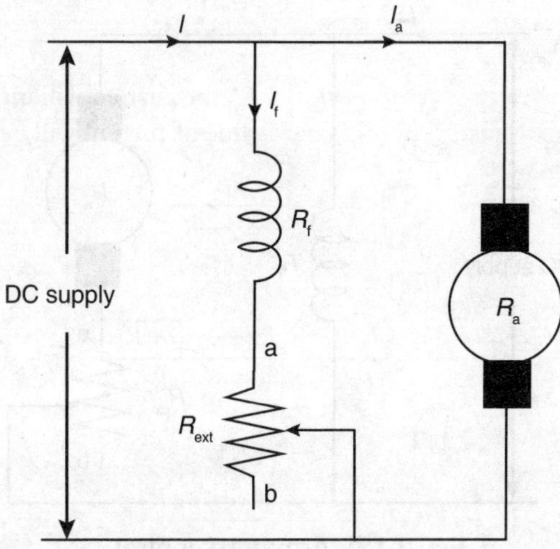

Fig. 3.16: Flux control

A variable resistanceR_{ext} called the shunt field resistance is connected in series with the field winding. Initially, it is kept at the minimum position (Point *a*). The speed of the motor at this position of external resistance is called the base speed of the motor. As the variable resistance is increased by moving the variable notch towards Position *b*, the field current (I_f) reduces. Since the field current, I_f, is directly proportional to the flux, φ, with reduction in field current, the flux φ reduces. From Eq. (3.21), as flux φ reduces, the speed of the motor increases. Thus, this method is used to increase the speed of the DC shunt motor above the base speed.

Since the field carries very less current,power wastage in the external resistance is also very less. This method of speed control is very simple and economical to use.

The advantages of this method are

i) It makes speed control very simple; also, very economical to use.
ii) Power loss is very less
iii) Speed control is independent of load on the machine.
iv) This method of speed control can be applied for DC compound motor also.

The disadvantages of this method are

i) A speed higher than the base speed can be obtained.
ii) It is not possible to get the creeping speed.

3.12.2 Armature Resistance Control or Armature Control Method

In this method, the speed of a DC shunt motor is controlled by varying the armature current, I_a, The circuit for this method is shown in Fig. 3.17.

A variable resistance, R_{ext}, is connected in series with the armature winding. Initially, it is kept at the maximum position (Point *b*). The speed of the motor at this position of external resistance is called the base speed of the motor.

As variable resistance is decreased by moving the variable notch towards Position *a*, the armature current increases. This increases the I_aR_a drop,because of which the difference $V - I_aR_a$ decreases and thus the speed decreases. So, this method is used to run the DC motor below the base speed.

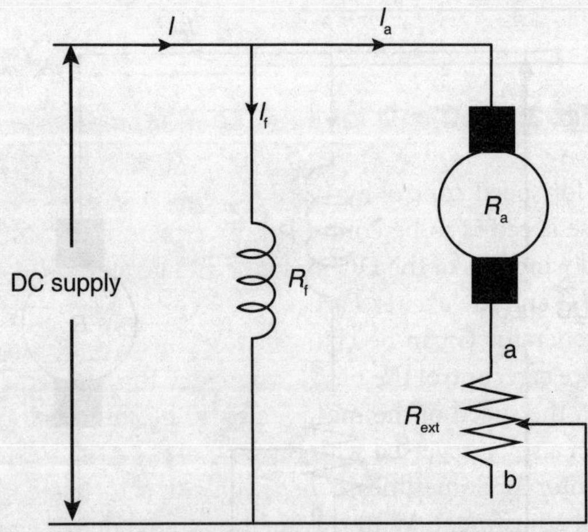

Fig. 3.17: Armature control

The advantages of this method are:

i) The creeping speed of the motor can be obtained
ii) Wide range of speed can be achieved
iii) The motor can develop any desired torque over its operating range.

The disadvantages of this method are:

i) Since the armature winding carries a large amount of current, a large amount of power is wasted in the external resistance
ii) Speed control is dependent on the load on the machine.
iii) Due to the loss in external resistance, the efficiency of the motor is reduced.

3.12.3 Voltage Control Method

With the advancement of solid-state devices, this method of speed control has become very versatile and attractive. Using solid-state devices, it is possible to obtain smooth voltage control. The speed of the DC motor can be varied by simply changing the applied voltage. This method is quite expensive and therefore, it is employed for large-size motors where efficiency is of great importance.

3.13 SPEED CONTROL BY WARD LEONARD METHOD

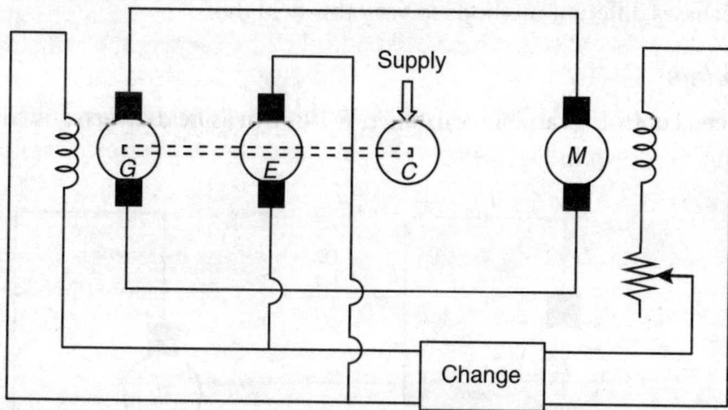

Fig. 3.18: Circuit for Ward Leonard method

A simple circuit diagram for speed control by Ward Leonard method is shown in Fig. 3.18. Here,*M* is the DC shunt motor whose speed is to be controlled. Its armature is connected with the separately excited generator *G*. The field windings of the DC motor *M* and generator *G* is fed by a constant voltage exciter *E*. Both the generator *G* and the exciter *E*are mounted on the same shaft and driven by a motor C. The field current of the generator *G* can be changed with the help of the voltage divider. Thus, by controlling the field current we can control the output voltage of the generator *G*. Since the generator is connected with the motor *M*, the speed of the motor can also be controlled. The direction of the field current of the generator *G* can also be reversed with the help of the switch 'change'. This will change the direction of rotation of the motor *M*. Sometimes, a field regulator is included in the field circuit of shunt motor *M* for additional speed adjustment. With this method, the motor may be operated at any speed up to its maximum speed.

The advantages of this method are:

i) Losses are less and there is high efficiency
ii) It can regenerate or return the motors energy of motion back to the main supply.
iii) The motor can be brought to a standstill quickly.

The disadvantages of this method are:

i) The set-up is costly
ii) The overall efficiency is low
iii) Special motor-generator set is required for each motor.

3.14 SPEED CONTROL OF DC SERIES MOTOR

There are two methods of speed control for DC series motor, namely,

i) Flux control method
ii) Armature control method
iii) Series–parallel control.

3.14.1 Flux Control Method

As the name suggests, in this method the speed of the DC series motor is controlled by varying the field flux. This section discusses different methods to vary this field flux.

By using field diverters

In this method of speed control, a variable resistance, R_d, known as field diverter, is connected in parallel with the field winding, as shown in Fig. 3.19.

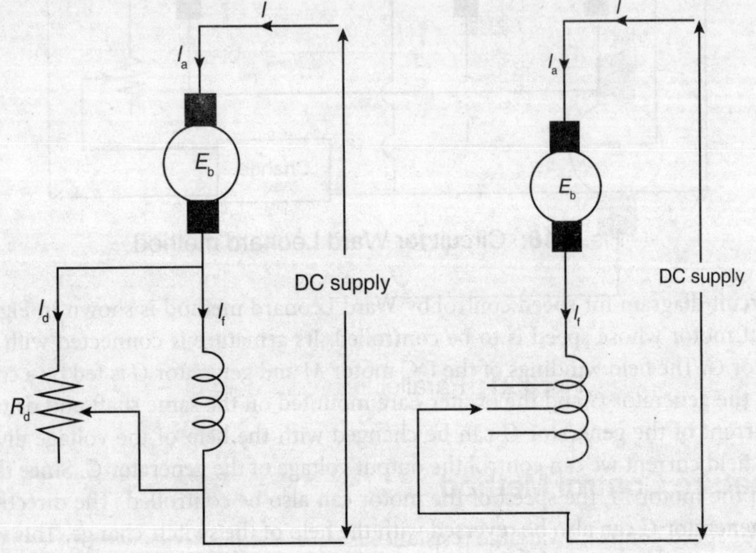

Fig. 3.19: Field diverters **Fig. 3.20:** Tapped field control

The armature current I_a, gets divided into two parts, I_d and I_f. Now, due to a parallel path the field current will get reduced. This will reduce the flux, φ, and thus the speed will increase. This method will give speed above the base speed. The series field diverter method is often employed in traction work.

Tapped field control

This method is also used to run the DC motor above its base speed. In this method, by decreasing the number of turns in the field winding, the flux is reduced. This, in turn, increases the speed of the motor. The circuit for this method is shown in Fig. 3.20.

Parallel field coils control method

This method is usually employed in the case of fan motors. By connecting the field coils in series– parallel combinations, as shown in Fig. 3.21, several fixed speeds can be obtained.

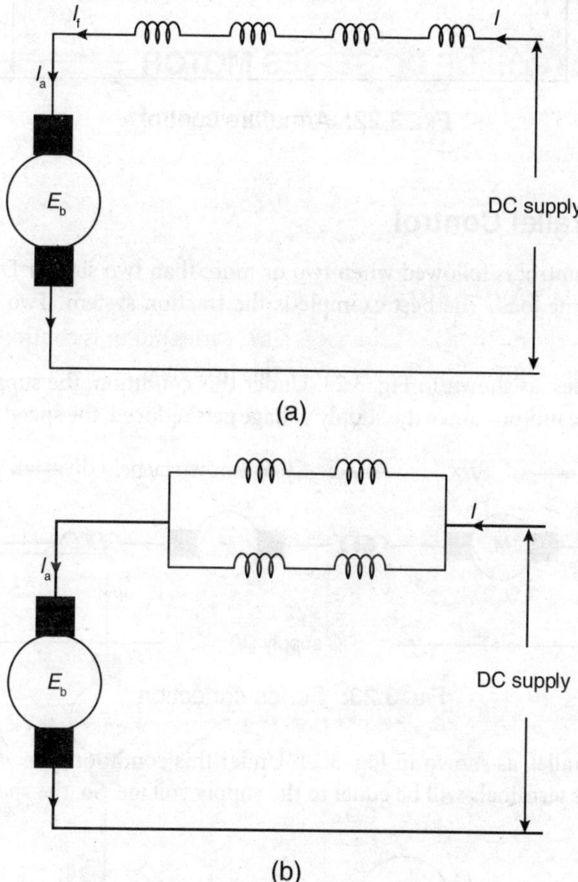

Fig. 3.21: Parallel field coils control

3.14.2 Armature Control Method

In this method of speed control, a variable resistance is connected in series with the field and armature windings, as shown in Fig. 3.22. This variable resistance will reduce the voltage across the armature and hence, the speed falls. By changing the value of variable resistance, any speed below the base speed can be obtained. This method of speed control is generally used for motors used in cranes, hoists and trains. However, it has poor speed regulation.

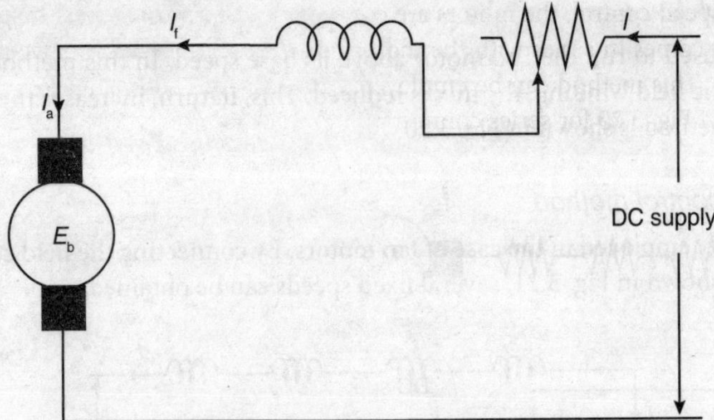

FIG. 3.22: Armature control

3.14.3 Series-Parallel Control

This method of speed control is followed when two or more than two similar DC motors are mechanically coupled to the same load. The best example is the traction system. Two or more than two DC motors can be,

i) Connected in series, as shown in Fig. 3.23. Under this condition, the supply voltage gets equally divided among the motors. Since the supply voltage gets reduced, the speed of the motor decreases.

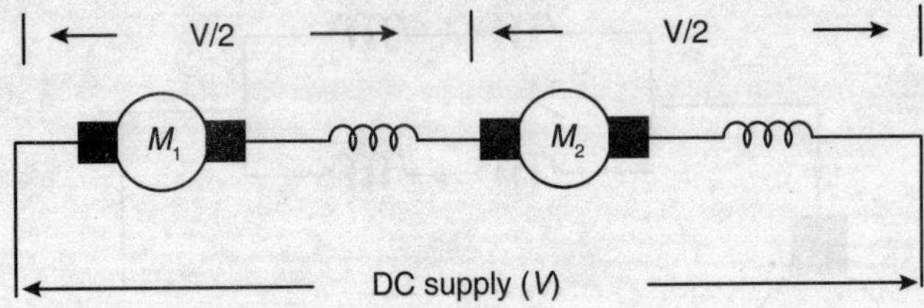

FIG. 3.23: Series conection

ii) Connected in parallel, as shown in Fig. 3.24. Under this condition, the voltage appearing across each of the motor terminals will be equal to the supply voltage. So, the speed of the motor will be high.

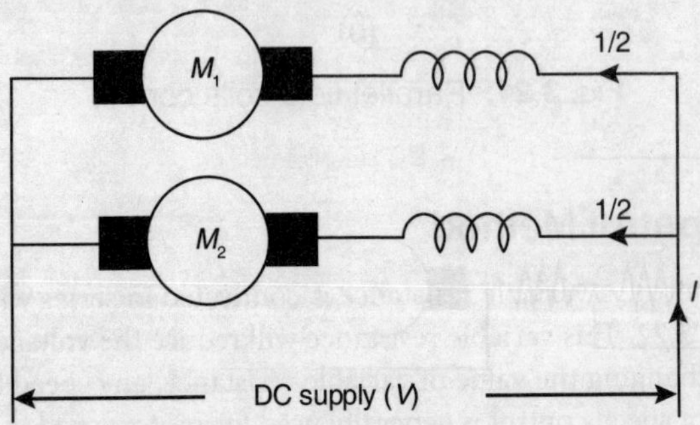

FIG. 3.24: Shunt connection

To obtain variable speed control, the motors are connected with a variable resistance. At the instant of starting, full resistance comes in series with the motors. As the motor picks up speed the resistances are slowly cut out in steps. This method can be employed for both series and parallel combination of motor connection as shown in Fig. 3.25 for series connection of motors, and Fig. 3.26 for parallel connection of motors.

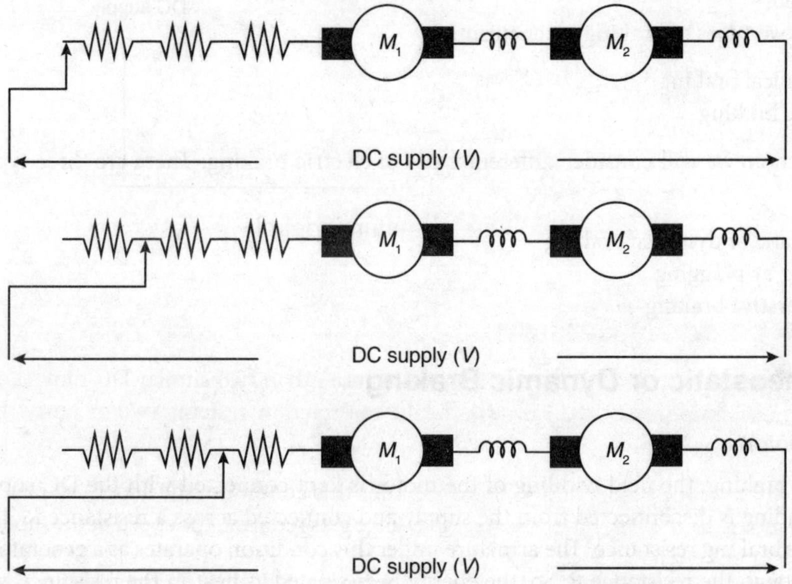

FIG. 3.25: Position of variable resistant for series motor connection

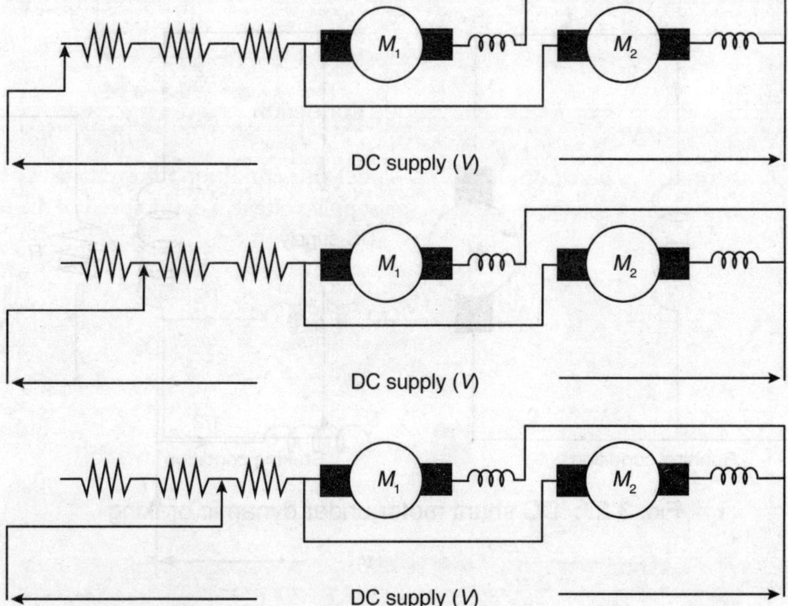

FIG. 3.26: Position of variable resistance for shunt motor connection

3.15 ELECTRIC BRAKING

Braking is used to stop the motor quickly. If a DC motor is being used for repeated operation, then it becomes necessary to stop the motor quickly so that it can be used again. If the supply from the motor is disconnected, it will not stop immediately but continue to run for some time due to inertia. However, the motor may sometimes have to be stopped quickly at a specific location. For stopping the motor quickly braking is used.

There are two types of braking mechanism:

i) Mechanical braking
ii) Electric braking

In this section, we will consider different types of electric braking. There are three types of electric braking.

i) Rheostatic or dynamic braking
ii) Braking by plugging
iii) Regenerative braking

3.15.1 Rheostatic or Dynamic Braking

DC shunt motor

For dynamic braking, the field winding of the motor is kept connected with the DC supply while the armature winding is disconnected from the supply and connected across a resistance *R*. This resistance *R* is called the braking resistance. The armature under this condition operates as a generator, completing its circuit through the resistance *R*. So, the energy is dissipated as heat in the resistance and ultimately the machine stops. Figure 3.27 shows the connection of a DC shunt motor during running and braking condition.

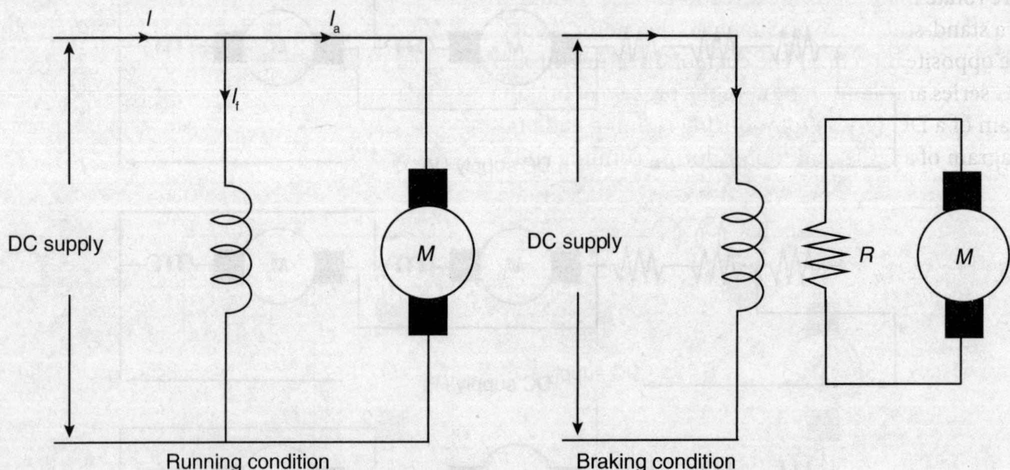

Fig. 3.27: DC shunt motor under dynamic braking

DC series motor

For DC series motor, the motor is disconnected from the supply and is connected across a braking resistance *R*. However, care should be taken that the field connection is reversed while connecting the motor with the braking resistance so that the current direction through the field winding remains same as before braking. If this is not done then the braking will not take place. Also, the value of the braking resistance must be less than the critical resistance. Figure 3.28 shows the connection diagram of a DC series motor during running and braking condition.

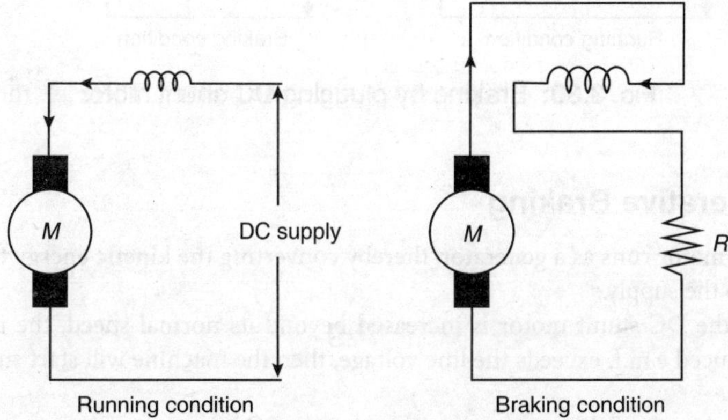

Fig. 3.28: DC series motor under dynamic braking

3.15.2 Braking by Plugging

In this method of braking, the armature connection with the supply is reversed. This will make the armature rotate in the opposite direction, thus providing the necessary braking effect. When the motor comes to a stand-still, the main supply to the motor is cut off. If this is not done the motor will start to rotate in the opposite direction. The current direction through the field winding should not change at all for both DC series and shunt motor, in the process of braking by plugging. Figure3.29 shows the connection diagram of a DC series motor during running and braking condition while Fig. 3.30 shows the connection diagram of a DC shunt motor during running and braking condition.

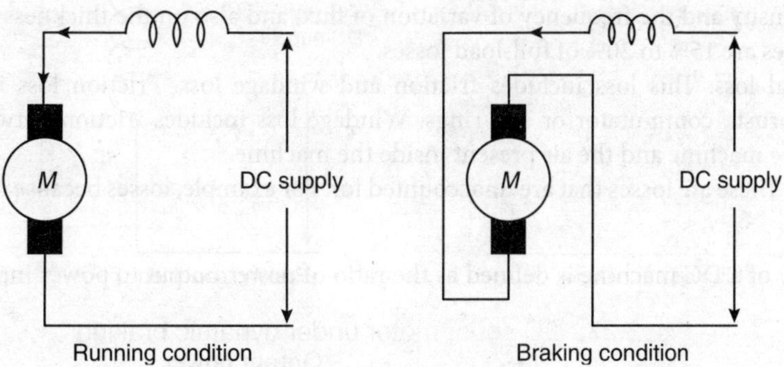

Fig. 3.29: Braking by plugging DC series motor

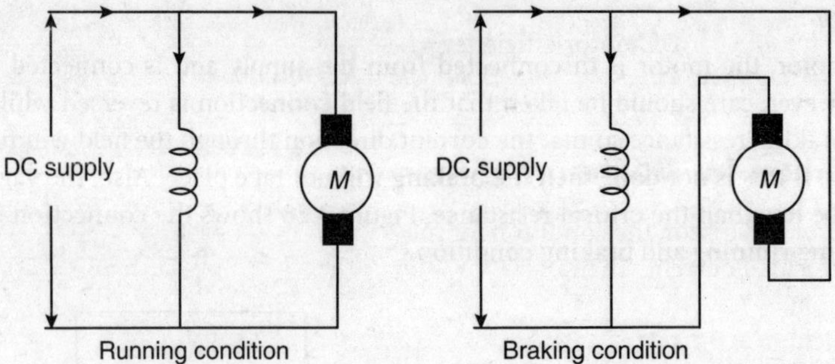

Fig. 3.30: Braking by plugging DC shunt motor

3.15.3 Regenerative Braking

In this method the motor runs as a generator, thereby converting the kinetic energy to electrical energy and feeds it back to the supply.

If the speed of the DC shunt motor is increased beyond its normal speed, the induced e.m.f. will increase. If this induced e.m.f. exceeds the line voltage, then the machine will start supplying current to the main.

The same can be done for field current also. That is, if the field current is increased to such a value that the induced e.m.f. exceeds the value of line voltage, then the motor starts supplying current to the main.

To brake the DC shunt motor, the field excitation in increased, due to which the speed of the motor reduces and the kinetic energy of the motor is fed back to the supply.

3.16 LOSSES AND EFFICIENCY OF A DC MACHINE

The different types of losses in a DC machine are categorised as follows:

i) Copper loss or electrical loss: It is given by I^2R and includes power loss in the armature conductor, in the field conductor, in the inter pole field, loss in the brush and loss in the compensating winding. This loss depends on the load and the armature current.

ii) Core loss or magnetic loss: This includes hysteresis loss and eddy current loss. The hysteresis loss depends on flux density and the frequency of variation of flux. The eddy current loss also depends on flux density and the frequency of variation of flux, and also on the thickness of the iron core. These losses are 15% to 20% of full-load losses.

iii) Mechanical loss: This loss includes friction and windage loss. Friction loss includes friction between brush, commutator or slip rings. Windage loss includes friction between the moving parts of the machine and the air present inside the machine.

iv) Stray loss: These are losses that are unaccounted for. For example, losses because of armature reaction.

The efficiency of a DC machine is defined as the ratio of power output to power input. It is given by the expression,

$$\text{DC generator efficiency} = \frac{\text{Output power}}{\text{Output power} + \text{Losses}}$$

$$\text{DCmotor efficiency} = \frac{\text{Input power} - \text{Losses}}{\text{Input power}}$$

3.16.1 Condition for Maximum Efficiency

For DC generator or DC motor, the condition for maximum efficiency is the same. Let V be the terminal voltage and I be the load current. Then,

Generated output= VI

Loss in the machine = $I^2R_a + P_{const}$

Where, P_{const} are the constant losses of the machine.

Here, the flux and the speed of the machine have been assumed to be constant. Under this condition, all the losses except the armature copper loss are constant.

Then,

$$\text{Efficiency } (\eta) = \frac{\text{Generated output}}{\text{Generated output} + \text{Loss in the machine}}$$

$$\text{or } \eta = \frac{VI}{VI + I^2R_a + P_{const}} \tag{3.22}$$

For maximum efficiency,

$$\frac{d\eta}{dI} = 0$$

$$\text{or, } \frac{d}{dI}\left(\frac{VI}{VI + I^2R_a + P_{const}}\right) = 0$$

$$\text{or, } \frac{\left(VI + I^2R_a + P_{const}\right)V - VI\left(V + 2IR_a\right)}{\left(VI + I^2R_a + P_{const}\right)^2} = 0$$

$$\text{or, } I^2 = P_{const} \tag{3.23}$$

Thus for maximum efficiency, variable loss must be equal to the constant loss.

3.17 BRAKE TEST ON DC MOTOR

This is a direct method of testing of DC motor. It is used to calculate the efficiency of small DC motors. In this method of testing, the DC motor shaft is coupled to a water-cooled pulley which is loaded by means of weight.

Let,

W_1 be the tension on the tight side of the brake in Kg.

W_2 be the tension on the slack side of the brake in Kg.

R be the radius of pulley in meter

N be the speed of the motor in r.p.m

V be the supply voltage in volts
I be the full load current in ampere.
Then,

$$\text{Output torque or shaft torque } (T_{sh}) = R(W_1 - W_2)\text{Kg-m}$$

$$\text{or, } T_{sh} = 9.81 \times R(W_1 - W_2)\text{N-m}$$

$$\text{Motor output} = \frac{2\pi N T_{sh}}{60}\text{ W}$$

$$= \frac{2\pi N \times 9.81 \times R(W_1 - W_2)}{60}\text{ W}$$

$$= 1.027(W_1 - W_2)RN\text{ W}$$

$$\text{Motor input} = VI$$

Then,

$$\text{Efficiency}(\eta) = \frac{\text{Motor output}}{\text{Motor input}} \times 100 = \frac{1.027(W_1 - W_2)RN}{VI} \times 100$$

3.18. SWINBURNE'S TEST

This is an indirect method of testing the efficiency of a DC machine. It is a no-load test and hence it cannot be conducted on a DC series motor. The motor is run on no-load at rated voltage and at rated speed. Since the motor is operated at no-load, the mechanical power output is zero. So, the power developed by the armature will supply frictional and windage loss, core loss and the copper loss in the armature. Let,
I_0 be the no load supply current.
I_f be the field current.
V be the supply voltage
I_a be the no-load armature current.
Then, the input power at no load = VI_0
Let us calculate the efficiency of the motor when the load current is I. Then,
For motoring action:

$$I_a = I - I_{sh}$$

$$\text{Input power} = VI$$

$$\text{Loss} = I_a^2 R_a + P_c = (I - I_{sh})^2 R_a + P_c$$

Where, P_c is the constant loss and is given by the expression

$$P_c = VI_0 - I_a^2 R_a = VI_0 - (I - I_{sh})^2 R_a$$

Then,

$$\text{Efficiency}\ (\eta) = \frac{\text{Input} - \text{Loss}}{\text{Input}} \times 100 = \frac{VI - \left[\left(I - I_{sh} \right)^2 R_a + P_c \right]}{VI} \times 100$$

For generating action:

$$I_a = I + I_{sh}$$

$$\text{Output power} = VI$$

$$\text{Loss} = I_a^2 R_a + P_c = \left(I + I_{sh} \right)^2 R_a + P_c$$

Where, P_c is the constant loss and is given by the expression

$$P_c = VI_0 - I_a^2 R_a = VI_0 - \left(I - I_{sh} \right)^2 R_a$$

Then

$$\text{Efficiency}\ (\eta) = \frac{\text{Output}}{\text{Output} + \text{Loss}} \times 100 = \frac{VI}{VI + \left[\left(I + I_{sh} \right)^2 R_a + P_c \right]} \times 100$$

Advantages of this test are:

i) It is economical because no-load input power is sufficient to perform the test
ii) Efficiency can be pre-determined

Disadvantages of this test are:

i) Effect of armature reaction is not considered here.
ii) Stray load loss is not considered.
iii) Temperature rise in the machine cannot be determined by this test.
iv) It is not applicable for DC series motor.

3.19 HOPKINSON'S TEST

It is called regenerative test or back-to-back test. It is carried out on two identical DC machines which are mechanically coupled to each other. In this, full-load test can be carried out on two identical shunt machines without wasting their outputs. One of the machines runs as a motor and the other runs as a generator. The mechanical output of the motor drives the generator and the generator's electrical output supplies the greater part of input to the motor. The motor is supplied from the mains only to compensate for losses. Due to these losses, the generator output is not sufficient to drive the motor.

Figure 3.31 shows the connection diagram for Hopkinson's test. The two shunt machines are connected in parallel. One of the machines is started as a motor. Initially, the switch S is kept open. The other machine which is coupled to the first will act as load (on the first, which is acting as motor). Thus,the second machine will act as a generator. The speed of the motor is adjusted to its rated value with the help of the field rheostat. The voltmeter reading is observed. The voltage of the generator is adjusted by its field rheostat so that voltmeter reading is zero. This will indicate that the generator voltage is having the same magnitude and polarity of that of the supply voltage and prevent heavy circulating current flowing in the local loop of armatures on closing the switch. Now, switch S is closed. The two machines can be put into any load by adjusting their field rheostats. The generator current I_2 can be adjusted to any value by increasing the excitation of the generator or by reducing the excitation of the motor. The various reading shown by different ammeters are noted for further calculations.

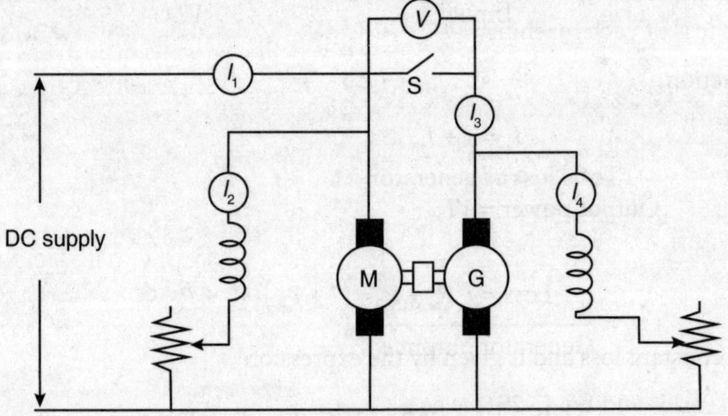

Fig. 3.31: Hopkinson's test circuit

From Fig. 3.31, we can write,

$$\text{Motor input} = V(I_1 + I_3) \tag{3.24}$$

$$\text{Generator output} = VI_3 \tag{3.25}$$

$$\text{Motor output} = \eta \times \text{Motor input} = \eta \times V(I_1 + I_3) \tag{3.26}$$

This output of the motor is the input for the generator. Then,

$$\text{Generator output} = \eta \times \text{Input to the generator} = \eta \times \eta \times V(I_1 + I_3) \tag{3.27}$$

Comparing Eq.(3.25) and Eq. (3.27), we get,

$$VI_3 = \eta^2 V(I_1 + I_3)$$

$$or,\ \eta = \sqrt{\frac{I_3}{I_1 + I_3}} \tag{3.28}$$

$$\text{Armature copper loss in motor} = (I_1 + I_3 - I_2)^2 R_{am}$$

$$\text{Shunt field loss in motor} = VI_2$$

$$\text{Armature copper loss in generator} = (I_3 + I_4)^2 R_{ag}$$

$$\text{Shunt field loss in generator} = VI_4$$

$$\text{Power taken from the supply} = VI_1$$

$$\text{Total iron loss} = P_c = VI_1 - \left[(I_1 + I_3 - I_2)^2 R_{am} + VI_2 + (I_3 + I_4)^2 R_{ag} + VI_4 \right]$$

Then, the iron loss of each machine $= \dfrac{P_c}{2}$

For generator

$$\text{Total loss of generator} = (I_3 + I_4)^2 R_{ag} + VI_4 + \frac{P_c}{2} \quad (3.29)$$

Then,

$$\eta_g = \frac{\text{Generator output}}{\text{Generator output} + \text{Total loss of generator}} \times 100$$

Substituting Eq.(3.25) and Eq. (3.29) we get,

$$\eta_g = \frac{VI_3}{VI_3 + \left[(I_3 + I_4)^2 R_{ag} + VI_4 + \dfrac{P_c}{2} \right]} \times 100 \quad (3.30)$$

For motor

$$\text{Total loss of motor} = (I_1 + I_3 - I_2)^2 R_{am} + VI_2 + \frac{P_c}{2} \quad (3.31)$$

Then,

$$\eta_m = \frac{\text{Motor input} - \text{Loss}}{\text{Motor input}} \times 100$$

Substituting Eq.(3.24) and Eq. (3.31) we get,

$$\eta_g = \frac{V(I_1 + I_3) - \left[(I_1 + I_3 - I_2)^2 R_{am} + VI_2 + \dfrac{P_c}{2} \right]}{V(I_1 + I_3)} \times 100 \quad (3.32)$$

Disadvantages of this test are:

1. It is very difficult to get two identical machines.
2. The iron losses in the two machines cannot be separated.
3. If the machines are not loaded equally, then the analysis will be very difficult.

SOLVED PROBLEMS

Ex.3.1 A DC shunt machine has the following parameters:
Resistance of the armature = 0.15 Ω.
Resistance of the field = 200 Ω.

Calculate the ratio of generated speed to motor speed, when the machine is connected across a 300 V supply with the line current of 56 A.

Solution

Given, $R_a = 0.15\ \Omega$, $R_f = 200\ \Omega$, V=300V, $I_L = 56$ A.

For a shunt generator:

$$I_f = \frac{V}{R_f} = \frac{300}{200} = 1.5 \text{ A}$$

$$I_a = I_L + I_f = 56 + 1.5 = 57.5 \text{ A}$$

$$\text{Back e.m.f.}\left(E_{bg}\right) = V + I_a R_a = 300 + 57.5 \times 0.15 = 308.63 \text{ A}$$

For a shunt motor:

$$I_a = I_L - I_f = 56 - 1.5 = 54.5 \text{ A}$$

$$\text{Back e.m.f.}\left(E_{bm}\right) = V - I_a R_a = 300 - 54.5 \times 0.15 = 291.83 \text{ A}$$

Now, let N_g be the speed of the machine during generation action and N_m be the speed of the machine during motoring action.

Again, we know that the back e.m.f. is given by the expression,

$$E = \frac{\varphi ZNP}{60A}$$

For a particular machine, φ, Z,P and A are constant. So we can now write,

$$E \propto N$$

Then for generation action,

$$E_{bg} \propto N_g \tag{i}$$

And, for motoring action,

$$E_{bm} \propto N_m \tag{ii}$$

Dividing, Eq. (i) byEq. (ii), we get,

$$\frac{N_g}{N_m} = \frac{E_{bg}}{E_{bm}} = \frac{308.63}{291.83} = 1.06$$

Ex.3.2 A wave-wound 8-pole DC motor has 500 conductors. The average flux density over one pole pitch is 1T. The pole shoe is 10 cm long and the armature diameter is 20 cm. Find the following:

(i) Gross mechanical power

(ii) Torque developed

Assume that the motor is running at 1200 r.p.m. and is drawing 15A current.

Solution

$$\text{Flux per pole}(\varphi) = \frac{\pi \times 20 \times 10^{-2} \times 10 \times 10^{-2} \times 1}{8} = 0.0078 \text{ Wb}$$

$$\text{e.m.f. induced}(E) = \frac{\varphi ZNP}{60A} = \frac{0.0078 \times 500 \times 1200 \times 8}{60 \times 2} = 312 \text{ V}$$

$$\text{Gross mechanical power}(P_m) = EI_a = 312 \times 15 = 4680 \text{ W}$$

$$\text{Torque developed} = \frac{P_m}{\frac{2\pi N}{60}} = \frac{4680}{\frac{2\pi \times 1200}{60}} = 37.24 \text{ Nm}$$

Ex.3.3 Consider a DC series motor with armature resistance of 0.8Ω and field resistance of 0.3Ω.It is running at 1200 r.p.m. and takes a current of 25A when connected across 250V supply. Find the speed of the motor when an external resistance of 0.3Ω is connected in parallel with the field winding. Assume that the flux at 12.5A current is 0.8 times of the flux at 25A current.

Solution

Given, $R_a = 0.8\Omega$, $R_f = 0.3\Omega$, $R_{ext} = 0.3\Omega$, N=1200 r.p.m., $I = 25$A, V=250V.
We know that,

$$E_b = V - I_a(R_a + R_f) = 250 - 25 \times (0.8 + 0.3) = 222.5\text{V}$$

Again,

$$E_b = k\varphi N$$

$$\text{or } k\varphi = \frac{E}{N} = \frac{222.5}{1200} = 0.185$$

The value of external resistance and the field resistance is the same,that is, 0.3Ω. So when they are connected in parallel, the current 25A gets equally divided in both the paths. That is, now the field current is 12.5A. Let, under this condition, the back e.m.f. be E_{b1}.
Then,

$$E_{b1} = V - I_a R_a - I_f R_f = 250 - 25 \times 0.8 - 12.5 \times 0.3 = 226.25 \text{ V}$$

Again

$$E_{b1} = k\varphi_1 N_1$$

$$k\varphi_1 = 0.8 \times 0.185$$

Then,

$$N_1 = \frac{E_{b1}}{k\varphi_1} = \frac{226.25}{0.8 \times 0.185} = 1529 \text{ r.p.m.}$$

Ex.3.4 A DC shunt motor, when fed from a DC source of 300 V, draws a current of 95 A and runs at 1200 r.p.m. Now, if suddenly the excitation is reduced to 60% of its previous value, calculate the new speed of the motor. Assume that the torque remains the same and the resistance of the armature is 0.5 Ω.

Solution

Given, $V = 300$ V, $I = 95$ A, N=1200 r.p.m., $R_a = 0.5\ \Omega$.
Let the initial flux be φ and the initial e.m.f. beE_b. Then,

$$E_b = V - I_a R_a = 300 - 95 \times 0.5 = 252.5\text{ V}$$

When the excitation is reduced by 60%, let the new flux be φ_1, new e.m.f. be E_{b1} and the new current be I_1. Then,

$$\varphi_1 = 0.6\varphi$$

Since the torque does not change, we can write,

$$I_1 \varphi_1 = I\varphi$$

$$I_1 = \frac{I\varphi}{\varphi_1} = \frac{95\varphi}{0.6\varphi} = 158.33\text{ A}$$

Then,

$$E_{b1} = V - I_1 R_a = 300 - 158.33 \times 0.5 = 220.835\text{ V}$$

Let the initial speed be N and the new speed when the excitation is changed be N_1. Then, we can write,

$$N = \frac{E_b}{\varphi} \qquad \text{(i)}$$

And

$$N_1 = \frac{E_{b1}}{\varphi_1} \qquad \text{(ii)}$$

Dividing Eq. (i) byEq. (ii), we get,

$$\frac{N}{N_1} = \frac{E_b}{E_{b1}} \times \frac{\varphi_1}{\varphi} = \frac{252.5}{220.835} \times 0.6 = 0.686$$

$$N_1 = \frac{N}{0.686} = \frac{1200}{0.686} = 1749.27\text{ r.p.m.}$$

Ex.3.5 A 230V DC shunt motor has a field and armature resistance of 150Ω and 0.3Ω respectively. The motor runs at 1500 r.p.m. and the armature current is 50A at full load. Calculate the following.

(i) Developed torque
(ii) If the variable field resistance is changed to 180Ω, what will be the new speed of the motor? Assume the armature current and torque is constant.

Solution

Given, V=230V, R_a = 0.3Ω, R_f = 150Ω, N=1500 r.p.m, I_a = 50A, R_{f_mod} = 180Ω.

$$E_a = V - I_a R_a = 230 - 50 \times 0.3 = 215 \text{ V}$$

$$\text{Power} = E_a I_a = 215 \times 50 = 10750 \text{ W}$$

(i) Developed torque $T_d = \dfrac{P}{\omega} = \dfrac{P}{\dfrac{2\pi N}{60}} = \dfrac{60P}{2\pi N \pi} = \dfrac{60 \times 10750}{2 \times \pi \times 1500} = 68.44$ Nm

(ii) Field current when the field resistance is 150Ω

$$I_{f1} = \frac{V}{R_f} = \frac{230}{150} = 1.53 \text{ A}$$

Field current when the field resistance is 180Ω

$$I_{f1} = \frac{V}{R_{f\_mod}} = \frac{230}{180} = 1.28 \text{ A}$$

Since the armature current has not changed.

$$I_{f1} N = I_{f2} N_2$$

$$or\ N_2 = \frac{I_{f1}}{I_{f2}} N = \frac{1.53}{1.28} \times 1500 = 1792 \text{ r.p.m.}$$

Ex.3.6 A 250V, 1400 r.p.m. DC shunt motor is started with the help of a starter. The starter resistances in the order in which they are cut out are 2Ω, 1.5Ω, 1Ω and 0.5Ω respectively. The armature has a resistance of 0.2Ω. Find the following:

(i) Speed at each position of the starter hand.
(ii) Initial and final value of armature current at each position of the starter hand.

Assume that, the starter is moved to its next position only when the armature current has come to its rated value of 30A.

Solution

Total resistance in the armature circuit at different starter positions are:

$$\text{Position} - 1: R_1 = 2 + 1.5 + 1 + 0.5 + 0.2 = 5.2\ \Omega$$

$$\text{Position} - 2: R_2 = 1.5 + 1 + 0.5 + 0.2 = 3.2\Omega$$

$$\text{Position} - 3: R_3 = 1 + 0.5 + 0.2 = 1.7\Omega$$

$$\text{Position} - 4: R_4 = 0.5 + 0.2 = 0.7\Omega$$

$$\text{Position} - 5: R_5 = 0.2\ \Omega$$

At Position-1: $R_1 = 5.2\ \Omega$
Back e.m.f. = 0 V .The starting current is,

$$I_{st} = \frac{V}{R_1} = \frac{250}{5.2} = 48.07\text{ A}$$

When the armature current drops to its rated value of 30 A, the back e.m.f. (E_{b1}) will be,

$$E_{b1} = V - I_a R_1 = 250 - 30 \times 5.2 = 94\text{ V}$$

The back e.m.f. at rated speed of 1400 r.p.m.

$$E_{b2} = V - I_a R_a = 250 - 30 \times 0.2 = 244\text{ V}$$

So the speed at e.m.f. E_{b1} will be,

$$N_1 = \frac{E_{b1}}{E_{b2}} \times 1400 = \frac{94}{244} \times 1400 = 539.34\text{ r.p.m.}$$

At Position-2: $R_2 = 3.2\ \Omega$

$$I_{st} = \frac{V - E_{b1}}{R_2} = \frac{250 - 94}{3.2} = 48.75\text{ A}$$

When the armature current has dropped to its rated value of 30 A,

$$E_{b3} = V - I_a R_2 = 250 - 30 \times 3.2 = 154\text{ V}$$

So the speed at e.m.f. E_{b3} will be,

$$N_3 = \frac{E_{b3}}{E_{b2}} \times 1400 = \frac{154}{244} \times 1400 = 883\text{ r.p.m.}$$

At Position-3: $R_3 = 1.7\ \Omega$

$$I_{st} = \frac{V - E_{b3}}{R_3} = \frac{250 - 154}{1.7} = 56.47\text{ A}$$

When the armature current has dropped to its rated value of 30 A,

$$E_{b4} = V - I_a R_3 = 250 - 30 \times 1.7 = 199\text{ V}$$

So the speed at e.m.f. E_{b4} will be,

$$N_4 = \frac{E_{b4}}{E_{b2}} \times 1400 = \frac{199}{244} \times 1400 = 1142\text{ r.p.m.}$$

At Position-4: $R_4 = 0.7\ \Omega$

$$I_{st} = \frac{V - E_{b4}}{R_4} = \frac{250 - 199}{0.7} = 72.86\text{ A}$$

When the armature current has dropped to its rated value of 30 A,

$$E_{b5} = V - I_a R_4 = 250 - 30 \times 0.7 = 229\ \text{V}$$

So the speed at e.m.f. E_{b5} will be,

$$N_5 = \frac{E_{b5}}{E_{b2}} \times 1400 = \frac{229}{244} \times 1400 = 1314\ \text{r.p.m.}$$

At Position-5: $R_4 = 0.2\ \Omega$

$$I_{st} = \frac{V - E_{b5}}{R_4} = \frac{250 - 229}{0.2} = 105\ \text{A}$$

When the armature current has dropped to its rated value of 30 A,

$$E_{b6} = V - I_a R_4 = 250 - 30 \times 0.2 = 244\ \text{V}$$

So the speed at e.m.f. E_{b6} will be,

$$N_6 = \frac{E_{b6}}{E_{b2}} \times 1400 = \frac{244}{244} \times 1400 = 1400\ \text{r.p.m.}$$

Ex.3.7 A DC shunt motor has the field resistance of 150 Ω and armature resistance of 0.2 Ω. When connected across a 230 V supply, it takes a no-load current of 3A. Calculate its efficiency at 30 A input current.

Solution

At no-load

$$\text{Input power} = 230 \times 3 = 690\ \text{W}$$

$$\text{Shunt field current at no-load}\left(I_{s0}\right) = \frac{230}{150} = 1.53\ \text{A}$$

$$\text{Armature current at no-load}\left(I_{a0}\right) = 3 - 1.53 = 1.47\ \text{A}$$

$$\text{Copper loss in the armature at no-load}\left(P_{cu0}\right) = I_{a0}^2 R_a = 1.47^2 \times 0.2 = 0.43\ \text{W}$$

$$\text{Constant loss}\left(P_c\right) = 690 - 0.43 = 689.57\ \text{W}$$

For an input current of 30 A

$$\text{Armature current}\left(I_a\right) = 30 - 1.53 = 28.47\ \text{A}$$

$$\text{Copper loss in the armature}\left(P_{cu}\right) = I_a^2 R_a = 28.47^2 \times 0.2 = 162.11\ \text{W}$$

$$\text{Total loss}\left(P_{loss}\right) = 689.57 + 162.11 = 851.68\ \text{W}$$

$$\text{Efficiency}(\eta) = \frac{\text{Input power} - \text{Total loss}}{\text{Input power}} \times 100 = \frac{(230\times 30) - 851.68}{(230\times 30)} \times 100 = 87.65\,\%$$

Ex.3.8 A DC shunt motor has the field and armature resistance of 220Ω and 0.4Ω respectively. When connected across a 230V supply, it has a no-load current of 3A and full-load current of 30A. Find the efficiency of the motor.

Solution

Given, $R_f = 220V$, $R_a = 0.4\Omega$, $V=230V$, $I_{a0} = 3A$, $I_{afl} = 30A$.

At no-load

$$\text{Input power} = 230\times 3 = 690\text{ W}$$

$$\text{Shunt field current at no-load}(I_{s0}) = \frac{230}{220} = 1.05\text{ A}$$

$$\text{Armature current at no-load}(I_{a0}) = 3 - 1.05 = 1.95\text{ A}$$

$$\text{Copper loss in the armature at no-load}(P_{cu0}) = I_{a0}^2 R_a = 1.95^2 \times 0.4 = 1.521\text{ W}$$

$$\text{Constant loss}(P_c) = 690 - 0.43 = 688.48\text{ W}$$

For an input current of 30 A

$$\text{Armature current}(I_a) = 30 - 1.05 = 28.95\text{ A}$$

$$\text{Copper loss in the armature}(P_{cu}) = I_a^2 R_a = 28.95^2 \times 0.4 = 335.241\text{ W}$$

$$\text{Total loss}(P_{loss}) = 688.48 + 335.241 = 1023.721\text{ W}$$

$$\text{Efficiency}(\eta) = \frac{\text{Input power} - \text{Total loss}}{\text{Input power}} \times 100 = \frac{(230\times 30) - 1023.721}{230\times 30} \times 100 = 85.16\,\%$$

Ex.3.9 Hopkinson's test is performed on two identical DC shunt machines. The data related to this test is as follows.

Supply voltage = 220 V.

Line current = 100 A.

Armature current of the motor = 410 A

Field current of the motor = 6A

Field current of the generator = 5A

Armature resistance of the motor = 0.02 Ω

Armature resistance of the generator = 0.02 Ω

Brush drop = 1V per brush

Calculate the efficiency of each machine.

Solution

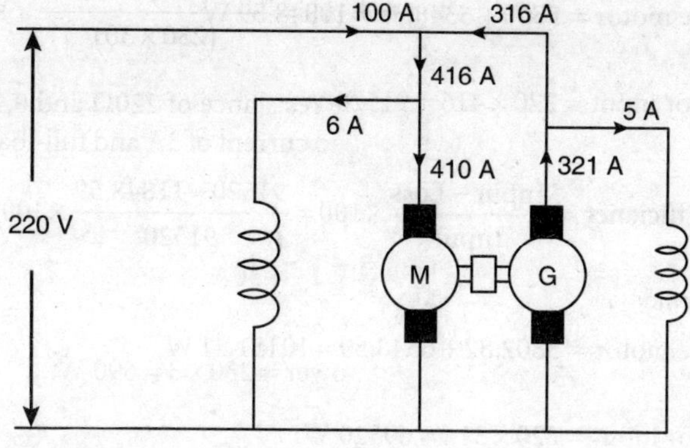

Fig. E3.9:

From Fig. E3.9, we can write,

Motor side

$$\text{Copper loss in the armature} = 410^2 \times 0.02 = 3362 \text{ W}$$

$$\text{Loss in brush} = 410 \times 2 = 820 \text{ W}$$

$$\text{Copper loss in the field} = 220 \times 6 = 1320 \text{ W}$$

$$\text{Total loss in motor} = 3362 + 820 + 1320 = 5500 \text{ W}$$

Generator side

$$\text{Copper loss in the armature} = 321^2 \times 0.02 = 2060.82 \text{ W}$$

$$\text{Loss in brush} = 321 \times 2 = 642 \text{ W}$$

$$\text{Copper loss in the field} = 220 \times 5 = 1100 \text{ W}$$

$$\text{Total loss in generator} = 2060.82 + 642 + 1100 = 3802.82 \text{ W}$$

$$\text{Total loss in the combined motor-generator set} = 5500 + 3802.82 = 9302.82 \text{ W}$$

$$\text{Input power} = 220 \times 100 = 22000 \text{ W}$$

$$\text{Stray loss} = 22000 - 9302.82 = 12697.18 \text{ W}$$

$$\text{Stray loss per machine} = \frac{12697.18}{2} = 6348.59 \text{ W}$$

Motor efficiency

Total loss in the motor = 5500 + 6348.59 = 11848.59 W

$$\text{Motor input} = 220 \times 416 = 91520 \text{ W}$$

$$\text{Efficiency} = \frac{\text{Input} - \text{Loss}}{\text{Input}} \times 100 = \frac{91520 - 11848.59}{91520} \times 100 = 87.05\%$$

Generator efficiency

Total loss in the motor = 3802.82 + 6348.59 = 10151.41 W

$$\text{Motor input} = 220 \times 316 = 69520 \text{ W}$$

$$\text{Efficiency} = \frac{\text{Output}}{\text{Output} + \text{Loss}} \times 100 = \frac{69520}{69520 + 10151.41} \times 100 = 87.26\%$$

Ex.3.10 Hopkinson's test is performed on two identical DC shunt machines. The data related to this test is as follows.

Supply voltage = 240 V.

Line current = 16 A.

Armature current of the generator = 56 A

Field current of the motor = 2A

Field current of the generator = 3A

Armature resistance of the motor = 0.2 Ω

Armature resistance of the generator = 0.2 Ω

Calculate the efficiency of each machine.

Solution

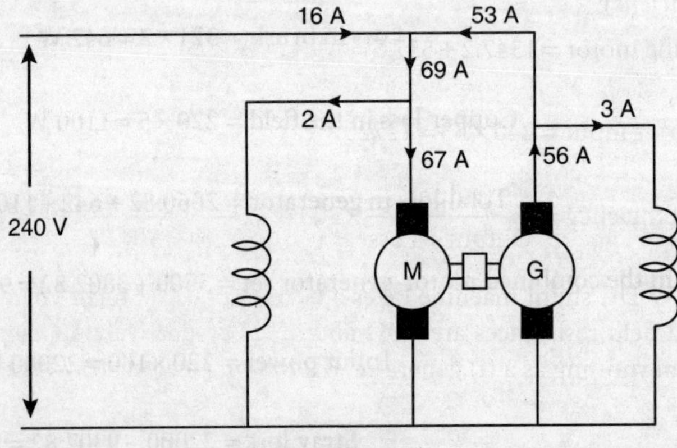

Fig. E3.10:

From Fig. E3.10, we can write,

Motor side

$$\text{Copper loss in the armature} = 67^2 \times 0.2 = 897.8 \text{ W}$$

$$\text{Copper loss in the field} = 240 \times 2 = 480 \text{ W}$$

$$\text{Total loss in motor} = 897.8 + 480 = 1377.8 \text{ W}$$

Generator side

$$\text{Copper loss in the armature} = 56^2 \times 0.2 = 627.2 \text{ W}$$

$$\text{Copper loss in the field} = 240 \times 3 = 720 \text{ W}$$

$$\text{Total loss in generator} = 627.2 + 720 = 1347.2 \text{ W}$$

$$\text{Total loss in the combined motor-generator set} = 1377.8 + 1347.2 = 2725 \text{ W}$$

$$\text{Input power} = 240 \times 16 = 3840 \text{ W}$$

$$\text{Stray loss} = 3840 - 2725 = 1115 \text{ W}$$

$$\text{Stray loss per machine} = \frac{1115}{2} = 557.5 \text{ W}$$

Motor efficiency

Total loss in the motor $= 1377.8 + 557.5 = 1935.3$ W

$$\text{Motor input} = 240 \times 69 = 16560 \text{ W}$$

$$\text{Efficiency} = \frac{\text{Input} - \text{Loss}}{\text{Input}} \times 100 = \frac{16560 - 1935.3}{16560} \times 100 = 88.31\%$$

Generator efficiency

Total loss in the motor $= 1347.2 + 557.5 = 1904.7$ W

$$\text{Motor input} = 240 \times 53 = 12720 \text{ W}$$

$$\text{Efficiency} = \frac{\text{Output}}{\text{Output} + \text{Loss}} \times 100 = \frac{12720}{12720 + 1904.7} \times 100 = 86.97\%$$

Ex.3.11 A 230V, 15 kW DC shunt machine takes a current of 4A at rated voltage and rated speed. The armature and field resistances are 0.3Ω and 250Ω respectively. Calculate the efficiency of the machine when running as a (i) generator, (ii) motor. Neglect stray losses.

Solution

As generator

$$I_L = \frac{15 \times 10^3}{230} = 65.22 \text{ A}$$

$$I_f = \frac{230}{250} = 0.92 \text{ A}$$

$$I_a = I_L + I_f = 66.14 \text{ A}$$

$$\text{Loss in armature} = I_a^2 R_a = 66.14^2 \times 0.3 = 1312.35 \text{ W}$$

$$\text{Loss in the field} = \frac{230^2}{250} = 211.6 \text{ W}$$

$$\text{Constant loss} = 230 \times 4 - 4^2 \times 0.3 = 915.2 \text{ W}$$

$$\text{Total loss} = 1312.35 + 915.2 + 211.6 = 2439.15 \text{ W}$$

$$\text{Efficiency} = \frac{15 \times 10^3}{15 \times 10^3 + 2439.15} \times 100 = 86\%$$

Motor

$$I_a = I_L - I_f = 64.3 \text{ A}$$

$$\text{Loss in armature} = I_a^2 R_a = 64.3^2 \times 0.3 = 1240.35 \text{ W}$$

$$\text{Loss in the field} = 211.6 \text{ W}$$

$$\text{Constant loss} = 915.2 \text{ W}$$

$$\text{Total loss} = 1240.35 + 211.6 + 915.2 = 2367.15 \text{ W}$$

$$\text{Efficiency} = \frac{15 \times 10^3 - 2367.15}{15 \times 10^3} \times 100 = 84\%$$

Ex.3.12 A DC shunt motor has the armature and field resistance as 0.2Ω and 100Ω respectively. When it is connected across a 200V supply, the no-load current is 3A and speed is 1400 r.p.m. Calculate the motor torque and motor speed at a full load current of 40A.

Solution

$$\text{Field current}(I_f) = \frac{200}{100} = 2 \text{ A}$$

$$\text{Armature current at no-load}(I_0) = 3 - 2 = 1 \text{ A}$$

$$\text{e.m.f. at no-load } (E_{bo}) = 200 - 1 \times 0.2 = 199.8 \text{ V}$$

$$\text{Armature current at full-load}(I_0) = 40 - 2 = 38 \text{ A}$$

$$\text{e.m.f. at full load } (E_{bf}) = 200 - 38 \times 0.2 = 192.4 \text{ V}$$

$$\text{No-load speed}(N_0) = 1400 \text{ r.p.m.}$$

Let the speed of the motor at full load be N_f. Then,

$$\frac{E_{b0}}{E_{bf}} = \frac{N_0}{N_f}$$

$$N_f = \frac{E_{bf} N_0}{E_{b0}} = \frac{192.4}{199.8} \times 1400 = 1348 \text{ r.p.m.}$$

$$\text{Loss at no-load} = 200 \times 3 - 2^2 \times 100 - 1^2 \times 0.2 = 199.8 \text{ W}$$

$$\text{Motor output} = 200 \times 40 - 2^2 \times 100 - 38^2 \times 0.2 - 199.8 = 7111.4 \text{ W}$$

$$\text{Again, motor output} = \frac{2\pi NT}{60}$$

$$T = \frac{60 \times \text{Motor output}}{2\pi N} = \frac{60 \times 7111.4}{2 \times \pi \times 1348} = 50.38 \text{ Nm}$$

Ex.3.13 Consider a DC machine with field and armature resistance of 250Ω and 1Ω respectively. When supplied with a 240V DC supply, it draws a current of 30A from the source (motoring action) running at 1500 r.p.m. Calculate the speed if the above condition is required to be achieved as a generator.

Solution

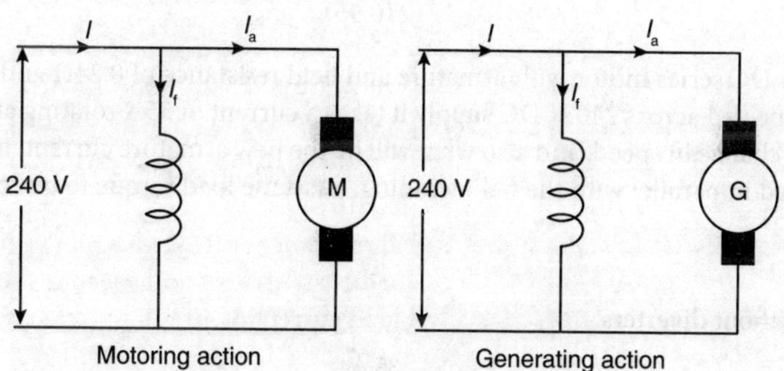

Fig. E3.13:

When working as a motor:

$$\text{Field current } (I_f) = \frac{V}{R_f} = \frac{240}{250} = 0.96 \text{ A}$$

$$I = 30\text{A}.$$

$$I_a = I - I_f = 30 - 0.96 = 29.04\text{A}$$

$$\text{Back e.m.f.}(E_b) = V - I_a R_a = 240 - 29.04 \times 1 = 210.96 \text{ A}$$

Again we can write,

$$E_b = I_f N_m \quad \text{(i)}$$

Where, N_m is the speed of rotation during motoring action.
When working as a generator:

$$\text{Field current } (I_f) = \frac{V}{R_f} = \frac{240}{250} = 0.96 \text{ A}$$

$$I = 30\text{A}.$$

$$I_a = I + I_f = 30 + 0.96 = 230.96 \text{ A}$$

$$\text{e.m.f.}\left(E_g\right) = V + I_a R_a = 240 + 30.96 \times 1 = 270.96 \text{ A}$$

Again we can write,

$$E_g = I_f N_g \tag{ii}$$

Where, N_g is the speed of rotation during generating action.
From Eq. (i) and Eq. (ii), we can write,

$$\frac{E_g}{E_b} = \frac{I_f N_g}{I_f N_m}$$

$$\text{or, } N_g = \frac{E_g N_m}{E_b} = \frac{270.96 \times 1500}{210.96} = 1927 \text{ r.p.m.}$$

Ex.3.14 Consider a DC series motor with armature and field resistance of 0.24Ω and 0.2Ω respectively. When connected across 240 V DC supply it takes a current of 35A rotating at 1200 r.p.m. What will be the change in speed, and also what will be the new armature current, if a diverter of 0.2Ω is connected in parallel with the field winding. (Assume load torque to be constant in both the cases).

Solution
Case-1: Without diverters

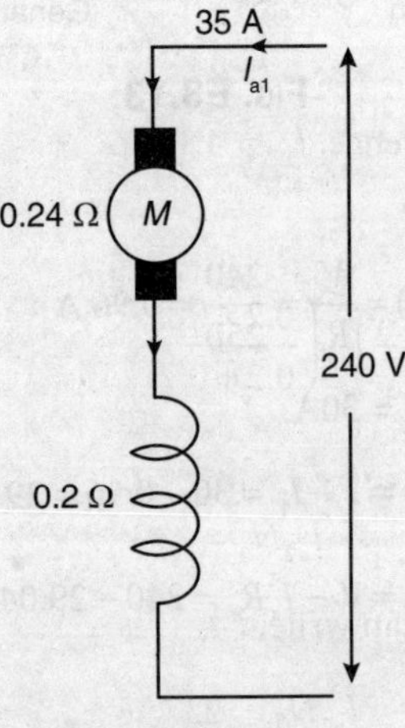

Fig. E3.14 (a):

$$E_{b1} = V - I_{a1}\left(R_a + R_f\right) = 240 - 35(0.2 + 0.24) = 224.6 \text{ V}$$

Again,

$$I_{a1}N_1 = 224.6 \text{ V} \tag{i}$$

Case-2: With diverters

$$I_{f2} = I_{a2}\frac{0.2}{0.2+0.2} = 0.5I_{a2}$$

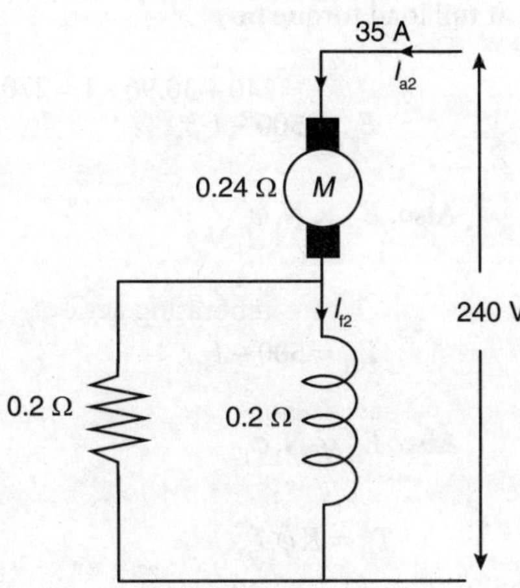

Fig. E3.14 (b):

Since the torque in both the cases is the same, we can write,

$$KI_{a1}^2 = KI_{a2}I_{f2}$$

$$\text{or, } KI_{a1}^2 = K \times 0.5 \times \times I_{a2}^2$$

$$\text{or, } I_{a2} = \sqrt{\frac{I_{a1}^2}{0.5}} = 8.36 \text{ A}$$

$$\text{Hence, } I_{f2} = 4.18 \text{ A}$$

Again,

$$E_{b2} = V - I_{a2}\left(\frac{0.2 \times 0.2}{0.2+0.2} + 0.24\right) = 237.15 \text{ V}$$

Again,

$$I_{a2}N_2 = 237.15 \text{ V} \tag{ii}$$

From Eq. (i) and Eq. (ii), we can write,

$$\frac{I_{a2}N_2}{I_{a1}N_1} = \frac{237.15}{224.6} = 1.06$$

$$\text{or, } N_2 = \frac{1.06 \times 35 \times 1200}{8.36} = 5325 \text{ r.p.m.}$$

Ex.3.15 A 500 V DC shunt motor has a speed of 2500 r.p.m. at no-load. At full-load torque, full-field voltage and reduced armature voltage, it runs at 900 r.p.m. Keeping the armature voltage and field voltage constant, if the load torque is reduced to 40% of rated value, the speed increased to 1000 r.p.m. Calculate the armature voltage drop at full load.

Solution

Let the armature current at full load torque be I_{a1}.

Then,

$$E_{a1} = 500 - I_{a1} r_a$$

$$\text{Also, } E_{a1} \propto N_1 \varphi_1$$

At 40 % torque,

$$E_{a1} = 500 - I_{a1} r_a$$

$$\text{Also, } E_{a1} \propto N_1 \varphi_1$$

$$T_{e1} = K \varphi_1 I_{a1}$$

Again,

$$0.4\, T_{e1} = K \varphi_2 I_{a2}$$

Since the field voltage is constant, $\varphi_1 = \varphi_2$

$$\frac{T_{e1}}{0.4 T_{e1}} = \frac{K \varphi_1 I_{a1}}{K \varphi_2 I_{a2}}$$

$$\text{or, } \frac{1}{0.4} = \frac{I_{a1}}{I_{a2}}$$

$$\text{or, } I_{a2} = 0.4 I_{a1}$$

At 40 % of rated torque

$$E_{a2} = 500 - I_{a2} r_a = 500 - 0.4 I_{a1} r_a$$

$$\text{Also, } E_{a2} \propto N_2 \varphi_2$$

$$\text{or, } E_{a2} \propto N_2 \varphi_1$$

Then,

$$\frac{E_{a1}}{E_{a2}} = \frac{N_1 \varphi_1}{N_2 \varphi_2} =$$

$$\frac{500 - I_{a1}r_a}{500 - 0.4I_{a1}r_a} = \frac{900 \times \varphi_1}{1000 \times \varphi_1} = 0.9$$

$$500 - I_{a1}r_a = 500 \times 0.9 - 0.4 \times 0.9 I_{a1}r_a$$

Solving the equation we get,

$$I_{a1}r_a = 78.125 \text{ V}$$

Ex.3.16 A DC series motor at rated speed and voltage drives an electric train. Calculate the change in supply voltage when the speed reduces by 0.5 p.u. Assume the power to be constant.

Solution

$$I_{a2} = I_{a1} \times \sqrt{\frac{\omega_1}{\omega_2}} = I_{a1} \times \sqrt{\frac{\omega_1}{0.5 \times \omega_1}}$$

$$\text{or, } I_{a2} = 1.414 I_{a1}$$

Since the power is constant,

$$V_{t2}I_{a2} = V_{t1}I_{a1}$$

$$V_{t2} = V_{t1}\frac{I_{a1}}{I_{a2}} = \frac{1}{1.414}V_{t1}$$

Ex. 3.17 A separately excited DC motor runs at 100 rpm supplying rated torque when the applied voltage is 50 volts and the armature current is 10A. The armature resistance is $R_a = 1$ ohm. A thyristor-controlled chopper circuit is used to decrease the applied voltage to 40 volts, at constant field flux, to drive a load at 75% of the rated torque. The conducting time T_{on} of this circuit is fixed at 0.2 seconds. Compute the speed of the motor at 40 volts and the blocking time T_{off} of the chopper circuit.

Solution

$$T_1 = k\varphi \cdot I_{a1}$$

$$T_2 = 0.75T_1 = k\varphi \cdot I_{a2} = 0.75(k\varphi \cdot I_{a1}) \Rightarrow I_{a2} = 0.75 I_{a1} = (0.75)(10A) = 7.5 \text{ A}$$

$$E_{a1} = V_{t1} - I_{a1}R_a = 50 - (10)(1) = 40 \text{ V}$$

$$E_{a2} = V_{t2} - I_{a2}R_a = 40 - (7.5)(1) = 32.5 \text{ V}$$

$$E_{a1} = k\varphi \cdot \omega_1, E_{a2} = k\varphi \cdot \omega_2 \Rightarrow \frac{E_{a1}}{\omega_1} = \frac{E_{a2}}{\omega_2} \Rightarrow \omega_2 = \frac{E_{a2}}{E_{a1}}\omega_1 = \frac{32.5}{40}(100) = 81.25 \text{ r.p.m}$$

$$\frac{T_{on}}{T_{on} + T_{off}} = \frac{V_{\text{Load, DC}}}{V} \Rightarrow \frac{0.2}{0.2 + T_{off}} = \frac{40}{50} \Rightarrow T_{off} = 0.05 \text{sec}$$

Ex.3.18 The armature and the field resistance of a DC shunt motor is 0.2 Ω and 220Ω respectively and it is connected across a 310 V supply and 50 A line current. Calculate the ratio of motor speed to generated speed.

Solution

Given, $R_a = 0.2\ \Omega$, $R_f = 22\ \Omega$, V=310V, $I_L = 50$ A.

For a shunt generator:

$$I_f = \frac{V}{R_f} = \frac{310}{220} = 1.41\text{ A}$$

$$I_a = I_L + I_f = 50 + 1.41 = 51.41\text{ A}$$

$$\text{Back e.m.f.}\left(E_{b1}\right) = V + I_a R_a = 310 + 51.41 \times 0.2 = 320.28\text{ A}$$

For a shunt motor:

$$I_a = I_L - I_f = 50 - 1.41 = 48.59\text{ A}$$

$$\text{Back e.m.f.}\left(E_{b2}\right) = V - I_a R_a = 310 - 48.59 \times 0.2 = 300.282\text{ A}$$

Now, let N_g be the speed of the machine during generation action and N_m be the speed of the machine during motoring action.

Again, we know that, the back e.m.f. is given by the expression,

$$E = \frac{\varphi ZNP}{60A}$$

For a particular machine, φ, Z,P and $A$are constant. So we can now write,

$$E \propto N$$

Then, for generation action,

$$E_{b1} \propto N_g \qquad \text{(i)}$$

And, for motoring action,

$$E_{b2} \propto N_m \qquad \text{(ii)}$$

Dividing, Eq. (i) by Eq. (ii), we get,

$$\frac{N_m}{N_g} = \frac{E_{b2}}{E_{b1}} = \frac{300.282}{320.28} = 0.937$$

Ex.3.19 Find the torque developed in a 4-pole lap-wound DC motor with 400 conductors, which is running at 1500 r.p.m and drawing 20 A current. The average flux density over one pole pitch is 2T. The pole shoe is 9 cm long and the armature diameter is 15 cm.

Solution

$$\text{Flux per pole}\left(\varphi\right) = \frac{\pi \times 15 \times 10^{-2} \times 9 \times 10^{-2} \times 2}{4} = 0.021\text{ Wb}$$

$$\text{e.m.f.induced}\left(E\right) = \frac{\varphi ZNP}{60A} = \frac{0.021 \times 400 \times 1500 \times 4}{60 \times 4} = 210\text{ V}$$

$$\text{Gross mechanical power}(P_m) = EI_a = 210 \times 20 = 4200 \text{ W}$$

$$\text{Torque developed} = \frac{P_m}{\frac{2\pi N}{60}} = \frac{4200}{\frac{2\pi \times 1500}{60}} = 26.74 \text{ Nm}$$

Ex.3.20 The armature and the field resistance of a DC series motor running at 1500 r.p.m. with a current of 20 A, when connected across 300 V supply, is 0.8Ω and 0.5Ω respectively. What will be the speed of the motor when an external resistance of 0.5 Ω is connected in parallel with the field winding? Assume that the flux at 10A current is 0.4 times of the flux at 20A current.

Solution

Given, $R_a = 0.8\Omega$, $R_f = 0.5\Omega$, $R_{ext} = 0.5\Omega$, N=1500 r.p.m., I = 20A, V=300V.

We know that,

$$E_b = V - I_a(R_a + R_f) = 300 - 20 \times (0.8 + 0.5) = 274 \text{ V}$$

Again,

$$E_b = k\varphi N$$

$$\text{or } k\varphi = \frac{E}{N} = \frac{274}{1500} = 0.183$$

The value of external resistance and the field resistance is same, that is, 0.3Ω. So, when they are connected in parallel, the current 20A gets equally divided in both the paths. That is, now the field current is 10A. Let, under this condition, the back e.m.f. beE_{b1}.

Then,

$$E_{b1} = V - I_a R_a - I_f R_f = 300 - 20 \times 0.8 - 10 \times 0.5 = 279 \text{ V}$$

Again

$$E_{b1} = k\varphi_1 N_1$$

$$k\varphi_1 = 0.4 \times 0.183$$

Then,

$$N_1 = \frac{E_{b1}}{k\varphi_1} = \frac{279}{0.4 \times 0.183} = 3811 \text{ r.p.m.}$$

Ex.3.21 A DC shunt motor when connected across 230V supply draws a current of 70 A running at 1000 r.p.m. The resistance of the armature is 0.3 Ω. Assuming that the torque remaining same, what will be the new speed of the motor if the excitation is reduced to 40% of its previous existing value?

Solution

Given, V = 230 V, I = 70 A, N=1000 r.p.m., R_a = 0.3 Ω.

Let the initial flux be φ and the initial e.m.f. beE_b. Then,

$$E_b = V - I_a R_a = 230 - 70 \times 0.3 = 209 \text{ V}$$

When the excitation is reduced by 40%, let the new flux be φ_1, the new e.m.f. be E_{b1} and the new current be I_1. Then,

$$\varphi_1 = 0.4\varphi$$

Since the torque does not change, we can write,

$$I_1 \varphi_1 = I\varphi$$

$$I_1 = \frac{I\varphi}{\varphi_1} = \frac{70\varphi}{0.4\varphi} = 175 \text{ A}$$

Then,

$$E_{b1} = V - I_1 R_a = 230 - 175 \times 0.3 = 177.5 \text{ V}$$

Let the initial speed be N and the new speed when the excitation is changed be N_1. Then, we can write,

$$N = \frac{E_b}{\varphi} \quad \text{(i)}$$

And

$$N_1 = \frac{E_{b1}}{\varphi_1} \quad \text{(ii)}$$

Dividing Eq. (i) by Eq. (ii), we get,

$$\frac{N}{N_1} = \frac{E_b}{E_{b1}} \times \frac{\varphi_1}{\varphi} = \frac{209}{177.5} \times 0.4 = 0.47$$

$$N_1 = \frac{N}{0.47} = \frac{1000}{0.47} = 2127 \text{ r.p.m.}$$

Ex.3.22 A DC shunt motor with field and armature resistance of 200Ω and 0.5Ω respectively, when connected across a 300V supply, runs at 1200 r.p.m. with armature current of 45A. If the variable field resistance is changed to 250Ω, what will be the new speed of the motor? Assume the armature current and torque is constant.

Solution

Given, V=300V, R_a = 0.5Ω, R_f = 200Ω, N=1200 r.p.m, I_a = 45A, R_{f_mod} = 250Ω.

Field current when the field resistance is 200Ω

$$I_{f1} = \frac{V}{R_f} = \frac{300}{200} = 1.5 \text{ A}$$

Field current when the field resistance is 250Ω

$$I_{f1} = \frac{V}{R_{f\_mod}} = \frac{300}{250} = 1.2 \text{ A}$$

Since the armature current has not changed.

$$I_{f1}N = I_{f2}N_2$$

$$\text{or } N_2 = \frac{I_{f1}}{I_{f2}}N = \frac{1.5}{1.2} \times 1200 = 1500 \text{ r.p.m.}$$

Ex.3.23 The armature and the field resistance of a DC shunt motor is 100Ω and 0.4Ω respectively. Its no-load current is 5 A when connected across a DC supply of 300V. Calculate its efficiency at 40 A input current.

Solution

At no-load

$$\text{Input power} = 300 \times 5 = 1500 \text{ W}$$

$$\text{Shunt field current at no-load}\left(I_{s0}\right) = \frac{300}{100} = 3 \text{ A}$$

$$\text{Armature current at no-load}\left(I_{a0}\right) = 5 - 3 = 2 \text{ A}$$

$$\text{Copper loss in the armature at no-load}\left(P_{cu0}\right) = I_{a0}^2 R_a = 3^2 \times 0.4 = 3.6 \text{ W}$$

$$\text{Constant loss}\left(P_c\right) = 1500 - 3.6 = 1496.4 \text{ W}$$

For an input current of 30 A

$$\text{Armature current}\left(I_a\right) = 40 - 3 = 37 \text{ A}$$

$$\text{Copper loss in the armature}\left(P_{cu}\right) = I_a^2 R_a = 37^2 \times 0.4 = 547.6 \text{ W}$$

$$\text{Total loss}\left(P_{loss}\right) = 547.6 + 1496.4 = 2044 \text{ W}$$

$$\text{Efficiency}\left(\eta\right) = \frac{\text{Input power} - \text{Total loss}}{\text{Input power}} \times 100 = \frac{\left(300 \times 40\right) - 2044}{\left(300 \times 40\right)} \times 100 = 82.96\ \%$$

Ex.3.24 A DC shunt motor has the field and armature resistance of 200Ω and 0.3Ω respectively. When connected across a 290V supply has a no-load current of 2A and full-load current of 45A. Find the efficiency of the motor.

Solution

Given, $R_f = 200\text{V}$, $R_a = 0.3\Omega$, V=290V, $I_{a0} = 2\text{A}$, $I_{afl} = 45\text{A}$.

At no-load

$$\text{Input power} = 290 \times 2 = 580 \text{ W}$$

$$\text{Shunt field current at no-load}\left(I_{s0}\right) = \frac{290}{200} = 1.45 \text{ A}$$

$$\text{Armature current at no-load}\left(I_{a0}\right) = 2 - 1.45 = 0.55 \text{ A}$$

$$\text{Copper loss in the armature at no-load}\left(P_{cu0}\right) = I_{a0}^2 R_a = 0.55^2 \times 0.3 = 0.091 \text{ W}$$

$$\text{Constant loss}\left(P_c\right) = 580 - 0.091 = 578 \text{ W}$$

For an input current of 45 A

$$\text{Armature current}\left(I_a\right) = 45 - 1.45 = 43.55 \text{ A}$$

$$\text{Copper loss in the armature}\left(P_{cu}\right) = I_a^2 R_a = 43.55^2 \times 0.3 = 568.98 \text{ W}$$

$$\text{Total loss}\left(P_{loss}\right) = 578 + 568.98 = 1146.98 \text{ W}$$

$$\text{Efficiency}\left(\eta\right) = \frac{\text{Input power} - \text{Total loss}}{\text{Input power}} \times 100 = \frac{\left(290 \times 45\right) - 1146.98}{290 \times 45} \times 100 = 91.21\,\%$$

Ex.3.25 A 300V, 20 kW DC shunt machine with armature resistance of 0.4Ω and field resistance of 300Ω, takes a current of 5 A at rated voltage and speed. What will be the efficiency of the machine as a generator and as a motor? (Neglect stray losses)

Solution

As generator:

$$I_L = \frac{20 \times 10^3}{300} = 66.67 \text{ A}$$

$$I_f = \frac{300}{300} = 1 \text{ A}$$

$$I_a = I_L + I_f = 66.67 + 1 = 67.67 \text{ A}$$

$$\text{Loss in armature} = I_a^2 R_a = 67.67^2 \times 0.4 = 1831.69 \text{ W}$$

$$\text{Loss in the field} = \frac{300^2}{300} = 300 \text{ W}$$

$$\text{Constant loss} = 300 \times 5 - 5^2 \times 0.4 = 1490 \text{ W}$$

$$\text{Total loss} = 1831.69 + 1490 + 300 = 3721.69 \text{ W}$$

$$\text{Efficiency} = \frac{20 \times 10^3}{20 \times 10^3 + 3721.69} \times 100 = 84.31\%$$

As motor:

$$I_a = I_L - I_f = 65.67 \text{ A}$$

$$\text{Loss in armature} = I_a^2 R_a = 65.67^2 \times 0.4 = 1725.02 \text{ W}$$

$$\text{Loss in the field} = 300 \text{ W}$$

$$\text{Constant loss} = 1490 \text{ W}$$

$$\text{Total loss} = 1725.02 + 300 + 1490 = 3515.02 \text{ W}$$

$$\text{Efficiency} = \frac{20 \times 10^3 - 3515.02}{20 \times 10^3} \times 100 = 82.42\%$$

Ex.3.26 The no-load speed of 450 V DC shunt motor is 300 r.p.m.. It runs at 1000 r.p.m. at reduced armature voltage, full-load torque and full-field voltage. If the load torque is reduced to 30% keeping the armature voltage and field voltage constant, the speed increases to 1300 r.p.m. what is the armature voltage drop at full-load?

Solution

Let the armature current at full load torque be I_{a1}.

Then,

$$E_{a1} = 450 - I_{a1} r_a$$

$$\text{Also, } E_{a1} \propto N_1 \varphi_1$$

At 30 % torque,

$$E_{a1} = 400 - I_{a1} r_a$$

$$\text{Also, } E_{a1} \propto N_1 \varphi_1$$

$$T_{el} = K \varphi_1 I_{a1}$$

Again,

$$0.3\, T_{el} = K \varphi_2 I_{a2}$$

Since the field voltage is constant, $\varphi_1 = \varphi_2$

$$\frac{T_{el}}{0.3 T_{el}} = \frac{K \varphi_1 I_{a1}}{K \varphi_2 I_{a2}}$$

$$\text{or,} \frac{1}{0.3} = \frac{I_{a1}}{I_{a2}}$$

$$\text{or,} I_{a2} = 0.3 I_{a1}$$

At 30 % of rated torque,

$$E_{a2} = 450 - I_{a2} r_a = 450 - 0.3 I_{a1} r_a$$

$$\text{Also, } E_{a2} \propto N_2 \varphi_2$$

$$\text{Or, } E_{a2} \propto N_2 \varphi_1$$

Then,

$$\frac{E_{a1}}{E_{a2}} = \frac{N_1 \varphi_1}{N_2 \varphi_2} = \frac{450 - I_{a1} r_a}{450 - 0.3 I_{a1} r_a} = \frac{1000 \times \varphi_1}{1300 \times \varphi_1} = 0.77$$

$$450 - I_{a1} r_a = 450 \times 0.77 - 0.3 \times 0.77 \times I_{a1} r_a$$

Solving the equation we get,

$$I_{a1} r_a = 134.59 \text{ V}$$

Ex.3.27 A separately excited DC motor, when connected across 300V, takes an armature current of 15 A at 1800 r.p.m. Find the torque developed when it draws an armature current of 10A. Assume armature resistance to be 0.3Ω.

Solution

$$E_b = 300 - 15 \times 0.3 = 295.5 \text{ V}$$

$$T = \frac{295.5 \times 15}{\frac{2 \times \pi \times 1800}{60}} = 23.51 \text{ Nm}$$

Torque when the armature current is 10A.

$$T_1 = 23.51 \times \frac{10}{15} = 15.67 \text{ Nm}$$

Ex.3.28 If the current drawn by a DC series motor is increased from 10 A to 15A, what will be the increase in torque expressed as a percentage of initial torque?

Solution

Let T_1 and T_2 be the torque of a DC series motor at 10A and 15A current respectively. Then,

$$T_2 = \left(\frac{15}{10}\right)^2 T_1$$

$$T_2 = 2.25 \times T_1$$

$$\text{Therefore, increase in torque} = \frac{T_2 - T_1}{T_1} \times 100 = \frac{2.25 - 1}{1} \times 100 = 125\%$$

Ex.3.29 The field and the armature resistance of a 300V, DC shunt motor is 200Ω and 0.5Ω respectively. Find the back e.m.f. when it takes a current of 35 A.

Solution

$$I_f = \frac{300}{200} = 1.5\,\Omega$$

$$I_a = 35 - 1.5 = 33.5 \text{ A}$$

$$\text{E.m.f.} = 300 - 33.5 \times 0.5 = 283.25 \text{ V}$$

Ex.3.30 The result of the Hopkinson's test conducted on two machines are as follows at full load.
Line current = 50A
Line voltage = 300V
Motor armature current = 400A
Field currents = 5 A and 4.5 A
What is the efficiency of the machine if the armature resistance of each machine is 0.04 Ω?

Solution

$$I_g = 400 - 50 = 350 \text{ A}$$

$$P_{input} = 300 \times 50 = 15000 \text{ W}$$

$$\text{Copper loss} = \left[400^2 + 350^2\right] \times 0.04 = 11300 \text{ W}$$

$$\text{Stray loss} = \frac{(15000 - 11300)}{2} = 6935 \text{ W (each machine)}$$

For motor:

$$P_m = (400 + 4.5) \times 300 = 121350 \text{ W}$$

$$P_{lm} = 6935 + 400^2 \times 0.04 + 300 \times 4.5 = 14685 \text{ W}$$

$$\text{Efficiency} = \frac{121350 - 14685}{121350} \times 100 = 87.89\%$$

For generator:

$$P_g = 300 \times 350 = 105000 \text{ W}$$

$$P_{lm} = 6935 + 350^2 \times 0.04 + 300 \times 5 = 13335 \text{ W}$$

$$\text{Efficiency} = \frac{105000 - 13335}{105000} \times 100 = 87.3\%$$

EXERCISES

1. A 300 V DC shunt motor takes a no-load current of 4A and full loads current of 45A. The armature and field resistance of the motor is 0.5W and 250W respectively. Find efficiency of the motor.
2. What is the principle of operation of DC motor?
3. What is back e.m.f? Give the expression for back e.m.f.
4. Derive the electromagnetic torque equation of a DCmotor.
5. Derive the condition for maximum power output in case of a DC motor.
6. Explain the commutation action in a DCmotor.
7. Explain the armature reaction in a DCmotor.
8. Give the classification of DC motor.
9. Obtain an expression for mechanical power in terms of armature current and back e.m.f. for a separately excited DC motor.
10. Derive an expression for mechanical power for a DC series motor.
11. Obtain an expression for mechanical power for a DC shunt motor.
12. How can you reverse the direction of rotation of a DC motor?
13. List out the factors on which the speed of a DC motor depends.
14. Explain the following three characteristics:
    (i) Speed versus armature current characteristics.
    (ii) Torque versus armature currentcharacteristics.
    (iii) Speed versus torque characteristics.

    For
    (i) DC shunt motor
    (ii) DC series motor
    (iii) DC compound motor
15. Why is a starter required to start a DC motor?
16. Draw and explain the operation of three-point starter.
17. Draw and explain the operation of four-point starter.
18. What are the different methods of speed control of DC shunt motor?
19. Draw and explain the shunt field control of controlling the speed of shunt DC motor.
20. What are the advantages and disadvantages of shunt field control method of speed control?
21. Draw and explain the armature control of controlling the speed of shunt DC motor.
22. What are the advantages and disadvantages of armature control method of speed control?
23. Explain the Ward Leonard method of speed control.
24. What are the advantages and disadvantages of Ward Leonard method of speed control?
25. What are the different methods of speed control of DC series motor?
26. Draw and explain the flux control method of speed control of DC series motor.
27. Draw and explain the armature control method of speed control of DC series motor.
28. Draw and explain the series parallel control method of speed control of DC series motor.

29. What is electric braking? What are the different methods of electric braking?
30. Explain the following braking mechanism in detail:
    (i) Rheostatic or Dynamic braking
    (ii) Braking by plugging
    (iii) Regenerative braking
31. What are the different types of losses in DC motor?
32. Obtain the condition for maximum efficiency for a DC motor
33. What is brake test on DC motor?
34. Explain Swinburne's Test
35. Explain Hopkinson's Test
36. A DC shunt machine has the following parameters: Resistance of the armature = 0.1Ω, Resistance of the field = 220 Ω. Calculate the ratio of generated speed to motor speed, when the machine is connected across a 240 V supply with the line current of 40 A.
37. A wave-wound 4-pole DC motor has 210 conductors. The average flux density over one pole pitch is 1.5T. The pole shoe is 9.7 cm long and the armature diameter is 23 cm. Find the mechanical power and torque developed. Assume that the motor is running at 1000 r.p.m. and is drawing 10A current.
38. A DC shunt motor, when fed from a DC source of 330 V, draws a current of 100 A and runs at 1000 r.p.m. Now, if the excitation is suddenly reduced to 50% of its previous value, calculate the new speed of the motor. Assume that the torque remains the same and the resistance of the armature is 0.7 Ω.
39. A 210V, a DC shunt motor has a field and armature resistance of 100Ω and 0.8Ω respectively. The motor runs at 1800 r.p.m. and the armature current is 80A at full load. Calculate the developed torque. If the variable field resistance is changed to 140Ω, what will be the new speed of the motor? Assume the armature current and torque is constant.
40. A 200V, 1500 r.p.m. DC shunt motor is started with the help of a starter. The starter resistances in the order in which they are cut out are 2Ω, 3Ω, 4Ω and 5Ω respectively. The armature has a resistance of 0.4Ω. Find the speed at each position of the starter hand, and the initial and final values of armature current at each position of the starter hand. Assume that the starter is moved to its next position only when the armature current has come to its rated value of 32A.
41. A DC shunt motor has the field resistance of 110 Ω and armature resistance of 0.5 Ω. When connected across a 200 V supply, it takes a no-load current of 10A. Calculate its efficiency at 70 A input current.
42. A DC shunt motor has the field and armature resistance of 200Ω and 0.8Ω respectively. When connected across a 230V supply, it has a no-load current of 9A and full-load current of 40A. Find the efficiency of the motor.
43. Hopkinson's test is performed on two identical DC shunt machines. The data related to this test is as follows.
    (i) Supply voltage = 200 V
    (ii) Line current = 110 A
    (iii) Armature current of the motor = 510 A
    (iv) Field current of the motor = 4A
    (v) Field current of the generator = 3A
    (vi) Armature resistance of the motor = 0.08 Ω
    (vii) Armature resistance of the generator = 0.09 Ω
    (viii) Brush drop = 1.2V per brush
    (ix) Calculate the efficiency of each machine.

44. Hopkinson's test is performed on two identical DC shunt machines. The data related to this test is as follows.

(i) Supply voltage = 200 V
(ii) Line current = 10 A
(iii) Armature current of the generator = 76 A
(iv) Field current of the motor = 8A
(v) Field current of the generator = 9A
(vi) Armature resistance of the motor = 0.2 Ω
(vii) Armature resistance of the generator = 0.2 Ω
(viii) Calculate the efficiency of each machine.

45. A 240V, a 25 kW DC shunt machine takes a current of 9A at rated voltage and rated speed. The armature and field resistances are 0.4Ω and 200Ω respectively. Calculate the efficiency of the machine when running as a (i) generator, (ii) motor. Neglect stray losses.

46. A DC shunt motor has the armature and field resistance as 0.1Ω and 90Ω respectively. When it is connected across 220V supply, the no-load current is 7A and the speed is 1500 r.p.m. Calculate the motor torque and motor speed at a full-load current of 50A.

CHAPTER 4

SINGLE-PHASE TRANSFORMER

4.1 INTRODUCTION

Transformer is an essential component in the field of electrical and electronic circuits since the 1830s. A transformer is a device that transfers electrical energy from one electrical circuit to another electrical circuit by electromagnetic induction (transformer action). There is no electrical link between the two circuits. The electrical energy is transferred without changing the frequency, but the voltage and current magnitude may change. Thus, transformers are commonly used in applications that require the conversion of AC voltage from one voltage level to another.

Transformers can be categorised into two broad categories:

i) *Electronic transformers*, which operate at very low power levels and are used in consumer electronic equipment like television sets, VCRs, CD players, personal computers, and many other devices.
ii) *Power transformers*, which are used in power generation, transmission and distribution systems to raise or lower the level of voltage to the desired levels.

The basic principle of operation of both types of transformers is the same.

4.2 CLASSIFICATION OF TRANSFORMER

A transformer can be classified into many different ways. Some of the classifications of transformers are as follows:

i) Classification based on application
   - Step-up transformer
   - Step-down transformer
ii) Classification based on construction
   - Core type transformer
   - Shell type transformer
iii) Classification based on number of phases
   - Single-phase transformer
   - Three-phase transformer
iv) Classification based on the location of transformer
   - Indoor type transformer
   - Outdoor type transformer
   - Station transformer

4.3 TRANSFORMER CONSTRUCTION

An electrical transformer has three main parts, namely,

i) The core, which makes a path for the magnetic flux.
ii) Transformer windings, which receives energy
iii) The transformer insulation

The different parts of the transformer are shown in Fig. 4.1.

4.3.1 Core of the Transformer

The transformer core can be either square or rectangular. The two vertical portions of the core, where the windings are wound is called the limb of the transformer. The top and the bottom portion of the core are called the yoke of the transformer. The core is made up of sheet steel laminations. The lamination is done to reduce eddy current losses in the transformer. These laminations are also insulated from each other by varnish insulated coating.

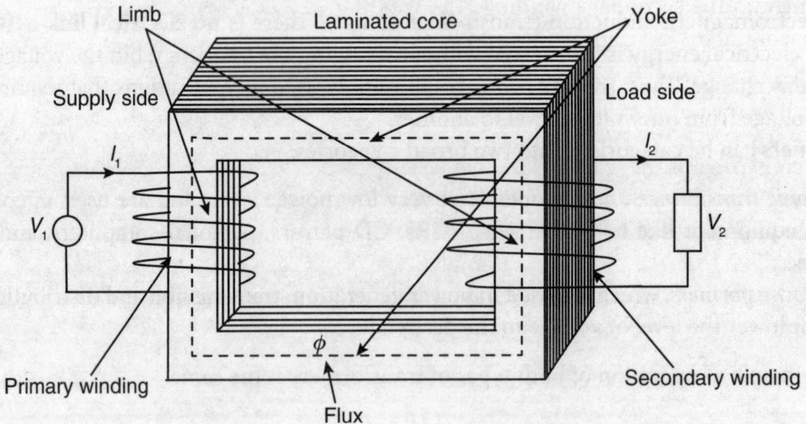

Fig. 4.1: Parts of a transformer

The transformer core should have the following properties.

i) Minimum iron losses.
ii) Maximum saturation flux density.
iii) Maximum permeability.

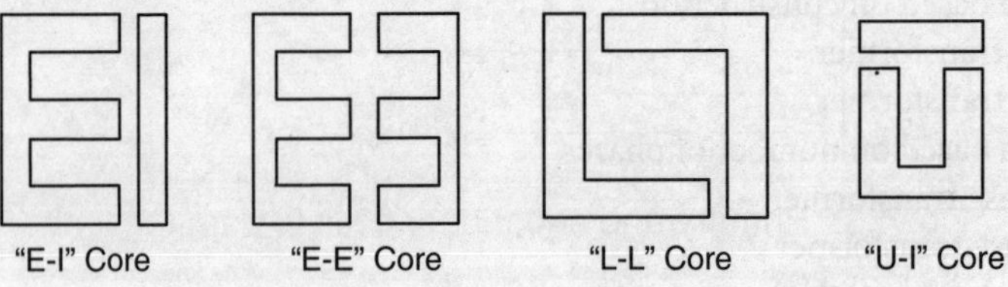

Fig. 4.2: Types of laminations for transformer core

The individual laminations of the core are stamped or punched out from larger steel sheets and formed into strips of thin steel resembling the letters "E's", "L's", "U's" and "I's" as shown in Fig. 4.2.

These lamination stampings are connected together to form the required core shape. For example, two "E" stampings plus two end-closing "I" stampings to give an E-I core forming one element of a standard shell-type transformer core. These individual laminations are tightly butted together during the transformers construction to reduce the reluctance of the air gap at the joints producing a highly saturated magnetic flux density.

Transformer core laminations are usually stacked alternately to each other to produce an overlapping joint with more lamination pairs being added to make up the correct core thickness. This alternate stacking of the laminations also gives the transformer the advantage of reduced flux leakage and iron losses. E-I core laminated transformer construction is mostly used in isolation transformers, step-up and step-down transformers as well as autotransformers.

4.3.2 Transformer Windings

A transformer mainly has two windings wound on the two limbs of the transformer, namely, the primary winding and the secondary winding. The winding which is connected with the supply is called the primary winding of the transformer and the limb is called the primary limb or the primary side of the transformer. The winding which is connected with the load, is called the secondary winding of the transformer and the limb is called the secondary limb or the secondary side of the transformer.

As shown in Fig. 4.1, the two windings of the transformer are wound on the two separate limbs. By this type of construction the leakage flux increases and the efficiency of the transformer decreases. To have maximum flux linkage, the two coils are brought close to each other by dividing both the coils in both the limbs as shown in the Fig. 4.3. Note that in the transformer shown in Fig. 4.3, the laminated core has not been shown to make the diagram simpler. These windings are made up of copper or aluminium wires having the property of maximum conductivity, good mechanical strength and less cost.

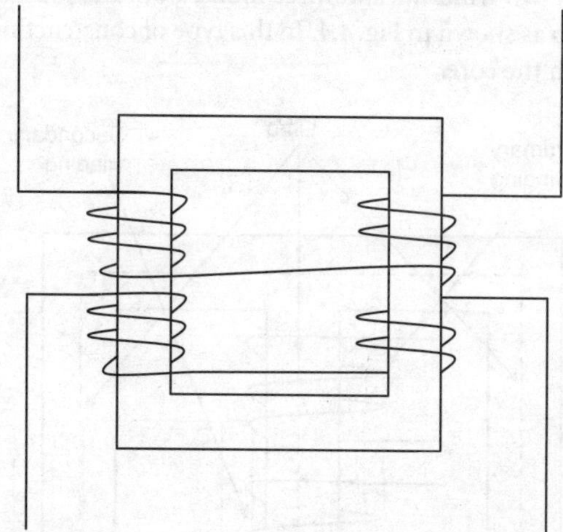

Fig. 4.3: Primary and secondary coils in both the limbs

4.3.3 Insulation

Insulation is a very important part of a transformer. The transformer windings are insulated from the core and also from each other by paper, cloth or mica insulation. To have maximum flux linkage both the primary and secondary coils are wound on both the limbs. Generally, the low voltage coil is placed nearer to the core so that it becomes easy to insulate it from the core. The high-voltage winding is placed after it.

4.4 CLASSIFICATION OF TRANSFORMER BASED ON CONSTRUCTION

Depending upon how the transformer core is designed, a single-phase transformer can be classified into the following categories.

- Core-type transformer
- Shell-type transformer
- Spiral-type transformer
- Berry-type transformer

4.4.1 Core-type Transformer

The transformer shown in Fig. 4.1 is a core-type transformer. A core-type transformer has a single path for the magnetic flux to flow in the transformer. The core can be in the form of a rectangle or a square and it is laminated to reduce eddy current loss. In this type of construction, the windings encircle the core. As stated earlier in this chapter, low-voltage coils are placed near the core with proper insulation and the high-voltage coil surrounds the low-voltage coil. For proper flux linkage, both the coils are present on both the limbs. Generally, cylindrical type windings are used for core-type transformers. These transformers are used for high-voltage applications. For small transformers, rectangular core with rectangular cylindrical coils are used. For large transformers, square core with circular cylindrical core is used.

4.4.2 Shell-type Transformer

A shell-type transformer has two windows and three limbs. Both the primary and secondary windings are placed on the central limb as shown in Fig. 4.4. In this type of construction the magnetic flux has two parallel paths to flow through the core.

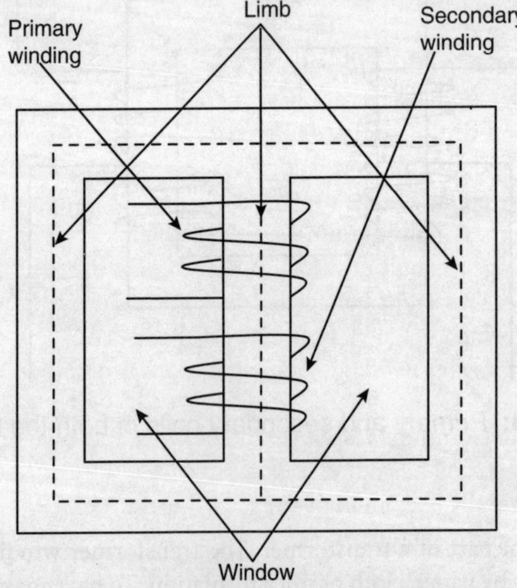

Fig. 4.4: Shell-type transformer

The coil used in this type of transformer is usually sandwich-type or multi-layer disc-type coils and they occupy the entire space of the windows. The core is made up of E-shaped or F-shaped laminated sheets. Low-voltage windings are placed near to the core and high-voltage windings are place above it with proper insulations. In this type of transformer, maintenance is difficult and natural cooling is not possible.

4.4.3 Difference between Core-type and Shell-type Transformer

The comparison between core-type and shell-type transformer is tabulated in Table 4.1

Table 4.1: Comparison between core-type and shell-type transformer

Core type transformer	Shell type transformer
It has a single window	It has two windows
It has a single path for the flux to flow	It has two parallel paths for the flux to flow
The core has two limbs	The core has three limbs
Cylindrical coils are used	Sandwich coils are used
Natural cooling is possible	Natural cooling is not possible
Used for low-voltage application	Used for high-voltage application
Maintenance is easy	Maintenance is difficult

4.5 OTHER PARTS OF A TRANSFORMER

Other than transformer core and its windings, some other important parts of a transformer are as follows:

i) *Bushings*: These are used to bring out the transformer terminals so that it can be connected with the external circuit. For smaller rating transformers, solid porcelain bushings are used while for large transformers oil-filled bushings are used. Solid porcelain bushings consist of high-grade porcelain cylinders that conductors pass through. Outside surfaces have a series of skirts to increase the leakage path distance to the grounded metal case. High-voltage bushings are generally oil-filled condenser type. Condenser-types have a central conductor wound with alternating layers of paper insulation and tin foil and are filled with insulating oil. This results in a path from the conductor to the grounded tank, consisting of a series of condensers. The layers are designed to provide approximately equal voltage drops between each condenser layer. A bushing has a 'conductor', (usually of copper or aluminium), surrounded by insulation, except for the terminal ends.

ii) *Tappings*: We can change the output voltage of the transformer by changing the turns ratio of the transformer. To achieve different voltage levels, tappings are provided at different places in the winding of the transformer.

iii) *Different types of transformer windings*: Some of the most commonly used transformer windings are:
   - Crossover winding: Commonly used on the high-voltage side of a small transformer. In this type of winding, each coil has number of layers with a specific number of turns per layer.
   - Disc winding: Generally used in high-capacity transformers. It is a stranded conductor of rectangular or circular cross section.

- Cylindrical conductor: In this, individual cylindrical layers are placed one on top of the other. It is used in transformers with high current and low voltage ratings.
- Sandwich winding: Used in shell-type transformers. High-voltage and low-voltage windings are wound together on the same limb. Low-voltage winding is put on the core first and then high-voltage winding is put with proper insulation.
- Helical winding or screw winding: The winding is wound in the form of a helix; it is used on the low-voltage side of the transformer. The windings are made up of large cross-sectional rectangular conductors.

iv) *Conservator and Breather*: The transformer main oil tank is always filled completely. During summer and heavy load conditions the oil expands and the conservator is provided for the expansion and contraction of oilv) *Breather*: During contraction of oil, the outside air gets into the transformer through breather. The breather has silica gel which absorbs the moisture. Also, the breather has a oil cap to stop entry of particles such as dust into the transformer

4.6 PRINCIPLE OF OPERATION OF TRANSFORMER

Transformer works on the principle of electromagnetic induction. Consider a single-phase transformer shown in Fig. 4.5. To make the diagram simple, the laminated core is not shown in the transformer circuit of Fig. 4.5.

The primary of the transformer having N_1 turns is fed from an AC supply of V_1 volts. The current I_1 will flow through the primary coil. The current through the primary will set up a flux φ in the core. This flux, when linked with the primary winding, will produce an induced e.m.f., E_1, in the primary. The flux φ will pass through the core and link with the secondary winding to induce an e.m.f., E_2, in the secondary winding. Because of this induced e.m.f., a current I_2 will flow through the load connected with the secondary winding. The load terminal voltage is V_2.

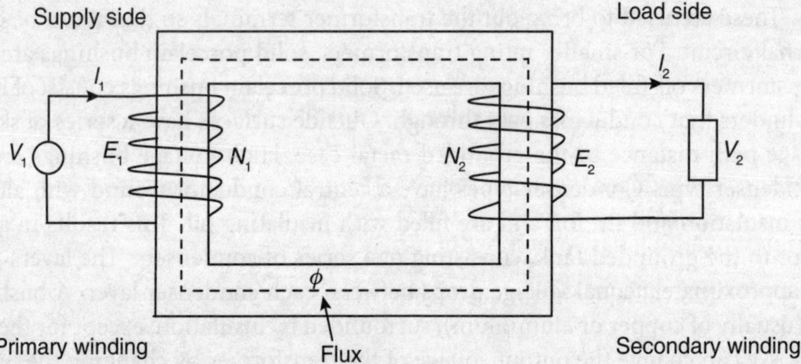

Fig. 4.5: Single-phase transformer

If the input voltage V_1 is greater than the output voltage V_2, then it is called the step-down transformer. If the input voltage V_1 is less than the output voltage V_2, then it is called the step-up transformer. If $V_1 = V_2$, it is called isolation transformer or one-to-one transformer. In the case of step-up transformer the number of turns of primary (N_1) is less than the number of turns of secondary (N_2). Similarly, in the case of step-down transformer, the number of turns of primary (N_1) is more than the number of turns of secondary (N_2).

4.7 EMF EQUATION OF A SINGLE-PHASE TRANSFORMER

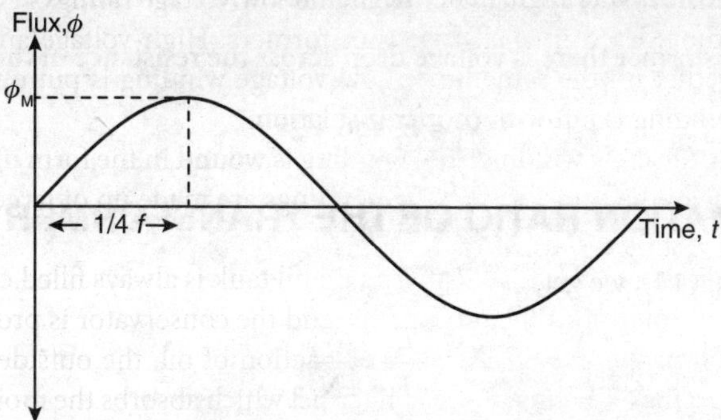

Fig. 4.6: Waveform of flux

Consider a single-phase transformer with N_1 turns on the primary side and N_2 turns on the secondary side. The primary is connected across an alternating sinusoidal voltage source, because of which an alternating flux, shown in Fig. 4.6, is set up in the core. This flux links with the primary and secondary windings. Let f be the frequency of supply and φ_m be the maximum value of the flux.

From the Fig. 4.6, we can write,

$$\text{Average rate of change of flux} = \frac{\varphi_m}{1/4f} = 4f\varphi_m$$

Since average e.m.f. induced in the coil per turn is equal to the average rate of change of flux, we can write,

$$\text{Average e.m.f. induced per turn} = 4f\varphi_m \text{ volt}$$

Again, for sinusoidal signal,

$$\frac{\text{RMS value}}{\text{Average value}} = \text{form factor} = 1.11$$

or, RMS value of the induced e.m.f. per turn $= 1.11 \times 4f\varphi_m = 4.44f\varphi_m$

Since the primary winding has N_1 number of turns,

$$\text{RMS value of the induced e.m.f. in primary} = E_1 = 4.44f\varphi_m N_1 \quad (4.1)$$

Similarly,

$$\text{RMS value of the induced e.m.f. in secondary} = E_2 = 4.44f\varphi_m N_2 \quad (4.2)$$

Equations (4.1) and (4.2) are called the e.m.f. equations of the transformer.

4.8 IDEAL TRANSFORMER

An ideal transformer is not possible to be constructed physically. However, the concept gives a powerful tool for the analysis of a practical transformer. An ideal transformer has the following properties.

- The resistance of the primary and secondary windings is zero.
- All the flux produced by the primary completely links with the secondary.

- Loss in the transformer is zero.
- Permeability of the core is very high, hence negligible current is required to establish flux in the core.

Since in an ideal transformer there is voltage drop across the resistance of the windings, we can say,

$$V_1 = E_1 \text{ and } V_2 = E_2. \tag{4.3}$$

4.9 TRANSFORMATION RATIO OF THE TRANSFORMER

Dividing Eq. (4.2) by Eq. (4.1), we get,

$$\frac{E_2}{E_1} = \frac{N_2}{N_1} \tag{4.4}$$

If we consider an ideal transformer, then from Eqs (4.3) and (4.4) we get,

$$\frac{E_2}{E_1} = \frac{N_2}{N_1} = \frac{V_2}{V_1} \tag{4.5}$$

In an ideal transformer, input power (VA) is equal to the output power (VA). That is,

$$V_1 I_1 = V_2 I_2 \tag{4.6}$$

$$\text{or, } \frac{V_2}{V_1} = \frac{I_1}{I_2} \tag{4.7}$$

From Eq. (4.5) and Eq. (4.7),

$$\frac{E_2}{E_1} = \frac{N_2}{N_1} = \frac{V_2}{V_1} = \frac{I_1}{I_2} = K \tag{4.8}$$

Equation (4.8) is called the transformation ratio of the transformer.

4.10 OPERATION OF IDEAL TRANSFORMER AT NO-LOAD

Under no-load condition, the secondary of the transformer is open-circuited, as shown in Fig. 4.7.

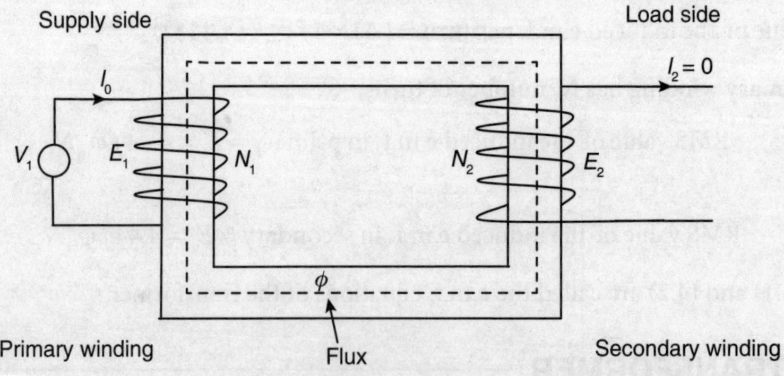

Fig. 4.7: Transformer at no-load

The primary is connected across an alternating voltage source (V_1), because of which some current flows through the primary winding. Since the secondary is open-circuited, no current will flow through the secondary winding (I_2 =0). Under this condition, the primary current will only be used to produce the flux (φ) in the core. Since the primary current is only being used to magnetise the core, this current is called the magnetisation current and is denoted by I_m.

For an ideal transformer, the resistances of the primary and secondary windings are assumed to be zero. Thus, if R_1 is the resistance of the primary and R_2 is the resistance of the secondary, then for an ideal transformer, $R_1 = R_2 = 0$. Or in other words, in an ideal transformer, the transformer coils are pure inductive coils. So under these assumptions, the current I_m will lag the supply voltage V_1 by 90^0. The alternating flux (φ) produced by I_m will be proportional to I_m and thus will be in phase with I_m.

This flux will link with both the primary and secondary windings and will produce an e.m.f. E_1 and E_2 in the primary and secondary respectively. These two e.m.f.s will oppose the cause of its production (Lenz's law), that is V_1, and hence they will be in anti-phase with V_1. The complete phasor of no-load operation of ideal transformer is shown in Fig. 4.8.

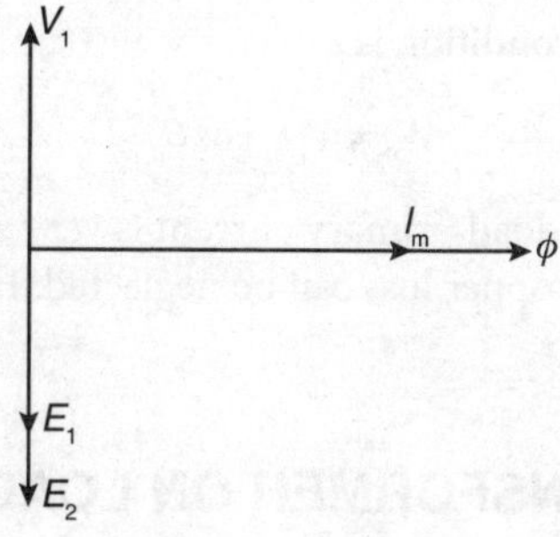

Fig. 4.8: Phasor of ideal transformer at no-load

4.11 OPERATION OF PRACTICAL TRANSFORMER AT NO-LOAD

A practical transformer will have losses in the core and also in the winding. This is due to the presence of resistance in the transformer windings and is given as i^2R. Thus, the primary no-load current will have two components.

i) The magnetisation component (I_m), as discussed in Section 4.10, which is responsible for the production of flux in the core.

ii) The power component (I_c), which will supply the total losses.

Hence, the no-load primary current (I_0) will be the vector sum of I_m and I_c.

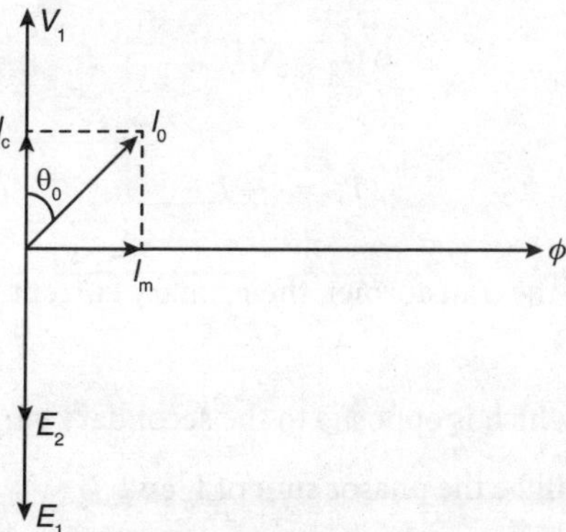

Fig. 4.9: Phasor diagram for no-load operation of practical transformer

Due to the presence of resistance in the transformer windings, the no-load current I_0 will not lag V_1 by 90^0, but by some angle θ, which is less than 90^0. The phasor diagram for no-load operation of practical transformer is shown in Fig. 4.9.

Resolving, I_0 into its x and y component, we get,

$$I_m = I_0 \sin \theta_0 \tag{4.9}$$

$$I_c = I_0 \cos \theta_0 \tag{4.10}$$

Then,

$$I_0 = \sqrt{I_m^2 + I_c^2} \tag{4.11}$$

Total power input under no-load condition is

$$W_0 = V_1 I_0 \cos \theta_0 \tag{4.12}$$

For a practical transformer the no-load primary current is very small, usually 4–5% of full-load current. Since the current is very small, copper loss can be neglected. Hence W_0 gives the iron loss component of the transformer only.

4.12 OPERATION OF TRANSFORMER ON LOAD

Under no-load condition, the primary current I_0 produces a flux φ in the core, which links with both the primary and secondary windings. Now, when the secondary of the transformer is loaded, the secondary is closed. Hence, the secondary current I_2 will now flow through the secondary winding (Fig. 4.5).

The magnitude and the phase of I_2 will depend upon the type of load connected across the secondary terminals. If the load is resistive, I_2 will be in phase with V_2. If the load is inductive, I_2 will lag V_2 and if the load is capacitive I_2 will lead V_2.

This secondary current, I_2, will set up its own flux φ_2 in the core. This flux φ_2 will oppose the main flux φ. Since φ_2 is opposing the flux φ, the net flux through the core will reduce, which in turn will reduce the induced e.m.f. in the primary. So the primary will draw more current from the supply. Let the additional primary current be I_2'. This additional primary current will produce an additional flux φ' in the core. The flux φ' will be equal in magnitude but opposite to φ_2. Since they are opposite, they will cancel each other and so the net flux through the core will remain the same.

Since φ_2 and φ' are equal in magnitude,

$$N_2 I_2 = N_1 I_2'$$

$$I_2' = \frac{N_2}{N_1} I_2 \tag{4.13}$$

So, under loaded condition of the transformer, the primary current (I_1), will have two components,

i) The no load component (I_0)

ii) The load component $\left(I_2'\right)$ which is opposite to the secondary current I_2

Thus the primary current I_1 will be the phasor sum of I_0 and I_2'.

4.13 PHASOR DIAGRAM OF TRANSFORMER ON LOAD

Phasor diagram of the transformer on load can be obtained from the phasor shown in Fig. 4.9.

i) Inductive load:

The secondary current (I_2) will lag $V_2 = E_2$ by an angle φ_2. The additional primary current I'_2, will be equal in magnitude to I_2 but will be opposite in direction to I_2. The primary current I_1 will be the phasor sum of I_0 and I'_2. The complete phasor is shown in Fig. 4.10.

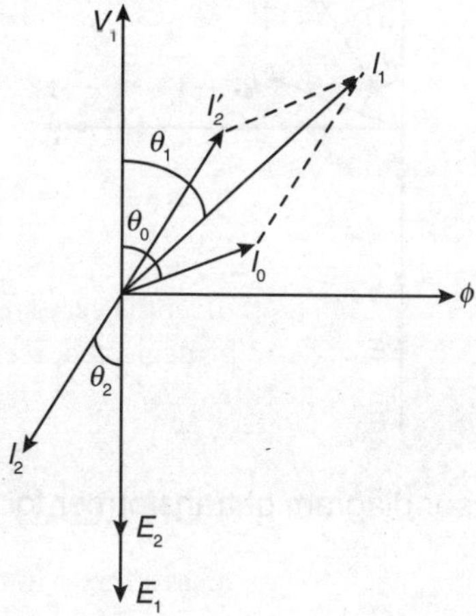

FIG. 4.10: Phasor diagram of transformer for inductive load

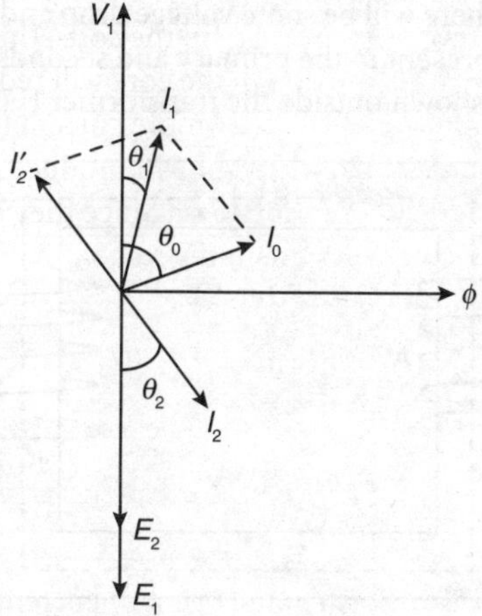

FIG. 4.11: Phasor diagram of transformer for capacitive load

ii) Capacitive load:

The secondary current (I_2) will lead $V_2 = E_2$ by an angle φ_2. The additional primary current I'_2, will be equal in magnitude to I_2 but will be opposite in direction to I_2. The primary current I_1 will be the phasor sum of I_0 and I'_2. The complete phasor is shown in Fig. 4.11.

iii) Resistive load:
The secondary current (I_2) will be in phase with V_2. The additional primary current I'_2, will be equal in magnitude to I_2 but will be opposite in direction to I_2. The primary current I_1 will be the phasor sum of I_0 and I'_2. The complete phasor is shown in Fig. 4.12.

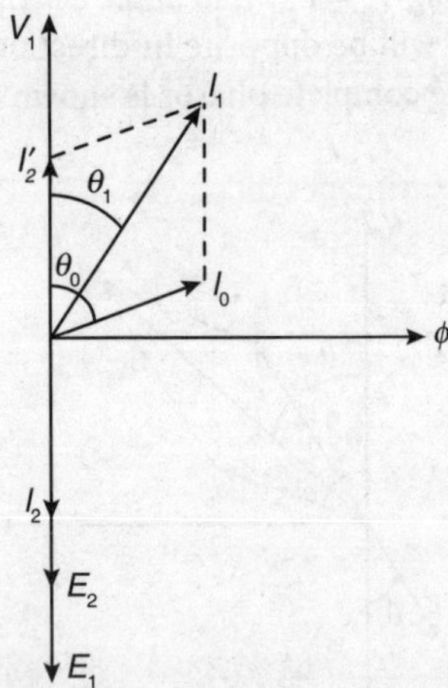

Fig. 4.12: Phasor diagram of transformer for resistive load

4.14 RESISTANCE AND LEAKAGE REACTANCE OF TRANSFORMER

In a practical transformer, the transformer windings are not purely inductive. The primary and secondary windings will always have some resistance in them. Because of the presence of these resistances in the primary and secondary windings, there will be some voltage drop and also I^2R loss in both the windings.

Let R_1 and R_2 be the resistances present in the primary and secondary windings respectively. For simplicity, these resistances have been shown outside the transformer (Fig. 4.13).

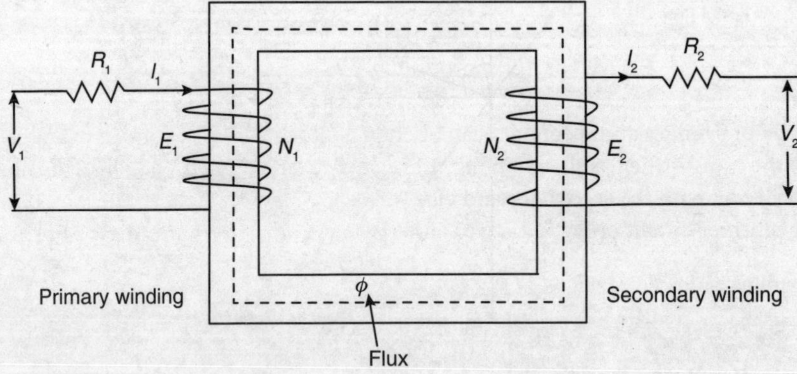

Fig. 4.13: Transformer with primary and secondary winding resistance.

From the transformer of Fig. 4.13, we can write,

$$\text{Primary side}: V_1 - I_1R_1 = E_1$$

$$\text{or } V_1 = E_1 + I_1R_1 \tag{4.14}$$

$$\text{Secondary side}: V_2 = E_2 - I_2R_2 \tag{4.15}$$

Again, in an ideal transformer, it was assumed that all the flux produced by the primary links with the secondary and vice versa. But in the case of practical transformer, this assumption is not true.

The total flux produced by the primary (φ_1) can be divided into two parts:

i) Flux (φ_{11}) which will link with the primary coil only, called the leakage flux.
ii) Flux (φ_{12}) which will pass through thc core and will link with the secondary winding, called the useful flux or mutual flux.

Similarly, total flux produced by the secondary (φ_2) can be divided into two parts:

i) Flux (φ_{22}) which will link with the secondary coil only, is called the leakage flux.
ii) Flux (φ_{21}), which will pass through the core and will link with the primary winding, is called the useful flux or mutual flux.

These fluxes are shown in Fig. 4.14.

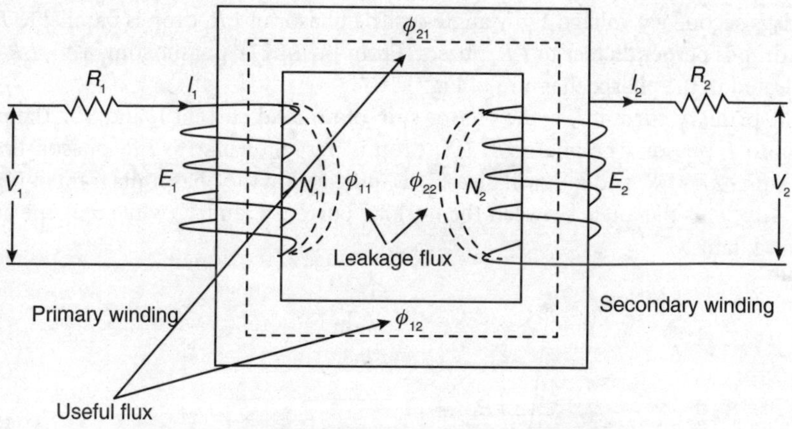

Fig. 4.14: Flux production in transformer

The leakage flux of primary and secondary can be shown outside the transformer by inductor coils on both the sides of the transformer, as shown in Fig. 4.15. Now, it can be assumed that all the flux produced by the primary will link with the secondary and vice versa.

From the transformer circuit of Fig. 4.15, we can write,

$$\text{Primary side}: V_1 - I_1(R_1 + jX_1) = E_1$$

$$\text{or } V_1 = E_1 + I_1(R_1 + jX_1) \tag{4.16}$$

$$\text{Secondary side}: V_2 = E_2 - I_2(R_2 + jX_2) \tag{4.17}$$

X_1 and X_2 are called the leakage reactance of the transformer.

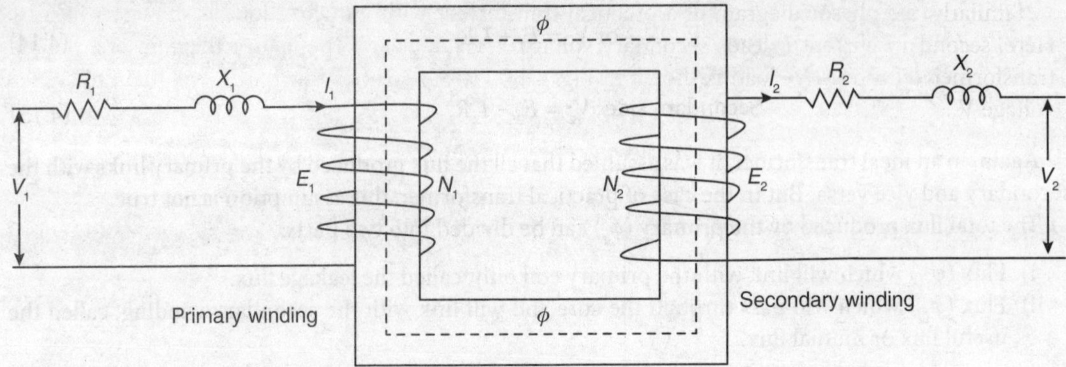

Fig. 4.15: Transformer with winding resistance and leakage reactance

4.15 PHASOR DIAGRAM OF TRANSFORMER ON LOAD WITH RESISTANCE AND LEAKAGE REACTANCE OF THE WINDINGS

The phasor diagram of a practical transformer with inductive load is shown in Fig. 4.16. Here, secondary current I_2 lags secondary voltage V_2 by angle φ_2. The phasor of I_2R_2 drop is parallel to I_2 phasor. The phasor of I_2X_2 drop is perpendicular to I_2R_2 phasor. From Eq. (4.17), phasor sum of V_2, I_2R_2 and I_2X_2 will give E_2, as indicated in the phasor diagram of Fig. 4.16.

Similarly, the primary current I_1 is the phasor sum of no-load current I_0 and I'_2. The phasor of I_1R_1 drop is parallel to I_1 phasor. The phasor of I_1X_1 drop is perpendicular to I_1R_1 phasor. From Equation (4.16), phasor sum of E_1, I_1R_1 and I_1X_1 will give V_1, as indicated in the phasor diagram of Fig. 4.16. In the phasor of Fig. 4.16, θ_0 is the angle between the no-load current I_0 and V_1 while θ_1 is the angle between primary current I_1 and V_1.

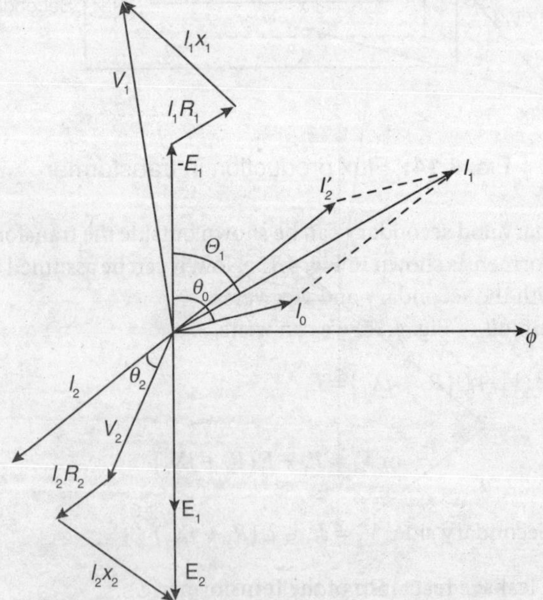

Fig. 4.16: Phasor diagram for inductive load

Similarly, the phasor diagram of a practical transformer with inductive load is shown in Fig. 4.17. Here, secondary current I_2 leads secondary voltage V_2 by angle φ_2. The phasor diagram of a practical transformer, with resistive load is shown in Fig. 4.18. Here, secondary current I_2 will be in phase with voltage V_2.

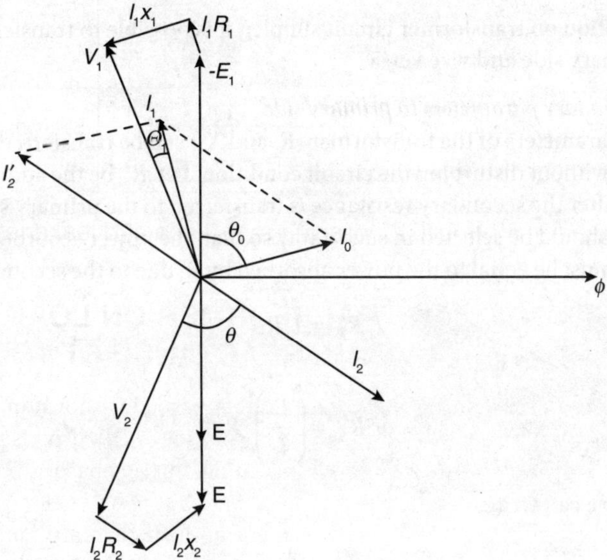

FIG. 4.17: Phasor diagram for capacitive load

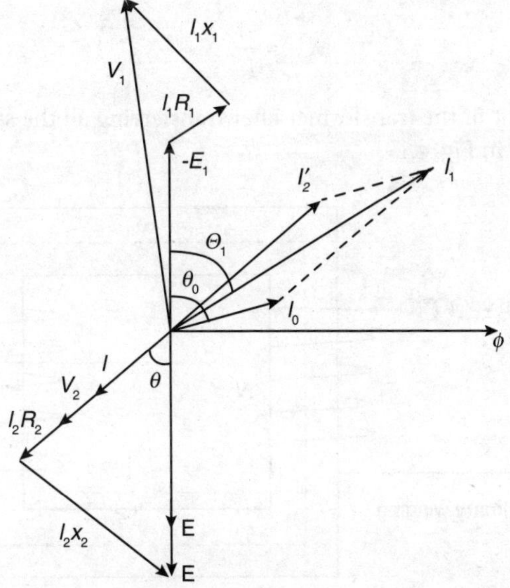

FIG. 4.18: Phasor diagram for resistive load

4.16 DETERMINATION EQUIVALENT RESISTANCE, REACTANCE AND IMPEDANCE OF TRANSFORMER WINDING

Refer to the transformer circuit shown in Fig. 4.15, where, R_1 and X_1 are the resistance and reactance of the primary winding respectively, and R_2 and X_2 are the resistance and reactance of the secondary winding respectively.

To make the calculation on transformer circuit simpler, it is possible to transfer the secondary circuit parameters to the primary side and vice versa.

i) *Transferring secondary parameters to primary side*

The secondary parameters of the transformer (R_2 and X_2) can be transferred to the primary side of the transformer without disturbing the circuit condition. Let R_2' be the additional resistance of the primary circuit after the secondary resistance is transferred to the primary side of the transformer. The value of R_2' should be selected in such a way so that, the power absorbed by R_2' due to the primary current I_1 must be equal to the power absorbed by R_2 due to the secondary current I_2. That is,

$$I_1^2 R_2' = I_2^2 R_2$$

$$or\ R_2' = \left(\frac{I_2}{I_1}\right)^2 R_2$$

From Eq. (4.8), we can write,

$$R_2' = \frac{1}{K^2} R_2 \tag{4.18}$$

In the similar manner, it is also possible to transfer the secondary reactance (X_2) to the primary side also. That is,

$$X_2' = \frac{1}{K^2} X_2 \tag{4.19}$$

The modified circuit of the transformer after transferring all the secondary parameters to the primary side is shown in Fig. 4.19.

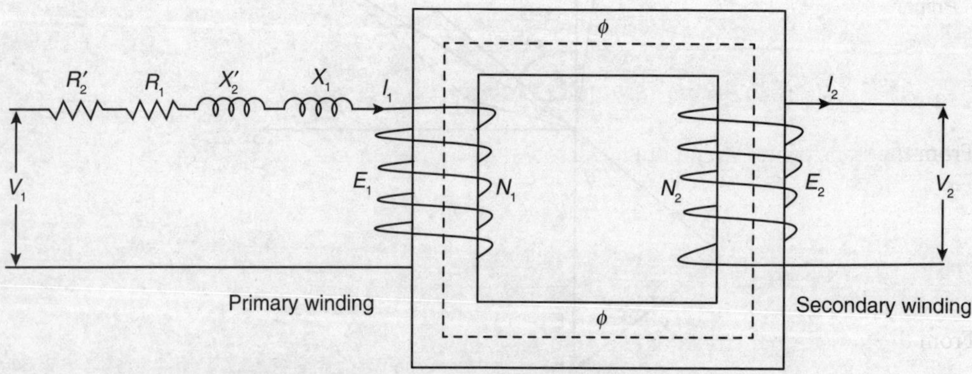

FIG. 4.19: Parameters referred to primary side

ii) *Transferring primary parameters to secondary side*

The primary parameters of the transformer (R_1 and X_1) can also be transferred to the secondary side without disturbing the circuit condition. Let R_1' be the additional resistance of the secondary circuit after the primary resistance is transferred to the secondary side of the transformer. The value of R_1' should be selected in such a way so that, the power absorbed by R_1' due to the secondary current I_2 must be equal to the power absorbed by R_1 due to the primary current I_1. That is,

$$\text{or } R_1' = \left(\frac{I_1}{I_2}\right)^2 R_1$$

From Eq. (4.8), we can write,

$$R_1' = R_2K^2 \tag{4.20}$$

In a similar manner, it is also possible to transfer the primary reactance (X_1) to the secondary side also. That is,

$$X_1' = X_2K^2 \tag{4.21}$$

The modified circuit of the transformer after transferring all the primary parameters to the secondary side is shown in Fig. 4.20.

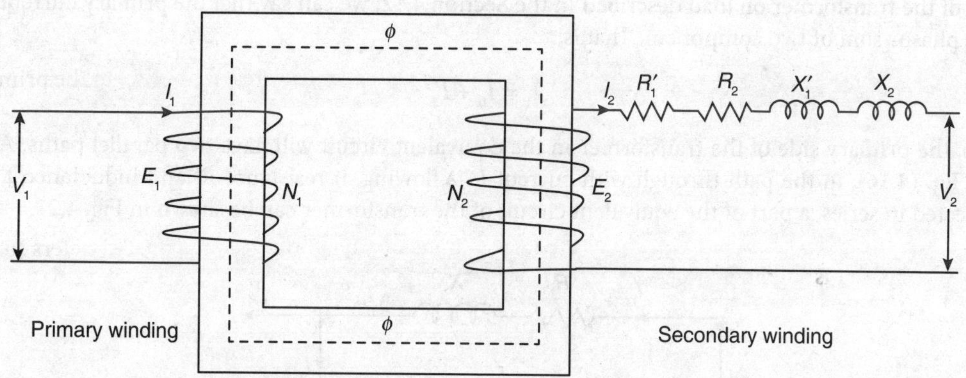

FIG. 4.20: Parameters referred to secondary side

From the transformer circuit of Fig. 4.19, we can write,

$$R_{e1} = R_1 + R_2' \tag{4.22}$$

$$X_{e1} = X_1 + X_2' \tag{4.23}$$

From the transformer circuit of Fig. 4 .20, we can write,

$$R_{e2} = R_2 + R_1' \tag{4.24}$$

$$X_{e2} = X_2 + X'_1 \tag{4.25}$$

Where, R_{e1} is called the equivalent resistance of the transformer referred to primary. X_{e1} is called the equivalent reactance of the transformer referred to primary. Similarly, R_{e2} is the equivalent resistance of the transformer referred to secondary and X_{e2} is called the equivalent reactance of the transformer referred to secondary.

Again, we can write,

$$Z_{e1} = R_{e1} + jX_{e1} \tag{4.26}$$

and

$$Z_{e2} = R_{e2} + jX_{e2} \tag{4.27}$$

Where, Z_{e1} is the equivalent impedance of the transformer referred to primary and Z_{e2} is the equivalent impedance of the transformer referred to secondary.

4.17 EQUIVALENT CIRCUIT OF TRANSFORMER

It is possible to represent the behaviour of electrical equipment in terms of an equivalent circuit. Equivalent circuit helps in predetermining the behaviour of any electrical equipment under various operating conditions. Equivalent circuit of a device can be obtained from the characteristic equations describing the performance of the device.

Equations (4.11), (4.16) and (4.17) give the characteristic equation of the transformer. From the operation of the transformer on load described in the Section 4.12, we can say that the primary current (I_1), is the phasor sum of two component. That is,

$$I_1 = I_0 + I'_2 \tag{4.28}$$

So, the primary side of the transformer in the equivalent circuit will have two parallel paths. Again from Eq. (4.16), in the path through with current I_1 is flowing, if resistance R_1 and inductance X_1 are connected in series, a part of the equivalent circuit of the transformer can be shown in Fig. 4.21.

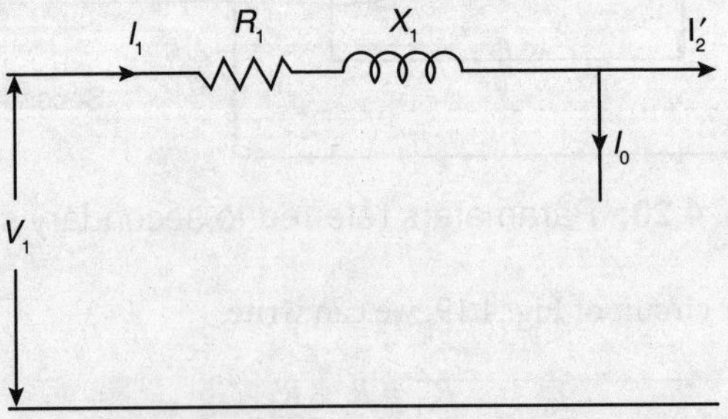

Fig. 4.21: Equivalent circuit of transformer

Again from Equation (4.11), the current I_0 is the phasor sum of I_m and I_c. Current I_c is in phase with V_1; hence, in this path a pure resistance is present. I_m lags V_1 by 90^0, so in this path pure inductor is present. The modified equivalent circuit of the primary side of the transformer is shown in Fig. 4.22.

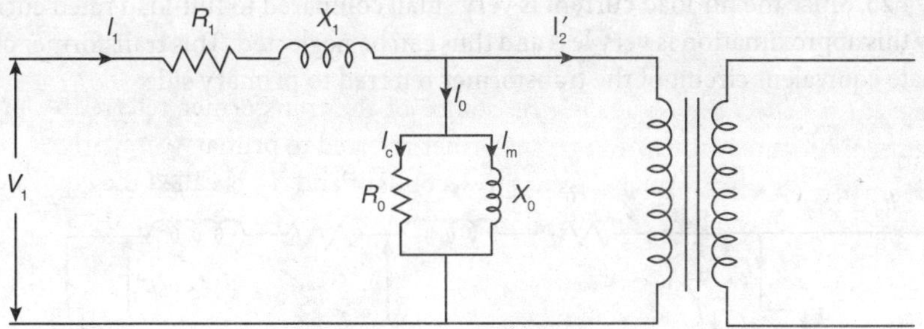

FIG. 4.22: Primary side equivalent circuit

Again from Eq. (4.17), in the secondary side of the transformer there is a resistance R_2 and inductance X_2 connected in series. So the complete equivalent circuit of the transformer is shown in Fig. 4.23. The transformer circuit shown in Fig. 4.23 is called the exact equivalent circuit of transformer.

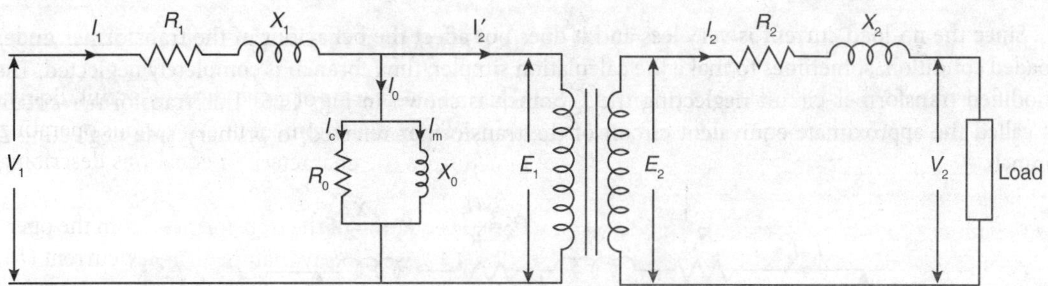

FIG. 4.23: Exact equivalent circuit

Now, as discussed in the earlier sections, if we transfer the secondary parameters to the primary side, the modified equivalent circuit of the transformer is shown in Fig. 4.24.

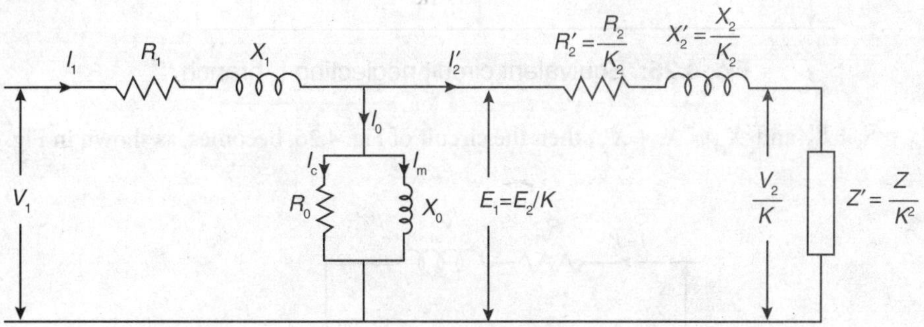

FIG. 4.24: Exact equivalent circuit referred to primary

The transformer circuit shown in Fig. 4.24 is called the exact equivalent circuit of the transformer referred to primary.

The no-load current I_0 is only 3% to 5% of the full-load rated current. So, the equivalent circuit diagram of the transformer can be further simplified by transferring the parallel R_0 and X_0 to the left, as shown in Fig. 4.25. Since the no-load current is very small compared to full-load rated current, the error introduced by this approximation is very less and thus can be neglected. This transformer circuit is called the approximate equivalent circuit of the transformer referred to primary side.

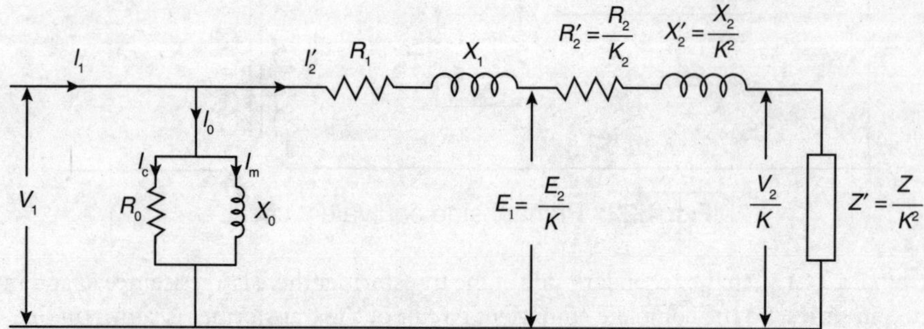

Fig. 4.25: Approximate equivalent circuit referred to primary

Since the no-load current is very less and it does not affect the behaviour of the transformer under loaded condition, sometimes to make the calculation simpler, the I_0 branch is completely neglected. The modified transformer circuit neglecting the I_0 branch is shown in Fig. 4.26. This transformer circuit is called the approximate equivalent circuit of the transformer referred to primary side neglecting I_0 branch.

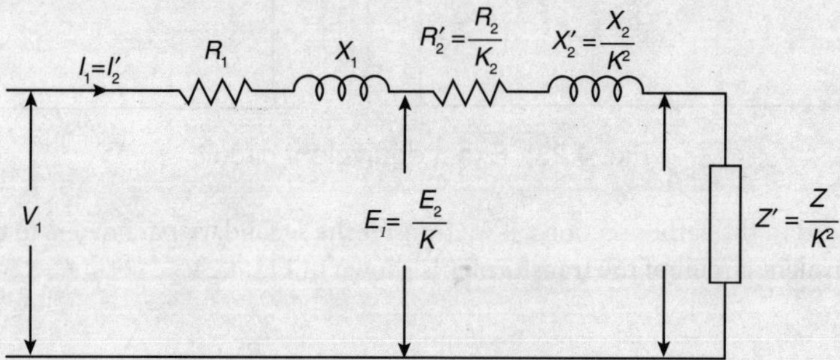

Fig. 4.26: Equivalent circuit neglecting I_0 branch

Let $R_{e1} = R_1 + R_2'$ and $X_{e1} = X_1 + X_2'$, then the circuit of Fig. 4.26, becomes, as shown in Fig. 4.27.

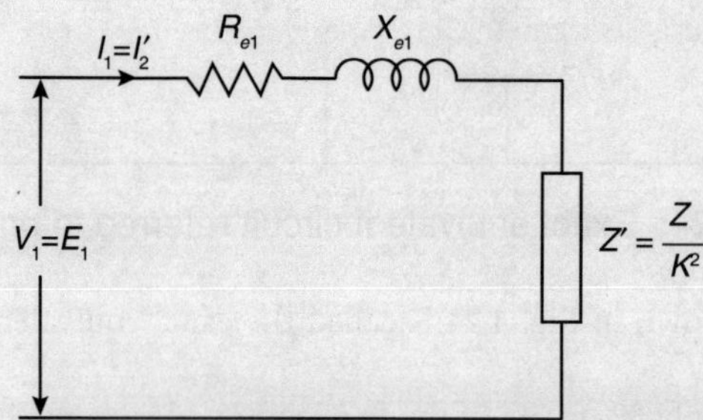

Fig. 4.27: Modified circuit

Similarly, Fig. 4.28 gives the approximate equivalent circuit of the transformer referred to the secondary side of the transformer.

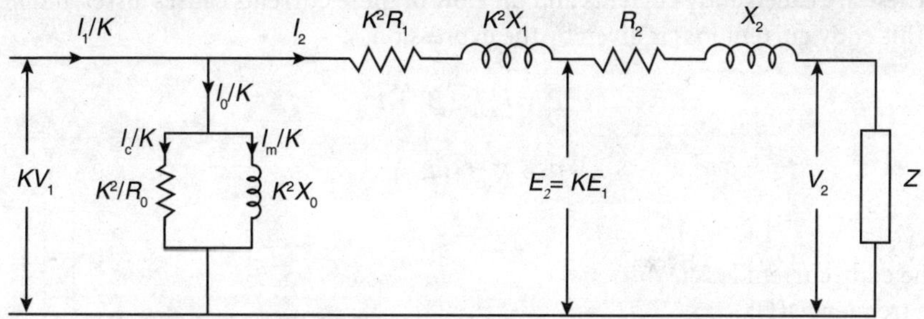

Fig. 4.28: Equivalent circuit referred to secondary

Let $R_{e2} = R_2 + K^2 R_1$ and $X_{e2} = X_2 + K^2 X_1$, and neglecting the parallel branch, the modified circuit of Fig. 4.28, is shown in Fig. 4.29.

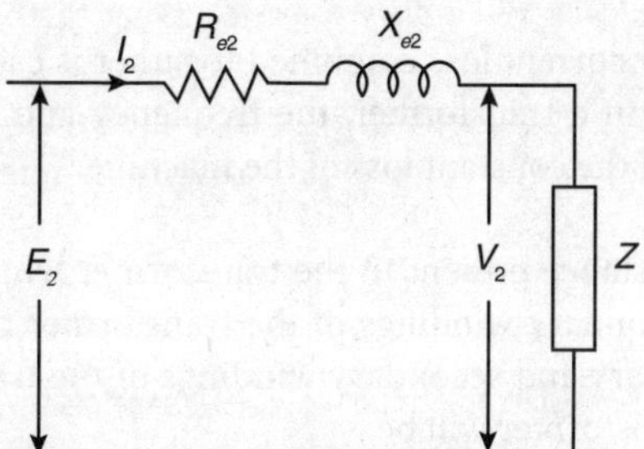

Fig. 4.29: Equivalent circuit without parallel branch

4.18 LOSSES IN TRANSFORMER

All the losses in transformer can be classified into two groups:

i) Iron loss or core loss

The iron loss in transformer includes hysteresis loss and eddy current loss. When a ferromagnetic material is magnetised, a hysteresis loop is formed. In the process of magnetisation and demagnetisation of ferromagnetic material, energy is stored and then released. This process is not completely reversible. This difference between energy storage and energy release is dissipated as heat. This dissipation of energy is known as hysteresis loss and is given by the expression

$$P_{\mathrm{h}} = \mathrm{K}_{\mathrm{h}} f \left(B_{\mathrm{m}}\right)^{n} \tag{4.29}$$

Where,

P_{h} is the hysteresis loss (W)

F is the frequency (Hz)

B_{m} is the maximum flux density (T)

K_{h} is a constant which depends on the material used.

The value of n varies from 1.5 to 2.5 depending on the type of material used.

If the transformer core is made up of thick laminated sheets, then the e.m.f. induced in the laminated sheet due to alternating magnetic field will produce circulating currents in the transformer core. These are called eddy currents and the flow of these currents causes losses and heating of the core. This eddy current loss is given by the expression.

$$P_e = k_e f^2 (B_m)^2 t^2 \quad (4.30)$$

$$\text{or, } P_e = K_e f^2 (B_m)^2 \quad (4.31)$$

Where,
P_e is the eddy current loss (W)
f is the frequency (Hz)
B_m is the maximum flux density (T)
t is the thickness of lamination.
k_e is a constant which depends on the material used.

$$K_e = k_e t^2 \text{ is a constant}$$

The hysteresis and eddy current loss combined together is called the iron loss or the core loss of the transformer. Since, in a transformer, the frequency and maximum flux density remains constant, core loss is called the constant loss of the machine.

ii) Copper loss

This loss is due to the resistance present in the transformer windings. If R_1 and R_2 are the resistances of primary and secondary windings of the transformer respectively and I_1 and I_2 are the currents through the primary and secondary windings of the transformer respectively, then, the total copper loss of the transformer will be,

$$P_c = I_1^2 R_1 + I_2^2 R_2 \quad (4.32)$$

Since the copper loss depends on the currents, it is not a constant loss; it varies according to the load condition of the transformer. If the copper loss at full load is P_c, then the copper loss at half load will be $\left(\frac{1}{2}\right)^2 P_c$. Similarly the copper loss at 1/4th of full load will be $\left(\frac{1}{4}\right)^2 P_c$. In general the copper loss at a load x times the full load will be $x^2 P_c$

4.19 TRANSFORMER OIL

The losses that occur in the transformer, gets dissipated in the form of heat. Transformer oil or insulating oil is a highly refined mineral oil that is stable at high temperatures and has excellent electrical insulating properties. Its functions are to insulate, and also to dissipate the heat of the transformer due to losses.

Generally, there are two types of transformer oil used in transformer,

1. Paraffin based transformer oil
2. Naphtha based transformer oil

Transformer oil has following characteristics:

1. It is colourless
2. It has low density
3. It has low viscosity

The properties of transformer oil can be categorised as follows:

1. Physical properties
2. Electrical properties
3. Chemical properties

Physical Properties

- Moisture content: For good transformer oil, its moisture content should be low.
- Flash point: The flash point should be high.
- Viscosity: Viscosity of the transformer oil should be very less.
- Pour point: Pour point should be low.

Electrical Properties

- Electrical breakdown voltage strength: Transformer oil should have high breakdown voltage strength.
- Resistivity: For good transformer oil, the resistivity should be high.
- Dielectric dissipation factor: Dielectric dissipation value should be low.

Chemical Properties

- Neutralisation value: The total acidity of the transformer oil should be low.
- Oxidation stability: Oxidation stability is a measure of neutralisation value and sludge after oil is aged by simulating the actual service conditions of the oil.
- Corrosive sulphur: Presence of corrosive sulphur should be very less.

4.20 COOLING SYSTEM OF TRANSFORMER

The losses that occur in the transformer get dissipated in the form of heat. Heat is generated because of copper loss in the transformer. The contribution of iron loss is comparatively less. If this heat is not dissipated properly, then the transformer may get damaged. To protect the transformer from damage due to excessive heat, different transformer cooling methods are available. The different transformer cooling methods are a follows:

i) For dry type transformers
   a) Air Natural (AN) cooling method
   b) Air Blast cooling method
ii) For oil immersed transformers
   a) Oil Natural Air Natural (ONAN) cooling
   b) Oil Natural Air Forced (ONAF) cooling
   c) Oil Forced Air Forced (OFAF) cooling
   d) Oil Forced Water Forced (OFWF)

Air Natural Cooling Method

This method of cooling is used in small transformers (up to 3 MVA). In this method, the transformer is cooled by the natural air flow surrounding the transformer.

Air Blast Cooling Method

For bigger transformers, this method of cooling is used. In this method filtered air is forced, with the help of a fan or blower, into the core and windings of the transformer. This air is filtered to prevent the accumulation of dust particles in ventilation ducts.

Oil Natural Air Natural (ONAN) cooling

ONAN stands for Oil Natural Air Natural cooling system. Here, the transformer is cooled by the natural convectional flow of hot oil in the transformer. When the transformer oil gets heated, it flows to the upper portion of the transformer tank and the lower portion is occupied by the cooled oil. Additional dissipating surface in the form of tubes or radiators are connected with the transformer tank so that the hot oil which comes to the upper side of the transformer tank, is able to dissipate the heat into the atmosphere by natural conduction, convection and radiation and thus become cold. These tubes or radiators are called the radiator of transformer.

Oil Natural Air Forced (ONAF) cooling

In the Oil Natural Air Forced (ONAF) method of cooling, heat dissipation from the hot oil is made faster by applying forced air flow on the dissipating surface. Fans blow air on to the cooling surface; this forced air takes away the heat from the surface of the radiator and provides better and faster cooling than natural air.

Oil Forced Air Forced (OFAF) cooling

In ONAF method of cooling, the natural conventional oil flow is used. That is, hot oil rises up and coil oil takes its place at the bottom. In OFAF system, the air circulation through the transformer tank is accelerated by applying some external force. In this method, cooling oil circulation in the tank is increased by means of oil pumps. This method makes the cooling faster and better. Its compact arrangement occupies less space as compared to the other cooling methods discussed earlier in this section.

Oil Force Water Forced (OFWF) cooling

Water is a better heat exchanging medium than air. In OFWF method of cooling, oil-to-water heat exchanger is used to dissipate the heat from the transformer oil. The hot oil is sent to the heat exchanger with the help of the oil pumps, where the oil is cooled by applying a shower of cold water on the oil pipes.

Oil Direct Air Forced (ODAF) Cooling

This method of cooling is generally applied to high-rating transformers. In this method, the cooled oil coming out from the radiator is passed through the transformer windings where gaps for oil flow or pre-decided oil flow paths between the insulated conductors are provided for faster and better cooling of the transformer.

Oil Direct Water Forced (OFWF) Cooling

It is same as ODAF cooling system; only instead of air, forced water is used to cool the heated transformer oil.

4.21 VOLTAGE REGULATION OF A TRANSFORMER

Voltage regulation of a transformer is defined as the ratio of change is secondary terminal voltage from no-load to full-load voltage. That is,

$$\text{Voltage regulation} = \frac{\text{No-load voltage} - \text{Full-load voltage}}{\text{Full-load voltage}}$$

$$\text{or Voltage regulation} = \frac{E_2 - V_2}{V_2}. \tag{4.33}$$

Consider the approximate equivalent circuit of the transformer referred to the secondary, shown in Fig. 4.29. The phasor diagram for lagging load and leading load is shown in Fig. 4.30 and Fig. 4.31 respectively.

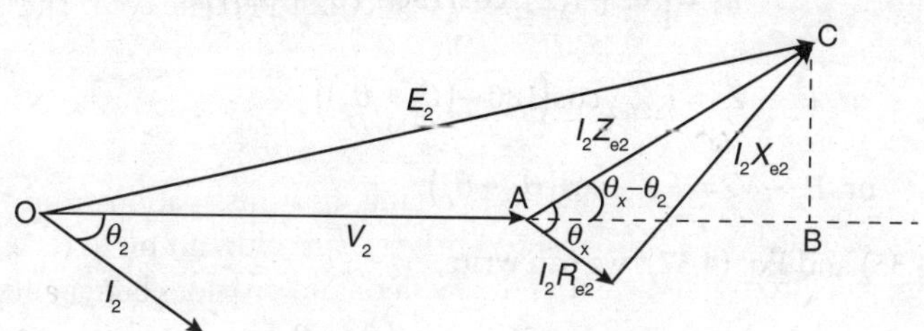

FIG. 4.30: For lagging load

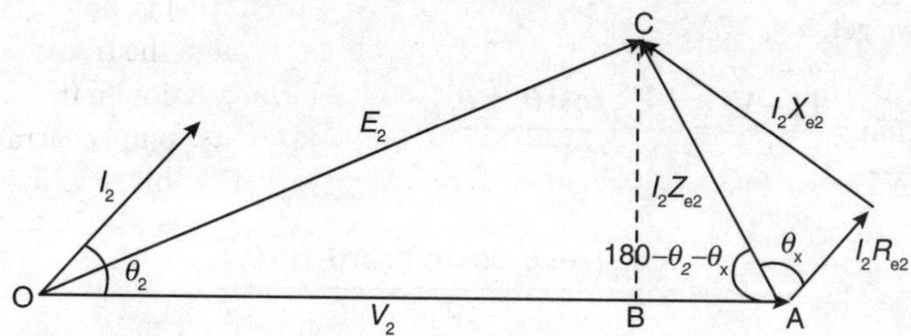

FIG. 4.31: For leading load

From the phasor of Fig. 4.30, we can write,

$$(OC)^2 = (OB)^2 + (BC)^2$$

$$\text{or } (OC)^2 = (OA + AB)^2 + (BC)^2$$

$$\text{or } E_2^2 = \left[V_2 + I_2 Z_{e2}\cos(\theta_x - \theta_2)\right]^2 + \left[I_2 Z_{e2}\sin(\theta_x - \theta_2)\right]^2 \tag{4.34}$$

For an ideal transformer, $I_2 Z_{e2}\sin(\theta_x - \theta_2)$ is very small as compared to E_2. So, Eq. (4.34) becomes,

$$\text{or } E_2^2 = \left[V_2 + I_2 Z_{e2}\cos(\theta_x - \theta_2)\right]^2$$

$$\text{or } E_2 - V_2 = I_2 Z_{e2}\cos(\theta_x - \theta_2) \tag{4.35}$$

From the phasor of Fig. 4.31, we can write,

$$(OC)^2 = (OB)^2 + (BC)^2$$

$$\text{or } (OC)^2 = (OA - AB)^2 + (BC)^2.$$

$$\text{or} \qquad E_2^2 = \left[V_2 - I_2 Z_{e2}\cos\left(180-(\theta_x+\theta_2)\right)\right]^2 + \left[I_2 Z_{e2}\sin\left(180-(\theta_x+\theta_2)\right)\right]^2 \tag{4.36}$$

For an ideal transformer, $I_2 Z_{e2}\sin\left(180-(\theta_x+\theta_2)\right)$ is very small as compared to E_2. So Eq. (4.36) becomes,

$$E_2^2 = \left[V_2 - I_2 Z_{e2}\cos\left(180-(\theta_x+\theta_2)\right)\right]^2$$

$$\text{or, } E_2 - V_2 = I_2 Z_{e2}\cos\left(180-(\theta_x+\theta_2)\right)$$

$$\text{or, } E_2 - V_2 = I_2 Z_{e2}\cos\left(\theta_x+\theta_2\right) \tag{4.37}$$

Combining Eq. (4.35) and Eq. (4.37), we can write,

$$E_2 - V_2 = I_2 Z_{e2}\cos\left(\theta_x \pm \theta_2\right) \tag{4.38}$$

Where,

– ve sign is for lagging power factor and +ve sign is for leading power factor load. Substituting Eq. (4.38) in Eq. (4.33) we get,

$$\text{Voltage regulation} = \frac{E_2 - V_2}{V_2} = \frac{I_2 Z_{e2}\cos\left(\theta_x \pm \theta_2\right)}{V_2}$$

$$= \frac{I_2 Z_{e2}\left(\cos\theta_x \cos\theta_2 \mp \sin\theta_x \sin\theta_2\right)}{V_2}$$

$$= \frac{\left(I_2 Z_{e2}\cos\theta_x\right)\cos\theta_2 \mp \left(I_2 Z_{e2}\sin\theta_x\right)\sin\theta_2}{V_2}$$

Again from the phasor of Fig. 4.30 and Fig. 4.31, we can write,

$$I_2 Z_{e2}\cos\theta_x = I_2 R_{e2}$$

$$\text{or, } I_2 Z_{e2}\sin\theta_x = I_2 X_{e2}$$

Substituting we get,

$$\text{Voltage regulation} = \frac{I_2 R_{e2}\cos\theta_2 \mp I_2 X_{e2}\sin\theta_2}{V_2} \times 100 \tag{4.39}$$

Where,

– ve sign is for leading power factor and +ve sign is for lagging power factor load.

4.21.1 Condition for Zero Voltage Regulation and Maximum Voltage Regulation

Consider the voltage regulation equation of (4.39). The regulation will be zero if,

$$I_2 R_{e2}\cos\theta_2 \mp I_2 X_{e2}\sin\theta_2 = 0$$

$$\text{or,}\quad \frac{\sin\theta_2}{\cos\theta_2} = \pm\frac{R_{e2}}{X_{e2}}$$

$$\text{or,}\ \tan\theta_2 = \pm\frac{R_{e2}}{X_{e2}} \tag{4.40}$$

Equation (4.40) gives the condition for zero voltage regulation, where −ve sign is for lagging power factor and +ve sign is for leading power factor load.

Again the voltage regulation will be maximum if,

$$\frac{d}{d\theta}\left[\frac{I_2R_{e2}\cos\theta_2 \mp I_2X_{e2}\sin\theta_2}{V_2}\right] = 0$$

$$\text{or,}\quad -\frac{I_2R_{e2}}{V_2}\sin\theta_2 \mp \frac{I_2X_{e2}}{V_2}\cos\theta_2 = 0$$

$$\text{or,}\quad \tan\theta_2 = \pm\frac{X_{e2}}{R_{e2}} \tag{4.41}$$

Equation (4.41), gives the condition for maximum voltage regulation, where −ve sign is for leading power factor and +ve sign is for lagging power factor load.

4.22 EFFICIENCY OF A TRANSFORMER

The efficiency (η) of a single phase transformer is defined as the ratio of output power to input power and is expressed as a percentage. That is,

$$\text{Efficiency}\,(\eta) = \frac{\text{Output Power}}{\text{Input Power}}\times 100 = \frac{\text{Output Power}}{\text{Output Power} + \text{Loss}}\times 100 \tag{4.42}$$

Where,

$$\text{Output Power} = V_2I_2\cos\theta_2$$

$$\text{Loss} = \text{Iron loss} + \text{copper loss}$$

$$\text{Iron loss}\,(P_i) = \text{Hysteresis loss} + \text{Eddy current loss}$$

Copper loss at full load is P_c.

So the efficiency of the transformer at full load will be.

$$\eta_{fl} = \frac{V_2I_2\cos\theta_2}{V_2I_2\cos\theta_2 + P_i + P_c} \tag{4.43}$$

The efficiency of the transformer at a load x times full load will be

$$\eta = \frac{xV_2I_2\cos\theta_2}{xV_2I_2\cos\theta_2 + P_i + x^2P_c} \tag{4.44}$$

4.22.1 Maximum Efficiency of a Single-phase Transformer

Since the secondary current of the transformer (I_2), changes with the load, the efficiency of the transformer also changes with the change in load. In this section, we will determine the condition for maximum efficiency of the transformer. Value of load current and KVA supplied at maximum efficiency.

The efficiency of a single phase transformer is given by the expression,

$$\eta = \frac{V_2 I_2 \cos\theta_2}{V_2 I_2 \cos\theta_2 + P_i + I_2^2 R_{e2}} \quad (4.45)$$

Where

V_2 is the voltage across the secondary terminal of the transformer.

I_2 is the current flowing in the secondary of the transformer.

$\cos\theta_2$ is the power factor of secondary.

R_{e2} is the equivalent resistance of the transformer

P_i is the constant loss of the transformer.

$P_c = I_2^2 R_{e2}$ is the copper loss of the transformer.

For maximum efficiency,

$$\frac{d\eta}{dI_2} = 0$$

Differentiating Eq. (E.45) with respect to I_2 and equation it to zero, we get,

$$\left(V_2I_2\cos\theta_2 + P_i + I_2^2R_{e2}\right)\frac{d}{dI_2}\left(V_2I_2\cos\theta_2\right) - \left(V_2I_2\cos\theta_2\right)\frac{d}{dI_2}\left(V_2I_2\cos\theta_2 + P_i + I_2^2R_{e2}\right) = 0$$

$$\text{or, } \left(V_2I_2\cos\theta_2 + P_i + I_2^2R_{e2}\right)\times\left(V_2\cos\theta_2\right) - \left(V_2I_2\cos\theta_2\right)\times\left(V_2\cos\theta_2 + 2I_2R_{e2}\right) = 0$$

$$\text{or, } V_2^2I_2\cos^2\theta_2 + P_iV_2\cos\theta_2 + I_2^2R_{e2}V_2\cos\theta_2 - V_2^2I_2\cos^2\theta_2 - 2V_2I_2^2R_{e2}\cos\theta_2 = 0$$

$$\text{or, } P_iV_2\cos\theta_2 - I_2^2R_{e2}V_2\cos\theta_2 = 0$$

$$\text{or, } V_2\cos\theta_2\left(P_i - I_2^2R_{e2}\right) = 0$$

$$\text{or, } P_i - I_2^2R_{e2} = 0$$

$$\text{or, } P_i = I_2^2R_{e2} \quad (4.46)$$

From Eq. (4.46), we get the condition for maximum efficiency of the transformer that is for maximum efficiency

Iron loss = Copper loss

Let the load current during maximum efficiency be I_{2m}. Then Eq. (4.46), can be written as,

$$P_i = I_{2m}^2 R_{e2}$$

$$\text{or, } I_{2m} = \sqrt{\frac{P_i}{R_{e2}}} \tag{4.47}$$

Equation (4.47) gives the load current at maximum efficiency.

Let I_{fl} be the current through the transformer at full load. Dividing Eq. (4.47) by I_{fl}, we get,

$$\frac{I_{2m}}{I_{fl}} = \frac{1}{I_{fl}}\sqrt{\frac{P_i}{R_{e2}}}$$

$$\text{or, } \frac{I_{2m}}{I_{fl}} = \sqrt{\frac{P_i}{I_{fl}^2 R_{e2}}} = \sqrt{\frac{P_i}{P_{cu}(fl)}}$$

Where,

$$P_{cu}(fl) = I_{fl}^2 R_{e2} = \text{Copper loss in the transformer at full load}$$

Then,

$$I_{2m} = I_{fl}\sqrt{\frac{P_i}{P_{cu}(fl)}} \tag{4.48}$$

Equation (4.48) gives the secondary current at maximum efficiency in terms of full load current of the transformer.

The KVA at maximum efficiency is given by,

$$(\text{KVA}) \text{ at maximum efficiency} = I_{2m} V_2$$

Substituting the expression of I_{2m} from equation (4.48), we get,

$$(\text{KVA}) \text{ at maximum efficiency} = V_2 I_{fl}\sqrt{\frac{P_i}{P_{cu}(fl)}} = \text{KVA} \times \sqrt{\frac{P_i}{P_{cu}(fl)}}$$

$$\text{or, } (\text{KVA}) \text{ at maximum efficiency} = \text{KVA} \times \sqrt{\frac{P_i}{P_{cu}(fl)}} \tag{4.49}$$

4.22.2 All-day Efficiency of a Transformer

The calculation of efficiency by the ratio of output power to input power is called the commercial efficiency of a transformer. This formula is useful provided the load and its power factor is constant throughout.

However in reality, as in the case of distribution transformer, the load varies widely over a day (24 hours). So the calculation of efficiency by using the formula of ratio of output power to input power will not give the exact efficiency of a transformer.

For distribution transformer all-day energy efficiency is calculated. In this method, the 24 hours in a day is divided into small segments or blocks. These segments or blocks are chosen in such a way so that in a particular segment or a block the load is constant.

The output power and the input power of each section are calculated and then the total output power and the total input power of a complete day is taken to calculate the efficiency of a transformer.

Consider the data given in Table 4.2

TABLE 4.2: Load data

Time Segment	KVA Loads	Degree of load	Power factor
T_1	S_1	X_1	$\cos\theta_1$
T_2	S_2	X_2	$\cos\theta_2$
T_3	S_3	X_3	$\cos\theta_3$
•	•	•	•
T_n	S_n	X_n	$\cos\theta_n$

From Table 4.2 we get,

$$\text{Output of the transformer in 24 Hr.} = \sum_{i=0}^{n} S_i T_i \cos\theta_i \tag{4.50}$$

$$\text{Total loss in the transformer in 24 Hr.} = 24P_i + \sum_{i=1}^{n} T_i P_{cu} x_i^2 \tag{4.51}$$

Where,

P_i is the iron loss of the transformer which is constant loss and does not change with the load.

P_{cu} is the copper loss of the transformer at full load.

x is a factor which is, $x = 1$ at full load, $x = 0.5$ at half load, $x = 0.25$ at 1/4th load and so on.

So the all day efficiency of the transformer is given by the expression,

$$\eta_{\text{all\_day}} = \frac{\sum_{i=0}^{n} S_i T_i \cos\theta_i}{\left(\sum_{i=0}^{n} S_i T_i \cos\theta_i\right) + \left(24P_i + \sum_{i=1}^{n} T_i P_{cu} x_i^2\right)} \tag{4.52}$$

4.23 TRANSFORMER TESTING

To obtain the parameters of the equivalent circuit of a transformer and also to calculate the efficiency and voltage regulation of a transformer at a particular load condition, two tests are carried out on a transformer. These tests are,

i) Open-circuit test or no-load test
ii) Short-circuit test or impedance test

These two tests give the parameters of the equivalent circuit of a transformer. Based on these test results the efficiency and the voltage regulation of the transformer at any load can be calculated without actually loading the transformer.

4.23.1 Open-circuit Test (O.C. Test)

The experimental set up for an open-circuit Test (O.C. Test) is shown in Fig. 4.32.

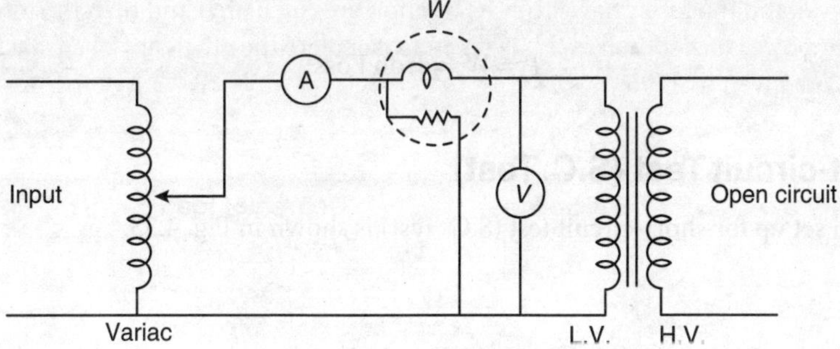

Fig. 4.32: Open-circuit test circuit

The open-circuit test (O.C. Test) is generally performed on the low-voltage (L.V.) side of the transformer and the high-voltage (H.V.) side is kept open. The test circuit is shown in Fig. 4.32.

All the measuring instruments are connected on the L.V. side of the transformer.

The steps for this test are:

i) The voltage of the low-voltage side of the transformer is slowly increased with the help of a variable autotransformer such as the Variac, until it becomes equal to the rated voltage. That is the voltmeter (V) will be reading the rated voltage of the low-voltage side of the transformer. For example, for a 230 V/440 V transformer, the rated voltage of the low-voltage side is 230V and the rated voltage of the high voltage side is 440V.
ii) At this rated voltage the readings of all measuring instruments (ammeter, voltmeter and wattmeter) are noted.

Let the voltmeter reading be V_0, ammeter reading be I_0 and the wattmeter reading be W_0.

Since the high voltage side is open-circuited, the ammeter reading I_0 is the no-load current of the transformer.

Then we can write,

$$W_0 = V_0 I_0 \cos\theta_0 \tag{4.53}$$

$$\cos\theta_0 = \frac{W_0}{V_0 I_0} \tag{4.54}$$

Again,

$$I_c = I_0 \cos\theta_0 \tag{4.55}$$

$$I_m = I_0 \sin\theta_0 \tag{4.56}$$

$$R_0 = \frac{V_0}{I_c} \tag{4.57}$$

$$X_0 = \frac{V_0}{I_m} \tag{4.58}$$

So, the open-circuit test gives the parameters of the parallel branch of the equivalent circuit of the transformer. Since the high voltage side is open-circuited, the wattmeter (W_0) will give the iron loss of the transformer.

That is,

$$P_i = W_o = \text{Iron Loss} \tag{4.59}$$

4.23.2 Short-circuit Test (S.C. Test)

The experimental set up for short-circuit test (S.C. Test) is shown in Fig. 4.33.

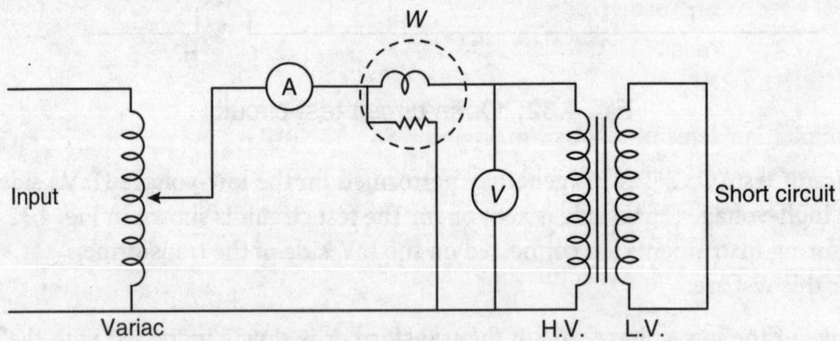

FIG. 4.33: Short-circuit test circuit

The short-circuit test (S.C. Test) is generally performed on the low-voltage (L.V.) side of the transformer and the low voltage (L.V.) side is short-circuited. The test circuit is shown in Fig. 4.33. All the measuring instruments are connected on the H.V. side of the transformer.

The steps for this test are:

i) The rated current of the H.V. side of the transformer is supplied with the help of autotransformer For example, if we consider a 2KVA, 110V/220V transformer, then the rated current of the H.V. side will be,

$$\frac{2\times10^3}{220} = 9.1 \text{ A}$$

ii) Once the ammeter reads the rated current, the reading of the other instruments is noted.

Let the voltmeter reading be V_{sc}, ammeter reading be I_{sc} and the wattmeter reading be W_{sc}. Then we can write,

$$W_{sc} = V_{sc} I_{sc} \cos\theta_{sc} \tag{4.60}$$

$$\cos\theta_{sc} = \frac{W_{sc}}{V_{sc} I_{sc}} \tag{4.61}$$

Again,

$$Z_e = \frac{V_{sc}}{I_{sc}} \tag{4.62}$$

$$R_e = \frac{W_{sc}}{I_{sc}^2} \tag{4.63}$$

$$X_e = \sqrt{Z_e^2 - R_e^2} \tag{4.64}$$

If we consider a step-down transformer, then the H.V. side will be the primary side of the transformer, so,

$$R_e = R_{e1}, X_e = X_{e1} \text{ and } Z_e = Z_{e1}.$$

Similarly, for a step-up transformer,

$$R_e = R_{e2}, X_e = X_{e2} \text{ and } Z_e = Z_{e2}.$$

So the short-circuit test gives the equivalent resistance, equivalent reactance and equivalent impedance of the transformer either referred to primary or secondary side. The wattmeter reading gives the total copper loss of the transformer.

4.23.3 Polarity Test

This test is conducted to determine the primary and secondary phasor polarities. AC voltage V_s is applied to the primary side of the transformer and the reading of the three voltmeters is noted.

If, $V_3 = V_1 \sim V_2$ then terminal A_1 and a_1 are of same polarity (Fig. 4.34).

If, $V_3 = V_1 + V_2$ then terminal A_1 and a_1 are of opposite polarity (Fig. 4.35).

This polarity test is very important for parallel operation of transformer.

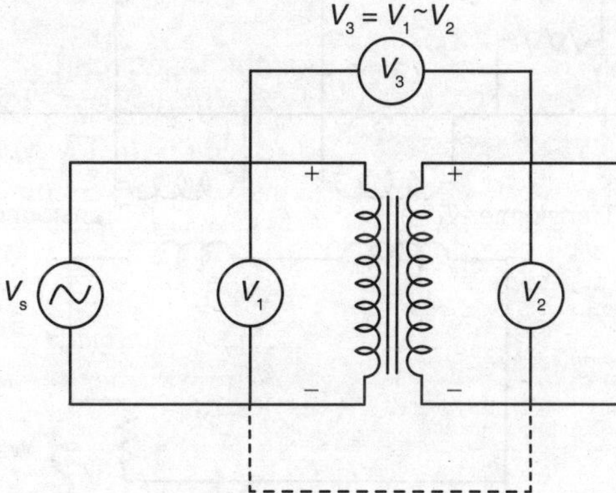

Fig. 4.34: Circuit when V_1 and V_2 are of same polarity

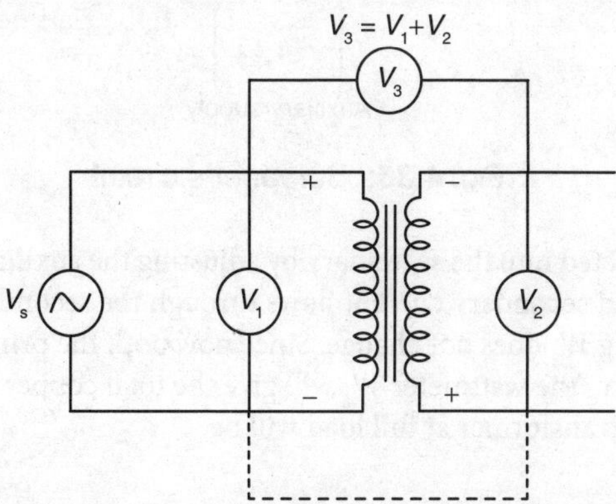

Fig. 4.35: Circuit when V_1 and V_2 are of opposite polarity

4.23.4 Sumpner's Test (Back-to-back Test)

This test is also called the load test. Using Sumpner's test we can determine the efficiency, voltage regulation and the maximum temperature rise of a transformer. In O.C. and S.C. test, the loading condition is absent and hence the result can be inaccurate. In Sumpner's test, the transformer is actually loaded and thus we can say that the result of Sumpner's test is much better and accurate than O.C. and S.C. test.

To perform this test, two identical transformers are required. The two primaries of the transformer are connected in parallel and the two secondaries are connected in series. The circuit connection of Sumpner's test is shown in Fig. 4.36. The series connection of secondary should be such that their polarities are in phase opposition, so that the induced e.m.f.'s in the two secondary opposes each other.

Initially, the supply from the auxiliary source connected across the secondary is zero. Now, rated voltage and frequency is applied from the supply connected across the primary. Since the transformers are identical, the e.m.f.'s induced in both the primary windings will be same. Again, since the secondary windings are connected in phase opposition, the induced e.m.f. will cancel out. This condition is similar to the O.C. test of the transformer and thus the wattmeter W_1 will give the total core loss of the two transformers.

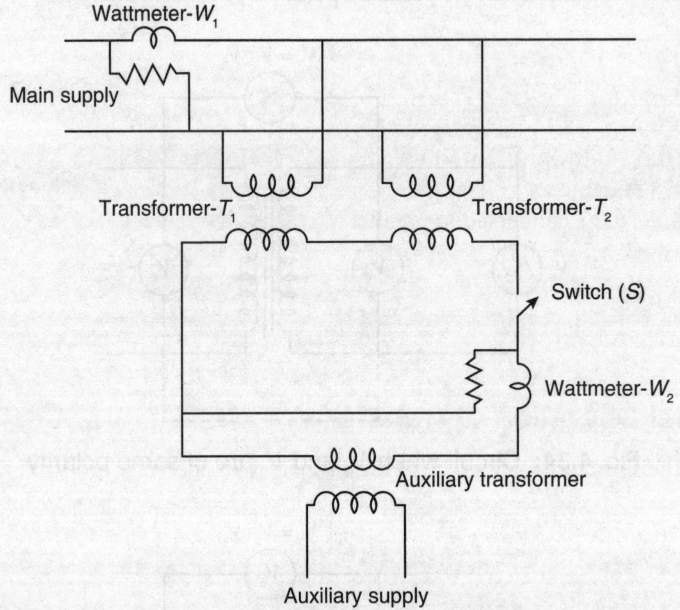

Fig. 4.36: Sumpner's circuit

Now, some voltage is injected into the secondary by adjusting the auxiliary supply. This auxiliary voltage is adjusted until the rated secondary current flows through the secondary winding. Under this condition, the wattmeter reading W_1 does not change. Since now both the primary and secondary windings are carrying the rated current, the wattmeter W_2 will give the total copper loss of the two transformers.

So the efficiency of each transformer at full load will be,

$$\text{Efficiency of each transformer} = \frac{\text{Output}}{\text{Output} + \frac{W_1}{2} + \frac{W_2}{2}} \times 100 \tag{4.65}$$

4.24 PARALLEL OPERATION OF TRANSFORMER

If the load on the transformer exceeds the rating of the given transformer, then there are two options that can be followed.

i) Replace the existing transformer with the next higher-rating transformer.
ii) Connect a second transformer in parallel with the first transformer.

It is always economical to install a number of small transformers in parallel rather than using one transformer of bigger rating. This is because,

- By running a number of transformers in parallel, we will get an option to switch ON only those transformers which will meet the demand by running it nearly at full load. In this way the efficiency of operation can be improved.
- If the transformers are connected in parallel, then the maintenance becomes easy as now total shut down is not required.
- Reliability of the system increases by connecting transformers in parallel. This is because in the case of breakdown of one transformer, the other transformer can share the load.

4.24.1 Condition for Parallel Operation of Transformers

Before connecting the transformers in parallel, the following conditions needs to be checked:

- The voltage rating of the primary and secondary windings of all the transformers to be connected in parallel must be same.
- The transformers to be connected in parallel must have the same polarity.
- Percentage impedance of all the transformers should be equal.
- X/R ratio of all the transformers must be same. Where X is the impedance and R is the resistance.
- The equivalent impedance must be inversely proportional to individual KVA rating, to avoid circulating currents.

4.24.2 Parallel Operation of Transformer with Equal Voltage Ratios

In Fig. 4.37, only the secondary side of the two transformers is shown to be connected in parallel.

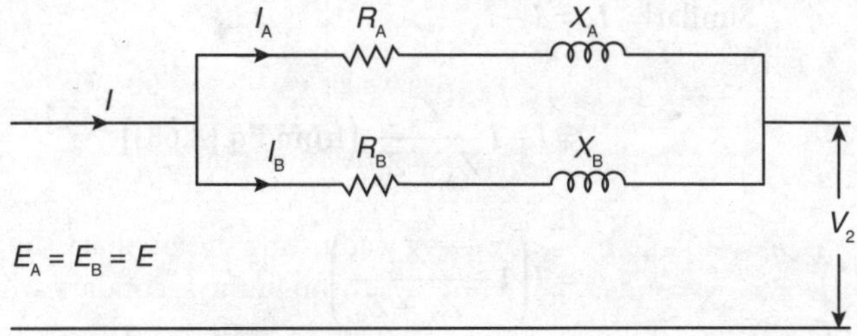

FIG. 4.37: Secondary side of the two transformers

Where,
R_A is the resistance of transformer A.
R_B is the resistance of transformer B.
X_A is the inductance of transformer A.

X_B is the inductance of transformer B.
V_2 is the secondary terminal voltage.
Since the transformer is having equal voltage ratio, we can write, $E_A = E_B = E$
Let,

$$Z_A = R_A + jX_A \text{ and } Z_B = R_B + jX_B$$

Then,

$$Z_{eq} = \frac{Z_A \times Z_B}{Z_A + Z_B}$$

Applying KVL in the circuit shown in Fig. 4.37, we get,

$$E_A - I_A R_A = V_2 \tag{4.66}$$

$$E_B - I_B R_B = V_2 \tag{4.67}$$

Since, $E_A = E_B = E$, from Eq. (4.66) and Eq. (4.67), we can write,

$$I_A Z_A = I_B Z_B$$

$$\text{or } \frac{I_A}{I_B} = \frac{Z_B}{Z_A}$$

$$\text{or } \frac{I_A}{I_A + I_B} = \frac{Z_B}{Z_A + Z_B}$$

$$\text{or } \frac{I_A}{I} = \frac{Z_B}{Z_A + Z_B} \text{ since } I = I_A + I_B$$

$$\text{or } I_A = I\frac{Z_B}{Z_A + Z_B} \tag{4.68}$$

$$\text{Similarly, } I_B = I - I_A$$

$$= I - I\frac{Z_B}{Z_A + Z_B} \text{(from Eq.(4.68))}$$

$$= I\left(1 - \frac{Z_B}{Z_A + Z_B}\right)$$

$$I_B = I\frac{Z_A}{Z_A + Z_B} \tag{4.69}$$

Thus the load shared by transformer A is

$$S_A = V_2 I_A = V_2 \times I\frac{Z_B}{Z_A + Z_B} \tag{4.70}$$

Similarly the load shared by transformer B will be,

$$S_B = V_2 I_B = V_2 \times I \frac{Z_A}{Z_A + Z_B} \tag{4.71}$$

So the total load on transformer is

$$S = S_A + S_B = V_2 I_A + V_2 I_B = V_2 (I_A + I_B) = V_2 I \tag{4.72}$$

Then,

$$S_A = S\left(\frac{Z_B}{Z_A + Z_B}\right) \text{and } S_B = S\left(\frac{Z_A}{Z_A + Z_B}\right) \tag{4.73}$$

4.24.3 Parallel Operation of Transformer with Unequal Voltage Ratios

For unequal voltage ratios E_A will not be now equal to E_B.

Applying KVL in the circuit shown in Fig. 4.37, we get,

$$E_A = V_2 + I_A Z_A \tag{4.74}$$

$$E_B = V_2 + I_B Z_B \tag{4.75}$$

If a load Z_L is connected across the secondary terminal, we can write,

$$V_2 = I Z_L = (I_A + I_B) Z_L \tag{4.76}$$

Substituting Eq. (4.76) in Eq. (4.75), we get,

$$E_B = (I_A + I_B) Z_L + I_B Z_B \tag{4.77}$$

Subtracting Eq. (4.75) from Eq. (4.74), we get,

$$E_A - E_B = I_A Z_A - I_B Z_B$$

$$\text{or } I_A = \frac{E_A - E_B + I_B Z_B}{Z_A} \tag{4.78}$$

Substituting Eq. (4.78) in Eq. (4.77), we get,

$$E_B = I_B Z_B + I_B Z_L + \left(\frac{E_A - E_B + I_B Z_B}{Z_A}\right) Z_L$$

Solving we get,

$$I_B = \frac{E_B Z_A - (E_A - E_B) Z_L}{Z_A Z_B + Z_L (Z_A + Z_B)} \tag{4.79}$$

Similarly we get,

$$I_A = \frac{E_A Z_B - (E_A - E_B) Z_L}{Z_A Z_B + Z_L (Z_A + Z_B)} \tag{4.80}$$

4.25 AUTOTRANSFORMER

In a two-winding transformer, there is no electrical connection between the two windings. They are magnetically coupled.

Change in voltage level can also be done effectively using a single coil. This is the concept behind an autotransformer.

4.25.1 Construction of Autotransformer

An autotransformer has only one winding wound on a laminated magnetic core. This single-winding serves as primary and secondary winding, both. The step-down and step-up autotransformer is shown in Fig. 4.38 and Fig. 4.39 respectively.

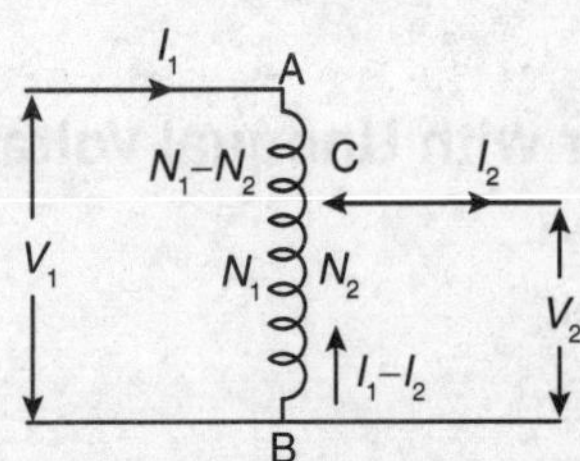

Fig. 4.38: Step-down autotransformer

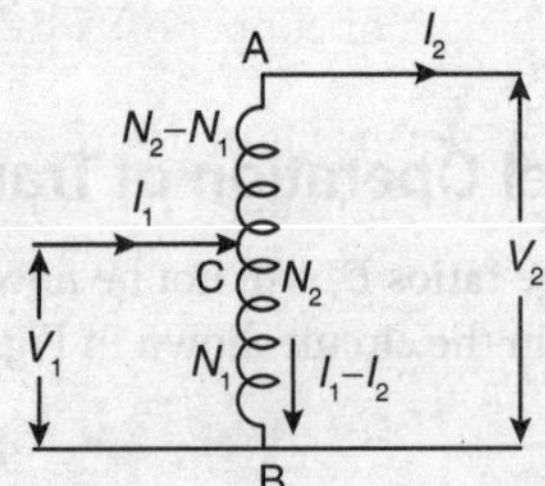

Fig. 4.39: Step-up autotransformer

As shown in Figs. 4.38 and 4.39, a part winding is common to both primary and secondary.

From the circuit shown in Fig. 4.38, we can say,

Primary turns = N_1

Secondary turns = N_2

Here $N_1 > N_2$ so $V_1 > V_2$, so it is a step-down transformer.

Similarly, from the circuit shown in Fig. 4.39, we can say,

Primary turns = N_1

Secondary turns = N_2

Here $N_1 < N_2$ so $V_1 < V_2$, so it is a step-up transformer.

The transformation ratio of an autotransformer is same as that of two-winding transformer (Eq. 4.8).

4.25.2 Copper Saving in an Autotransformer as Compared to Two-winding Transformer

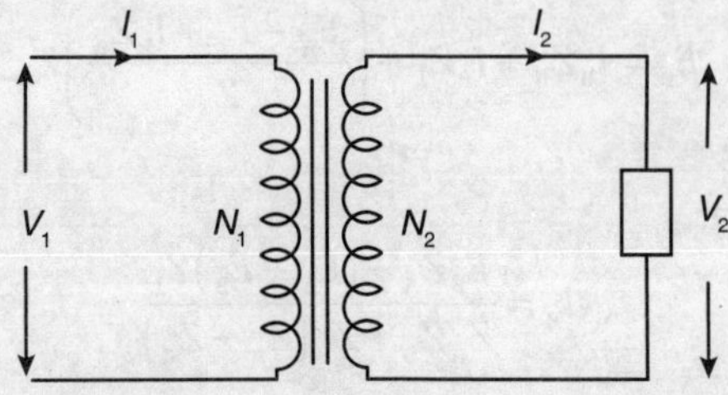

Fig. 4.40: Two-winding transformer

Consider a two-winding transformer as shown in Fig. 4.40, where V_1, I_1 and N_1 are the primary supply voltage, primary current and primary number of turns respectively. V_2, I_2 and N_2 are the secondary voltage, secondary current and secondary number of turns respectively.

For any transformer the weight of copper is proportional to the product of the number of turns (N) and current (I) through the winding.

Thus, for a two-winding transformer, shown in Fig. 4.40,

$$\text{Weight of copper of primary side} \propto N_1 I_1$$

$$\text{Weight of copper of secondary side} \propto N_2 I_2$$

$$\text{Total weight of copper of two-winding transformer } (W_{2T}) \propto N_1 I_1 + N_2 I_2 \tag{4.81}$$

Consider the step-down autotransformer shown in Fig. 4.38.
Section AB is having N_1 number of turns.
Section BC is having N_2 number of turns.
Section AC is having $N_1 - N_2$ number of turns.

$$\text{Weight of copper of section } AC \propto (N_1 - N_2) I_1$$

$$\text{Weight of copper of section } BC \propto N_2 (I_2 - I_1)$$

$$\text{Total weight of copper of autotransformer } (W_{aut}) \propto (N_1 - N_2) I_1 + N_2 (I_2 - I_1) \tag{4.82}$$

Dividing Eq. (4.82) by Eq. (4.81), we get,

$$\frac{W_{aut}}{W_{2T}} = \frac{(N_1 - N_2) I_1 + N_2 (I_2 - I_1)}{N_1 I_1 + N_2 I_2}$$

$$= \frac{N_1 I_1 - N_2 I_1 + N_2 I_2 - N_2 I_1}{N_1 I_1 + N_2 I_2}$$

$$= \frac{N_1 I_1 - 2N_2 I_1 + N_2 I_2}{N_1 I_1 + N_2 I_2}$$

$$= 1 - \frac{2N_2 I_1}{N_1 I_1 + N_2 I_2}$$

$$= 1 - \frac{2\left(\dfrac{N_2 I_1}{N_2 I_1}\right)}{\dfrac{N_1 I_1}{N_2 I_1} + \dfrac{N_2 I_2}{N_2 I_1}} = 1 - \frac{2}{\dfrac{N_1}{N_2} + \dfrac{I_2}{I_1}}$$

$$= 1 - \frac{2}{\dfrac{1}{K} + \dfrac{1}{K}} = 1 - K$$

$$\text{or } W_{aut} = (1-K)W_{2T} \tag{4.83}$$

$$\text{Saving of copper} = W_{2T} - W_{aut} = W_{2T} - (1-K)W_{2T} = KW_{2T} \tag{4.84}$$

For the step-up transformer shown in Fig. 4.39, we can say,

$$\text{Weight of copper of section } AC \propto (N_2 - N_1)I_2$$

$$\text{Weight of copper of section } BC \propto N_1(I_1 - I_2)$$

$$\text{Total weight of copper of autotransformer } (W_{aut}) \propto (N_2 - N_1)I_2 + N_1(I_1 - I_2) \tag{4.85}$$

Dividing Eq. (4.85) by Eq. (4.81), we get,

$$\frac{W_{aut}}{W_{2T}} = \frac{(N_2 - N_1)I_2 + N_1(I_1 - I_2)}{N_1 I_1 + N_2 I_2}$$

$$= \frac{N_2 I_2 - N_1 I_2 + N_1 I_1 - N_1 I_2}{N_1 I_1 + N_2 I_2}$$

$$= \frac{N_1 I_1 - 2N_1 I_2 + N_2 I_2}{N_1 I_1 + N_2 I_2}$$

$$= 1 - \frac{2N_1 I_2}{N_1 I_1 + N_2 I_2}$$

$$= 1 - \frac{2\left(\frac{N_1 I_2}{N_1 I_2}\right)}{\frac{N_1 I_1}{N_1 I_2} + \frac{N_2 I_2}{N_1 I_2}} = 1 - \frac{2}{\frac{N_2}{N_1} + \frac{I_1}{I_2}}$$

$$= 1 - \frac{2}{K + K} = 1 - \frac{1}{K}$$

$$\text{or } W_{\text{aut}} = \left(1 - \frac{1}{K}\right)W_{2\text{T}} \tag{4.86}$$

$$\text{Saving of copper} = W_{2\text{T}} - W_{\text{aut}} = W_{2\text{T}} - \left(1 - \frac{1}{K}\right)W_{2\text{T}} = \frac{1}{K}W_{2\text{T}} \tag{4.87}$$

4.25.3 Advantages of Autotransformer

Some of the advantages of autotransformers are

- Its efficiency is higher than two-winding transformer.
- It is smaller in size
- It is available at low cost
- It has a better voltage regulation.

4.25.4 Disadvantages of Autotransformer

Some of the disadvantages of an autotransformer are

- Low-voltage side is subjected to high-voltage stress and should be insulated from high voltage.
- Under fault condition, severe short circuit current flows through the windings.
- It is not possible to earth autotransformer on one side. Both sides are either earthed or insulated.

4.25.5 Application of Autotransformer

Some of the applications of autotransformer are

- It can be used as a starter to start electrical machines
- It can be used as a booster to give boost to distribution cables
- It can be used as a variac.

4.26 CONVERTING TWO-WINDING TRANSFORMER TO AUTOTRANSFORMER

Consider a two-winding transformer as shown in Fig. 4.41, with the polarity indicated by the dot.

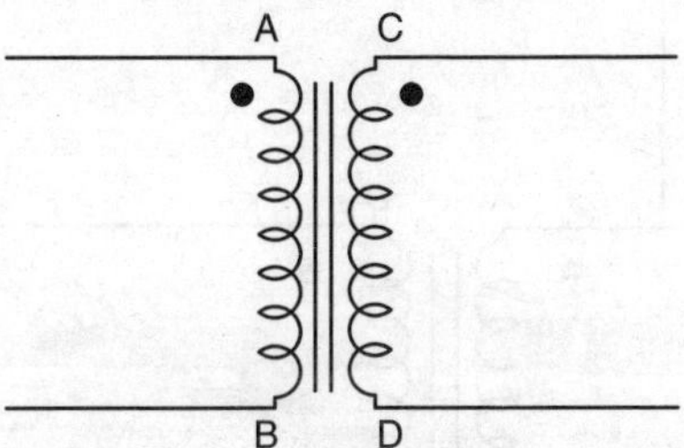

Fig. 4.41: Two-winding transformer

To convert this two-winding transformer to a step-up autotransformer, the primary and secondary windings are connected in series with additive polarity, as shown in Fig. 4.42.

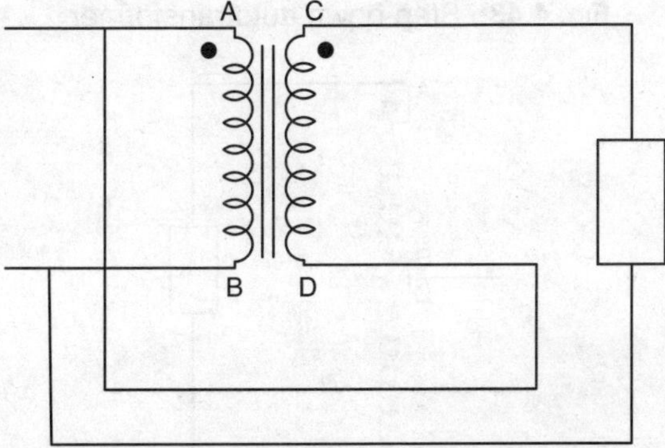

Fig. 4.42: Step-up autotransformer

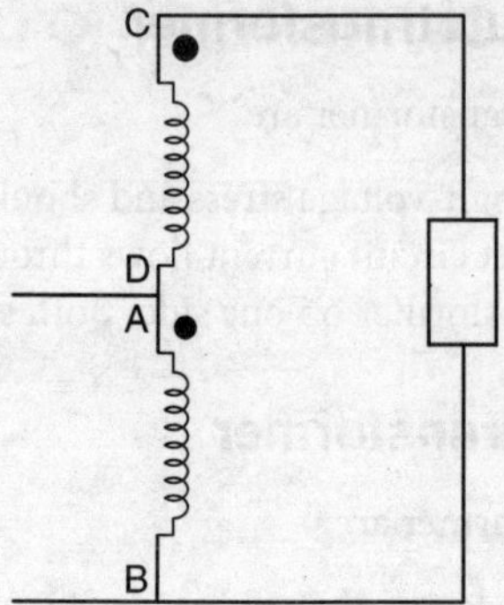

Fig. 4.42 (a): Equivalent circuit of step-up autotransformer shown in Fig. 4.42

The transformer circuit of Fig. 4.42 can be redrawn as shown in Fig. 4.42(a).

The connection diagram of a step-down autotransformer is shown in Fig. 4.43 and its equivalent in Fig. 4.43 (a).

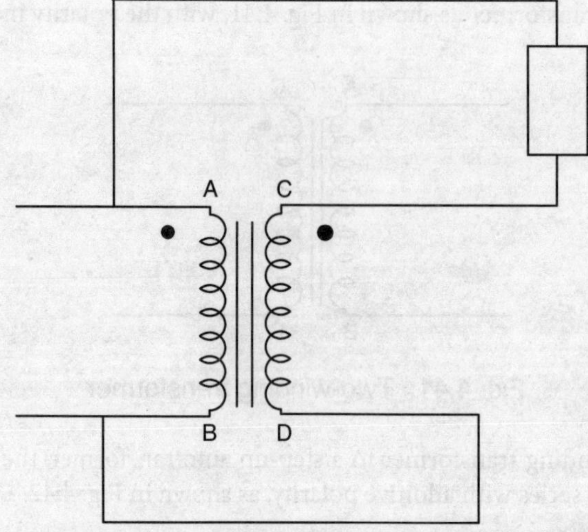

Fig. 4.43: Step-down autotransformer

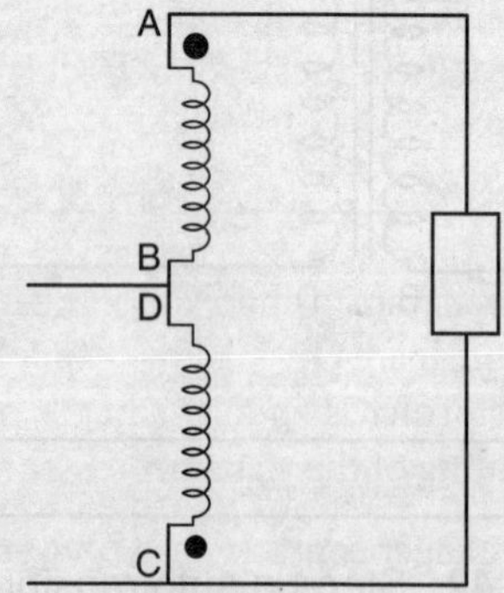

Fig. 4.43 (a): Equivalent circuit of step-down autotransformer shown in Fig. 4.43

4.27 VOLTAGE RATIO AND CURRENT RATIO OF AUTOTRANSFORMER

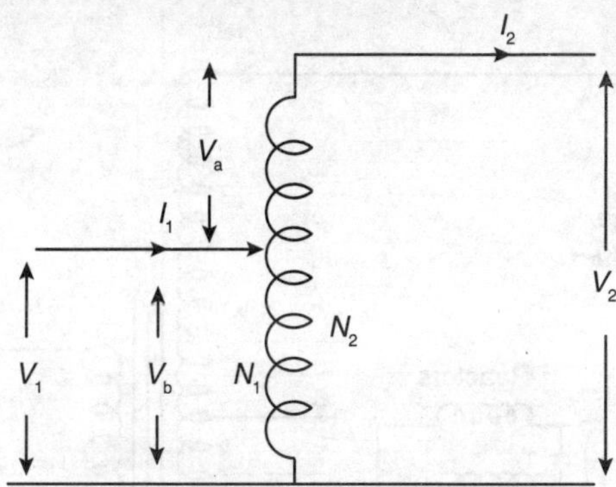

Fig. 4.44: Autotransformer

Consider the autotransformer shown in Fig. 4.44. The ratio of output over input is given by,

$$\frac{V_2}{V_1} = \frac{V_a + V_b}{V_b} = \frac{V_b + \frac{N_2}{N_1} V_b}{V_b} = 1 + K \tag{4.88}$$

$$\frac{I_2}{I_1} = \frac{I_2}{I_1 + I_2} = \frac{I_2}{I_2 + \frac{N_2}{N_1} I_2} = \frac{1}{1+K} \tag{4.89}$$

Comparing Eq. (4.88) and Eq. (4.89), we get,

$$\frac{V_2}{V_1} = \frac{I_1}{I_2} \tag{4.90}$$

4.28 DIMMERSTATS

An autotransformer can be used as a dimmerstat by changing the position of the variable contact. When the position of the variable contact is changed, it alters the output voltage of the autotransformer. This property is used in dimmerstat to control the intensity of light. It is mainly used in cinema halls.

4.29 TAP CHANGER

Tap changers are generally fitted to power transformers for regulation of the output voltage to required levels. Tap changers can be on-load or off-load. On-load tap changers, also called the circuit tap changers, allows tap changing when the transformer is on load (that is, on operation). On-load tap changing option is given on the high-voltage winding of the transformer. This is because,

- On the high-voltage side, the tap changer contacts, leads, etc can smaller because of low magnitude of current.

- Generally, the high-voltage winding is present outside the low-voltage winding. Thus, it is easy to get the tapping connections out to the tap changer.

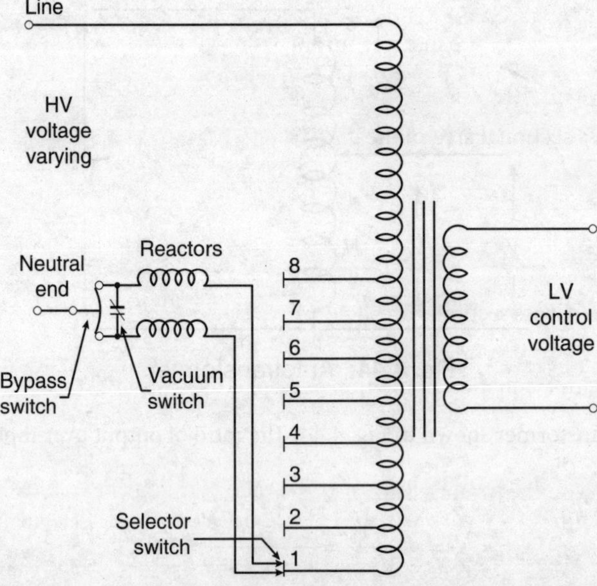

Fig. 4.45: Tap changer

The connection diagram of an on-load tap change is shown in Fig. 4.45. It has mainly four parts

- Selector Switches: It is used to select the physical tap position on the transformer winding.
- Reactors: During each tap change, there is an interval where two voltage taps are spanned. Reactors are used in the circuit to increase the impedance of the selector circuit which limits the circulating current due to this voltage difference. Under normal load conditions, equal load current flows in both halves of the reactor windings and the fluxes balance out giving no resultant flux in the core.
- Vacuum switch: It makes and breaks the current during the tap changing sequence.
- Bypass switch: It operates during the tap changing sequence but, at no time does it make or break load current, though it does make before break each connection.

SOLVED PROBLEMS

Ex.4.1 Consider a 100 kVA, 33/11 kV, 50 Hz transformer. If the e.m.f. per turn is 5V and the maximum flux density is 2T, calculate the following:

(i) Cross sectional area of the core

(ii) Number of turns on the primary and secondary sides

Solution

Given, KVA rating = 100 kVA, $E_1 = V_1 = 33$kV, $E_2 = V_2 = 11$kV, f = 50Hz, e.m.f. per turn = 5V, $B_m = 2$T.

We know that,

$$\text{e.m.f. per turn} = 4.44\varphi_m f$$

$$\text{or, } \varphi_m = \frac{\text{e.m.f. per turn}}{4.44 \times f} = \frac{5}{4.44 \times 50} = 0.225 \text{ Wb}$$

Let A be the cross sectional area of the core. Then, from the equation,

$$\varphi_m = B_m \times A$$

We can write,

$$\text{Cross sectional area } (A) = \frac{\varphi_m}{B_m} = \frac{0.225}{2} = 0.1125 \text{ m}^2$$

Again from the e.m.f. equation of the transformer,

$$E_1 = 4.44 f \varphi_m N_1$$

$$\text{Number of turns on the primary side of the transformer } (N_1) = \frac{E_1}{4.44 f \varphi_m} = \frac{33 \times 10^3}{4.44 \times 50 \times 0.225}$$

$$= 660 \text{ turns}$$

Similarly,

$$\text{Number of turns on the secondary side of the transformer } (N_2) = \frac{E_2}{4.44 f \varphi_m} = \frac{11 \times 10^3}{4.44 \times 50 \times 0.225}$$

$$= 220 \text{ turns}$$

Ex.4.2 For the transformer of example, **Ex.4.1**, calculate the value of primary and secondary currents.

Solution

$$\text{Primary current } (I_1) = \frac{\text{KVA rating}}{V_1} = \frac{100 \times 10^3}{33 \times 10^3} = 3.03 \text{ A}$$

$$\text{Secondary current } (I_2) = \frac{\text{KVA rating}}{V_2} = \frac{100 \times 10^3}{11 \times 10^3} = 9.09 \text{ A}$$

Ex.4.3 The core dimension of a 10 kVA, 230/440V, 50Hz transformer is 30 cm × 30 cm. If the maximum flux density is 1T, calculate the following:

(i) Maximum value of flux in the core
(ii) Number of turns on the primary and secondary sides
(iii) Value of currents on primary and secondary sides.

Solution

Given, KVA rating = 10 kVA, $V_1 = E_1 = 230$V, $V_2 = E_2 = 440$V, f=50Hz, dimensions of the core = 30 cm × 30 cm, B_m = 1T.

$$\text{Cross sectional area of the core } (A) = 0.3 \times 0.3 = 0.09 \text{ m}^2$$

$$\text{Maximum value of flux in the core } (\varphi_m) = B_m \times A = 1 \times 0.09 = 0.09 \text{ Wb}$$

$$\text{Number of turns on the primary } (N_1) = \frac{E_1}{4.44 \times \varphi_m \times f} = \frac{230}{4.44 \times 50 \times 0.09} = 11.51$$

$$\text{Number of turns on the secondary } (N_2) = \frac{E_2}{4.44 \times \varphi_m \times f} = \frac{440}{4.44 \times 50 \times 0.09} = 22$$

$$\text{Current in the primary } (I_1) = \frac{\text{KVA rating}}{V_1} = \frac{10 \times 10^3}{230} = 43.48 \text{ A}$$

$$\text{Current in the secondary } (I_2) = \frac{\text{KVA rating}}{V_2} = \frac{10 \times 10^3}{440} = 22.73 \text{ A}$$

Ex.4.4. In a 400/230 V, 50 Hz, single-phase transformer, the no-load current is 3A at 0.5 power factor lag. The primary is having 200 turns. Calculate the following:

(i) Magnetising component of no-load current
(ii) Working component of no-load current
(iii) Core loss of transformer
(iv) Secondary turns
(v) Maximum value of flux in the core

Solution

Given, $V_1 = 400\text{V}$, $V_2 = 230\text{V}$, $f = 50\text{Hz}$, $I_0 = 3\text{A}$, $\cos\theta_0 = 0.5$, $N_1 = 200$.

$$\cos\theta_0 = 0.5$$

$$\theta_0 = \cos^{-1}(0.5) = 60$$

$$\text{Magnetisation component of current } (I_m) = I_0 \sin\theta_0 = 3 \times \sin 60 = 3 \times 0.866 = 2.59\text{A}$$

$$\text{Working component of current } (I_c) = I_0 \cos\theta_0 = 3 \times \sin 60 = 3 \times 0.5 = 1.5 \text{ A}$$

$$\text{Core loss of the transformer} = V_1 I_0 \cos\theta_0 = 400 \times 3 \times 0.5 = 600 \text{ W}$$

From Eq. (4.5), we can write,

$$\text{Secondary turns } (N_2) = \frac{V_2}{V_1} N_1 = \frac{230}{440} \times 200 = 105 \text{ turns}$$

From the e.m.f. equation of the transformer,

$$E_1 = 4.44 \times f \times \varphi_m \times N_1$$

$$\text{Maximum value of flux } (\varphi_m) = \frac{E_1}{4.44 \times f \times N_1} = \frac{440}{4.44 \times 50 \times 200} = 0.0099 \text{ Wb}$$

Ex.4.5 A single-phase transformer has a no-load current of 2.5 A at 0.5 power factor lag. Calculate the primary current and the power factor when a load connected across the secondary draws a current of 100 A at 0.8 power factor lag. The primary and secondary turns of the transformer are 500 and 1000 respectively.

Solution

Given, $N_1 = 500$, $N_2 = 1000$, $I_0 = 2.5A$, $\cos\theta_0 = 0.5$, $I_2 = 100A$, $\cos\theta_2 = 0.8$.

$$\theta_0 = \text{Cos}^{-1}0.5 = 60^0$$

$$\theta_2 = \text{Cos}^{-1}0.8 = 36.869^0$$

$$I_2' = I_2 \times \frac{N_2}{N_1} = 100 \times \frac{1000}{500} = 200 \text{ A}$$

$$I_m = I_0 \sin\theta_0 = 2.5 \times \sin 60 = 2.165 \text{ A}$$

$$I_c = I_0 \cos\theta_0 = 2.5 \times \cos 60 = 1.25 \text{ A}$$

$$\text{No-load current } (I_0) = I_c + jI_m = 1.25 + j2.165 \text{ A} = 2.499\angle 59.99 \text{ A}$$

$$I_2' = 200 \times \cos 36.869 + j200 \times \sin 36.869$$

$$= 200 \times 0.8 + j200 \times 0.599$$

$$= 160 + j119.8\, A = 199.88\angle 36.82 \text{ A}$$

$$\text{Primary current } (I_1) = I_0 + I_2' = 1.25 + j2.165 + 160 + j119.8$$

$$= 161.25 + j121.965\, A = 202.181\angle 37.10 \text{ A}$$

Ex.4.6 A 440/230 V, 50Hz, single-phase transformer has the following results:
No-load current: 1.8 A at 0.4 power factor lag
On-load current: 28A at 0.8 power factor lag.
Calculate the primary current and primary power factor and draw the complete phasor.

Solution

Given, $V_1 = 440$, $V_2 = 230$, $I_0 = 1.8A$, $\cos\theta_0 = 0.4$, $I_2 = 28A$, $\cos\theta_2 = 0.8$.

$$\theta_0 = Cos^{-1}0.4 = 66.42^0$$

$$\theta_2 = Cos^{-1}0.8 = 36.869^0$$

$$I_2' = I_2 \times \frac{V_2}{V_1} = 28 \times \frac{230}{440} = 14.636 \text{ A}$$

$$I_m = I_0 \sin\theta_0 = 1.8 \times \sin 66.42 = 1.6497 \text{ A}$$

$$I_c = I_0 \cos\theta_0 = 1.8 \times \cos 66.42 = 0.72 \text{ A}$$

$$\text{No-load current } (I_0) = I_c + jI_m = 0.72 + j1.6497 \text{ A} = 1.799\angle 66.42 \text{ A}$$

$$I_2' = 14.636 \times \cos 36.869 + j\,14.636 \times \sin 36.869$$

$$= 14.636 \times 0.8 + j14.636 \times 0.599$$

$$= 11.709 + j8.78 \text{ A} = 14.635\angle 36.82 \text{ A}$$

$$\text{Primary current } (I_1) = I_0 + I_2' = 0.72 + j1.6497 + 11.709 + j8.78$$

$$= 12.429 + j10.4297 \text{ A} = 16.225\angle 40 \text{ A}$$

The phasor diagram is shown in Fig. E4.6. Where, $\theta_1 = 40^0$

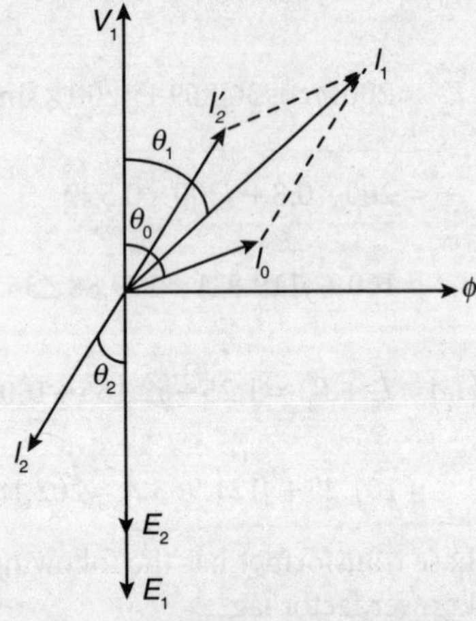

Fig. E4.6:

Ex.4.7 Consider a 100 KVA, 2000/200 V two-winding transformer as shown Fig. E4.7(a). It has been converted to an autotransformer by connecting the two windings in the manner as shown in Fig. E4.7(b). Calculate the following.

(i) HV and LV side voltages of the autotransformer
(ii) The KVA rating of the autotransformer.
(iii) The efficiency of the transformer at full load at 0.8 power factor lag if the loss in the transformer is 700 W

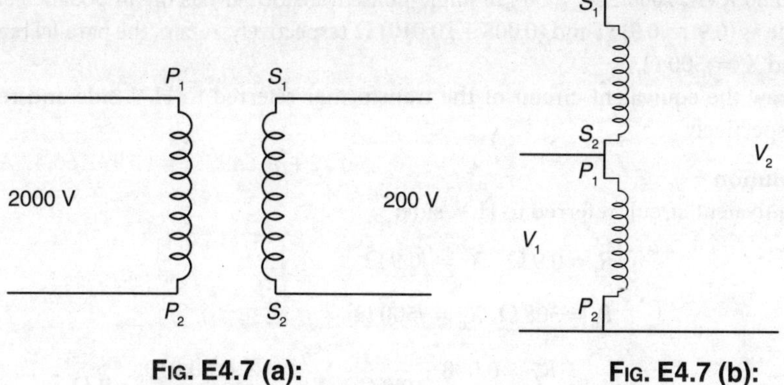

FIG. E4.7 (a): FIG. E4.7 (b):

Solution

For two-winding transformer:

$$I_p = \frac{100000}{2000} = 50 \text{ A } I_s = \frac{100000}{200} = 500 \text{ A}$$

These currents are indicated in the Fig.E4.7 (c) and Fig.E4.7 (d).

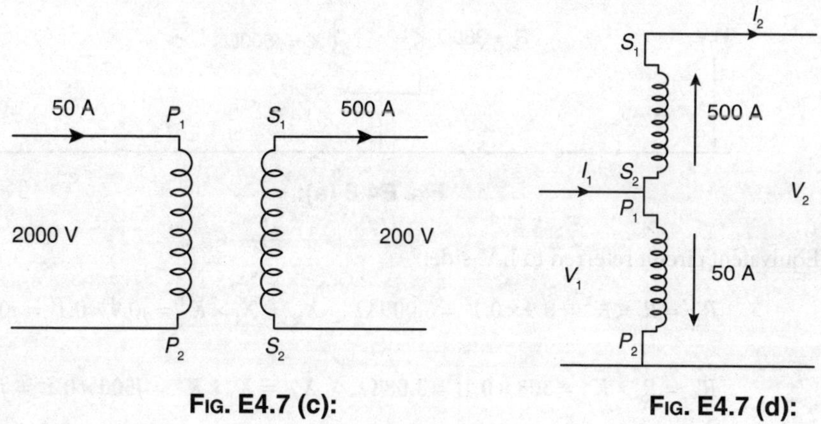

FIG. E4.7 (c): FIG. E4.7 (d):

For autotransformer:

$$V_1 = 2000 \text{ V } V_2 = 2000 + 200 = 2200 \text{ V}$$

$$I_1 = 500 + 50 = 550 \text{ A } I_2 = 500 \text{ A}$$

$$\text{KVA rating} = V_1 I_1 = 2000 \times 550 = 1100000 \text{ VA} = 1100 \text{ KVA}$$

Or

$$\text{KVA rating} = V_2 I_2 = 2200 \times 500 = 1100000 \text{ VA} = 1100 \text{ KVA}$$

$$\text{Efficiency } (\eta) = \frac{1100 \times 0.8}{1100 \times 0.8 + \left(\frac{700}{1000}\right)} = 99.9\ \%$$

Ex.4.8 An 80 KVA, 2000/200 V, 50 Hz single-phase transformer has the impedance of H.V. side and L.V. side as (0.9 + *j* 0.9) Ω and (0.008 + *j* 0.010) Ω respectively. Again, the parallel branch has $R_0 = 308\ \Omega$ and $X_0 = j500\ \Omega$.

Draw the equivalent circuit of the transformer referred to H.V. side and referred to L.V. side respectively.

Solution

Equivalent circuit referred to H.V. side:

$$R_1 = 0.9\ \Omega \quad X_1 = j0.9\ \Omega$$

$$R_0 = 308\ \Omega \quad X_0 = j500\ \Omega$$

$$R_2' = \frac{R_2}{K^2} = \frac{0.008}{0.1^2} = 0.8\ \Omega \quad X_2' = \frac{X_2}{K^2} = \frac{0.010}{0.1^2} = j1\ \Omega$$

The equivalent circuit referred to H.V. side is shown in Fig. E4.8 (a)

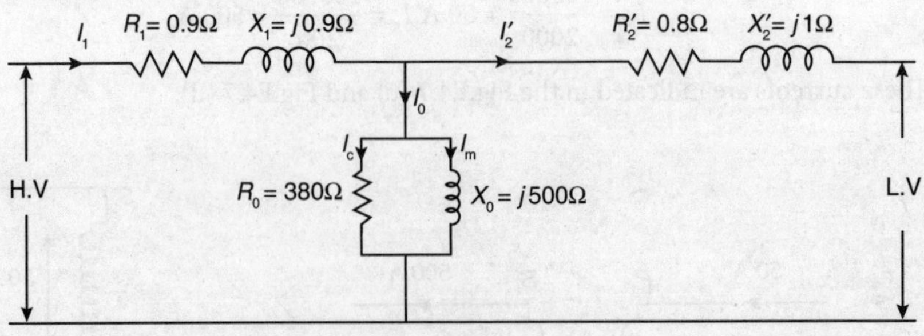

Fig. E4.8 (a):

Equivalent circuit referred to L.V. side:

$$R_1' = R_1 \times K^2 = 0.9 \times 0.1^2 = 0.009\ \Omega \quad X_1' = X_1 \times K^2 = j0.9 \times 0.1^2 = j0.009\ \Omega$$

$$R_0' = R_0 \times K^2 = 308 \times 0.1^2 = 3.08\ \Omega \quad X_0' = X_0 \times K^2 = j500 \times 0.1^2 = j5\ \Omega$$

$$R_2 = 0.008\ \Omega \quad X_2 = j0.010\ \Omega$$

The equivalent circuit referred to L.V. side is shown in Fig. E4.8 (b)

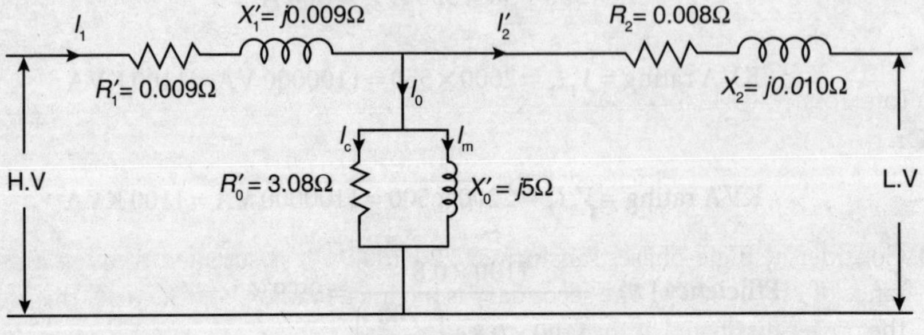

Fig. E4.8 (b):

Ex.4.9 The open-circuit (O.C.) and short-circuit (S.C.) tests conducted on a 230/115 V, 1000VA transformer gave the following readings:

O.C. Test:	230 V	0.45 A	30 W
S.C. Test:	19.1 V	8.7 A	42.3 W

Find the parameters of the equivalent circuit and the efficiency of the transformer at 0.8 power factor and full load.

Solution

From O.C. test:

$$\cos\theta_0 = \frac{30}{230\times0.45} = 0.289$$

$$\sin\theta_0 = \sin\left(cos^{-1}(0.289)\right) = 0.957$$

$$I_c = 0.45\times0.289 = 0.13 \text{ A}$$

$$I_m = 0.45\times0.957 = 0.431 \text{ A}$$

$$R_0 = \frac{230}{0.13} = 1769.23\ \Omega$$

$$X_0 = \frac{230}{0.431} = 533.64\ \Omega$$

From S.C. test:

$$\cos\theta_s = \frac{42.3}{19.1\times8.7} = 0.255$$

$$Z_e = \frac{19.1}{8.7} = 2.195\ \Omega$$

$$R_e = \frac{42.3}{8.7^2} = 0.559\ \Omega$$

$$X_e = \sqrt{Z_e^2 - R_e^2} = \sqrt{2.195^2 - 0.559^2} = 2.122\ \Omega$$

Total loss = 30+42.3=72.3 W

$$\text{Efficiency}\left(\eta\right) = \frac{1000\times0.8}{1000\times0.8+72.3}\times100 = 91.7\ \%$$

Ex.4.10 Consider a single-phase transformer 220/1000V. It is supplied through a cable (having resistance of 2Ω) at 220 V. The secondary is having a load of 3Ω. Calculate the primary current and the power dissipated at the load.

Solution

$$K = \frac{V_2}{V_1} = \frac{1000}{220} = 4.55$$

$$R_1' = 3 \times 4.55^2 = 62.11$$

$$R_{pri} = 2 + 62.11\,\Omega = 64.11\,\Omega$$

$$\text{Primary current}\,(I_1) = \frac{220}{64.11} = 3.43\text{ A}$$

$$\text{Secondary current}\,(I_2) = \left(\frac{V_1}{V_2}\right) I_1 = \left(\frac{220}{1000}\right) \times 3.43 = 0.755\text{ A}$$

$$\text{Power dissipated} = 3 \times 0.755^2 = 1.71\text{ W}$$

Ex.4.11 A 15 KVA, 2200/500 V transformer has $R_1 = 6\Omega$, $R_2 = 0.4\Omega$, $X_1 = 15\Omega$, $X_2 = 0.5\Omega$. Calculate the secondary terminal voltage when supplying 7A at 0.8 power factor lag.

Solution

$$K = \frac{V_2}{V_1} = \frac{500}{2200} = 0.227$$

$$R_1' = R_1 K^2 = 6 \times 0.227^2 = 0.309\,\Omega$$

$$X_1' = X_1 K^2 = 15 \times 0.227^2 = 0.773\,\Omega$$

$$R_{eq} = R_2 + R_1' = 0.4 + 0.309 = 0.709\,\Omega$$

$$X_{eq} = X_2 + X_1' = 0.5 + 0.773 = 1.273\,\Omega$$

$$\text{Drop} = I_2\left(R_{eq}\cos\theta + X_{eq}\sin\theta\right)$$

$$= 7\left(0.709 \times 0.8 + 1.273 \times 0.6\right) = 9.317\text{ V}$$

$$\text{Secondary terminal voltage} = 500 - 9.317 = 490.68\text{ V}$$

Ex.4.12 The open-circuit and short-circuit test result of a 5KVA, 500/250 V, 50Hz, single-phase transformer is as follows:

O.C. test:	500 V	1 A	50 W
S.C. test:	25 V	10 A	60 W

Calculate the following:

(i) Parameters of the equivalent circuit of the transformer.
(ii) Efficiency at full load 0.8 power factor lag.
(iii) Efficiency at 60% of full load at 0.8 power factor lag.
(iv) Voltage regulation at 0.8 power factor lead at full load.

Solution

From O.C. test:

$$\cos\theta_0 = \frac{50}{500\times 1} = 0.1$$

$$\text{Sin}\theta_0 = \sin\left(\cos^{-1}(0.289)\right) = 0.957$$

$$I_c = 1\times 0.1 = 0.1 \text{ A}$$

$$I_m = 1\times 0.995 = 0.995 \text{ A}$$

$$R_0 = \frac{500}{0.1} = 5000\ \Omega$$

$$X_0 = \frac{500}{0.995} = 502.51\ \Omega$$

From S.C. test:

$$Z_e = \frac{25}{10} = 2.5\ \Omega$$

$$R_e = \frac{60}{10^2} = 0.6\ \Omega$$

$$X_e = \sqrt{Z_e^2 - R_e^2} = \sqrt{2.5^2 - 0.6^2} = 2.43\ \Omega$$

Total loss = 50 + 60 = 110 W

$$\text{Efficiency at full load } (\eta) = \frac{5000\times 0.8}{5000\times 0.8 + 110}\times 100 = 97.32\ \%$$

$$\text{Efficiency at 60 \% full load } (\eta) = \frac{5000\times 0.8\times 0.6}{5000\times 0.8\times 0.6 + 50 + 0.6^2\times 60}\times 100 = 97.10\ \%$$

$$\%\text{ voltage regulation} = \frac{10\times 0.6\times 0.8 - 10\times 2.43\times 0.6}{500}\times 100 = -1.95\ \%$$

Ex.4.13 The open-circuit and short-circuit test result of a 11KVA, 500/250 V, 50Hz, single-phase transformer is as follows:

O.C. test:	250 V	3 A	200 W
S.C. test (L.V. side):	15 V	30 A	300 W

Calculate the parameters of the equivalent circuit of the transformer, efficiency and voltage regulation at full-load at 0.8 power factor lag.

Solution

From O.C. test:

$$\cos\theta_0 = \frac{200}{250\times 3} = 0.267$$

$$\sin\theta_0 = \sin\left(cos^{-1}(0.267)\right) = 0.964$$

$$I_c = 3\times 0.267 = 0.8 \text{ A}$$

$$I_m = 3\times 0.964 = 2.89 \text{ A}$$

$$R_0 = \frac{250}{0.8} = 312.5\,\Omega$$

$$X_0 = \frac{250}{2.89} = 86.51\,\Omega$$

From S.C. test:

$$Z_e = \frac{15}{30} = 0.5\,\Omega$$

$$R_e = \frac{300}{30^2} = 0.333\,\Omega$$

$$X_e = \sqrt{Z_e^2 - R_e^2} = \sqrt{0.5^2 - 0.333^2} = 0.373\,\Omega$$

Full-load current on L.V. side is

$$I_{fl} = \frac{11\times 10^3}{250} = 44 \text{ A}$$

This shows that the short-circuit test has not been conducted at rated current, so the watt meter reading will not give the total copper loss.

$$\text{Total copper loss} = \left(\frac{44}{30}\right)^2 \times 300 = 645.33 \text{ W}$$

Total loss 645.33 + 200 = 845.33 W

$$\text{Efficiency at full load}\left(\eta\right)=\frac{11\times10^3\times0.8}{11\times10^3\times0.8+845.33}\times100=91.23\,\%$$

$$\%\text{ voltage regulation}=\frac{40\times0.333\times0.8-40\times0.373\times0.6}{250}\times100=0.68\,\%$$

Ex.4.14 The iron loss and the copper loss of a 600 KVA distribution transformer are 4KW and 6KW respectively. The 24-hour loading of the transformer is as follows:

Number of Hours	Load (kW)	Power Factor
7	410	0.8
9	298	0.79
5	150	0.75
3	50	0.8

Calculate the all-day efficiency.

Solution

Number of Hours	Load (kW)	Power Factor	Load (KVA)
7	410	0.8	= 410/0.8 = 512.5
9	298	0.79	= 298/0.79 = 377.22
5	150	0.75	= 150/0.75 = 200
3	50	0.8	= 50/0.8 = 62.5

Copper loss at 600 KVA = 6 KW

$$\text{Copper loss at 512.5 KVA}=6\times\left(\frac{512.5}{600}\right)^2=4.38\text{ KW}$$

$$\text{Copper loss at 377.22 KVA}=6\times\left(\frac{377.22}{600}\right)^2=2.37\text{ KW}$$

$$\text{Copper loss at 200 KVA}=6\times\left(\frac{200}{600}\right)^2=0.667\text{ KW}$$

$$\text{Copper loss at 62.5 KVA}=6\times\left(\frac{62.5}{600}\right)^2=0.065\text{ KW}$$

$$\text{Total copper loss}=4.38\times7+2.37\times9+0.667\times5+0.065\times3=55.52\text{ KWh}$$

$$\text{Total iron loss}=24\times4=96\text{ KWh}$$

$$\text{Output of the transformer} = 410\times7+298\times9+150\times5+50\times3 = 6452 \text{ KWh}$$

$$\text{All-day efficiency } \left(\eta_{\text{all}_{\text{day}}}\right) = \frac{6452}{6452+55.52+96}\times100 = 97.71\,\%$$

Ex.4.15 A transformer with 500 turns on the primary takes a current of 5A and power 95 W, when connected across 200V, 50Hz supply. If the primary resistance is 0.45 Ω, calculate the core loss, no-load power factor, and maximum value of flux.

Solution

$$\text{Core loss } (P_c) = 95-0.45\times5^2 = 83.75 \text{ W}$$

$$\text{Power factor } (\cos\theta_0) = \frac{95}{1.4\times200} = 0.34$$

$$\text{Maximum value of flux } (\varphi_m) = \frac{200}{4.44\times50\times500} = 0.0018 \text{ Wb}$$

Ex.4.16 Consider a 5 KVA, 440/220 V single-phase transformer. When a voltage of 35 V is applied to the H.V. side with L.V. side short-circuited, rated current flows through the windings, and the power input is 100 W. Find the resistance and reactance of both the windings under the condition that power flow and the ratio of reactance to resistance is same in both the windings.

Solution

$$I_{\text{L.V.}}(\text{Rated}) = \frac{5\times10^3}{220} = 22.73 \text{ A}$$

$$I_{\text{H.V.}}(\text{Rated}) = \frac{5\times10^3}{440} = 11.36 \text{ A}$$

$$R = \frac{100}{11.36^2} = 0.774\,\Omega$$

$$Z = \frac{35}{11.36} = 3.08\,\Omega$$

$$X = \sqrt{3.08^2-0.774^2} = 2.98\,\Omega$$

$$\frac{X}{R} = \frac{2.98}{0.774} = 3.85$$

Since the loss is equal in the L.V. and H.V. winding
H.V. side:

$$r = \frac{0.774}{2} = 0.387\,\Omega$$

$$x = 3.85 \times 0.387 = 1.49\ \Omega$$

L.V. side:

$$r = 0.387 \times \left(\frac{220}{440}\right)^2 = 0.097\ \Omega$$

$$x = 1.49 \times \left(\frac{220}{440}\right)^2 = 0.373\ \Omega$$

Ex.4.17 A 10 KVA transformer has 400-Watt iron losses and 600-Watt copper losses. Determine the load at which maximum efficiency takes place. Also determine maximum efficiency of the transformer at 0.8 power factor lag.

Solution

The load at which the maximum efficiency will occur $= \sqrt{\frac{400}{600}} \times 10 = 8.165$ KVA

At maximum efficiency,

$$\text{Iron loss} = \text{Copper loss}$$

$$\text{Total loss} = 400 + 400 = 800\ \text{W}$$

$$\text{Maximum efficiency} = \frac{\left(\sqrt{\frac{400}{600}}\right) \times 10 \times 0.8}{\left(\sqrt{\frac{400}{600}}\right) \times 10 \times 0.8 + 800 \times 10^{-3}} \times 100 = 89.1\%$$

Ex.4.18 In a 2KVA, 500/150V single-phase transformer, calculate the rated primary and secondary currents.

Solution

$$\text{Primary current } (I_1) = \frac{2 \times 10^3}{500} = 4\ \text{A}$$

$$\text{Secondary current } (I_2) = \frac{2 \times 10^3}{150} = 13.33\ \text{A}$$

Ex.4.19 A voltage of $V = 155.5 \sin 377t + 15.5 \sin 1131t$ Volts is applied to the primary of a single phase transformer with 200 turns on primary. Calculate the instantaneous and r.m.s. values of the core flux.

Solution

$$\varphi_{\text{inst}} = \frac{1}{200}\int (155.5 \sin 377t + 15.5 \sin 1131t)\,dt = -2.05 \cos 377t - 0.068 \cos 1131t\ \text{mWb}$$

$$\varphi_{\text{rms}} = \sqrt{\left(\frac{2.05}{\sqrt{2}}\right)^2 + \left(\frac{0.068}{\sqrt{2}}\right)^2} = 1.45\ \text{mWb}$$

Ex.4.20 Calculate the number of turns of primary and secondary side for an 11 KV/230 V, 10 KVA transformer having maximum flux density of 1T and e.m.f. per turn of 6V. Also calculate the currents flowing through both the windings.

Solution

Given, KVA rating = 10 kVA, $E_1 = V_1 = 11000$, $E_2 = V_2 = 230$, $f = 50\text{Hz}$, e.m.f. per turn = 6V, $B_m = 1\text{T}$.

We know that,

$$\text{e.m.f. per turn} = 4.44\varphi_m f$$

$$\text{or, } \varphi_m = \frac{\text{e.m.f. per turn}}{4.44 \times f} = \frac{6}{4.44 \times 50} = 0.027 \text{ Wb}$$

Let A be the cross sectional area of the core. Then, from the equation,

$$\varphi_m = B_m \times A$$

We can write,

$$\text{Cross sectional area } (A) = \frac{\varphi_m}{B_m} = \frac{0.027}{1} = 0.027 \text{ m}^2$$

Again from the e.m.f. equation of the transformer,

$$E_1 = 4.44 f \varphi_m N_1$$

$$\text{Number of turns on the primary side of the transformer } (N_1) = \frac{E_1}{4.44 f \varphi_m} = \frac{11000}{4.44 \times 50 \times 0.027}$$

$$= 1835 \text{ turns}$$

Similarly,

$$\text{Number of turns on the secondary side of the transformer } (N_2) = \frac{E_2}{4.44 f \varphi_m}$$

$$= \frac{230}{4.44 \times 50 \times 0.027} = 38 \text{ turns}$$

$$\text{Primary current } (I_1) = \frac{\text{KVA rating}}{V_1} = \frac{10 \times 10^3}{11 \times 10^3} = 0.909 \text{ A}$$

$$\text{Secondary current } (I_2) = \frac{\text{KVA rating}}{V_2} = \frac{10 \times 10^3}{230} = 43.47 \text{ A}$$

Ex.4.21 A 10 KVA, 33/11 KV, 50 Hz transformer has a core dimension of 25 cm × 20 cm. The maximum flux density is 3T. Calculate the number of turns of primary and secondary side and also the current flowing through them

Solution

$$\text{Cross-sectional area of the core } (A) = 0.25 \times 0.2 = 0.05 \text{ m}^2$$

$$\text{Maximum value of flux in the core } (\varphi_m) = B_m \times A = 3 \times 0.05 = 0.15 \text{ Wb}$$

$$\text{Number of turns on the primary } (N_1) = \frac{E_1}{4.44 \times \varphi_m \times f} = \frac{33000}{4.44 \times 50 \times 0.15} = 991$$

$$\text{Number of turns on the secondary } (N_2) = \frac{E_2}{4.44 \times \varphi_m \times f} = \frac{11000}{4.44 \times 50 \times 0.15} = 330$$

$$\text{Current in the primary } (I_1) = \frac{\text{KVA rating}}{V_1} = \frac{10 \times 10^3}{33000} = 0.303 \text{ A}$$

$$\text{Current in the secondary } (I_2) = \frac{\text{KVA rating}}{V_2} = \frac{10 \times 10^3}{11000} = 0.909 \text{ A}$$

Ex.4.22 In a 440/220 V, 50 Hz, single-phase transformer, the no-load current is 5A at 0.7 power factor lag. The primary is having 400 turns. Calculate the following:

(i) Magnetising component of no-load current
(ii) Working component of no-load current
(iii) Core loss of transformer

Solution

$$\cos\theta_0 = 0.7$$

$$\theta_0 = \text{Cos}^{-1}(0.7) = 45.57$$

$$\text{Magnetisation component of current } (I_m) = I_0 \sin\theta_0 = 5 \times \sin 45.57 = 3.57 \text{ A}$$

$$\text{Working component of current } (I_c) = I_0 \cos\theta_0 = 5 \times \sin 45.57 = 5 \times 0.7 = 3.5 \text{ A}$$

$$\text{Core loss of the transformer} = V_1 I_0 \cos\theta_0 = 440 \times 5 \times 0.7 = 1540 \text{ W}$$

Ex.4.23 A single-phase transformer has a no-load current of 3A at 0.7 power factor lag. Calculate the primary current and the power factor when a load connected across the secondary draws a current of 120A at 0.8 power factor lag. The primary and secondary turns of the transformer are 1000 and 1300 respectively.

Solution

$$\theta_0 = \cos^{-1} 0.7 = 45.57^0$$

$$\theta_2 = \cos^{-1} 0.8 = 36.869^0$$

$$I_2' = I_2 \times \frac{N_2}{N_1} = 120 \times \frac{1300}{1000} = 156 \text{ A}$$

$$I_m = I_0 \sin\theta_0 = 3\times\sin 45.57 = 2.14\ \text{A}$$

$$I_c = I_0 \cos\theta_0 = 3\times\cos 45.57 = 2.1\ \text{A}$$

$$\text{No-load current}\left(I_0\right) = I_c + jI_m = 2.1 + j2.14\ \text{A}$$

$$I_2' = 120\times\cos 36.869 + j120\times\sin 36.869$$

$$= 120\times 0.8 + j120\times 0.599$$

$$= 96 + j71.88\ \text{A}$$

$$\text{Primary current}\left(I_1\right) = I_0 + I_2' = 2.1 + j2.14 + 96 + j71.88$$

$$= 98.1 + j74.02\ \text{A}$$

Ex.4.24 A 66KV/11KV, 50Hz, single phase transformer no load current and power factor as 5A at 0.5 lag respectively and on load current and power factor as 45A at 0.8 lag respectively Calculate the primary current and primary power factor

Solution

$$\theta_0 = \cos^{-1} 0.5 = 60^0$$

$$\theta_2 = \cos^{-1} 0.8 = 36.869^0$$

$$I_2' = I_2\times\frac{V_2}{V_1} = 45\times\frac{11}{66} = 0.1667\ \text{A}$$

$$I_\text{m} = I_0 \sin\theta_0 = 0.1667\times\sin 60 = 0.144\ \text{A}$$

$$I_\text{c} = I_0 \cos\theta_0 = 0.1667\times\cos 60 = 0.083\ \text{A}$$

$$\text{No-load current}\left(I_0\right) = I_c + jI_\text{m} = 0.083 + j0.144\ \text{A}$$

$$I_2' = 0.1667\times\cos 36.869 + j\,0.1667\times\sin 36.869$$

$$= 0.133 + j0.1\ \text{A}$$

$$\text{Primary current}\left(I_1\right) = I_0 + I_2' = 0.083 + j0.144 + 0.133 + j0.1$$

$$= 0.216 + j10.244\ \text{A} = 0.325\angle 48.48\ \text{A}$$

Ex.4.25 Consider a 1000 KVA, 5000/2000 V two winding transformer as shown Fig. E4.25(a). It has been converted to an autotransformer by connecting the two windings in the manner as shown in Fig. E4.25(b). Calculate the following: the efficiency of the transformer at full load at 0.8 power factor lag if the loss in the transformer is 800 W

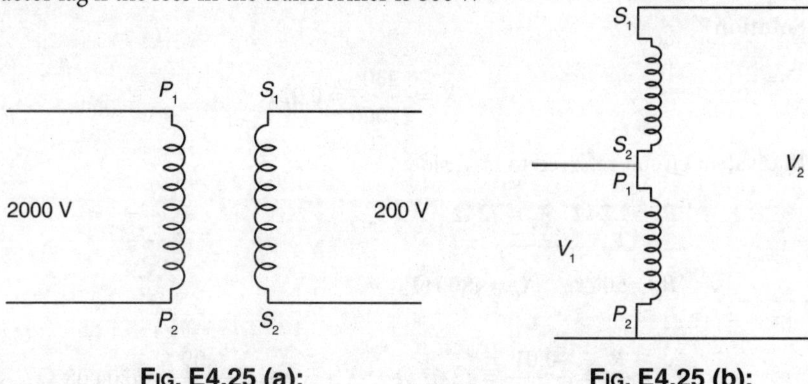

FIG. E4.25 (a): **FIG. E4.25 (b):**

Solution

For a two-winding transformer:

$$I_P = \frac{1000000}{5000} = 200\text{ A} \quad I_s = \frac{1000000}{2000} = 500\text{ A}$$

These currents are indicated in Fig.E4.7 (c) and Fig.E4.7 (d) respectively.

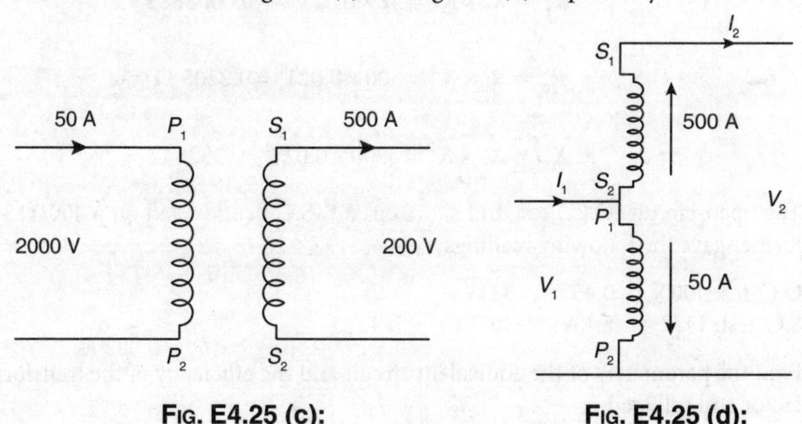

FIG. E4.25 (c): **FIG. E4.25 (d):**

For autotransformer:

$$V_1 = 5000\text{ V } V_2 = 5000 + 2000 = 7000\text{ V}$$

$$I_1 = 200 + 500 = 700\text{ A} \quad I_2 = 500\text{ A}$$

$$\text{KVA rating} = V_1 I_1 = 5000 \times 700 = 3.5\text{ MVA}$$

$$\text{Efficiency}\,(\eta) = \frac{3.5 \times 10^6 \times 0.8}{3.5 \times 10^6 \times 0.8 + 800} = 99.9\,\%$$

Ex.4.26 The high-voltage and low-voltage side impedance of a 60 KVA, 11Kv/230 V, 50 Hz single-phase transformer are (1.2+*j* 2) Ω and (0.01 + *j*0.09) Ω respectively. The parallel branch has $R_0 = 500\ \Omega$ and $X_0 = j800\ \Omega$. Obtain the parameters of equivalent circuit of the transformer referred to H.V. side and referred to L.V. side respectively.

Solution

$$K = \frac{230}{11000} = 0.021$$

Equivalent circuit referred to H.V. side:

$$R_1 = 1.2\ \Omega \quad X_1 = j2\ \Omega$$

$$R_0 = 500\ \Omega \quad X_0 = j800\ \Omega$$

$$R_2' = \frac{R_2}{K^2} = \frac{0.01}{0.021^2} = 22.67\ \Omega \qquad X_2' = \frac{X_2}{K^2} = \frac{0.09}{0.021^2} = j204.08\ \Omega$$

The equivalent circuit referred to H.V. side is shown in Fig. E4.8 (a)
Equivalent circuit referred to L.V. side:

$$R_1' = R_1 \times K^2 = 1.2 \times 0.021^2 = 0.00053\ \Omega$$

$$X_1' = X_1 \times K^2 = j2 \times 0.021^2 = j0.00088\ \Omega$$

$$R_0' = R_0 \times K^2 = 500 \times 0.021^2 = 0.2205\ \Omega$$

$$X_0' = X_0 \times K^2 = j800 \times 0.021^2 = j3528\ \Omega$$

Ex.4.27 The open-circuit (O.C.) test and short-circuit (S.C.) conducted on a 400/115 V, 100VA transformer gave the following readings:

O.C. test:400 V	0.45 A	32 W
S.C. test: 11 V	8.1 A	40 W

Find the parameters of the equivalent circuit and the efficiency of the transformer at 0.8 power factor and full load.

Solution
From O.C. test:

$$\cos\theta_0 = \frac{32}{400 \times 0.45} = 0.1778$$

$$\sin\theta_0 = \sin\left(\cos^{-1}(0.1778)\right) = 0.984$$

$$I_c = 0.45 \times 0.1778 = 0.08\ \text{A}$$

$$I_m = 0.45 \times 0.984 = 0.443 \text{ A}$$

$$R_0 = \frac{400}{0.08} = 5000\,\Omega$$

$$X_0 = \frac{400}{0.443} = 9021.9\,\Omega$$

From S.C. test:

$$\cos\theta_s = \frac{40}{11 \times 8.1} = 0.448$$

$$Z_e = \frac{11}{8.1} = 1.36\,\Omega$$

$$R_e = \frac{40}{8.1^2} = 0.609\,\Omega$$

$$X_e = \sqrt{Z_e^2 - R_e^2} = \sqrt{1.36^2 - 0.609^2} = 1.216\,\Omega$$

Total loss = 32 + 40 = 72 W

$$\text{Efficiency}\,(\eta) = \frac{100 \times 0.8}{100 \times 0.8 + 72} \times 100 = 82.23\,\%$$

Ex.4.28 Calculate the primary current and the power dissipated in a 440/230 V, 50Hz single-phase transformer connected across a 440 V supply, with secondary having a load of 4Ω

Solution

$$K = \frac{V_2}{V_1} = \frac{230}{440} = 0.523$$

$$R_1' = 4 \times 0.523^2 = 1.09$$

$$R_{\text{pri}} = 1.09\,\Omega$$

$$\text{Primary current}\,(I_1) = \frac{440}{1.09} = 403.67 \text{ A}$$

$$\text{Secondary current}\,(I_2) = \left(\frac{V_1}{V_2}\right) I_1 = \left(\frac{440}{230}\right) \times 403.67 = 772.24 \text{ A}$$

$$\text{Power dissipated} = 4 \times 772.24^2 = 2.38 \text{ MW}$$

Ex.4.29 A 1100/440 V transformer has $R_1 = 5\Omega, R_2 = 0.3\Omega, X_1 = 10\Omega, X_2 = 0.3\Omega$. Calculate the secondary terminal voltage when supplying 10A at 0.8 power factor lag.

Solution

$$K = \frac{V_2}{V_1} = \frac{440}{1100} = 0.4$$

$$R_1' = R_1 K^2 = 5 \times 0.4^2 = 0.8\,\Omega$$

$$X_1' = X_1 K^2 = 10 \times 0.4^2 = 1.6\,\Omega$$

$$R_{eq} = R_2 + R_1' = 0.3 + 0.8 = 1.1\,\Omega$$

$$X_{eq} = X_2 + X_1' = 0.3 + 1.6 = 1.9\,\Omega$$

$$\text{Drop} = I_2\left(R_{eq}\cos\theta + X_{eq}\sin\theta\right)$$

$$= 10(1.1 \times 0.8 + 1.9 \times 0.6) = 20.2\text{ V}$$

$$\text{Secondary terminal voltage} = 440 - 20.2 = 419.8\text{ V}$$

Ex.4.30 The open-circuit and short-circuit test result of a 10 KVA, 600/300V, 50Hz, single-phase transformer is as follows:

O.C. test: 600 V 1 A 50 W
S.C. test: 30 V 16.67 A 60 W

Calculate the following:

(i) Parameters of the equivalent circuit of the transformer.
(ii) Efficiency at full load at 0.8 power factor lag.
(iii) Efficiency at 60% of full load at 0.8 power factor lag.

Solution
From O.C. test:

$$\cos\theta_0 = \frac{50}{600 \times 1} = 0.08$$

$$\sin\theta_0 = \sin\left(cos^{-1}(0.08)\right) = 0.996$$

$$I_c = 1 \times 0.08 = 0.08\text{ A}$$

$$I_m = 1 \times 0.996 = 0.996\text{ A}$$

$$R_0 = \frac{600}{0.08} = 7500\ \Omega$$

$$X_0 = \frac{600}{0.996} = 602.41\ \Omega$$

From S.C. test:

$$Z_e = \frac{30}{16.67} = 1.799\ \Omega$$

$$R_e = \frac{60}{16.67^2} = 0.215\ \Omega$$

$$X_e = \sqrt{Z_e^2 - R_e^2} = \sqrt{1.799^2 - 0.215^2} = 1.786\ \Omega$$

Total loss = 50 + 60 = 110 W

$$\text{Efficiency at full load } (\eta) = \frac{10000 \times 0.8}{10000 \times 0.8 + 110} \times 100 = 98.64\ \%$$

$$\text{Efficiency at 80 \% full load } (\eta) = \frac{10000 \times 0.8 \times 0.8}{10000 \times 0.8 \times 0.8 + 50 + 0.8^2 \times 60} \times 100 = 98.63\ \%$$

EXERCISES

1. What is the function of a transformer?
2. What is the difference between electronic transformer and power transformer?
3. Give the classification of transformer based on its application.
4. Classify transformers based on the following parameters:
   (i) Construction
   (ii) Number of phases
   (iii) Location
5. List the different parts of a transformer.
6. What is the difference between limb and the yoke in a transformer?
7. Explain why transformer core is laminated?
8. List the properties of a transformer core.
9. What are the materials used for transformer winding?
10. How can you insulate the transformer winding from the core?
11. Differentiate between core-type and shell-type transformer.
12. Draw and explain the operation of core-type and shell-type transformer.
13. What is the function of bushings in a transformer?
14. What is the use of tappings in a transformer?
15. List the different types of windings used in a transformer.
16. Write a note on:

(i) Crossover winding
(ii) Disc winding
(iii) Cylindrical conductor
(iv) Sandwich winding
(v) Helical winding

17. Draw and explain the principle of operation of a single-phase transformer.
18. Obtain the e.m.f. equation of a single phase transformer.
19. What is an ideal transformer? What are its properties?
20. Obtain the transformation ratio of a single-phase transformer.
21. Explain the operation of an ideal transformer at no load.
22. Explain the operation of a practical transformer at no load.
23. Explain the operation of a transformer at load.
24. Draw and explain the phasor diagram of a transformer for
(i) Inductive load
(ii) Capacitive load
(iii) Resistive load

25. Draw and explain the phasor diagram of a transformer on load with resistance and leakage reactance of the windings for inductive load.
26. What is the procedure to transform the secondary parameters to primary side?
27. What is the procedure to transform the primary parameters to secondary side?
28. Step by step, obtain the equivalent circuit of a transformer.
29. What are the losses in a transformer?
30. List the characteristics of transformer oil.
31. Explain the physical, electrical and chemical properties of transformer oil.
32. What are the different types of cooling methods available in the case of a transformer?
33. Explain the following cooling methods:
(a) Air Natural (AN) cooling method
(b) Air Blast (AB) cooling method
(c) Oil Natural Air Natural (ONAN) cooling
(d) Oil Natural Air Forced (ONAF) cooling
(e) Oil Forced Air Forced (OFAF) cooling
(f) Oil Forced Water Forced (OFWF) cooling

34. Define voltage regulation. Obtain an expression for voltage regulation.
35. Derive the condition for zero voltage regulation and maximum voltage regulation
36. Define efficiency of a transformer. Obtain an expression for it.
37. Obtain the condition for maximum efficiency of a transformer.
38. What is all-day efficiency?
39. Draw and explain the open-circuit test of a single-phase transformer.
40. Draw and explain the short-circuit test of a single-phase transformer.
41. Explain how polarity test is done on a transformer.
42. Explain Sumpner's Test.
43. What are the advantages of running transformers in parallel?
44. List the condition for parallel operation of transformers
45. Explain the parallel operation of a transformer with equal voltage ratios.
46. Explain the parallel operation of a transformer with unequal voltage ratios.
47. Draw and explain the construction and operation of autotransformer.
48. Derive the copper saving in an autotransformer as compared to two-winding transformer.

49. List the advantages and disadvantages of an autotransformer.
50. Give the application of autotransformer.
51. How does one convert a two-winding transformer into an autotransformer?
52. Derive the voltage ratio and current ratio of an autotransformer
53. The O.C. and S.C. test results conducted on a 25 KVA, 440/220 V , 50 Hz transformer are as follows:
    O.C. test (L.V. side): 220 V   9.6 A   710 W
    S.C. test (H.V. side): 42 V   57 A   1030 W
    Calculate the following:
    (a) The parameters of the equivalent circuit of the transformer.
    (b) The voltage regulation of the transformer.
    (c) The efficiency of the transformer at 0.8 power factor lag at full load.
    (d) The efficiency of the transformer at 0.8 power factor lag at one-fourth of full load.
54. The O.C. and S.C. test result conducted on a 11 KVA, 220/110 V, 50 Hz transformer is as follows:
    O.C. test (H.V. side): 220 V   3.16 A   500 W
    S.C. test :   65 V   10 A   400 W
    Calculate the following:
    (a) The parameters of the equivalent circuit of the transformer
    (b) The voltage regulation of the transformer
    (c) The efficiency of the transformer at 0.8 power factor lag at full load
    (d) The efficiency of the transformer at 0.8 power factor lag at one-fourth of full load.
55. A 2500/250V, 200 KVA, 50 Hz transformer has the following parameters:
    $R_1 = 0.4\Omega$, $R_2 = 0.002\Omega$, $X_1 = 0.5\Omega$, $X_2 = 0.005\Omega$, $R_0 = 1500\Omega$, $X_0 = 1000\Omega$
    Calculate the voltage regulation and efficiency at full load at 0.8 power factor lag.
56. A transformer has the following parameters:
    $R_1 = 0.8\ \Omega$, $R_2 = 0.03\ \Omega$, $X_1 = 4\ \Omega$, $X_2 = 0.4\ \Omega$, $R_0 = 3500\Omega$, $X_0 = 1000\Omega$
    If the turns ratiois 8, find the equivalent circuit of the transformer referred to
    (a) Primary   (b) Secondary
57. Consider a 1000 KVA, 330/110 kV, 50 Hz transformer. If the e.m.f. per turn is 10V and the maximum flux density is 1T, calculate the following:
    (a) Cross sectional area of the core
    (b) Number of turns on the primary and secondary sides.
58. For the transformer of Exercise 57, calculate the value of primary and secondary currents.
59. The core dimension of an 11 KVA, 220/440V, 50Hz transformer is 20 cm × 20 cm. If the maximum flux density is 2T, calculate the following:
    (a) Maximum value of flux in the core
    (b) Number of turns on the primary and secondary sides
    (c) Value of currents on primary and secondary sides.
60. In a 430/110 V, 50 Hz, single phase transformer, the no-load current is 2A at 0.3 power factor lag. The primary has 300 turns. Calculate the following:
    (a) Magnetising component of no-load current
    (b) Working component of no load current
    (c) Core loss of transformer
    (d) Secondary turns
    (e) Maximum value of flux in the core.
61. A single-phase transformer has a no-load current of 2 A at 0.6 power factor lag. Calculate the primary current and the power factor when a load connected across the secondary draws a current

of 110 A at 0.8 power factor lag. The primary and secondary of the transformer are 300 and 900 respectively.

62. A 410/220 V, 50Hz, single phase transformer has the following results:
No-load current: 2 A at 0.5 power factor lag
On-load current: 32A at 0.8 power factor lag.
Calculate the primary current and primary power factor and draw the complete phasor.

63. Consider a 300 KVA, 1100/220 V two-winding transformer as shown Fig. Q63 (a). It has been converted to an autotransformer by connecting the two windings in the manner as show in Fig. Q63 (b). Calculate the following.
(a) HV and LV side voltages of the autotransformer
(b) The KVA rating of the autotransformer.
(c) The efficiency of the transformer at full-load at 0.8 power factor lag if the losses in the transformer is 900 W.

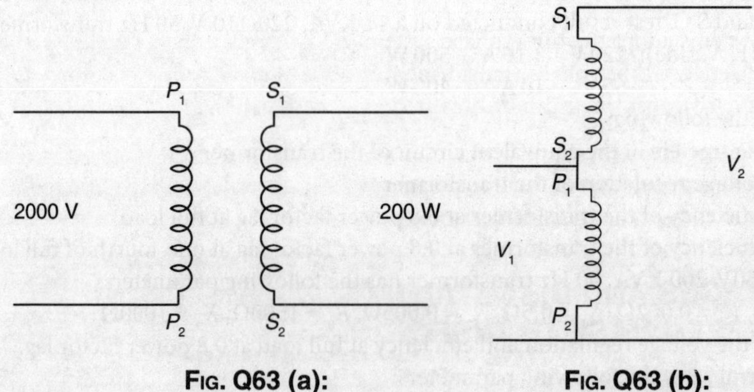

Fig. Q63 (a): Fig. Q63 (b):

64. A 90 KVA, 1100/210 V, 50 Hz single-phase transformer has the impedance of H.V. side and L.V. side as $(1 + j\,2)\ \Omega$ and $(0.018 + j\,0.019)\ \Omega$ respectively. Again, the parallel branch has $R_0 = 3080\ \Omega$ and $X_0 = j\,5000\ \Omega$.
Draw the equivalent circuit of the transformer referred to H.V. side and referred to L.V. side respectively.

65. The open-circuit (O.C.) test and short-circuit (S.C.) conducted on a 230/115 V, 1000VA transformer gave the following readings:
O.C. Test: 230 V 0.5 A 45 W
S.C. Test: 19.1 V 8.7 A 52 W
Find the parameters of the equivalent circuit and the efficiency of the transformer at 0.8 power factor and full load.

66. The iron loss and the copper loss of an 800 KVA distribution transformer are 5KW and 8KW respectively. The 24-hour loading of the transformer is as follows:

Number of Hours	Load (kW)	Power Factor
8	400	0.8
8	200	0.76
6	190	0.7
2	40	0.85

Calculate the all-day efficiency.

CHAPTER 5

THREE-PHASE TRANSFORMER

5.1 INTRODUCTION

Modern power systems adopt the three-phase system for generation, transmission and distribution of electrical power. Hence, a three-phase transformer is required to step up or step down the three phase voltage. A three-phase transformer can be obtained in two different ways.

- Three numbers of identical single-phase transformers can be suitably connected to make a three-phase transformer. Such three-phase transformers are called a *bank of three-phase transformer.*
- Alternatively, a three-phase transformer can be constructed as a single unit.

5.2 ADVANTAGES AND DISADVANTAGES OF A SINGLE UNIT THREE-PHASE TRANSFORMER

Some of the advantages and disadvantages of a single unit three-phase transformer as compared with a bank of three-phase transformer are listed below:

Advantages

1. It occupies less space
2. Its weight is less
3. Its cost is comparatively less
4. It is much easier to transport
5. It is more efficient
6. Core size is comparatively small.

Disadvantages

When one of the phases of a single-unit three-phase transformer becomes faulty, the entire unit of three-phase transformer needs to be removed from the supply for repair.

But in the case of a bank of three-phase transformer, only the faulty transformer is removed from the supply for repair while the other two transformers remain connected in the system without interrupting the supply completely.

5.3 CONSTRUCTION OF A THREE-PHASE TRANSFORMER

Like single-phase transformer, a three-phase transformer can also be categorised into two categories.

1. Core-type three-phase transformer
2. Shell-type three-phase transformer

5.3.1 Core-type Three-phase Transformer

A most common three-phase core-type transformer is shown in Fig. 5.1. It is called a three-limb core transformer. In the three-limb arrangement the cross-sectional area of the limbs and the cross-sectional area of the yoke is same. The three windings are wound on the three limbs. Each limb will have the primary and secondary windings of one phase. These windings are arranged concentrically. The magnetic circuit is completed through the two yokes, of which one is at the top and the other is at the bottom.

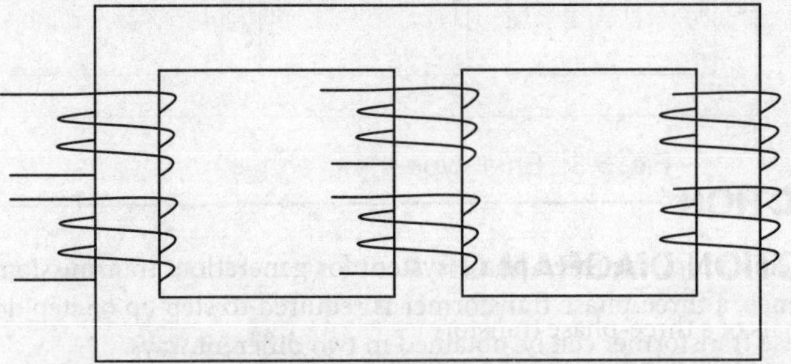

Fig. 5.1: Three-phase core-type transformer

A five-limb core arrangement is shown in Fig. 5.2. Here, the cross-sectional area of the yoke is half the cross-sectional area of the limbs where the three windings are placed. In the five-limb arrangement there are two unused limbs (at the extreme ends). The cross-sectional area of the unused limbs is half the cross-sectional area of the limbs where the windings are wound. A five-limb core has less height and weight than a three-limb core and is most commonly used.

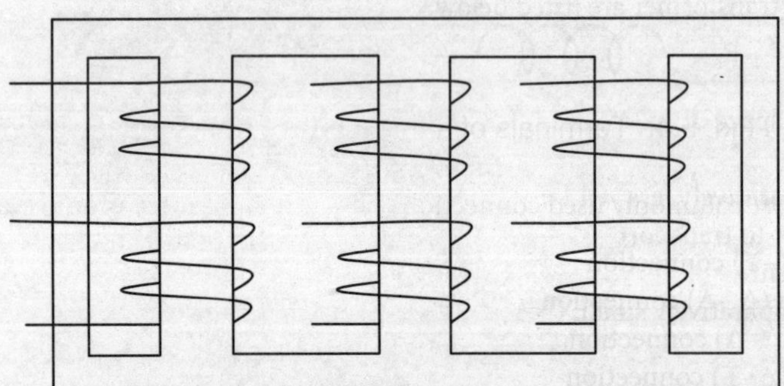

Fig. 5.2: Five-limb core-type transformer

5.3.2 Shell-type Three-phase Transformer

A three-phase shell-type transformer is shown in Fig. 5.3. In this type of construction, each phase has an independent magnetic circuit. The phase magnetic circuits are in parallel. Shell-type three-phase transformers are rarely used since it does not have any design advantage. They are used only for very large transformers.

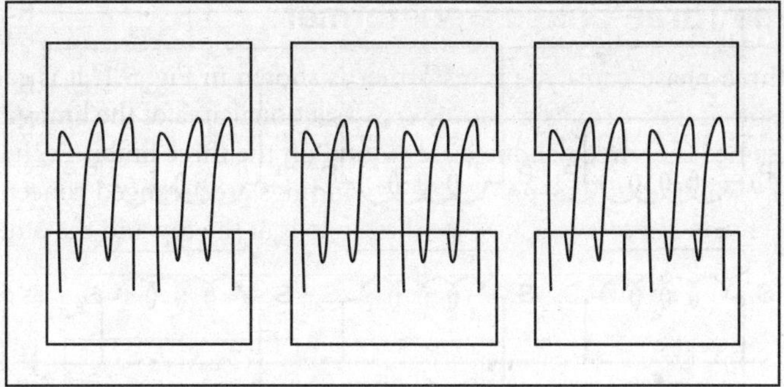

FIG. 5.3: Shell-type three-phase transformer

5.4 CONNECTION DIAGRAM OF A THREE-PHASE TRANSFORMER

The twelve terminals of a three-phase transformer winding are shown in Fig. 5.4. Out of these twelve terminals, six are of primary winding and the remaining six are of secondary winding. P_1 to P_6 are primary windings terminals and S_1 to S_6.

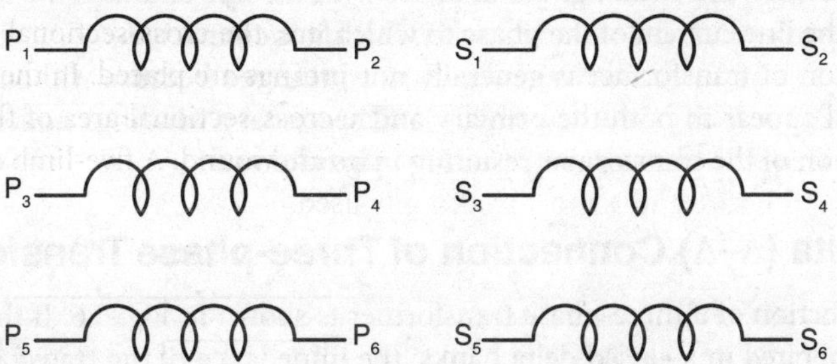

FIG. 5.4: Terminals of a three-phase transformer winding

Some of the most commonly used connections for a three-phase transformer are

i) Star–Star (Υ–Υ) connection
ii) Delta–Delta (Δ –Δ) connection
iii) Star–Delta (Υ–Δ) connection
iv) Delta–Star (Δ–Υ) connection
v) Open Delta (V–V) connection
vi) Scott (T–T) connection
vii) Zigzag connection

5.4.1 Star–Star (Υ–Υ) Connection of Three-phase Transformer

In this type of connection, both the primary and the secondary windings of the three-phase transformer are connected in star form. The connection diagram of the three-phase transformer windings in star–star connection is shown in Fig. 5.5.

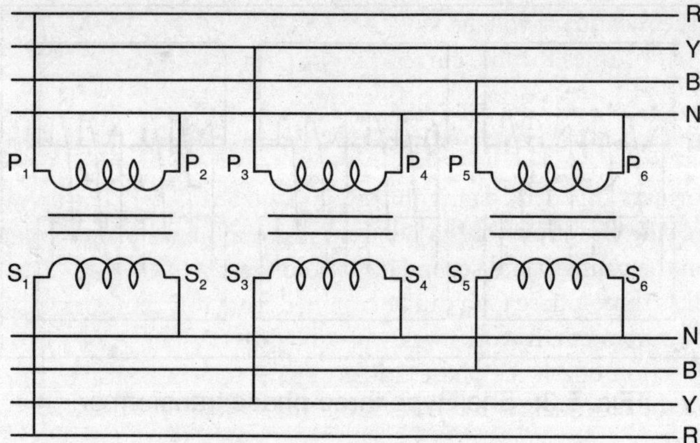

Fig. 5.5: Star–Star connection

This connection is preferred if there is a need to supply large single-phase loads together with three-phase loads. In this type of transformer, the neutral of the primary and the neutral of the secondary are connected together electrically if the transformer is operated in unbalanced load condition. For a star-connected transformer, the windings are in series with the line and thus the current carried by the windings is equal to the line current of the phase to which it is connected.

Star–star connection of transformer is generally not preferred because without neutral, large third harmonic voltage will appear in both the primary and secondary windings under unbalanced load. It will stress the insulation of the transformer resulting in breakdown.

5.4.2 Delta–Delta (Δ–Δ) Connection of Three-phase Transformer

The delta–delta connection of a three-phase transformer is shown in Fig. 5.6. If the three transformers are connected and operated in a closed delta banks, the impedance of the transformers must be same. If it is not same then the difference in voltages due to this unequal voltage drops will cause circulating currents to flow in the delta, thus heating the transformer. This problem also exists in a single three-phase transformer unit. For a delta-connected transformer, the line voltage is equal to the phase voltage.

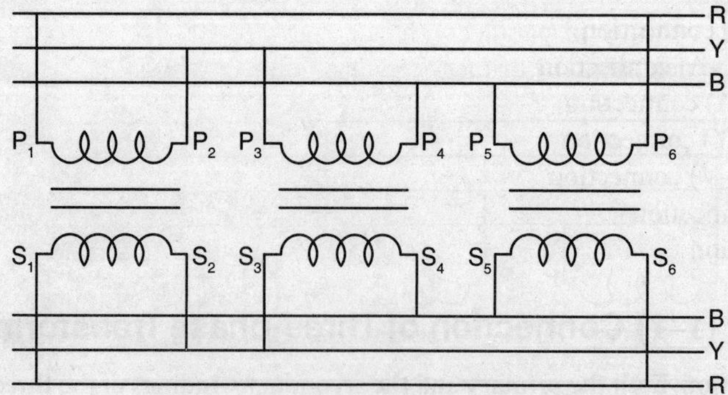

Fig. 5.6: Delta–Delta connection

The delta–delta connection has the following advantages:

- This connection permits unbalanced loads
- It provides a closed path for the circulating current to flow

- No distortion in the secondary voltage
- Phase current is $1/\sqrt{3}$ times the line current.

5.4.3 Delta–Star (Δ –Υ) Connection of Three-phase Transformer

In large power transformers this delta–star connection is preferred for step-up transformers. The star side is the high voltage side and it provides a neutral for grounding the high-voltage circuit. There is a 30^0 phase shift between the primary and secondary line voltages.

Single-phase as well as three-phase load can be supplied from this transformer. There is no distortion due to third harmonic component. It works well with unbalanced load.

The delta–star connection of a three phase transformer is shown in Fig. 5.7. Terminals P_1 to P_6 are connected in Delta and the terminals S_1 to S_6 are connected in star.

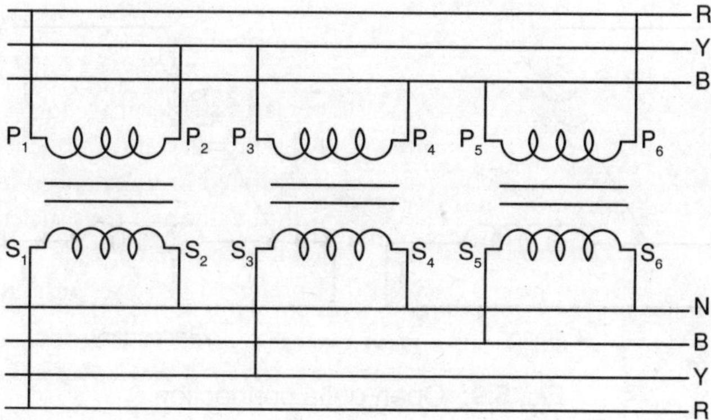

Fig. 5.7: Delta–Star connection

5.4.4 Star–Delta (Υ–Δ) Connection of Three-phase Transformer

As in the case of a delta–star transformer, star-delta transformer also has a 30^0 phase shift between primary and secondary line voltages. The neutral is now present on the primary side of the transformer. This type of transformer is mainly used in the distribution side, to step down the voltage. It is not possible to connect the transformer in parallel with star–star or delta–delta transformers. It can handle unbalanced load and the neutral of the primary can be grounded to avoid distortion.

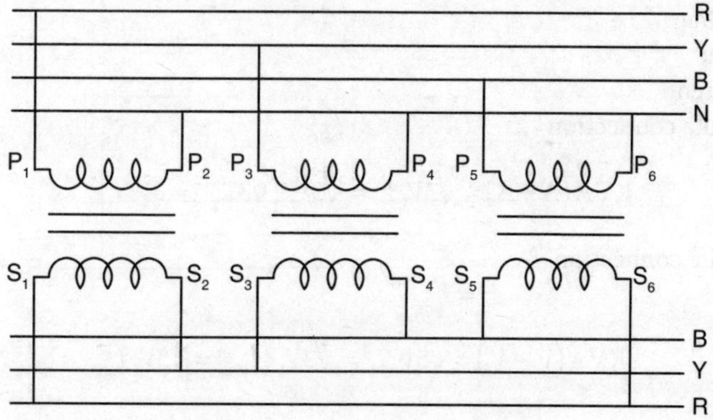

Fig. 5.8: Star–Delta connection

The star–delta connection of a three-phase transformer is shown in Fig. 5.8. Terminals P_1 to P_6 are connected in star and the terminals S_1 to S_6 are connected in delta

5.4.5 Open Delta (V–V) Connection of Three-phase Transformer

As discussed earlier in this chapter, a three-phase transformer can be made by considering three identical single-phase transformers. Thus a three-phase delta–delta transformer can be constructed by considering three single-phase identical delta-connected transformers.

If one of the single-phase delta-connected transformer is removed from the supply for the purpose of maintenance and if it is required to supply the load with the remaining two transformers at reduced efficiency, then we get an open delta connection.

Open delta transformer is shown in Fig. 5.9

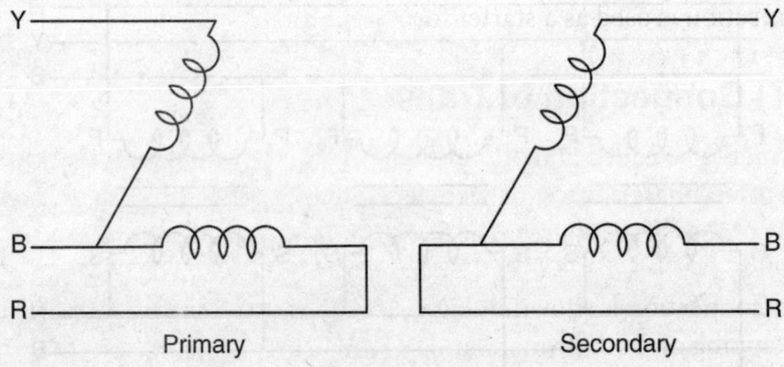

Fig. 5.9: Open delta connection

We know that, for a delta connection

$$V_L = V_{Ph}$$

And

$$I_L = \sqrt{3} I_{Ph}$$

Where,
V_L is the line voltage
V_{ph} is the phase voltage
I_L is the line current
I_{ph} is the phase current
Then, for delta–delta connection

$$\text{KVA}(\Delta - \Delta) = \sqrt{3} V_L I_L = \sqrt{3} V_L \left(\sqrt{3} I_{Ph}\right) = 3 V_L I_{ph} \tag{5.1}$$

Then, for open delta connection

$$I_L = I_{ph}$$

$$\text{KVA}(V - V) = \sqrt{3} V_L I_L = \sqrt{3} V_L \left(I_{Ph}\right) = \sqrt{3} V_L I_{ph} \tag{5.2}$$

Dividing Eq. (5.2) by Eq. (5.1), we get,

$$\frac{\text{KVA}(V-V)}{\text{KVA}(\Delta-\Delta)} = \frac{\sqrt{3}V_{\text{L}}I_{\text{ph}}}{3V_{\text{L}}I_{\text{ph}}} = 0.577 \tag{5.3}$$

From Eq. (5.3) we can conclude that the open delta connection will be able to carry only 57.7% of the original load.

Some of the features of open delta connection are

- The power handling capacity of open delta circuit is 0.577 times that of delta–delta circuit.
- The secondary terminal voltage becomes unbalanced with increase in load.
- The power factor is 86.6% of the balanced load factor.
- The two transformers in open delta connection will be operated at different power factor.

The open delta connection is used as a starter for AC motors.

5.4.6 Scott (T–T) Connection of Transformer

A **Scott T–T transformer** (also called a **Scott connection**) is a type of circuit used to derive two-phase electric power from a three-phase source, or vice-versa. The Scott connection evenly distributes a balanced load between the phases of the source. The Scott three-phase transformer was invented by Charles F. Scott, in the late 1890s.

A three-phase Scott connection is shown in Fig. 5.10. It has two transformers. The main transformer is provided with 50% tapping on both primary and secondary windings. The second transformer, called the teaser transformer, has 86.6% tap and is connected to centre tap on main transformer on primary and secondary side. Three-phase supply is given at the other end of the teaser transformer and the two ends of the main transformer.

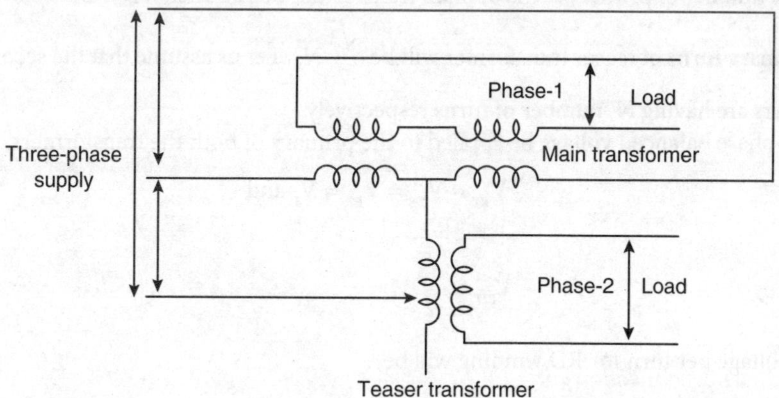

Fig. 5.10: Scott connection

5.5 CONVERSION FROM THREE-PHASE TO TWO-PHASE

Nowadays, the transmission of electrical power is in three-phase. However, there are some applications which require two-phase supply (e.g. electrical furnace) and thus it becomes necessary to have two-phase supply. By using Scott connection of transformers, it is possible to convert a three-phase supply to a two-phase supply and vice versa.

The connection diagram of two single-phase transformers for three-phase to two-phase conversion is shown in Fig. 5.11

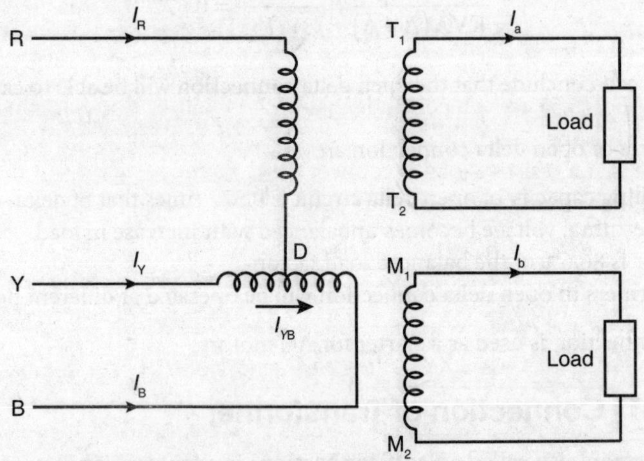

FIG. 5.11: Connection diagram for three-phase to two-phase conversion

In the circuit shown in Fig. 5.11, the primary side is a three-phase supply and the secondary side is two-phase output connected across a load. In the connection shown in Fig. 5.11, there are two transformers – main transformer and teaser transformer. For teaser transformer, the primary is RD and the secondary side is T_1T_2. For main transformer, the primary side is YB and the secondary side is M_1M_2. D is the midpoint of primary of main transformer, which is connected to the primary of teaser transformer.

Let the total number of primary turns of main transformer be N_1. Then $YD = BD = \frac{N_1}{2}$ and the total number of primary turns of teaser transformer will be $\frac{\sqrt{3}}{2}N_1$. Let us assume that the secondary of both the transformers are having N_2 number of turns respectively.

Let a three-phase balanced voltage be applied to the primary of both the transformers. So,

$$V_{RY} = V_{YB} = V_{BR} = V_L \text{ and}$$

$$V_{RD} = \frac{\sqrt{3}}{2}V_L$$

Then, the voltage per turn for RD winding will be

$$\frac{\frac{\sqrt{3}}{2}V_L}{\frac{\sqrt{3}}{2}N_1} = \frac{V_L}{N_1} \tag{5.4}$$

Again, since, $V_{YB} = V_L$, then

$$V_{YD} = V_{BD} = \frac{V_L}{2}$$

Then the voltage per turn of YB winding will be

$$\frac{V_L/2}{N_1/2} = \frac{V_L}{N_1} \tag{5.5}$$

From Eq. (5.4) and Eq. (5.5), we can conclude that the voltage per turn for primary winding of main and teaser transformer is same.

Now, let us assume that a resistive load is connected across the secondary of main and teaser transformer. The phasor of the primary and secondary side is shown in Fig. 5.12

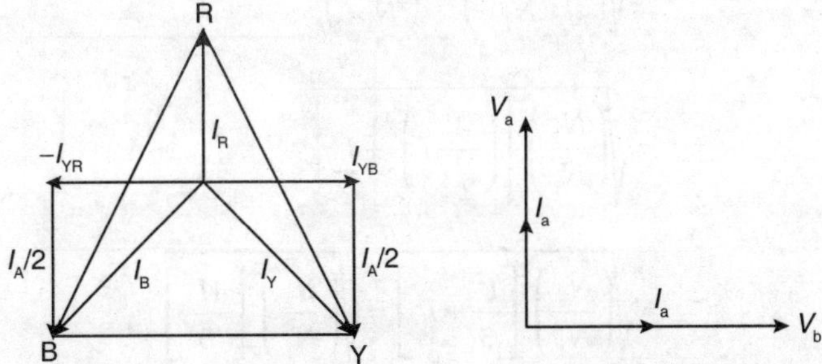

Fig. 5.12: Phasor with resistive load

Let the current of the secondary side of the main and teaser transformers be I_b and I_a respectively. Since the load is a resistive load, the power factor is unity and hence the currents are in phase with the respective voltages as shown in the Fig. 5.12, where V_a is the voltage of secondary side of teaser transformer and V_b is the voltage of secondary side of main transformer.

V_a and V_b will be equal in magnitude but displaced in phase by 90⁰.

Again,

$$\frac{I_R}{I_a} = \frac{N_2}{\frac{\sqrt{3}}{2}N_1} = \frac{2}{\sqrt{3}}\frac{N_2}{N_1}$$

$$\text{or } I_R = \frac{2}{\sqrt{3}}\frac{N_2}{N_1}I_a \tag{5.6}$$

For balanced load, let,

$$I_a = I_b = I$$

Then Eq. (5.6) becomes,

$$I_R = \frac{2}{\sqrt{3}}\frac{N_2}{N_1}I \tag{5.7}$$

Again,

$$\frac{I_{YB}}{I_b} = \frac{N_2}{N_1}$$

$$I_{YB} = \frac{N_2}{N_1} I \tag{5.8}$$

From the phasor of Fig. 5.12, we can write,

$$I_B = \sqrt{\left(\frac{I_R}{2}\right)^2 + (I_{YB})^2}$$

$$= \sqrt{\left(\frac{1}{\sqrt{3}}\frac{N_2}{N_1} I\right)^2 + \left(\frac{N_2}{N_1} I\right)^2}$$

$$= \sqrt{\left(\frac{N_2}{N_1}\right)^2 \left[\left(\frac{1}{\sqrt{3}} I\right)^2 + I^2\right]}$$

$$= \sqrt{\left(\frac{N_2}{N_1}\right)^2 \left[\frac{I^2}{3} + I^2\right]} = \sqrt{\left(\frac{N_2}{N_1}\right)^2 \left[\frac{4I^2}{3}\right]}$$

$$I_B = \frac{N_2}{N_1} \frac{2}{\sqrt{3}} I \tag{5.9}$$

The above expression shows that when the two-phase load is balanced, then the three-phase side is also balanced.

5.6 CONVERSION FROM THREE-PHASE TO SIX-PHASE

By dividing the secondary winding of a three-phase transformer into two parts and connecting them for currents 180^0 opposite to each other, it is possible to get six-phase from three-phase supply. Six or more phases of supply are required for supplying power rectifiers.

Some of the most commonly used schemes for three-phase to six-phase conversion are:

i) Double star connection
ii) Double delta connection
iii) Six-phase star connection
iv) Diametric connection

5.6.1 Double Star Connection

In this type of connection scheme, three identical single-phase transformers are used. The primaries of these three transformers are connected in delta. The secondary of each transformer is divided into two equal parts. By taking one part of the secondary winding of each transformer, a star connection is formed. Further, by taking the other part of the secondary winding of each transformer, a second star connection is formed. Since two star connections are present on the secondary side, it is called double star connection. The double star-connected transformer is shown in Fig. 5.13.

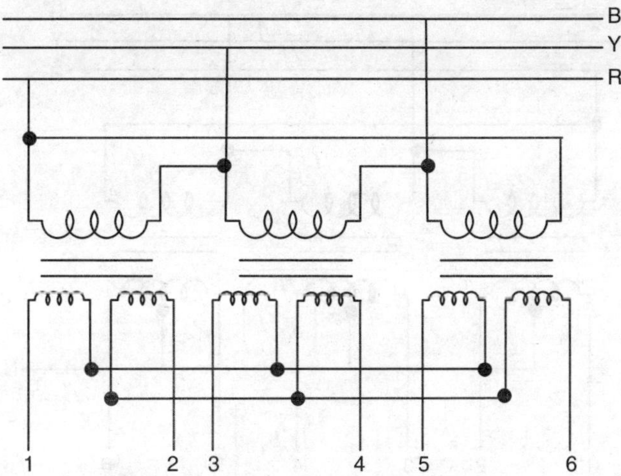

Fig. 5.13: Double star connection

5.6.2 Double Delta Connection

It is same as double star connection. The only difference is that now the secondary windings are connected in delta instead of star. The connection is shown in Fig. 5.14

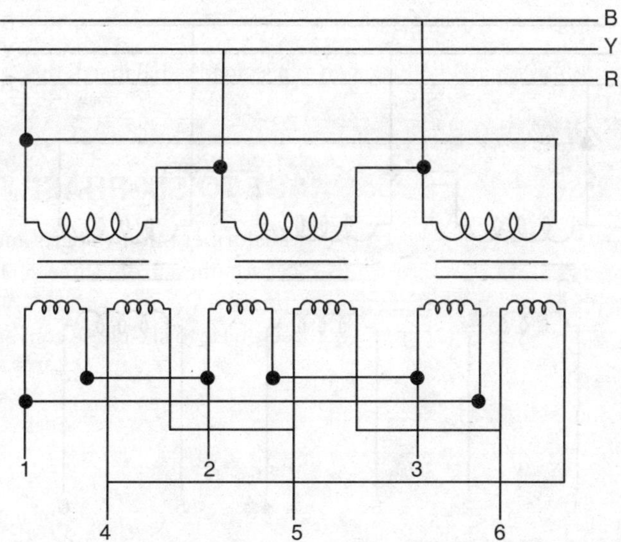

Fig. 5.14: Double delta connection

5.6.3 Six-phase Star Connection

In this type of connection scheme, three identical single-phase transformers are used. The primaries of these three transformers are connected in delta. In this type of connection, the secondary is not divided into two equal parts. Instead, the secondary is centre-tapped, and the centre-tap of each secondary is connected to neutral. The connection is shown in Fig. 5.15

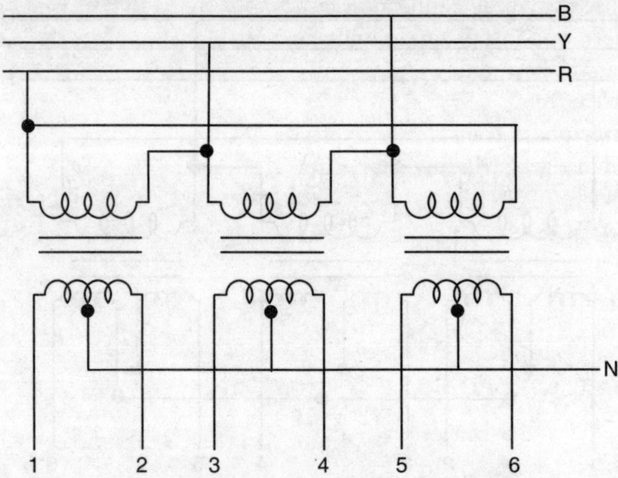

FIG. 5.15: Six-phase star connection

5.6.4 Diametric Connection

It is the same as six-phase star connection, but without the centre-tap on the secondary side. The connection is shown in Fig. 5.16

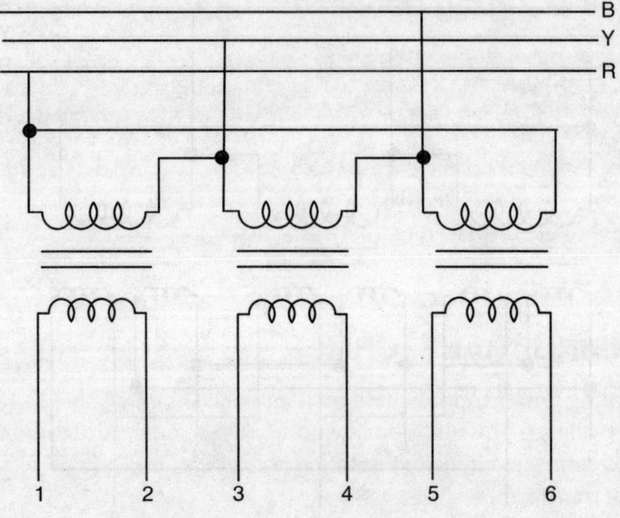

FIG. 5.16: Diametric Connection

5.7 ZIGZAG CONNECTION

A zigzag transformer is used in connection with three-phase and is made up of six coils connected in a "Y" manner. Each leg of the "Y" is made up of a coil on a different phase leg of the transformer. The neutral formed by the zigzag connection is very stable. Therefore, this type of transformer, or in some cases an autotransformer, lends itself very well for establishing a neutral for an ungrounded three-phase system. You can also use the zigzag connection in power systems to trap triple *n*-harmonic (3rd, 9th, 15th, etc.) currents. Here, you install zigzag units near loads that produce large triplen harmonic currents. The windings trap the harmonic currents and prevent them from travelling upstream, where they can produce undesirable effects.

A nine-winding three-phase transformer, typically having 3 primary and six identical secondary windings which is used in zigzag winding connection as shown in Fig. 5.17

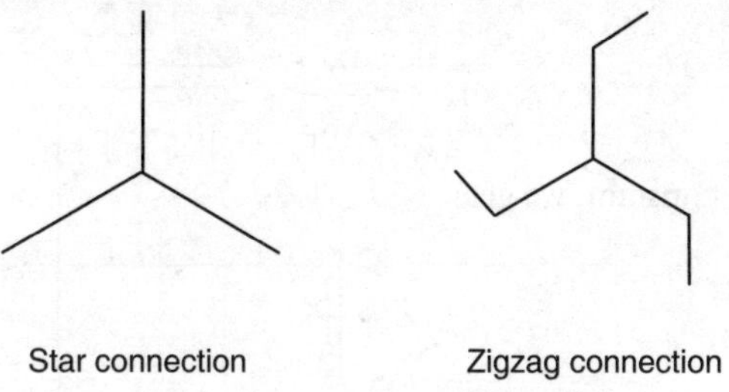

Fig. 5.17: Zigzag connection

The first coil on each zigzag winding core is connected contrariwise to the secondary coil on the next core. The second coils are then all tied together to form the neutral and the phases are connected to the primary coils. Each phase, therefore, couples with each other phase and the voltages cancel out. As such, there would be negligible current through the neutral point, which can be tied to ground

5.7.1 Advantages of Zigzag Connection

Some of the advantages of zigzag connection are as follows,

i) The cost of zigzag transformer is very less as compared to other simple three-phase star- or delta-connected transformer.
ii) This transformer suppresses triple harmonic
iii) A zigzag transformer is more effective for grounding purposes
iv) There is no phase angle displacement between the primary and the secondary circuit with this connection.

5.8 THREE-WINDING TRANSFORMER

In some of the high-rating power transformers, an additional winding is used apart from the usual primary and secondary windings. This additional winding is called the **tertiary winding of transformer**. The tertiary winding is connected in delta formation. Because of this tertiary winding, the transformer is called three-winding transformer.

Let us consider a transformer with N turns and let the reluctance of the magnetic path be R_L, then,

$$\text{m.m.f.} = NI = \varphi R_L \tag{5.10}$$

Where I is the current and φ is the flux in the transformer.We know that,

$$\text{Induced voltage}\left(E\right) = 4.44\varphi fN \tag{5.11}$$

Where, f is the frequency of supply.

Again,

$$E \propto \varphi$$

$$\text{or, } \varphi = KV \tag{5.12}$$

Now, from Eqs. 5.11 and 5.12, we get,

$$NI = KVR_L$$

$$\text{or, } \frac{V}{I} = \frac{N}{KR_L}$$

Assuming, N and K to be constant, we get,

$$Z \propto \frac{1}{R_L} \tag{5.13}$$

From Eq. 5.13, we can say that impedance is inversely proportional to reluctance. The impedance offered by the return path of unbalanced load current is very high. Hence, it will provide a low reluctance path for the unbalanced flux.

Again, we know that the unbalanced current in a three-phase system can be divided into positive sequence, negative sequence and zero sequence components. If the value of the zero sequence current in each line is I_o, then total current that flows through the neutral of secondary side of transformer is $I_n = 3.I_o$. This current cannot be balanced by primary current as the zero sequence current cannot flow through the isolated neutral star connected primary. So, this current will set up a magnetic flux in the core, which will increase the impedance of the zero sequence current. On the other hand, a delta-connected tertiary winding of transformer permits the circulation of zero sequence current in it. This circulating current in delta winding balances the zero sequence component of unbalance load, thus preventing unnecessary development of unbalanced zero-sequence flux in the transformer core. In a few words, it can be said that placement of tertiary winding in star–star neutral transformer considerably reduces the zero sequence impedance of transformer.

5.8.1 Advantages of using tertiary winding in transformer

Some of the advantages of tertiary winding transformer are:

i) This transformer has the ability to redistribute the flow of fault current
ii) It reduces the unbalancing in the primary due to unbalancing in three-phase load.
iii) Auxiliary load at different voltage levels from the tertiary winding can be supplied from this transformer in addition to its main secondary load.

5.9 THREE-PHASE AUTOTRANSFORMER

Three similar single-phase autotransformers can be connected in the manner as shown in Fig. 5.18, to form a three-phase autotransformer. The theory of operation is same as that of a single-phase autotransformer. Three-phase autotransformers are used to start a three-phase induction motor at reduced voltage. Apart from being economical, autotransformers have less leakage flux; hence, there is improved regulation. However its one major disadvantage is that it cannot provide isolation of HV and LV sides.

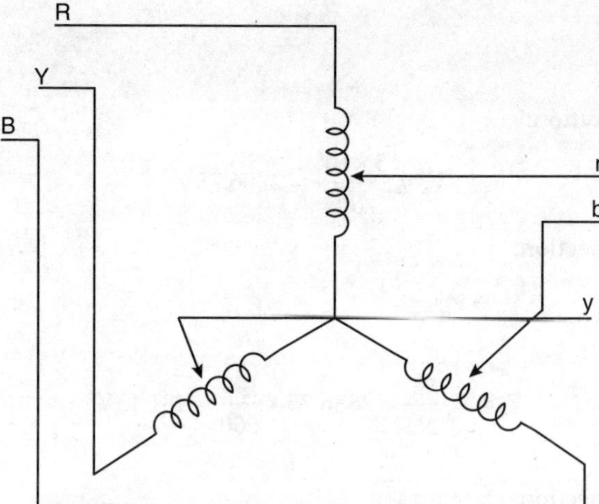

FIG. 5.18: Three-phase autotransformer

5.10 HARMONICS IN TRANSFORMER

Electromagnetic devices such as transformers act as source of harmonics. Transformer magnetic material characteristic is non-linear; this non-linearity is the main source of harmonics during excitation. Sources of harmonics in transformers can be classified under the following categories.

- **Normal excitation:** The normal excitation current in the transformer is non-sinusoidal. The distortion is due to the zero sequence and the third sequence component present in the excitation current. Presence of high reluctance electrical path like air, oil or tank, reduces the zero sequence component. Delta connection of poly-phase transformer can be used to reduce third harmonic component, provided the three phases are balanced.
- **Symmetric over-excitation:** Saturation of magnetic core of the transformer due to rise in voltage, with symmetrical magnetising current, generates odd harmonics. This harmonic current can be absorbed by the delta connection in the balanced system.
- **Inrush current harmonic:** When the transformer is switched off and switched on again, sometimes due to residual flux density in the core, when the transformer is switched ON, the flux density can reach twice the flux density or more. This will cause the magnetisation current to increase drastically. This is called the inrush current. This causes the generation of second order harmonic component in the transformer current.
- **DC magnetisation:** Under magnetic imbalance, the shape of the magnetising characteristic and the excitation current are different from those under no-load condition. Under this condition, the transformer excitation current contains both odd and even harmonic components.

SOLVED PROBLEMS

Ex.5.1 A three-phase transformer has 600 turns on the primary and 100 turns on secondary. The supply voltage is 5KV. Find the secondary line voltage at no-load when the windings are connected as,

(i) Star–Delta

(ii) Delta–Star

Solution

(i) For star connection:

$$V_P = \frac{5\times10^3}{\sqrt{3}} = 2886.75\text{V}$$

For delta connection:

$$V_L = V_P$$

$$V_s = V_P\frac{N_s}{N_P} = 2886.75\times\frac{100}{600} = 481.13\text{V}$$

(ii) For delta connection:

$$V_L = V_P = 5\times10^3\text{V}$$

$$V_s = V_P\frac{N_s}{N_P} = 5000\times\frac{100}{600} = 833.33\text{V}$$

For star connection:

$$V_L = \sqrt{3}V_s = 1443.37\text{V}$$

Ex.5.2 Two Scott-connected transformers supply a balanced load of 550V, 40KVA, from a three-phase supply of 4000 V. Calculate the voltage, current and the KVA rating of the main and the teaser transformer.

Solution

Primary of the main transformer will have a voltage of 4000 V

Primary of the teaser transformer will have a voltage of $0.866\times4000 = 3464$ V

Secondary of the main transformer will have a voltage of 550 V

Secondary of the teaser transformer will have a voltage of $0.866\times550 = 476.3$ V

$$\text{Primary Line current across the main and the teaser transformer} = \frac{40\times10^3}{\sqrt{3}\times4000}$$

$$= 5.77\text{ A}$$

$$\text{Secondary Line current across the main and the teaser transformer} = 5.77\times\frac{550}{4000}$$

$$= 0.7933\text{A}$$

Ex.5.3 Two identical transformers each of rating 200KVA, 1000/2500 V, 50Hz are connected in open delta on primary and secondary sides. Calculate the following:

(i) Total load this transformer arrangement can supply.
(ii) If 300 KVA, 200 V three-phase delta connected load is connected to this transformer bank, calculate the line current on H.V. side.

Solution

(i) Total load on open delta connection $= \sqrt{3} \times 200 = 346.4$ KVA

(ii) For delta connection:

$$I_{AB} = I_{BC} = I_{CA} = \frac{1}{3}\left(\frac{300 \times 10^3}{200}\right) = 500\,\text{A}$$

then,

$$I_A = I_{AB} - I_{CA} = 2 \times \frac{\sqrt{3}}{2} \times 500 = 866\,\text{A}$$

$$\text{Line current on H.V.side} = 866 \times \left(\frac{2500}{1000}\right) = 2165\text{A}$$

Ex.5.4 Consider a three-phase star–delta transformer of rating 25000 KVA, 440/220 KV. A balanced delta connected resistive load of 5000 KW is connected on the L.V. side of the transformer. Find:

(i) The resistive load in each phase measured from line to neutral on the H.V. side of the transformer.
(ii) What will be the value of this resistance if the load is now star-connected.

Solution

(i) $I_L = \dfrac{5000}{\sqrt{3} \times 220} = 13.12\text{A}$

$$R = \frac{440 \times 10^3 / \sqrt{3}}{13.12} = 19362\ \Omega$$

(ii) Value of R when connected in star

$$R = 3 \times 19362 = 58086\ \Omega$$

Ex.5.5 A three-phase transformer was made by connecting three identical single-phase transformer of rating 11 KV/220V, 50KVA. The leakage reactance per transformer is 0.08 Ω. A balanced star connected load of 25 Ω per phase is connected across the secondary of the three-phase transformer. Determine the equivalent per phase impedance as seen from the primary side of the transformer, when the transformer is connected in,

(i) Star–Star connection
(ii) Star–Delta connection
(iii) Delta–Star connection
(iv) Delta–Delta connection

Solution

(i) Star–Star connection

$$V_L(\text{Primary}) = \sqrt{3}\times 11\times 10^3 = 19052 \text{ V}$$

$$V_L(\text{Secondary}) = \sqrt{3}\times 220 = 381 \text{ V}$$

$$R'_L = \frac{25}{\left(\dfrac{220\times\sqrt{3}}{\sqrt{3}\times 11\times 10^3}\right)^2} = 62500\ \Omega$$

$$Z'_L = R'_L + j0.08 = 62500 + j0.08\ \Omega$$

(ii) Star–Delta connection

$$V_L(\text{Primary}) = \sqrt{3}\times 11\times 10^3 = 19052 \text{ V}$$

$$V_L(\text{Secondary}) = 220 \text{ V}$$

$$R'_L = \frac{25}{\left(\dfrac{220}{\sqrt{3}\times 11\times 10^3}\right)^2} = 187500\ \Omega$$

$$Z'_L = R'_L + j0.08 = 187500 + j0.08\ \Omega$$

(iii) Delta–Star connection

$$V_L(\text{Primary}) = 11\times 10^3 \text{ V}$$

$$V_L(\text{Secondary}) = \sqrt{3}\times 220 = 381 \text{ V}$$

$$R'_L = \frac{25}{\left(\dfrac{220\times\sqrt{3}}{11\times 10^3}\right)^2} = 20833\ \Omega$$

$$Z'_L = R'_L + j0.08 = 20833 + j0.08\ \Omega$$

(iv) Delta–Delta connection

$$V_L(\text{Primary}) = 11\times 10^3 \text{ V}$$

$$V_L(\text{Secondary}) = 220\,\text{V}$$

$$R'_L = \frac{25}{\left(\frac{220}{11\times10^3}\right)^2} = 62500\,\Omega$$

$$Z'_L = R'_L + j0.08 = 62500 + j0.08\,\Omega$$

Ex.5.6 Three, 50 KVA, 500/100 V single-phase, 50 Hz transformer are connected in delta–delta manner. The total load connected is 80 KVA. Due to some fault, one transformer is removed and the other two is operated in open delta manner. Calculate the maximum power this new connection can take.

Solution

For the transformer under delta–delta connection, Power = 50 × 3 = 150 KVA.
For open delta connection, 0.866 × 150 = 130 KVA

Ex.5.7 Two transformers are connected in open delta manner. A balanced three-phase load takes 80 KW, at 200 V, 0.8 power factor lag. The input voltage to the transformer is 500V. Determine the maximum power rating of each transformer.

Solution

$$P = \sqrt{3}VI\cos,$$

$$I = \frac{P}{\sqrt{3}V\cos\theta} = \frac{80\times10^3}{\sqrt{3}\times200\times\times0.8} = 288.67\,\text{A}$$

$$\text{Power rating} = 200\times288.67 = 57734\ \text{VA} = 57.73\ \text{KVA}.$$

Ex.5.8 Three single-phase transformers are connected in delta–star way to make a three-phase transformer. Draw the phasor diagram for the same.

Solution

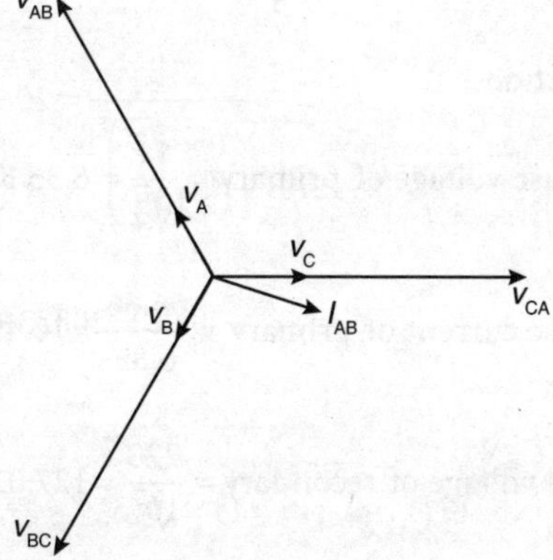

Ex.5.9 Two transformers, each of rating 150 KVA, 11/2.2 KV, 50 Hz are connected in open delta on primary and secondary sides.

(i) What is the total load that can be supplied from this transformer bank?
(ii) A 200 KVA, 2.2 KV, 0.866 power factor lagging three-phase delta-connected load is connected to the transformer bank. What will be the line current on the H.V. side?

Solution

(i) Load on open delta $= 150 \times \sqrt{3} = 259.8$ KVA

(ii) Load is delta connected:

$$I_{AB} = I_{BC} = I_{CA} = \frac{1}{3} \times \frac{200 \times 10^3}{2200} = 30.3 \text{ A}$$

$$I_A = I_{AB} - I_{BC} = \left(2 \times \frac{\sqrt{3}}{2} \times 30.3\right) = 52.48 \text{ A}$$

Ex.5.10 It is proposed to transmit the power generated by a 200 MVA, 11 KV, 50 Hz, three-phase generator to a three-phase 220 kV transmission line using a bank of three single-phase transformers. Find the turns-ratio and the voltage and current ratings for each single-phase transformer when the connections are

(a) Star–Star,
(b) Star–Delta,
(c) Delta–Star, and
(d) Delta–Delta.

Solution

$$V_p = 11 \text{ KV}$$

$$V_s = 220 \text{ KV}$$

$$\text{MVA per phase} = \frac{200}{3} = 66.66 \text{ MVA}$$

(a) For Star–Star connection

$$\text{Phase voltage of primary} = \frac{11}{\sqrt{3}} = 6.35 \text{ KV}$$

$$\text{Phase current of primary} = \frac{66.66}{6.35} = 10.49 \text{ A}$$

$$\text{Phase voltage of secondary} = \frac{220}{\sqrt{3}} = 127.02 \text{ KV}$$

$$\text{Phase current of secondary} = \frac{66.66}{127.02} = 0.52 \text{ A}$$

(b) For Star–Delta connection

$$\text{Phase voltage of primary} = \frac{11}{\sqrt{3}} = 6.35 \text{ KV}$$

$$\text{Phase current of primary} = \frac{66.66}{6.35} = 10.49 \text{ A}$$

$$\text{Phase voltage of secondary} = 220 \text{ KV}$$

$$\text{Phase current of secondary} = \frac{66.66}{220} = 0.3 \text{ A}$$

(c) For Delta–Star connection

$$\text{Phase voltage of primary} = 11 \text{ KV}$$

$$\text{Phase current of primary} = \frac{66.66}{11} = 6.06 \text{ A}$$

$$\text{Phase voltage of secondary} = \frac{220}{\sqrt{3}} = 127.02 \text{ KV}$$

$$\text{Phase current of secondary} = \frac{66.66}{127.02} = 0.52 \text{ A}$$

(d) For Delta–Delta connection

$$\text{Phase voltage of primary} = 11 \text{ KV}$$

$$\text{Phase current of primary} = \frac{66.66}{11} = 6.06 \text{ A}$$

$$\text{Phase voltage of secondary} = 220 \text{ KV}$$

$$\text{Phase current of secondary} = \frac{66.66}{220} = 0303 \text{ A}$$

Ex.5.11 The identical impedances of $10\angle -15\ \Omega$ are star-connected to balance three-phase line voltage of 208 V. Specify all the line and phase voltages and the currents for the above system.

Solution

$$|V_{\text{Line}}| = 208 \text{ V}$$

$$|V_{\text{Phase}}| = \frac{208}{\sqrt{3}} = 120 \text{ V}$$

Then the line voltages will be,

$$V_{ab} = 208\angle 0 \text{ V}$$

$$V_{bc} = 208\angle 120 \text{ V}$$

$$V_{ca} = 208\angle 240 \text{ V}$$

Here, V_{ab} has been assumed to the reference,
Similarly, the phase voltages will be,

$$V_{an} = 120\angle -30 \text{ V}$$

$$V_{bn} = 120\angle 90 \text{ V}$$

$$V_{cn} = 120\angle 210 \text{ V}$$

Then the phase currents will be,

$$I_{an} = \frac{V_{an}}{Z} = \frac{120\angle -30}{10\angle 15} = 12\angle -15 \text{ A}$$

$$I_{bn} = \frac{V_{bn}}{Z} = \frac{120\angle 90}{10\angle 15} = 12\angle 105 \text{ A}$$

$$I_{cn} = \frac{V_{cn}}{Z} = \frac{120\angle 210}{10\angle 15} = 12\angle 225 \text{ A}$$

Ex.5.12 A star-connected of three identical 100 KVA 7967/277 volt transformers are supplied with power directly from a large constant-voltage bus. In the short-circuit test, the recorded values on the high-voltage side for one of these transformers are

$$V_{SC} = 560 \text{ V } I_{sc} = 12.6 \text{ A } P_{sc} = 3300 \text{ W.}$$

(a) If this bank delivers a rated load at 0.88 PF lagging and rated voltage, what is the line-to-line voltage on the primary of the transformer bank?
(b) What is the voltage regulation under these conditions?

Solution
Since the transformer is star-connected, the per phase values of the reading will be,

$$V_{sc}\left(\text{per phase}\right) = \frac{560}{\sqrt{3}} = 323.32 \text{ V}$$

$$I_{sc}\left(\text{per phase}\right) = 12.6 \text{ A}$$

$$P_{sc}\left(\text{per phase}\right) = \frac{3300}{3} = 1100 \text{ W}$$

$$Z\left(\text{per phase}\right) = \frac{323.32}{12.6} = 25.66\ \Omega$$

$$\theta = \cos^{-1}\left(\frac{1100}{323.32 \times 12.6}\right) = 74.3°$$

Then,

$$Z = 25.66\angle 74.3\ \Omega = 6.94 + j24.7\ \Omega$$

So,

$$R = 6.94\ \Omega \quad X = 24.7\ \Omega$$

(a) Power $(P) = \dfrac{100}{3} = 33.33$ VA

The equivalent current flowing in the secondary of one transformer referred to the primary side is,

$$I_{sc}(\text{p}) = \frac{33.33}{7967} = 4.184\text{ A}$$

$$\text{or, } I_{sc}(\text{p}) = 4.184\angle -28.36\text{ A}$$

$$V(\text{p}) = 7967 + (4.184\angle -28.36) \times (6.94 + j24.7) = 8042\angle 0.55\text{V}$$

(b) $\%\text{ regulation} = \dfrac{8042 - 7967}{7967} \times 100 = 0.94\ \%$

Ex.5.13 Two transformers of 66/11 KV, star–delta configuration each is operated in parallel and is supplying a load of 45 MVA, at 11 KV, 0.8 p. f. lag. The ratings of the transformers are as follows:

Transformer-1: 25MVA, $X = 0.09$ p.u.
Transformer-2: 20 MVA, $X = 0.07$ p.u.

Find the magnitude of the current in per unit through each transformer. Find:

(i) The MVA output of each transformer
(ii) The MVA to which the total load must be limited so that none of the transformers are overloaded.

Take base of 45 MVA, 11KV on L.V. side.

Solution

$$X_1 = 0.09 \times \frac{45}{25} = 0.162\text{ p.u.}$$

$$X_2 = 0.07 \times \frac{45}{20} = 0.158\text{ p.u.}$$

$$S_1 = \frac{0.158}{0.158 + 0.162} \times 45 = 22.22\text{ MVA}$$

$$S_2 = \frac{0.162}{0.158 + 0.162} \times 45 = 22.78\text{ MVA}$$

Transformer-2 is overloaded, so the load must be reduced to $= \dfrac{20}{22.78} = 39.51$ MVA

Ex.5.14 For the transformer of Ex. 5.13, let the taps on the Transformer-1 are set at 55 KV to give a 4% boost in the voltage towards the L.V. side of that transformer compared to Transformer-2 which remains at 66 KV tap. Calculate the MVA output of each transformer for the original 45 MVA load and maximum MVA of the total load which will not overload the transformer. Use the same base as in Ex.5.13.

Solution

$$I_1 = \frac{22.22}{45}(0.8 - j0.6) = 0.395 - j0.297$$

$$I_2 = \frac{22.78}{45}(0.8 - j0.6) = 0.405 - j0.304$$

For 4% boost

$$I_c = \frac{0.04}{j0.162 + j0.158} = -j0.152$$

$$I_1 + I_c = 0.395 - j0.449 = 0.598\angle -48.66$$

$$I_2 + I_c = 0.405 - j0.456 = 0.609\angle -48.38$$

$$S_1 = 0.598 \times 1 \times 45 = 26.91 \text{ MVA}$$

$$S_2 = 0.609 \times 1 \times 45 = 27.41 \text{ MVA}$$

Both the transformers are overloaded.

$$\text{Transformer-1} = \frac{26.91}{25} = 1.08$$

$$\text{Transformer-2} = \frac{27.41}{20} = 1.37$$

$$\text{Reduce the load by} = \frac{45}{1.37} = 32.85 \text{ MVA}$$

Ex.5.15 Consider a three-winding transformer as shown in Fig. E 5.15. Where, $N_1 = 2N_2 = 3N_3$ and load of resistance R is connected across Coil 2 and Coil 3. Find the input impedance Z.

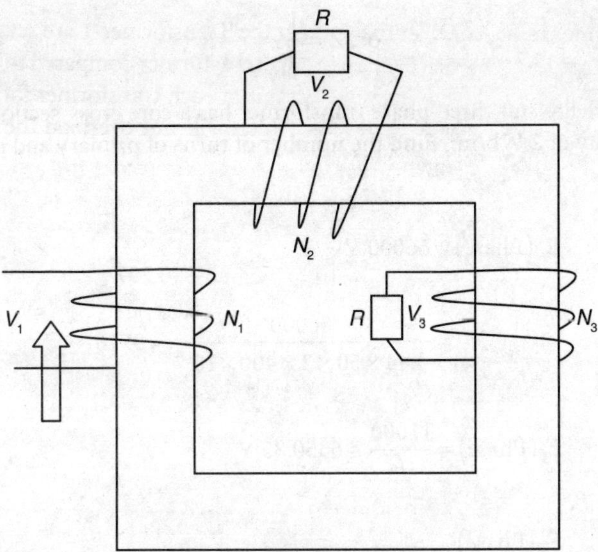

FIG. E5.15:

Solution

Let I_1, I_2 and I_3 be the current through N_1, N_2 and N_3 respectively. Then from MMF balance equation we can write,

$$N_1I_1 = N_2I_2 + N_3I_3$$

$$\text{or } N_1I_1 = \frac{1}{2}N_1I_2 + \frac{1}{3}N_1I_3$$

$$\text{or } I_1 = \frac{1}{2}I_2 + \frac{1}{3}I_3 \tag{i}$$

$$\text{Again, } V_1 = 2V_2 = 3V_3$$

$$\text{So, } I_2 = \frac{V_2}{R} = \frac{V_1}{2R}$$

$$I_3 = \frac{V_3}{R} = \frac{V_1}{3R}$$

Substituting in Eq. (i), we get,

$$I_1 = \frac{1}{2} \times \frac{V_1}{2R} + \frac{1}{3} \times \frac{V_1}{3R}$$

$$\text{or, } I_1 = \frac{13}{36R}V_1$$

$$\text{or, } Z = \frac{V_1}{I_1} = \frac{36}{13}R$$

Ex.5.16 A 66/11 KV, delta–star three-phase transformer has a core cross-sectional area of 400 cm² and the flux density of 2 Wb/m². Find the number of turns of primary and secondary.

Solution

$$E_p(\text{Phase}) = 66000 \text{ V}$$

$$N_p = \frac{66000}{4.44 \times 50 \times 2 \times 400 \times 10^{-4}} = 3716$$

$$E_s(\text{Phase}) = \frac{11000}{\sqrt{3}} = 6350.85 \text{ V}$$

$$\frac{E_p(\text{Phase})}{E_s(\text{Phase})} = \frac{N_p}{N_s}$$

$$N_s = \frac{E_s(\text{Phase})}{E_p(\text{Phase})} \times N_p = \frac{6350.85 \times 3716}{66000} = 358$$

Ex.5.17 Three single-phase 11KV/220 V, 200 KVA, transformers connected in delta–delta form supply a load of 500 KVA at 0.8 p.f. lag. If one of the transformers is removed for repair, then calculate the following:

(i) For each transformer, find the load in KVA
(ii) Percentage load shared by each transformer.
(iii) Total KVA rating of the transformer in open delta configuration
(iv) V–V bank to delta–delta transformer ratio
(v) Load increase in percentage for each transformer.

Solution

$$\text{Load in KVA for each transformer} = \frac{500}{\sqrt{3}} = 288.68 \text{ KVA}$$

$$\text{Percentage load shared by each transformer} = \frac{288.68}{200} \times 100 = 144.34\ \%$$

$$\text{Total KVA rating} = \sqrt{3} \times 200 = 346 \text{ KVA}$$

$$\frac{\text{V} - \text{V Bank}}{\Delta - \Delta \text{Bank}} = \frac{346}{3 \times 200} = 57.67\%$$

$$\text{Load KVA per transformer} = \frac{500}{3} = 166.67\,\text{KVA}$$

$$\text{Percentage increase in load} = \frac{288.63}{166.67} \times 100 = 173\ \%$$

Ex.5.18 A Scott-connected transformer is fed from 7000 V , two-phase network and supplies three-phase power at 600V (Line) on a four-wire system. If there are 650 turns per phase on the three-phase side, find the number of turns in the L.V. side and the position of the tapping of the neutral wirc.

Solution

$$N_{\text{lv}} = \frac{600 \times 650}{7000} = 56$$

$$N_{\text{ps}} = \frac{\sqrt{3}}{2} \times 56 = 49$$

$$N_{\text{pn}} = \frac{2}{3} \times 49 = 33$$

The neutral point is located at 33rd turn from the point A

Ex.5.19 Two transformers are connected in open delta supply with a 300 KVA balanced load and operating at 0.8 p.f. lag. If the voltage is 500V obtain the KVA rating of each transformer.

Solution

$$\theta = \cos^{-1} 0.8 = 36.86$$

The KVA rating of each transformer will be,

$$\text{KVA}_1 = \frac{300/2}{0.8} \times \cos(30 + 36.86) = 73.68\ \text{KVA}$$

$$\text{KVA}_2 = \frac{300/2}{0.8} \times \cos(30 - 36.86) = 186.16\ \text{KVA}$$

Ex.5.20 A three-phase star connected autotransformer supplies a balanced three-phase load of 200 KW at 400 V, 0.8 p.f. lag. If the supply voltage is 500V, determine the current in the winding as well as in the input and output lines.

Solution

$$I_1 = \frac{200 \times 10^3}{\sqrt{3} \times 400 \times 0.8} = 360.84\ \text{A}$$

$$I_{\text{h}} = \frac{400}{500} \times 360.84 = 288.5\ \text{A}$$

EXERCISES

1. What are the different ways of obtaining a three-phase transformer?
2. What are the advantages and disadvantages of a single unit three-phase transformer?
3. Explain the construction of core-type three-phase transformer.
4. Explain the construction of shell-type three-phase transformer.
5. What are the most commonly used connections for a three-phase transformer?
6. Draw and explain the following three phase transformer connection.
   (i) Star–Star (Y–Y) Connection
   (ii) Delta–Delta (Δ–Δ) Connection
   (iii) Star–Delta (Y–Δ) Connection
   (iv) Delta–Star (Δ–Y) Connection
7. Explain the open delta connection in a three-phase transformer.
8. Prove that the open delta connection can carry only 57.7% of the original load.
9. What are the features of an open delta connection?
10. Draw and explain the Scott connection in transformer.
11. How can you obtain two-phase electric power from a three-phase source?
12. How can you convert three-phase to two-phase?
13. How can you convert three-phase to six-phase?
14. What are the most commonly used schemes for three-phase to six-phase conversion?
15. Draw and explain the following connections.
    (i) Double star connection
    (ii) Double delta connection
    (iii) Six-phase star connection
    (iv) Diametric connection
16. Explain zigzag connection.
17. What are the advantages of zigzag connection?
18. Explain the operation of a three-winding transformer.
19. What are the advantages of using tertiary winding in transformer?
20. A three-phase transformer has 800 turns on the primary and 200 turns on secondary. The supply voltage is 10KV. Find the secondary line voltage at no-load when the windings are connected as,
    (i) Delta–Delta
    (ii) Star–Star
21. Two Scott-connected transformers supplies a balanced load of 800V, 90KVA, from a three phase supply of 6000V. Calculate the voltage, current and the KVA rating of the main and the teaser transformer.
22. Two identical transformers, each of rating 900KVA, 10000/2500 V, 50Hz are connected in open delta on primary and secondary sides. Calculate the following:
    (i) Total load this transformer arrangement can supply.
    (ii) If 250 KVA, 300V three-phase star connected load is connected to this transformer bank, calculate the line current on H.V. side.
23. Consider a three-phase star–delta transformer of rating 2500 KVA, 440/220 KV. A balanced delta connected resistive load of 500 KW is connected on the L.V. side of the transformer. Find:
    (i) The resistive load in each phase measured from line to neutral on the H.V. side of the transformer.
    (ii) The value of this resistance if the load is now star-connected.

24. A three-phase transformer was made by connecting three identical single-phase transformers of rating 22 KV/220V, 100KVA. The leakage reactance per transformer is 8 Ω. A balanced star-connected load of 65 Ω per phase is connected across the secondary of the three-phase transformer. Determine the equivalent per-phase impedance as seen from the primary side of the transformer, when the transformer is connected in,

(i) Star–Star connection
(ii) Star–Delta connection
(iii) Delta–Star connection
(iv) Delta–Delta connection

25. Three, 80 KVA, 300/200 V single-phase, 50 Hz transformers are connected in delta delta manner. The total load connected is 40 KVA. Due to some fault, one transformer is removed and the other two are operated in open delta manner. Calculate the maximum power this new connection can take.

26. Two transformers are connected in open delta manner. A balanced three-phase load takes 90 KW, at 240V, 0.8 power factor lag. The input voltage to the transformer is 600V. Determine the maximum power rating of each transformer.

CHAPTER 6

SPECIAL TRANSFORMERS

6.1 INTRODUCTION

Sometimes, transformers are manufactured for special applications in the industry or power sector. Transformers that are used for such special operations are called special transformers. In this chapter, we shall study a few such special transformers.

6.2 INSTRUMENT TRANSFORMER

In electrical power system, the range of current and voltage is very high and so it is not possible to measure these values directly by any meter of reasonable size and construction. So, there is a need to step down these voltages or currents to such a value that can be easily measured.

Such transformers which are used with the measuring instruments are called instrument transformers. A transformer used for the measurement of current is called the current transformer or CT. Similarly, a transformer used for measurement of voltage is called the potential transformer or PT.

6.2.1 Advantages of Instrument Transformer

Some of the advantages of instrument transformer are:

- Their readings do not depend upon circuit constants such as R, L and C.
- It is possible to standardise the instrument around circuit constant ratings.
- The measuring circuit is isolated from the power circuit
- There is low power consumption in the metering circuit
- Several instruments can be operated from a single instrument transformer.

6.2.2 Definitions

- Transformation ratio (R):
  For a current transformer or CT, it is defined as the ratio of primary winding current to secondary winding current. That is,

$$\text{Transformation ratio }(R)\text{ for CT} = \frac{\text{Primary winding current}}{\text{Secondary winding current}} \quad (6.1)$$

  For a potential transformer or PT, it is defined as the ratio of primary winding voltage to secondary winding voltage. That is,

$$\text{Transformation ratio } (R) \text{ for PT} = \frac{\text{Primary winding voltage}}{\text{Secondary winding voltage}} \tag{6.2}$$

- Nominal Ratio (K_n):
  For a current transformer or CT, it is defined as the ratio of rated primary winding current to rated secondary winding current. That is,

$$\text{Nominal ratio } (K_n) \text{ for CT} = \frac{\text{Rated primary winding current}}{\text{Rated secondary winding current}} \tag{6.3}$$

For a potential transformer or PT, it is defined as the ratio of rated primary winding voltage to rated secondary winding voltage. That is,

$$\text{Nominal ratio } (K_n) \text{ for PT} = \frac{\text{Rated primary winding voltage}}{\text{Rated secondary winding voltage}} \tag{6.4}$$

- Turns Ratio (n):
  For a current transformer or CT, it is defined as the ratio of number of secondary winding turns to number of primary winding turns. That is,

$$\text{Turns ratio } (n) \text{ for CT} = \frac{\text{Number of secondary winding turns}}{\text{Number of primary winding turns}} \tag{6.5}$$

For a potential transformer or PT, it is defined as the ratio of number of primary winding turns to number of secondary winding turns. That is,

$$\text{Nominal ratio } (K_n) \text{ for PT} = \frac{\text{Number of primary winding turns}}{\text{Number of secondary winding turns}} \tag{6.6}$$

6.2.3 Burden of an Instrument Transformer

The secondary load of a CT is generally called the burden to distinguish it from the load of the current whose current is being measured. The burden in a CT metering circuit is the impedance presented to its secondary winding. The typical burden ratings for IEC CTs are, 1.5 VA, 3 VA, 5 VA, 10 VA, 15 VA, 20 VA, 30 VA, 45 VA, 60 VA. For ANSI/IEEE burden ratings are B-0.1, B-0.2, B-0.5, B-1.0, B-2.0 and B-4.0. This means a CT with a burden rating of B-0.2 can tolerate up to 0.2Ω of impedance in the metering circuit before its secondary accuracy falls outside the accuracy specification. The most common source of excess burden is the conductor between the meter and the CT.

6.2.4 Current Transformer

The primary winding of a CT is connected in series with the line whose current needs to be measured. The primary winding has less number of turns and so the voltage drop across it is also less. The secondary winding, on the other hand, has large number of turns. The meters (ammeter or the current coil of the wattmeter) are connected directly across the secondary winding. Thus, we can say that the secondary side of the CT is operated under nearly short circuit condition. Since the secondary is nearly short-circuited, one of the terminals of the secondary winding is earthed to protect the equipment and also the personnel in the case of an insulation breakdown in CT. Figure 6.1 shows the circuit connection with CT for measurement purpose.

6.2.4.1 Relationships in a CT

The equivalent circuit of a CT is shown in Fig. 6.2.

Where,

R_1 is the resistance of the primary winding.
X_1 is the reactance of the primary winding.
R_2 is the resistance of the secondary winding.

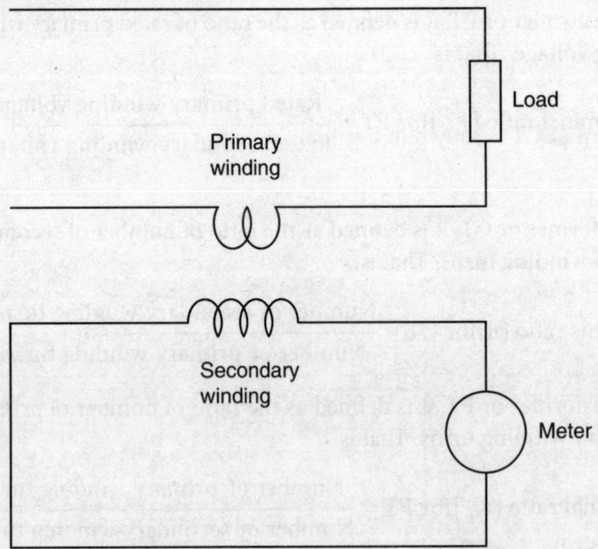

Fig. 6.1: CT for measurement purpose

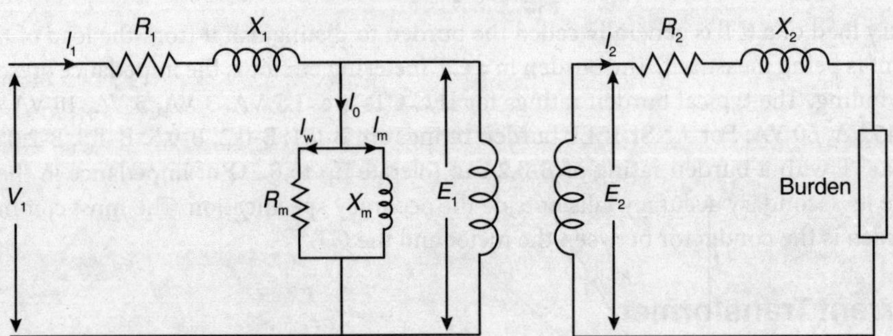

Fig. 6.2: Equivalent circuit of CT

X_2 is the reactance of the secondary winding.
N_1 is the number of turns of primary.
N_2 is the number of turns of secondary
R_L is the resistance of the burden.
X_L is the reactance of the burden.

E_1 is the primary induced e.m.f.
E_2 is the primary induced e.m.f.
I_1 is the primary current.
I_2 is the secondary current.
I_0 is the excitation current.
I_m is the magnetisation component of excitation current.
I_w is the loss component of excitation current.
α is the angle between I_0 and flux φ.
θ is the phase angle of the transformer.
δ angle between the secondary current and the secondary induced e.m.f.

$$K = \frac{N_2}{N_1}$$

The phasor diagram of CT is shown in Fig. 6.3

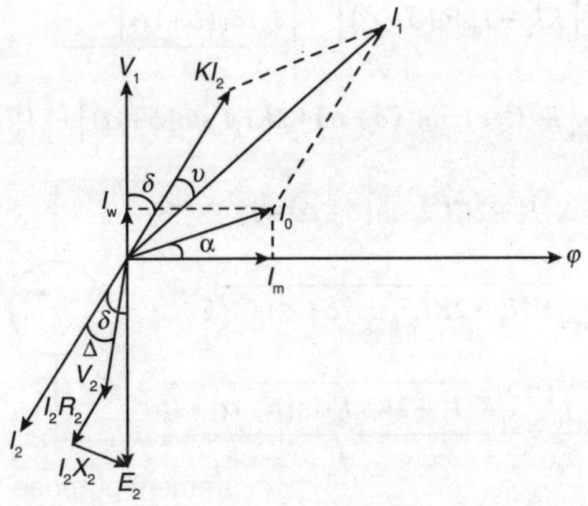

FIG. 6.3: Phasor of CT

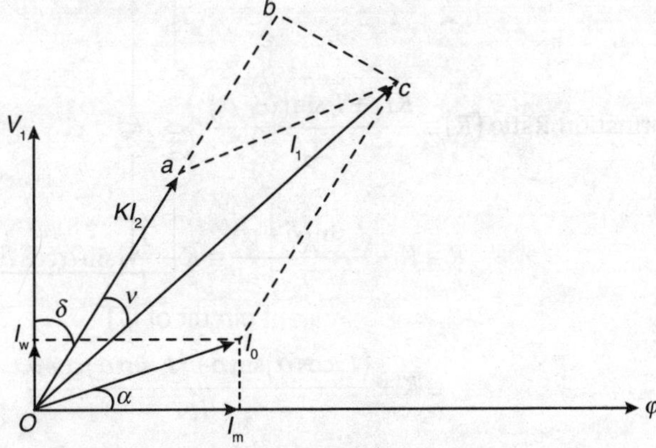

FIG. 6.4: Phasor of CT

The upper portion of the phasor shown in Fig. 6.3 has been redrawn again in Fig. 6.4 for the purpose of analysis. From the phasor of Fig. 6.4, we can write,

$$\angle bac = 90 - \delta - \alpha \qquad ac = I_0 \qquad oa = KI_2 \qquad oc = I_1$$

Then,

$$bc = I_0 \cos(\delta + \alpha) \quad ab = I_0 \sin(\delta + \alpha)$$

Again,

$$(oc)^2 = (ob)^2 + (bc)^2$$

$$= (oa + ab)^2 + (bc)^2$$

$$I_1^2 = [KI_2 + I_0 \sin(\delta + \alpha)]^2 + [I_0 \cos(\delta + \alpha)]^2$$

$$= [K^2 I_2^2 + I_0^2 sin^2(\delta + \alpha) + 2KI_2 I_0 sin(\delta + \alpha)] + [I_0^2 cos^2(\delta + \alpha)]$$

$$= K^2 I_2^2 + 2KI_2 I_0 sin(\delta + \alpha) + I_0^2$$

$$I_1 = \sqrt{K^2 I_2^2 + 2KI_2 I_0 sin(\delta + \alpha) + I_0^2} \tag{6.7}$$

$$\text{Transformation Ratio } (R) = \frac{I_1}{I_2} = \frac{\sqrt{K^2 I_2^2 + 2KI_2 I_0 \sin(\delta + \alpha) + I_0^2}}{I_2} \tag{6.8}$$

Generally, $I_0 \ll KI_2$ and $I_1 \cong KI_2$ so,

$$\text{Transformation Ratio } (R) \cong \frac{\sqrt{K^2 I_2^2 + 2KI_2 I_0 \sin(\delta + \alpha) + I_0^2 \sin^2(\delta + \alpha)}}{I_2}$$

$$\text{Transformation Ratio } (R) \cong \frac{KI_2 + I_0 \sin(\delta + \alpha)}{I_2}$$

$$R \cong K + \frac{I_0 \sin(\delta + \alpha)}{I_2} \cong K + \frac{I_0}{I_2}[\sin\alpha\cos\delta + \cos\alpha\sin\delta]$$

$$\cong K + \frac{(I_0 \cos\alpha)\sin\delta + (I_0 \sin\alpha)\cos\delta}{I_2}$$

$$R \cong \frac{I_m \sin\delta + I_w \cos\delta}{I_2} \tag{6.9}$$

6.2.4.2 Phase Angle

The angle by which the secondary phasor, when reversed, differs in phase from primary current phasor is called phase angle. The angle is positive if the reversed secondary current phasor leads the primary current phasor and the angle is negative if the reversed secondary current phasor lags the primary current phasor.

From the phasor of Fig. 6.4,

$$\tan\theta = \frac{bc}{ob} = \frac{bc}{oa+ab} = \frac{I_0\cos(\alpha+\delta)}{KI_2 + I_0\sin(\alpha+\delta)}$$

Generally θ is very small, so we can write,

$$\theta = \frac{I_0\cos(\alpha+\delta)}{KI_2 + I_0\sin(\alpha+\delta)} = \frac{I_0\cos\alpha\cos\delta - I_0\sin\alpha\sin\delta}{KI_2 + I_0\sin(\alpha+\delta)} \tag{6.10}$$

Since I_0 is very small when compared with KI_2, the component $I_0\sin(\alpha+\delta)$ in the denominator can be neglected.

$$\theta = \frac{I_0\cos\alpha\cos\delta - I_0\sin\alpha\sin\delta}{KI_2} \tag{6.11}$$

$$\theta = \frac{I_m\cos\delta - I_w\sin\delta}{KI_2} \tag{6.12}$$

6.2.4.3 Causes of Errors in CT

The causes of errors in CT are

- The flux density in the core is not a linear function of the magnetising force.
- There is always a magnetic leakage.
- Core loss and copper loss cannot be completely removed.
- The transformer always draws some magnetisation current.

6.2.4.4 Classification of Current Transformers

Current transformer can be classified on the basis of its construction

1) **Bar Type:** They are usually bolted to the current carrying device and are insulated for the operating voltage of the system. They have a permanent bar installed as a primary conductor.
2) **Wound CT:** Wound CT has a primary and secondary winding like a normal transformer. They are designed to measure currents from 1 amp to 100 amps. Wound primary CT are available in ratios from 2.5:5 to 100:5. The wound type provides excellent performance under a wide operating range.
3) **Window:** They do not have any primary winding and so are installed around the primary conductor. The electric field created by current flowing through the primary conductor interacts with the CT core to transform the current to the appropriate secondary output. Window Current Transformers are the most commonly used.
4) **Split-core CT:** Split-core CT have one end removable so that the load conductor or bus bar does not have to be disconnected to install the CT. There are used for measuring currents.

6.2.4.5 Installation of CT

Normally, CT should not be installed on live services. The power should be disconnected before the installation of CT. Measurements must have the same polarity to keep the power factor and direction of power flow measurements accurate and consistent. Usually, CTs are labelled to show which side of the CT should face either the source or the load. Generally, the primary side of the CT is marked with "dot" or with the symbol H1 and H2. The "dot" symbol of H1 gives the direction of flow of current into the CT. Therefore, H1 or the "dot" side of the CT will face the power source and H2 side will face the load. Similarly the secondary side of the CT is marked with the symbol X1 and X2. X1 represents the input side and it is also the polarity terminal. The polarity marks of a current transformer indicate that when a primary current enters at the polarity mark (H1) of the primary, a current in phase with the primary current and proportional to it in magnitude will leave the polarity terminal of the secondary (X1).

6.2.5 Potential Transformer

In a potential transformer (PT), the primary is connected across the line whose voltage needs to be measured and the voltage circuit is connected across the secondary winding. Potential transformer uses large core and conductor sizes in comparison to conventional power transformer. Shell-type configuration is used for low-voltage and core-type configuration is used for high-voltage transformers.

Potential transformer operates on the same principle as a power or distribution transformer. However, the capacity of potential transformer is smaller as compared to a power transformer. Potential transformer has ratings from 100 to 500 VA. LV side is usually wound for 115 or 1200 V. It has highly accurate ratios between primary and secondary voltage values (error is less than 0.4%). Power transformers are not designed for such high accuracy.

6.2.5.1 Relationships in a PT

Potential transformer is very much similar to power transformer. The only difference is that, in a potential transformer, since the loading is very small, the order of excitation current is same as that of the secondary winding current. But in a power transformer, excitation current is only a fraction of secondary winding current. The equivalent circuit of a potential transformer is same as that of a current transformer, as is shown in Fig. 6.2. The phasor diagram of a potential transformer is shown in Fig. 6.5. The phasor diagram referred to primary is shown in Fig. 6.6.

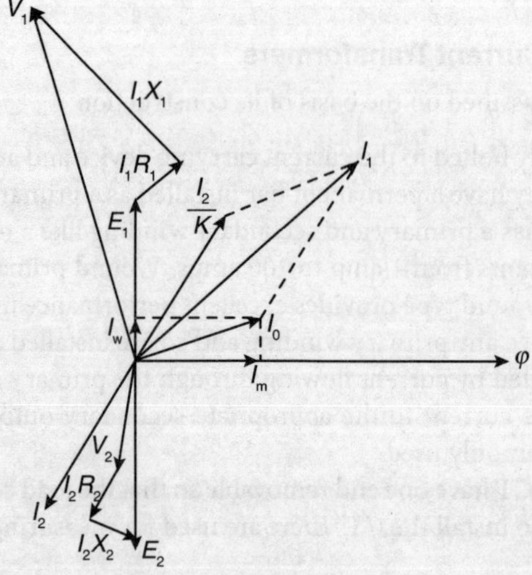

Fig. 6.5: Phasor of PT

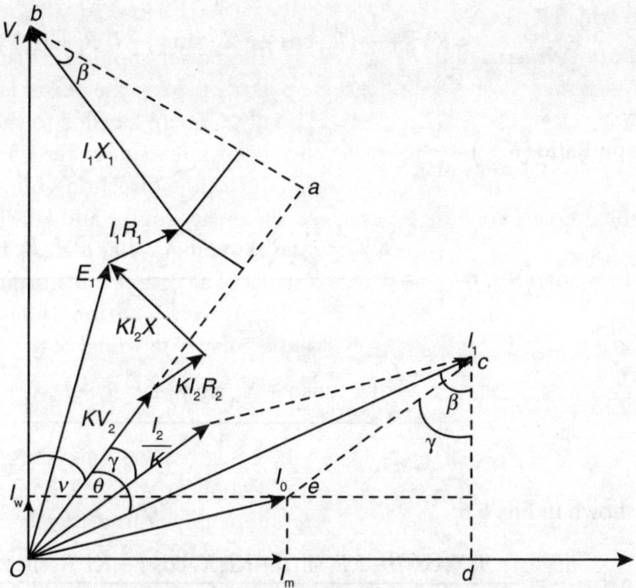

Fig. 6.6: Phasor of PT referred to primary

Actual Transformation Ratio

For the phasor of Fig. 6.6,

$$oa = V_1 \cos\theta = KV_2 + KI_2R_2 \cos\gamma + KI_2X_2 \sin\gamma + I_1R_1 \cos\beta + I_1X_1 \sin\beta$$

$$= KV_2 + KI_2\left(R_2 \cos\gamma + X_2 \sin\gamma\right) + I_1R_1 \cos\beta + I_1X_1 \sin\beta \tag{6.13}$$

Since θ is very small, $\angle ocd \approx \beta \angle ecd \approx \gamma \text{ and } \cos\theta \approx 1$

So, Eq. (6.13) becomes,

$$V_1 = KV_2 + KI_2\left(R_2 \cos\gamma + X_2 \sin\gamma\right) + \left(I_w + \frac{I_2}{K}\cos\gamma\right)R_1 + \left(I_m + \frac{I_2}{K}\sin\gamma\right)X_1$$

$$= KV_2 + I_2 \cos\gamma\left(KR_2 + \frac{R_1}{K}\right) + I_2 \sin\gamma\left(KX_2 + \frac{X_1}{K}\right) + I_wR_1 + I_mX_1$$

$$= KV_2 + \frac{I_2}{K}\cos\gamma\left(K^2R_2 + R_1\right) + \frac{I_2}{K}\sin\gamma\left(K^2X_2 + X_1\right) + I_wR_1 + I_mX_1$$

Let, $R_{e1} = K^2R_2 + R_1$ and $X_{e1} = K^2X_2 + X_1$

Then,

$$V_1 = KV_2 + \frac{I_2}{K}\cos\gamma\left(R_{e1}\right) + \frac{I_2}{K}\sin\gamma\left(X_{e1}\right) + I_wR_1 + I_mX_1$$

$$= KV_2 + \frac{I_2}{K}(R_{e1}\cos\gamma + X_{e1}\sin\gamma) + I_w R_1 + I_m X_1$$

Actual Transformation Ratio $(K_{act}) = \frac{V_1}{V_2}$

$$\text{or, } (K_{act}) = \frac{V_1}{V_2} = \frac{KV_2 + \frac{I_2}{K}(R_{e1}\cos\gamma + X_{e1}\sin\gamma) + I_w R_1 + I_m X_1}{V_2}$$

$$= K + \frac{\frac{I_2}{K}(R_{e1}\cos\gamma + X_{e1}\sin\gamma) + I_w R_1 + I_m X_1}{V_2} \tag{6.14}$$

From the phasor shown in Fig. 6.6,

$$\tan\theta = \frac{ab}{oa} = \frac{I_1X_1\cos\beta - I_1R_1\sin\beta + KI_2X_2\cos\gamma - KI_2R_2\sin\gamma}{KV_2 + KI_2R_2\cos\gamma + KI_2X_2\sin\gamma + I_1R_1\cos\beta + I_1X_1\sin\beta} \tag{6.15}$$

In the denominator, I_1 and I_2 are very small and hence can be neglected. So Eq. (6.15), becomes,

$$\tan\theta = \frac{ab}{oa} = \frac{I_1X_1\cos\beta - I_1R_1\sin\beta + KI_2X_2\cos\gamma - KI_2R_2\sin\gamma}{KV_2}$$

$$= \frac{X_1\left(I_w + \frac{I_2}{K}\cos\gamma\right) - R_1\left(I_m + \frac{I_2}{K}\sin\gamma\right) + KI_2X_2\cos\gamma - KI_2R_2\sin\gamma}{KV_2}$$

$$= \frac{I_2\cos\gamma\left(\frac{X_1}{K} + KX_2\right) - I_2\sin\gamma\left(\frac{R_1}{K} + KR_2\right) + I_wX_1 - I_mR_1}{KV_2}$$

$$= \frac{\frac{I_2\cos\gamma}{K}(X_1 + K^2X_2) - \frac{I_2\sin\gamma}{K}(R_1 + K^2R_2) + I_wX_1 - I_mR_1}{KV_2}$$

Let, $X_{e1} = X_1 + K^2X_2$ and $R_{e1} = R_1 + K^2R_2$

$$= \frac{\frac{I_2\cos\gamma}{K}(X_{e1}) - \frac{I_2\sin\gamma}{K}(R_{e1}) + I_wX_1 - I_mR_1}{KV_2} \tag{6.16}$$

Since, θ is very small,

$$\tan\theta \approx \theta$$

So equation (6.16), becomes,

$$\theta = \frac{\frac{I_2}{K}\left(X_{e1}\cos\gamma - R_{e1}\sin\gamma\right) + I_w X_1 - I_m R_1}{KV_2} \qquad (6.17)$$

6.2.6 Difference between CT and PT

Some of the differences between the current transformer and potential transformer are listed in Table.6.1.

Table 6.1: Difference between CT and PT

Current transformer (CT)	Potential transformer (PT)
Secondary must be kept under short-circuit condition always.	Secondary is open circuited.
Windings carry full line current.	Windings have full line voltage.
Primary current is independent of secondary.	Primary is dependent on secondary.
It can be considered as series transformer under short-circuit condition.	It can be considered as parallel transformer under open-circuit condition.
Small voltage exists across its terminals.	Full line voltage appears across its terminals.

6.2.7 Construction of Potential Transformer

Potential transformers have large core and conductor sizes as compared with the power transformer. The output of potential transformer is very small and the size is comparatively large; so there is no problem of temperature with the potential transformer. The core of the transformer is made up of material having high permeability; this is to reduce iron losses, ratio error and phase angle error. Potential transformer is basically a two-winding, step-down transformer. The primary is connected to high voltage source and has large number of turns of windings, while the secondary is connected to low voltage and has lesser number of turns. The core of the potential transformer can be of shell-type or core-type construction. Normally, shell-type core is used for potential transformer. The primary and secondary windings are arranged co-axially to reduce leakage reactance. Usually, cotton tape and varnish are used as insulation. Oil filled bushings are usually used for oil-filled potential transformer to reduce the size of the transformer.

6.2.8 Capacitive Potential Transformer

Capacitor voltage transformers are used for metering and protection in high-voltage network systems. In the case of inductive-type transformers, insulation of windings at high voltages is the biggest problem. Moreover, losses in the iron core at high voltage increase to a greater extent, which may affect the accuracy of the transformer.

A capacitive voltage transformer works on the principle of voltage division, by using two capacitances, one many times larger than the other. Consider two capacitance C1 and C2, where C2 >> C1), the voltage across the larger capacitance C2 will be V2 = V*C1/ (C1+C2) ~. Thus, we have effectively reduced the voltage level to a fraction of the original without using step-down transformers (which, for large voltages, will be bulky and expensive). CVTs, on the other hand, are cheaper and smaller.

6.3 EARTHING TRANSFORMER

The main purpose of an earthing transformer is to pass the faulty current during earth fault condition. It provides a neutral point for grounding purpose and carries no useful load. Low-impedance ground path can be obtained with the help of three pairs of concentric coils. There coils are connected in the manner so that they oppose the mmf or ampere turns. Because of this, very less magnetisation current flows through the transformer under normal condition. However, the earth fault current gets a path of very low impedance to flow, and hence it is used as an earthing transformer. Moreover, in terms of cost, it is very economical to use. It provides a stable neutral and it is free from harmonic. It can be used with both star- or delta-connected windings to feed the load. It keeps the zero sequence impedance almost constant under load condition and constant even when the auxiliary winding is under load. The most important property of this transformer is that the fault current is not reflected on the secondary side.

6.3.1 Function of Earthing Transformers

The main function is to provide a path for the earth fault current to flow easily. Some of the other function of earthing transformer is as follows.

i) Unbalanced load current can circulate in the neutral
ii) Neutral shift is kept within acceptable range
iii) Provides better earthing to the system
iv) Limits the current during faults

6.3.2 Rating of an Earthing Transformer

Earthing transformer works on load and supplies current only if one of the lines becomes grounded. Since it works on almost no load, it should have low iron losses. The KVA rating of a three-phase earthing transformer or a bank is the product of normal line-to-neutral voltage (KV) and the neutral or ground amperes that the transformer is designed to carry under fault conditions for a specified time.

6.4 PULSE TRANSFORMER

A pulse transformer is a special transformer that is designed for producing rectangular electrical pulses. Pulses are mainly used in electronic circuits. Special high-voltage pulse transformers are also used to generate high-power pulses for radar, particle accelerators, or other high-energy pulsed power applications. A pulse transformer generally has low values of leakage inductance and distributed capacitance, and a high open-circuit inductance. A pulse transformer is characterised by the product of the peak pulse voltage and the duration of the pulse. In AC power transformers, magnetic fluxes alternate between positive and negative values, but in case of pulse transformer, this phenomenon is not applicable.

6.4.1 Types of Pulse Transformers

Pulse transformers are classified as signal-type pulse transformer and power-type pulse transformer

i) Signal-type pulse transformer: This type of transformer is small in size and is used for low-power applications. The amplitude of the pulses from this type of transformer is also small. This type of pulse transformer is mainly used in telecommunication services. Signal-type pulse transformers are also used in digital logic applications, gate controlling, turning ON and turning OFF of SCRs, etc.

ii) Power-type pulse transformer: The size of this type of pulse transformer is large enough as compared to single pulse-type transformer. The amplitude of the pulse of this transformer will also be very high. So, sufficiently good insulations are provided for this type of transformer. These transformers are used in power transmission and distribution sectors. They are also used to combine low-voltage circuits and high-voltage semiconductor gates.

6.4.2 Construction of Pulse Transformer

Shell-type core is used in the design of pulse transformer. Shell-type cores are mechanically strong and hence, this type of transformer can withstand high voltage levels and its eddy current loss is comparatively less. Also, duc to shell type construction the capacitance effect in between the primary and secondary winding is reduced. The rising and falling process is very rapid in pulse series; thus, the core is subjected to rapid magnetisation and demagnetisation. To withstand with this rapid change in magnetisation, the core (frame) is usually made with high permeability material. Instead of silicon steel, they are constructed with Permalloy or ferrite materials as they offer more permeability and inductance. Sufficient insulation is provided for the core and copper windings, which ensures the electrical isolation for high-voltage pulse transformers.

6.4.3 Working of Pulse Transformer

For pulse transformer, the primary is supplied with a square pulse signal or a distorted sinusoidal signal. This distorted signal randomises the flux distribution in primary coil. The high reluctance path of the core confines those fluxes within it and thus, reduces the leakage fluxes. The distortion of pulses is also reduced as the core is made of high permeability material.

Then, the flux linkages in secondary winding leads to the development of another pulse series, which is nothing but the dV / dt of the input signal. Amplitude of the pulses are dependent upon the turns ratio. So, it gives a distortion-less and modified pulse series as the output signal.

The magnetic flux in a typical AC transformer core alternates between positive and negative values but the magnetic flux in a typical pulse transformer does not. The typical pulse transformer operates in "unipolar" mode (flux density may meet but does not cross zero).

A fixed DC current could be used to create a biasing DC magnetic field in the transformer core, thereby forcing the field to cross over the zero line. Pulse transformers usually (not always) operate at high frequency necessitating use of low loss cores (usually ferrites).

6.5 FURNACE TRANSFORMER

Furnace transformers are somewhat different from other power transformers due to their application, operating conditions and their close proximity to hot furnaces.

Furnace transformers are used to step down from voltages between 11 kV and 33 kV to levels of several hundred volts only. This results in massive secondary currents. As an example a 30 MVA unit at 150 V would result in a secondary current of 115 kilo Ampere. For such high secondary currents, special bushings are required to connect to the busbars. These bushings are specified with very specific arrangements to suit the busbar arrangement and cooling system. Furnace busbars are mostly water cooled.

Due to the high secondary currents and resistive losses, the furnace layout is such as to limit the burbar length. The furnace transformers are located close to the furnace itself and if they are single-phase units, they are arranged in a triangle around the furnace. To reduce the fire risk the transformers are enclosed in rooms, which adds to the high ambient temperature.

Furnace transformers are very much in a production environment. Loading of these transformers is therefore very close to rated values and even beyond and hence, the reliability of these transformers has to be very high.

Due to the nature of the process, furnace transformers are specified with large tapping ranges. It is common to have as many as thirty tap positions. Adding to the wide tapping range is the utilisation of the tap changer. Some users require up to 800 operations of the tap changer per day. This demands high maintainability and efforts to increase the maintenance intervals. On-line tap changer oil filters are thus essential. To reduce downtime further, plug-in type diverters are specified. This allows a quick changeover of the diverter switch and an overhaul in a workshop environment with more time at hand.

Another aspect of the process is the large number of short circuits that these transformers are subjected to every day. Transformers associated with open-arc furnaces can be subjected to a number of short circuits per melt as the material being melted collapses across the electrodes. A very robust design in terms of the transformers' ability to withstand the dynamic effects of repeated short circuits is required. Minimum impedance values for furnace transformers of the core-type are in the order of 4 – 5 %. To achieve lower values, one would need a shell-type transformer. Upper levels for impedance could be any value from 10 to 24 % depending on the configuration and tapping range.

Connections and ratings are of utmost importance. The production capacity of the furnace is directly connected to this rating. The wide tapping range makes it expensive to have the full rating across the complete tapping range. Reduced ratings at lower secondary voltages are allowed to reduce size and costs.

Another factor that can make a furnace transformer expensive is the step size of the different LV tap voltages, as well as the regulating range, the ratio of range to lowest LV voltage and the existence and level of the knee-point voltage. If equal LV voltage steps are specified, the internal connection of the transformer requires a dual active part, one being the regulator and the other the furnace transformer. The regulating and the furnace transformer have different magnetic flux levels and are on separate cores, but in the same tank. For the regulating transformer an auto connection is preferred.

6.6 RECTIFIER TRANSFORMER

Rectifier transformers are used to supply rectifier bridges for DC applications such as DC arc furnaces. These transformers step down the distribution voltages to the level used by between 500 V and 3300 V depending on the configuration of the power electronics.

The bridge rectifiers can have different arrangements requiring vector groups with phase displacements other than 30° intervals. Zigzag and extended delta vectors are not uncommon. Depending on the number of pulses used in the bridge rectifier, there could be a requirement for two active parts in one tank.

Furthermore, the power electronics generate DC and harmonics which influence the transformer design. The design flux density is kept low to prevent core saturation due to DC components and the winding conductor and cooling design need to cater to the additional losses due to harmonic currents.

SOLVED PROBLEMS

Ex.6.1 The reading of an ammeter connected with 440/10 CT is 5A. Calculate the line current.

Solution

$$\frac{I_1}{I_2} = \frac{440}{10}$$

$$\text{or, } \frac{I_1}{5} = \frac{440}{10}$$

$$\text{So, } I_1 = 220\,\text{A}$$

Ex.6.2 A 30VA, 110/10 CT has iron loss of 0.5W and magnetisation current of 2A, at its rated load. Calculate the ratio error and the phase angle when supplying rated output to a meter having a ratio of resistance to reactance of 8.

Solution

$$K = n = \frac{110}{10} = 11$$

Given,

$$\frac{R_2}{X_2} = 8$$

$$I_p = 110\,\text{A and } I_s = 10\,\text{A}$$

Again,

$$\delta = \tan^{-1}\frac{X_2}{R_2} = \tan^{-1}\frac{1}{8} = 7.125^0$$

$$\cos\delta = \cos 7.125 = 0.922$$

$$\sin\delta = \sin 7.125 = 0.124$$

Also,

$$E_p I_p = 30.$$

$$E_p = \frac{30}{I_p} = \frac{30}{110} = 0.273\,\text{V}$$

Then,

$$I_c = \frac{0.5}{0.273} = 1.832\,\text{A}$$

$$I_m = 2\,\text{A}(\text{Given})$$

$$\text{Actual ratio }(R) = 11 + \frac{1.832 \times 0.992 + 2 \times 0.124}{10} = 11.207$$

$$\text{Ratio error} = \frac{11 - 11.207}{11.207} \times 100 = -1.85$$

$$\text{Phase angle error} = \frac{180}{\pi}\left(\frac{2 \times 0.992 - 1.832 \times 0.124}{11 \times 10}\right) = 0.915 \text{ degree}$$

Ex.6.3 A CT has 2 turns on primary and 80 turns on secondary. The secondary is connected to an ammeter having a resistance of 0.2Ω. The resistance of the secondary winding is 0.25Ω. If the primary current is 400 A, find the reading of the ammeter, potential difference across the ammeter and VA on the secondary side.

Solution

$$\text{Ammeter reading }(I_2) = I_1 \frac{N_1}{N_2} = 400 \frac{2}{80} = 10\,\text{A}$$

$$\text{Potential difference across ammeter} = 10 \times 0.2 = 2\,\text{V}$$

$$\text{Total resistance of secondary} = 0.25 + 0.2 = 0.45\ \Omega$$

$$\text{Induced e.m.f.} = 10 \times 0.45 = 4.5\,\text{V}$$

$$\text{Total VA} = 4.5 \times 10 = 45\ \text{VA}$$

Ex.6.4 Consider the network shown in Fig.E6.4

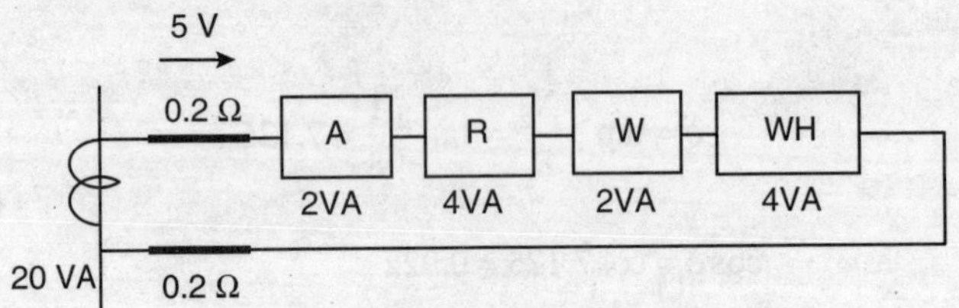

Fig. E6.4:

Is 20VA, 5A CT is suitable for this network?

Solution

$$\text{Total instrument burden} = 2 + 4 + 2 + 4 = 12\ \text{VA}$$

$$\text{Total lead resistance} = 0.2 + 0.2 = 0.4\ \Omega$$

$$\text{Total drop across the lead} = 0.4 \times 5 = 2\,\text{V}$$

$$\text{Total burden on the lead} = 2 \times 5 = 10\,\text{VA}$$

$$\text{Total burden on CT} = 12 + 10 = 22\,\text{VA}$$

Since the total burden is 22VA so a 20 VA CT is not sufficient for the circuit.

Ex.6.5 The nominal secondary voltage of a 4000/80 single-phase PT is 50V. Calculate the ratio error and the phase angle error of the CT when the burden on it is 90 + *j*150 Ω. The equivalent resistance and reactance referred to secondary is 4Ω and 2Ω respectively.

Solution

Given,

$$R_e = 4\,\Omega \quad X_e = 2\,\Omega\; V_p = 4000\,\text{V} \quad V_s = 80\,\text{V}$$

$$K = \frac{4000}{80} = 50$$

$$Z_b = \sqrt{90^2 + 150^2} = 175\,\Omega$$

$$\delta = \tan^{-1}\frac{150}{90} = 50.04^0$$

$$R = 50 + \frac{50(4\times\cos\delta + 2\sin\delta)}{175} = 50 + \frac{50(4\times 0.5144 + 2\times 0.8575)}{175}$$

$$= 51.077$$

$$\%\text{ratio error} = \frac{50 - 51.077}{51.077}\times 100 = -2.11\%$$

$$\theta = \frac{1}{175}(2\times 0.5144 - 4\times 0.8575) = -0.013 \text{ radian}$$

Ex.6.6 To achieve low PT error what should be the value of burden?

Solution

In PT, the burden should be in permissible range for error-less measurement.

Ex.6.7 A bat-type CT has 1 turn on its primary and 160 turns on its secondary. It is to be used with an ammeter that has an internal resistance of 0.2Ω. The ammeter is required to give a full-scale deflection when the primary current is 800A. Calculate the maximum secondary current and secondary voltage across the ammeter.

Solution

$$I_s = 800\times\frac{1}{160} = 5\text{A}$$

$$V_s = 5\times 0.2 = 1\text{V}$$

Ex.6.8 A CT has 1 turn on its primary and 100 turns on its secondary. The secondary has a purely resistive burden of 1.5W, which takes a current of 6A. The magnetising ampere turn is 60 AT. The supply frequency is 50 Hz and the cross-sectional area is 800mm². Find the following:

(i) Ratio and phase angle error
(ii) Flux density

Solution

$$E_s = 5\times1.5 = 9\text{V}$$

$$I_m = \frac{60}{1} = 60\,\text{A}$$

$$I_s = 6\text{A}$$

$$\text{turns ratio}\,(n) = \frac{100}{1} = 100$$

$$I_s' = 100\times6 = 600\text{A}$$

$$I_p = \sqrt{600^2 + 60^2} = 603\,\text{A}$$

$$\text{Actual transformation ratio} = \frac{603}{6} = 100.5$$

$$\text{Phase angle} = \tan^{-1} = \frac{60}{600} = 5.7$$

$$\varphi_m = \frac{9}{4.44\times50\times100} = 0.41\text{mWb}$$

$$B_m = \frac{0.41\times10^{-3}}{800\times10^{-6}} = 0.51\,\text{Wb/m}^2$$

Ex.6.9 The impedance of a bar-type CT having 500 turns on its secondary is (2 + j2) Ω. Calculate the ratio of phase angle error if the secondary current is 7A, magnetising m.m.f. is 90 A and iron loss is 1.5 W

Solution

$$\text{Turns ratio} = \frac{500}{1} = 500$$

$$Z_s = \sqrt{2^2 + 2^2} = 2.83\,\Omega$$

$$\sin\delta = \cos\delta = \frac{2}{2.83} = 0.707$$

$$E_s = 2.83\times7 = 19.81\text{V}$$

$$E_{\mathrm{p}} = \frac{19.81}{500} = 0.0396\,\mathrm{V}$$

$$I_{\mathrm{c}} = \frac{1.5}{0.0396} = 37.88\,\mathrm{A}$$

$$I_{\mathrm{m}} = \frac{90}{1} = 90\,\mathrm{A}$$

$$\text{Transformation ratio} = \frac{90 \times 0.707 + 37.88 \times 0.707}{7} + 500 = 512.92$$

$$\%\text{ratio error} = \frac{500 - 512.92}{512.92} \times 100 = -2.51\%$$

$$\text{Phase angle error} = \frac{180}{\pi}\left(\frac{90 \times 0.707 - 37.88 \times 0.707}{500 \times 7}\right) = 0.603$$

EXERCISES

1. For what purpose are instrument transformers used?
2. What are the applications of instrument transformers?
3. What are the advantages of using an instrument transformer?
4. Define the following terms:
   (i) Transformation ratio
   (ii) Nominal ratio
   (iii) Turns ratio
5. What do you mean by burden on an instrument transformer?
6. What is the purpose of current transformer?
7. Explain why one of the terminals of the secondary winding of CT is earthed.
8. Derive an expression for transformation ratio for CT.
9. Derive an expression for phase angle for CT.
10. List the causes of errors in CT.
11. How are CT classified? Explain each of them.
12. What is the procedure for the installation of CT.?
13. What is the purpose of potential transformer?
14. Derive an expression for transformation ratio for PT.
15. Derive an expression for phase angle for PT.
16. Differentiate between CT and PT
17. Explain the construction of a potential transformer.
18. Explain capacitive potential transformer.
19. Write a note on earthing transformer.
20. What are the functions of earthing transformer?

21. What is the rating of earthing transformer?
22. Write a note on pulse transformer.
23. Give the classification of pulse transformer.
24. Explain the construction of pulse transformer.
25. Explain the working of pulse transformer.
26. Write a note on furnace transformer.
27. Explain rectifier transformer.
28. A CT has 400 turns on secondary and a single turn on primary. The resistance and the reactance of the secondary are 4Ω and 3Ω respectively. The current through the secondary is 8A, the magnetising m.m.f. is 150A and iron loss is 1.2W. Find the ratio error and phase angle error.
29. A CT has single primary and 300 secondary turns. 8A current flows through the secondary to a burden of 2Ω. It has an m.m.f. of 100A. Find the ratio error and phase angle error.
30. The reading of an ammeter connected with 500/8 CT is 8A. Calculate the line current.
31. A 70VA, 150/5 CT has iron loss of 0.9W and magnetisation current of 5A, at its rated load. Calculate the ratio error and the phase angle when supplying rated output to a meter having a ratio of resistance to reactance of 10.

CHAPTER 7

THREE-PHASE INDUCTION MOTOR

7.1 INTRODUCTION

The most common type of AC machine that is being throughout the world is the induction motor. In the field of conversion of electrical energy to mechanical energy, 80% of all the motors being used for such conversion is the induction motor. Induction motor has so much importance because of the following reasons:

- It can be considered to be the cheapest motor.
- It is rugged and requires less maintenance.
- It is simple in design.
- It gives reliable operation.
- Its efficiency is very high.
- It is easy to control
- It runs at constant speed from zero to full load

7.2 CLASSIFICATION OF INDUCTION MOTOR

Depending on the rotor construction, induction motor can be classified into two categories:

- Squirrel-cage induction motor.
- Slip-ring induction motor or wound rotor induction motor.

Depending on the number of phases it can be classified as:

- Single-phase induction motor
- Three-phase induction motor

7.3 CONSTRUCTION OF THREE-PHASE INDUCTION MOTOR

As stated in the earlier sections, a three-phase induction motor has mainly two parts, namely, the stator and the rotor. The stator is the stationary part of the machine. It is a hollow, cylindrical core made up of thin laminations of silicon steel. The lamination is provided to reduce hysteresis and eddy current losses. A number of evenly spaced slots are provided on the inner periphery of the laminations. Three-phase windings are provided for a definite number of poles as per requirement of speed. These windings are 120^0 electrically and are either star or delta connected. The leads of these windings are brought out to the terminal box mounted on the motor frame. The insulation between the windings is generally varnish or oxide coated. When a three-phase supply is given to the stator winding of the induction motor, a rotating magnetic field of constant magnitude is produced. This rotating field induces currents in the rotor by electromagnetic induction. The rotor of a three-phase induction motor is a hollow cylinder mounted on a shaft. It is also laminated and has slots for winding in the outer periphery. Depending on how these rotor windings are placed, the rotor of a three-phase induction motor can be classified as,

i) Squirrel-cage type induction motor
ii) Wound-type induction motor

Squirrel-cage induction motor

This type of induction motor has parallel slots in the outer periphery of the rotor cylinder and copper or aluminium bar is placed in each of these slots. All these bars are joined at the end with the help of metal rings called end rings This type of construction looks like a cage and hence the name, squirrel-cage. The slots in the periphery of the rotor are not made parallel to each other, but a bit skewed as the skewing prevents magnetic locking of stator and rotor teeth and makes the working of motor more smooth and quieter. As the bars are permanently shorted by end rings, the rotor resistance is very small and it is not possible to add external resistance from outside (as the bars are permanently shorted). The absence of slip ring and brushes make the construction of squirrel-cage three phase induction motor very simple and robust and hence, widely used three-phase induction motor.

Wound rotor induction motor

This is also called slip-ring type induction motor. The rotor winding, in this case, is uniformly distributed in the slots and is usually star-connected. The three open ends of rotor winding are brought out and joined to three insulated slip rings mounted on the rotor shaft with one brush for each slip-ring. The three brushes are then connected to a three-phase star-connected rheostat. When the motor is started, these external resistances are kept at maximum position to give a large starting torque. As the motor picks up speed, these resistances are slowly cut out from the circuit.

7.4 DIFFERENCE BETWEEN SLIP-RING AND SQUIRREL-CAGE INDUCTION MOTOR

The difference between the slip-ring induction motor and squirrel-cage induction motor is listed in the Table .7.1.

TABLE 7.1 Difference between slip-ring and squirrel-cage induction motor

Slip-ring or phase-wound Induction motor	Squirrel-cage induction motor
Due to presence of slip-ring and brushes, the construction is complicated.	Construction is very simple as no slip-rings are required.
The rotor winding is similar to the stator winding.	Rotor bars are placed in the rotor, which is shorted with the help of end rings.
External resistance can be added in the rotor circuit.	It is not possible to add external resistance in the rotor circuit.
Has high starting torque.	Staring torque is low.
Slip-ring and brushes are present.	Slip-ring and brushes are absent.

Table 7.1 (continued)

Slip-ring or phase-wound Induction motor	Squirrel-cage induction motor
Requires regular maintenance.	Less maintenance is required
Rarely used.	Due to its simple construction and low cost, the squirrel-cage induction motor is widely used.
Efficiency is less	Efficiency is high.
Speed control by rotor resistance method is possible	Speed control by rotor resistance method is not possible.
Slip-ring induction motor is used where high starting torque is required, i.e., in hoists, cranes, elevator etc	Squirrel-cage induction motor is used in lathes, drilling machine, fan, blower printing machines etc

7.5 PRODUCTION OF ROTATING MAGNETIC FIELD IN THREE-PHASE INDUCTION MOTOR

The stator of a three-phase induction motor has a three-phase winding connected either in star or delta connection. When this three-phase stator winding is fed from a balanced three-phase supply, three-phase current displaced by 120⁰ (electrical degree) flows through the winding. The currents in all the three phases will produce their own flux, displaced from each other by 120⁰. Mathematically, it can be represented by the expression,

$$\varphi_R = \varphi_m \sin \omega t \quad (7.1)$$

$$\varphi_B = \varphi_m \sin(\omega t + 120) \quad (7.2)$$

$$\varphi_Y = \varphi_m \sin(\omega t - 120) \quad (7.3)$$

Where, φ_R, φ_B and φ_Y are the flux produced in the three phases R, B and Y of the induction motor. φ_m is the maximum value of the flux.

From Eqs. (7.1), (7.2) and (7.3), we can say that flux φ_R leads flux φ_Y by 120⁰, and flux φ_Y leads flux φ_B by 120⁰. The phasor of the three fluxes is shown in Fig. 7.1.

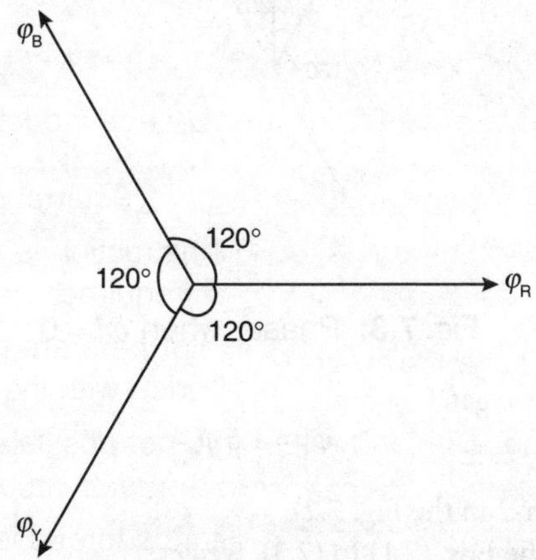

Fig. 7.1: Phasor of three-phase flux

The sinusoidal waveform of all the three fluxes is shown in Fig. 7.2.

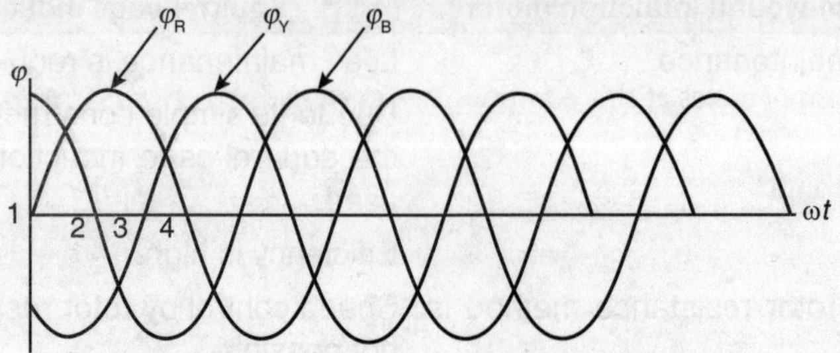

Fig. 7.2: Waveform of three-phase flux

Now, we will see the resultant effect of all these three fluxes at different values of ωt. The resultant flux at any instant is given by the expression,

$$\bar{\varphi}_T = \bar{\varphi}_R + \bar{\varphi}_B + \bar{\varphi}_Y \tag{7.4}$$

Case-1: When $\omega t = 0$ (Position-1 in Fig. 7.2)
Substituting the value of ωt in the equation (7.1) to equation (7.3), we get,

$$\varphi_R = 0$$

$$\varphi_B = \varphi_m \sin 120 = 0.866\varphi_m$$

$$\varphi_Y = \varphi_m \sin -120 = -0.866\varphi_m$$

The phasor of the three fluxes at this condition is shown in Fig. 7.3.

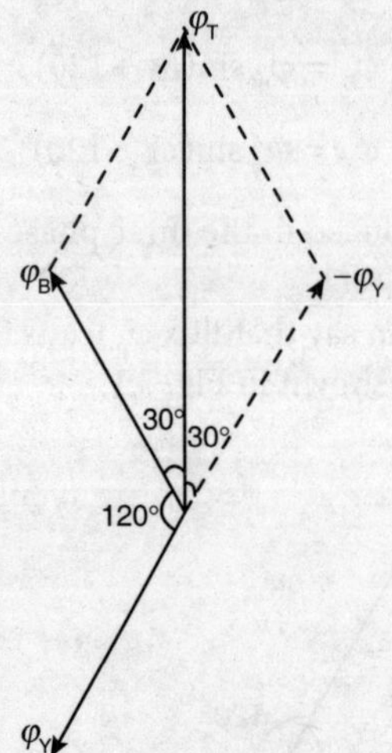

Fig. 7.3: Phasor when $\omega t = 0$

From the phasor of Fig. 7.3, we get,

$$\varphi_T = 1.5\,\varphi_m$$

Case-2: When $\omega t = 60$ (Position-2 in the Fig. 7.2)
Substituting the value of ωt in the Eqs. (7.1) to (7.3), we get,

$$\varphi_B = 0$$

$$\varphi_R = \varphi_m \sin 60 = 0.866\varphi_m$$

$$\varphi_Y = \varphi_m \sin -60 = -0.866\varphi_m$$

The phasor of the three fluxes at this condition is shown in Fig. 7.4.

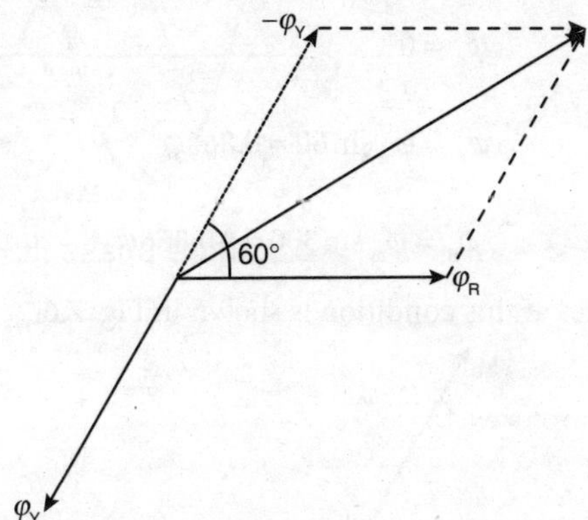

Fig. 7.4: Phasor when $\omega t = 60$

From the phasor of Fig. 7.4, we get,

$$\varphi_T = 1.5\varphi_m$$

Case-3: When $\omega t = 120$ (Position-3 in the Fig. 7.2)
Substituting the value of ωt in the Eq. (7.1) in Eq. (7.3), we get,

$$\varphi_Y = 0$$

$$\varphi_R = \varphi_m \sin 120 = 0.866\varphi_m$$

$$\varphi_B = \varphi_m \sin -240 = -0.866\varphi_m$$

The phasor of the three fluxes at this condition is shown in Fig. 7.5.

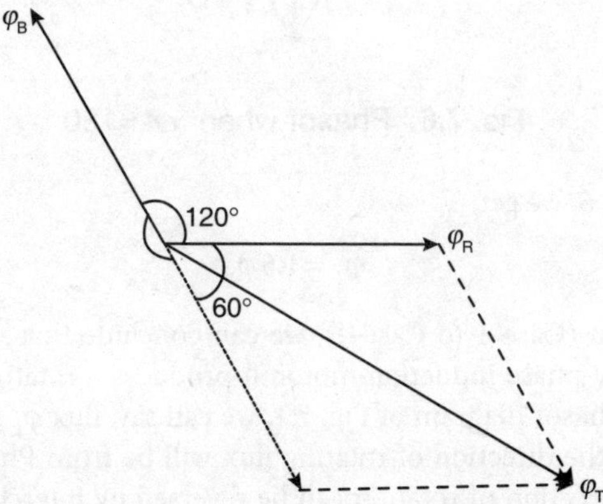

Fig. 7.5: Phasor when ωt =120

From the phasor of Fig. 7.5, we get,

$$\varphi_T = 1.5\varphi_m$$

Case-4: When ωt = 180 (Position-4 in the Fig. 7.2)
Substituting the value of ωt in Eq. (7.1) to Eq. (7.3), we get,

$$\varphi_R = 0$$

$$\varphi_Y = \varphi_m \sin 60 = 0.866\varphi_m$$

$$\varphi_B = \varphi_m \sin 300 = -0.866\varphi_m$$

The phasor of the three fluxes at this condition is shown in Fig. 7.6.

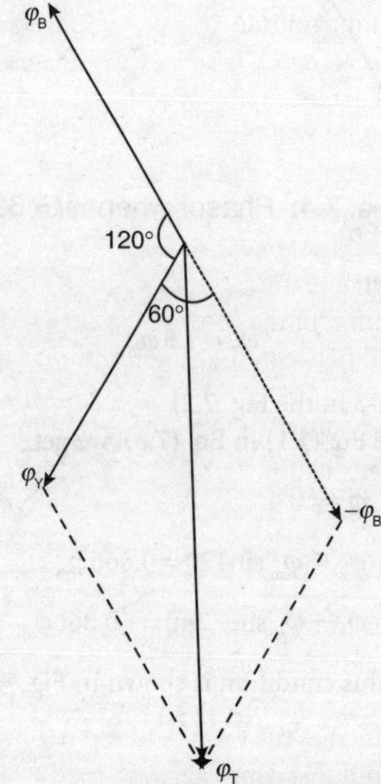

Fig. 7.6: Phasor when ωt =180

From the phasor of Fig. 7.6, we get,

$$\varphi_T = 1.5\varphi_m$$

From the phasor diagram (Case-1 to Case-4), we can conclude that when a three-phase supply is given to the stator of a three-phase induction motor, it produces a rotating magnetic field of constant magnitude. Also from the phasor diagram of Fig. 7.1, we can say, flux φ_R leads flux φ_Y by 120^0 and flux φ_Y leads flux φ_B by 120^0; so the direction of rotating flux will be from Phase R to Phase Y to Phase B. (clockwise rotation). This direction of rotation can be reversed by interchanging any two terminals of the three-phase winding.

The speed at which this magnetic field rotates is called the synchronous speed of the three-phase induction motor and it is given by the expression,

$$N_s = \frac{120f}{P} \tag{7.5}$$

Where,
N_s is the synchronous speed (speed of rotation of magnetic flux).
f is the frequency of supply
P is the number of poles in the stator

7.6 WORKING PRINCIPLE OF THREE-PHASE INDUCTION MOTOR

When a balanced three-phase supply is given to the stator of a three-phase induction motor, it produces a rotating magnetic field of constant magnitude, rotating with a speed of N_s. This rotating field is cut by the stationary rotor conductors; thus an e.m.f. is induced in the rotor. Since the rotor is close-circuited, this induced e.m.f. will cause some current to flow through the rotor.

As per Lenz`s law, this rotor current will oppose the cause of its production; that is, it opposes the relative speed between the rotating magnetic field and stationary rotor conductors. So to oppose this relative speed or to reduce this relative speed, the rotor starts rotating in the same direction of the rotating magnetic field, trying to catch up with the speed of the rotating magnetic field.

If the rotor speed (N) becomes equal to the rotating magnetic field (N_s), then the relative speed will be zero. So, there will be no induced e.m.f. in the rotor and hence no rotor current and the rotor will stop rotating. However, the relative speed between the stator and the rotor soon becomes appreciable again and as a result, the motor restarts.

The rotor of the induction motor always rotates at a speed slightly less than the synchronous speed. Since the rotor never attains the synchronous speed, it continues to rotate without stopping.

7.7 SLIP OF AN INDUCTION MOTOR

When a three-phase supply is given to the stator of an induction motor, it produces a magnetic field rotating at synchronous speed (N_s) and the rotor rotates with the speed (N), which is less than the synchronous speed (N_s).

Slip of an induction motor is defined as the ratio of difference between synchronous speed and speed of the rotor to synchronous speed, and it is denoted by s. That is,

$$\text{Slip}(s) = \frac{N_s - N}{N_s} \tag{7.6}$$

$$\text{or,}\ sN_s = N_s - N$$

$$\text{or,}\ N = (1-s)N_s \tag{7.7}$$

At starting of the motor $N = 0$ and so $s=1$.
Percentage slip is given by the expression,

$$\%s = \frac{N_s - N}{N_s} \times 100 \tag{7.8}$$

7.8 EFFECT OF SLIP ON ROTOR PARAMETER

In this section, we will study in detail the effect of slip on various rotor parameters like,

i) Effect of slip on rotor frequency
ii) Effect of slip on rotor induced e.m.f.
iii) Effect of slip on rotor resistance, reactance and impedance
iv) Effect of slip on rotor current.

From these parameters, we will obtain the torque equation of the induction motor in Section 7.10

Effect of slip on rotor frequency

Equation (7.5) gives the speed of rotation of stator flux. When the rotor is at standstill, the rotor speed (*N*) is zero and the value of slip is 1. When the rotor starts rotating with the speed *N*, the relative speed between the stator and the rotor will be ($N_s - N$). Let the rotor frequency at the rotor speed *N* be f_r. Then, from Eq. (7.5), we can write,

$$N_s - N = \frac{120 f_r}{P} \tag{7.9}$$

Dividing Eq. (7.9) by Eq. (7.5), we get,

$$\frac{N_s - N}{N_s} = \frac{f_r}{f} \tag{7.10}$$

Again from Eq. (7.6), we can write,

$$s = \frac{f_r}{f}$$

$$\text{or, } f_r = sf \tag{7.11}$$

From Eq. (7.11), we can say that when the rotor is rotating with speed *N*, the rotor frequency will be *s* times the supply frequency, or the frequency of the stator.

Effect of slip on rotor-induced e.m.f.

Let,

E_2 be the rotor-induced e.m.f. at standstill (when the rotor is not rotating).
E_{r2} be the rotor-induced e.m.f. during running condition.
Since at standstill $N = 0$,

$$E_2 \propto N_s \tag{7.12}$$

Similarly, during running condition,

$$E_{r2} \propto (N_s - N) \tag{7.13}$$

Dividing Eq. (7.13) by Eq. (7.12), we get,

$$\frac{N_s - N}{N_s} = \frac{E_{r2}}{E_2}$$

$$\text{or}, s = \frac{E_{r2}}{E_2}$$

$$\text{or}, E_{r2} = sE_2 \tag{7.4}$$

From Eq. (7.14), we can say that when the rotor is rotating with speed *N*, the rotor induced e.m.f will be *s* times the rotor e.m.f. at standstill.

Effect of slip on rotor resistance, reactance and impedance

Let,

R_2 and X_2 be the rotor resistance per phase and reactance per phase respectively at standstill.

R_{r2} and X_{r2} be the rotor resistance per phase and reactance per phase respectively during running condition. Since the resistance *R* does not depends of frequency,

$$R_2 = R_{r2} \tag{7.15}$$

Again, we can write,

$$X_2 = 2\pi f L (\text{at standstill}) \tag{7.16}$$

$$X_{r2} = 2\pi f_r L (\text{during running}) \tag{7.17}$$

Dividing Eq. (7.17) by Eq. (7.16), we get,

$$\frac{X_{r2}}{X_2} = \frac{f_r}{f}$$

$$\text{or}, X_{r2} = sX_2 \tag{7.18}$$

Reactance of the rotor during running condition is *s* times the reactance of the rotor at standstill. The rotor impedance at stand still will be,

$$Z_2 = \sqrt{R_2^2 + X_2^2} \tag{7.19}$$

The rotor impedance at running condition will be,

$$Z_2 = \sqrt{R_2^2 + X_{r2}^2} \tag{7.20}$$

$$\text{or}, Z_2 = \sqrt{R_2^2 + (sX_2)^2} \tag{7.21}$$

Effect of slip on rotor current

Let, I_2 be the rotor current, then,

$$I_2 = \frac{E_2}{Z_2} = \frac{E_2}{\sqrt{R_2^2 + X_2^2}} \text{at standstill} \tag{7.22}$$

$$I_{r2} = \frac{E_{r2}}{Z_2} = \frac{sE_2}{\sqrt{R_2^2 + (sX_2)^2}} \text{at running condition} \tag{7.23}$$

7.9 INDUCTION MOTOR AND TRANSFORMER

We know that transformer is a device in which two windings are magnetically coupled and when one winding is excited by AC supply of certain frequency, the e.m.f. gets induced in the second winding having same frequency as that of supply given to the first winding. The winding to which supply is given is called primary winding while winding in which e.m.f. gets induced is called secondary winding. The induction motor can be regarded as the transformer.

There are several differences between an induction motor and a normal transformer:

1. The normal transformer is an alternating flux transformer while induction motor is a rotating flux transformer.
2. The normal transformer has no air gap, whereas an induction motor has a distinct air gap between its stator and rotor.
3. In an alternating flux transformer, the frequency of induced e.m.f. and current in primary and secondary is always same. However in the induction motor, the frequency of e.m.f. and current on the stator side remains same but the frequency of rotor e.m.f. and current depends on the slip, which in turn depends on load on the motor. So, we have a variable frequency on the rotor side. But it is important to remember that at start when $N = 0$, the value of slip is unity ($s = 1$). Thus, the frequency of supply to the stator and the induced e.m.f. in the rotor is same. The effect of slip on the rotor parameters is already discussed in the previous section.
4. In case of the alternating flux transformer, the entire energy present in the secondary circuit is in the electrical form. However in an induction motor, part of its energy in the rotor circuit is in electrical form and the remaining part is converted into mechanical form.

In general, an induction motor can be treated as a generalised transformer. In this, the slip ring induction motor with star connected stator and rotor is shown.

So if E_1 = Stator e.m.f. per phase in volts.

E_2 = Rotor induced e.m.f. per phase in volts at start when motor is at standstill.

Then, according to transformer theory there exists a fixed relation between E_1 and E_2 called transformer ratio.

$\therefore$ At start when $N = 0, s = 1$

and we get,

$$\frac{E_2}{E_1} = K = \frac{\text{Rotor turns per phase}}{\text{Stator turns per phase}} \tag{7.24}$$

7.10 TORQUE EQUATION OF A THREE-PHASE INDUCTION MOTOR

The torque of a three-phase induction motor depends on the following factors:

- Rotor current (I_2)
- Power factor of the rotor circuit ($\cos\theta_2$)
- Flux which links with the rotor (φ)

So, the generalised torque equation will be,

$$\text{Torque}(T) \propto \varphi I_2 \cos\theta_2 \tag{7.25}$$

Where,

The flux that is produced in the stator is directly proportional to the stator-induced e.m.f. (E_1). That is,

$$\varphi \propto E_1 \tag{7.26}$$

From Eq. (7.24), we can write,

$$E_2 = KE_1$$

$$\text{or } E_2 = K\varphi$$

$$\text{or } E_2 \propto \varphi \tag{7.27}$$

Substituting Eq. (7.27) in Eq. (7.25) we get,

$$\text{Torque}\left(T\right) \propto E_2 I_2 \cos\theta_2 \tag{7.28}$$

7.10.1 Torque Equation for Induction Motor in Running Condition

From Eq. (7.23), the rotor current under running condition is

$$I_{r2} = \frac{E_{r2}}{Z_2} = \frac{sE_2}{\sqrt{R_2^2 + \left(sX_2\right)^2}}$$

And power factor under running condition is

$$\cos\theta_2 = \frac{R_2}{Z_2} = \frac{R_2}{\sqrt{R_2^2 + \left(sX_2\right)^2}} \tag{7.29}$$

From Eq. (7.14), the rotor e.m.f. under running condition is sE_2. Substituting Eq. (7.14), Eq. (7.23) and Eq. (7.29) in Eq. (7.28), we get the expression for torque under running condition.

$$T_{\text{run}} \alpha sE_2 \times \frac{sE_2}{\sqrt{R_2^2 + \left(sX_2\right)^2}} \times \frac{R_2}{\sqrt{R_2^2 + \left(sX_2\right)^2}}$$

$$\text{or } T_{\text{run}} = \frac{KsE_2^2 R_2}{R_2^2 + \left(sX_2\right)^2} \tag{7.30}$$

Equation (7.30), is the torque equation of induction motor under running condition, Where,

$$\text{Constant}\left(K\right) = \frac{3}{2\pi N_2} \tag{7.31}$$

7.10.1.1 Condition for maximum running torque

The condition for maximum running torque can be obtained by differentiating Eq. (7.30) with respect to s and equating it to zero. That is,

$$\frac{dT_{\text{run}}}{ds} = 0$$

$$\text{or } \frac{d}{ds}\left[\frac{KsE_2^2 R_2}{R_2^2 + \left(sX_2\right)^2}\right] = 0$$

$$\text{or} \quad \frac{KR_2E_2^2\left(R_2^2+(sX_2)^2\right)-KsR_2E_2^2\left(2sX_2^2\right)}{\left(R_2^2+(sX_2)^2\right)^2}=0$$

$$\text{or} \quad R_2^2-s^2X_2^2=0$$

$$\text{or} \quad R_2=sX_2 \tag{7.32}$$

Equation (7.32) gives the condition for maximum running torque.

So, the slip at maximum torque will be,

$$s_m=\frac{R_2}{X_2}$$

7.10.2 Starting Torque of Three-phase Induction Motor

At the instant of starting, the rotor is stationary and so the value of slip $s = 1$. So, the expression for starting torque can be obtained by substituting $s = 1$ in Eq. (7.30).

$$T_{st}=\frac{KE_2^2R_2}{R_2^2+X_2^2} \tag{7.33}$$

7.10.2.1 *Condition for maximum starting torque of three-phase induction motor*

The condition for maximum starting torque can be obtained by differentiating Eq. (7.33) with respect to R_2 and equating it to zero. That is,

$$\frac{dT_{st}}{dR_2}=0$$

$$\text{or} \quad \frac{d}{dR_2}\left[\frac{KE_2^2R_2}{R_2^2+X_2^2}\right]=0$$

$$\text{or} \quad \frac{KE_2^2\left(R_2^2+X_2^2\right)-KE_2^2R_2\left(2R_2\right)}{\left(R_2^2+X_2^2\right)^2}=0$$

$$\text{or} \quad R_2=X_2 \tag{7.34}$$

Equation (7.34) gives the condition for maximum starting torque.

7.11 TORQUE–SLIP CHARACTERISTICS AND TORQUE–SPEED CHARACTERISTICS OF THREE-PHASE INDUCTION MOTOR

We know that when slip $s = 0$, $N_r = N_s$. Under this condition, the motor stops. So the torque (T) at this value of s is zero. This shows that the torque–slip characteristics starts from the origin.

From equation (7.30), we can say that the torque of an induction motor depends on slip (s). The plot of the variation of torque with the change in slip is called the torque–slip characteristics of induction motor.

Consider the torque Eq. (7.30),

$$T_{run} = \frac{KsE_2^2R_2}{R_2^2 + (sX_2)^2}$$

For small value of slip $(sX_2)^2$ is negligible as compared to R_2^2. Again for a particular induction motor, the supply E_2 is constant. So the torque Eq. (7.30) can be written as

$$T_{run} \propto \frac{sR_2}{R_2^2}$$

$$\text{or } T_{run} \propto \frac{s}{R_2} \tag{7.35}$$

Again for an induction motor, R_2 is constant. So we can say for small value of slip,

$$T_{run} \propto s\sqrt{a^2 + b^2} \tag{7.36}$$

So, we can say for smaller values of slip, the torque is directly proportional to the slip. Hence the characteristic is a straight line with an angle of 45⁰ with the x-axis (slip axis).

Again, for higher values of slip $(sX_2)^2$ now will be larger than R_2^2. So, Eq. (7.30) becomes

$$T_{run} \propto \frac{sR_2}{(sX_2)^2}$$

Again for an induction motor, R_2 and X_2 is constant. So we can say for higher value of slip,

$$T_{run} \propto \frac{1}{s} \tag{7.37}$$

Thus, for larger values of slip, the torque is inversely proportional to slip. Hence, the characteristic will be rectangular hyperbola. The complete torque–slip characteristic of an induction motor is shown in Fig. 7.7.

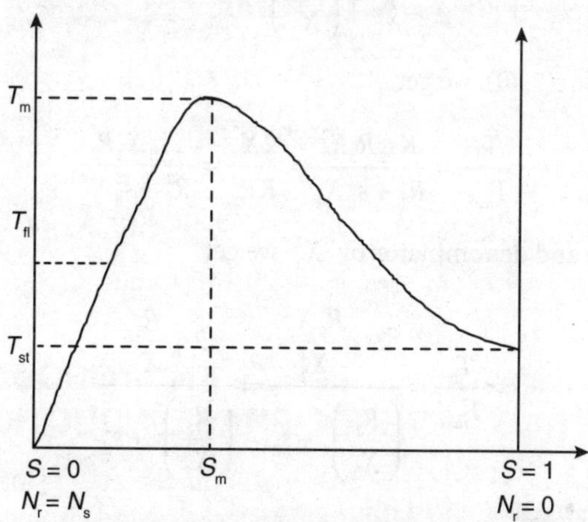

Fig. 7.7: Torque–slip characteristic

When the load on the motor increases, the torque increases as does the slip; but the speed of the motor decreases. The increased torque demand results in drawing more current from the supply. The load which the motor can drive safely, that is, the current drawn by the motor within safe limit, is called the full load of the motor. The torque produced at this load is called the full-load torque (T_{fl}) of the motor (Fig. 7.7).

If the load on the motor is increased beyond this point, the current drawn by the motor will also increase and so does the heat produced in the motor, which may damage the winding insulation of the motor, if the motor is run for long period of time. So, the load on the motor can be increased beyond the full-load point up to maximum torque for a short duration of time. In the region between the full-load torque and the maximum torque, the motor can be used for a short duration of time but not for continuous operation. For continuous operation the motor is used with in full-load torque.

The direction of increasing slip is opposite to the direction of increasing rotor speed. The magnitude of torque at zero rotor speed is called the starting torque of the motor.

7.12 RATIO BETWEEN FULL-LOAD TORQUE AND MAXIMUM TORQUE

Let the value of slip at full-load torque (T_{fl}) be s_{fl}. Then, from Eq. (7.30), we get the expression for the fill-load torque as,

$$T_{fl} = \frac{Ks_{fl}R_2E_2^2}{R_2^2 + s_{fl}^2X_2^2} \tag{7.38}$$

We know that the slip at maximum torque is,

$$s_m = \frac{R_2}{X_2} \tag{7.39}$$

Substituting Eq. (7.39) in Eq. (7.30), we get,

$$T_{max} = \frac{K\dfrac{R_2}{X_2}R_2E_2^2}{R_2^2 + \left(\dfrac{R_2}{X_2}\right)^2 X_2^2} = K\frac{E_2^2}{2X_2} \tag{7.40}$$

Dividing Eq. (7.38) by Eq. (7.40), we get,

$$\frac{T_{fl}}{T_{max}} = \frac{Ks_{fl}R_2E_2^2}{R_2^2 + s_{fl}^2X_2^2} \times \frac{2X_2}{KE_2^2} = \frac{2s_{fl}X_2R_2}{R_2^2 + s_{fl}^2X_2^2}$$

Dividing the numerator and denominator by X_2^2 we get,

$$\frac{T_{fl}}{T_{max}} = \frac{2s_{fl}\dfrac{R_2X_2}{X_2^2}}{\left(\dfrac{R_2}{X_2}\right)^2 + s_{fl}^2} = \frac{2s_{fl}\dfrac{R_2}{X_2}}{\left(\dfrac{R_2}{X_2}\right)^2 + s_{fl}^2}$$

Substituting Eq. (7.39), we get,

$$\frac{T_{fl}}{T_{max}} = \frac{2s_{fl}s_m}{s_m^2 + s_{fl}^2} \tag{7.41}$$

7.13 RATIO BETWEEN STARTING TORQUE AND MAXIMUM TORQUE

From Eq. (7.33)

$$T_{st} = \frac{KE_2^2 R_2}{R_2^2 + X_2^2}$$

From Eq. (7.40)

$$T_{max} = K\frac{E_2^2}{2X_2}$$

Dividing Eq. (7.33) by Eq. (7.40), we get,

$$\frac{T_{st}}{T_{max}} = \frac{KE_2^2 R_2}{R_2^2 + X_2^2} \times \frac{2X_2}{KE_2^2} = \frac{2R_2 X_2}{R_2^2 + X_2^2}$$

Dividing the numerator and denominator by X_2^2 we get,

$$\frac{T_{st}}{T_{max}} = \frac{2\dfrac{R_2}{X_2}}{\left(\dfrac{R_2}{X_2}\right)^2 + 1}$$

Substituting Eq. (7.39), we get,

$$\frac{T_{st}}{T_{max}} = \frac{2s_m}{s_m^2 + 1} \qquad (7.42)$$

7.14 EFFECT OF SUPPLY VOLTAGE ON STARTING TORQUE

From Eq. (7.33)

$$T_{st} = \frac{KE_2^2 R_2}{R_2^2 + X_2^2}$$

For a particular induction motor R_2 and X_2 are constant. So we can write now,

$$T_{st} \propto E_2^2 \qquad (7.43)$$

At standstill,

$$E_2 \propto \varphi \text{ and } \varphi \propto V$$

So we can say, at standstill,

$$E_2 \propto V$$

Where,

φ is the flux and V is the supply voltage.

So Eq. (7.43) becomes,

$$T_{st} \propto V^2 \quad (7.44)$$

From Eq. (7.44), it can be stated that the starting torque is directly proportional to the square of the supply voltage.

7.15 EFFECT OF ROTOR RESISTANCE ON TORQUE–SLIP CHARACTERISTIC OF INDUCTION MOTOR

From Eq. (7.30) for torque,

$$T_{run} = \frac{KsE_2^2R_2}{R_2^2 + (sX_2)^2}$$

If E_2 is taken as constant, we can say that if the rotor resistance R_2 is large as compared to sX_2, then the torque is inversely proportional to the rotor resistance R_2. Again, if the rotor resistance is small as compared to sX_2, then the torque is proportional to the rotor resistance R_2. Figure 7.8 shows the variation of torque–slip characteristics for different values of rotor resistance.

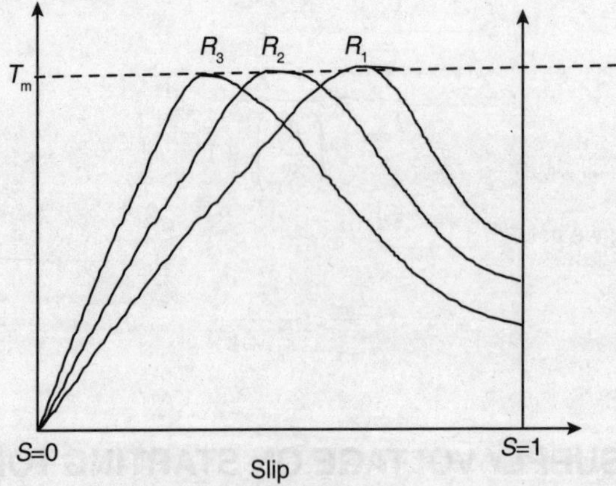

Fig. 7.8: Effect of resistance on torque–slip characteristic

The plot in Fig. 7.8 shows that with increase in rotor resistance, the value of the maximum torque Tm does not change but the slip at which this maximum torque takes place changes. Again, with the increase in rotor resistance, the starting torque decreases.

7.16 POWER OUTPUT OF A THREE-PHASE INDUCTION MOTOR

Let V_1, I_1 and $\cos\theta_1$ be the per-phase value of stator voltage, stator current and stator power factor respectively. Then,

$$\text{Input power to induction motor}\left(P_{in}\right) = 3V_1I_1\cos\theta_1$$

From the input power after subtracting the stator core loss and stator copper loss, we get air gap power (P_{ag}).

$$\text{Air gap power}\left(P_{ag}\right) = P_{in} - 3I_1^2R_1 - \text{Stator core loss}$$

Again, after subtracting the rotor copper loss from the air gap power, we get the mechanical power (P_m) of induction motor.

$$\text{Mechanical power}\left(P_m\right) = P_{ag} - \text{Rotor copper loss}$$

Some part of this mechanical power is wastage towards friction loss, windage loss and stray losses. After subtracting these losses from mechanical power we get the power output (P_o) of induction motor.

$$P_0 = P_m - \text{Friction loss} - \text{Windage loss} - \text{Stray loss}$$

7.17 EQUIVALENT CIRCUIT OF INDUCTION MOTOR

As in the case of transformer, it is also possible to represent induction motor as an equivalent circuit under sinusoidal steady-state operating condition. The equivalent circuit helps in quantitative predictions about the behaviour of the induction machine under various operating conditions.

The equivalent circuit of induction motor for only one phase, with rotor open-circuited is shown in Fig. 7.9.

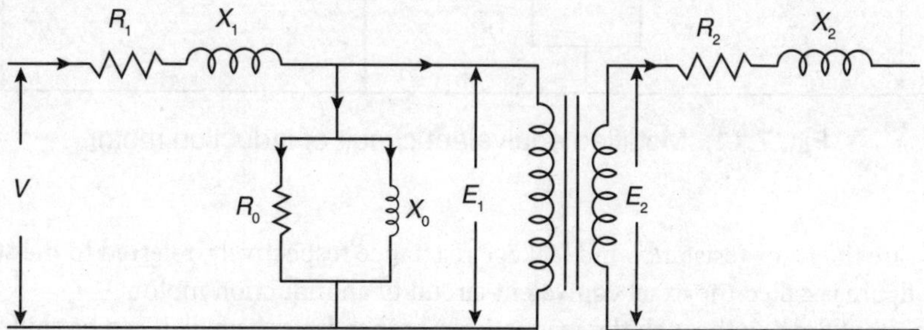

Fig. 7.9: Equivalent circuit of induction motor

Where,

R_1 and X_1 are the stator resistance and stator leakage reactance per phase respectively. R_2 and X_2 are the rotor resistance and standstill rotor leakage reactance per phase respectively. V is the supply voltage. R_0 and X_0 are the no-load resistance and reactance per phase respectively.

Now, let us consider that the rotor is short-circuited (Fig. 7.10). Under this condition, the rotor will start to rotate with a speed (N) less than the synchronous speed (N_s).

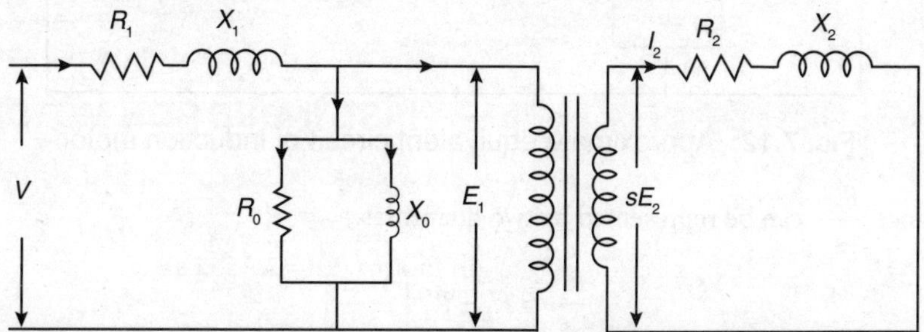

Fig. 7.10: Equivalent circuit of induction motor with rotor short-circuited

The rotor current (I_2), under this condition, is given by the expression,

$$I_2 = \frac{sE_2}{\sqrt{R_2^2 + sX_2^2}}$$

$$\text{or } I_2 = \frac{E_2}{\sqrt{\left(\frac{R_2}{s}\right)^2 + X_2^2}} \quad (7.45)$$

In the expression given by Eq. (7.45), the slip no longer multiplies the leakage reactance. Hence, the rotor current I2 is now proportional to E2 not sE2. Therefore, the transformer in Fig. 7.10 can also be removed and the modified circuit will be as shown in Fig. 7.11. .

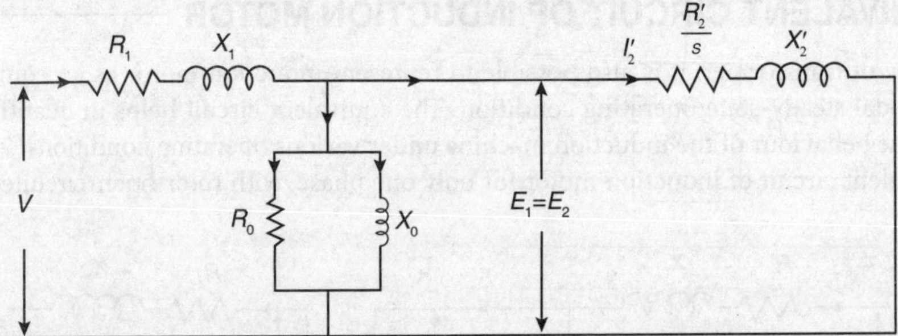

Fig. 7.11: Modified equivalent circuit of induction motor

Where

R'_2 and X'_2 are the rotor resistance and leakage reactance respectively, referred to the stator side.

The above figure is called the exact equivalent circuit of an induction motor.

Since the current flowing through the magnetising branch is very small, it can be shifted towards the left. This does not affect the performance of the circuit much. The modified equivalent circuit is shown in Fig. 7.12. This is called the approximate equivalent circuit of the induction motor.

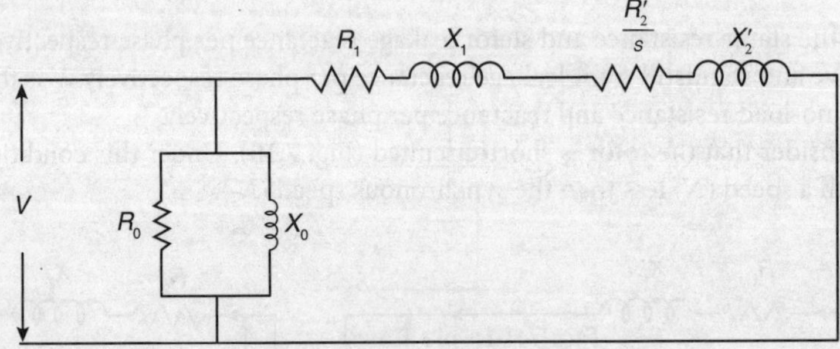

Fig. 7.12: Approximate equivalent circuit of induction motor

The resistance $\frac{R'_2}{s}$ can be represented by two quantities.

$$\frac{R'_2}{s} = R'_2 + R'_2\left(\frac{1-s}{s}\right) \quad (7.46)$$

Where,

$\frac{R_2'}{s}$ is called the air gap power.

R_2' is called the rotor resistance.

$R_2'\left(\frac{1-s}{s}\right)$ is called the mechanical output induction motor. It is also called the electrical equivalent of mechanical load.

So the equivalent circuit can be redrawn as shown in Fig. 7.13

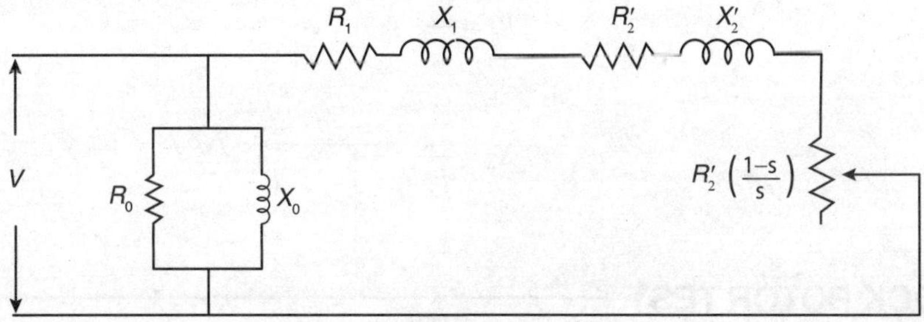

Fig. 7.13: Equivalent circuit of induction motor with electrical equivalent of mechanical load

7.18 NO-LOAD TEST

This test is carried out on an induction motor to determine the rotational and core loss of the motor. By this test, it is also possible to obtain the value of R_0 and X_0 of the equivalent circuit of the induction motor. The test is similar to the open-circuit test of a transformer. In this test, rated voltage is given to the three-phase stator winding. The rotor is rotated freely, without any load.

The circuit arrangement for no-load test is shown in Fig. 7.14.

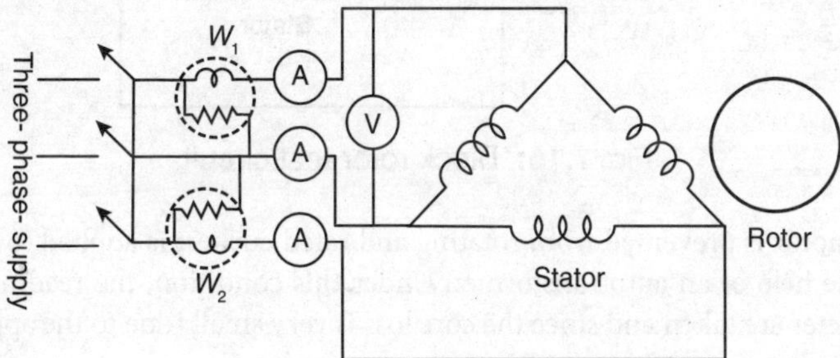

Fig. 7.14: No-load test circuit

The total power input (P_0) to the motor is measured by two wattmeter method. The input voltage to the stator (V_0) and the line current (I_0) is measured by the voltmeter and ammeter respectively, as shown in Fig. 7.14.

Since the motor is not connected to any load, this no-load current I_0 is very small and hence the copper losses in this case are negligible. Again, since the motor is running at no-load, the power factor would be very low.

From the readings of the wattmeter, voltmeter and ammeter, the following calculations can be done:

$$\cos\theta_0 = \frac{P_0}{\sqrt{3}V_0 I_0} \quad (7.47)$$

$$I_w = I_0 \cos\theta_0 \quad (7.48)$$

$$I_m = I_0 \sin\theta_0 \quad (7.49)$$

$$R_0 = \frac{V_0}{I_w} \quad (7.50)$$

$$X_0 = \frac{V_0}{I_m} \quad (7.51)$$

7.19 BLOCK ROTOR TEST

This test is similar to the short-circuit test conducted on the transformer. The circuit connection for this test is same as for no-load test and is shown in Fig. 7.15. This test gives the total winding resistance and leakage reactance of the motor.

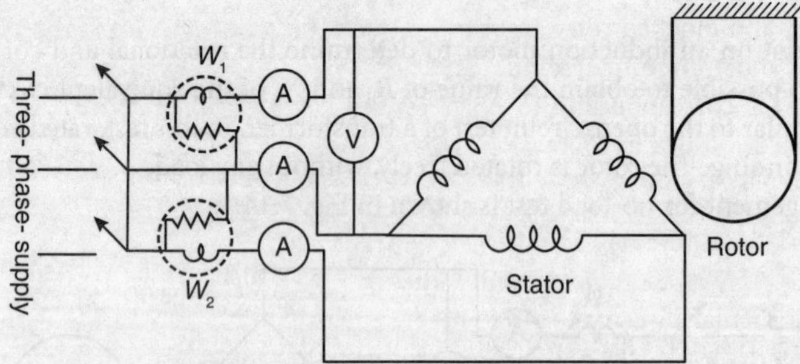

Fig. 7.15: Block rotor test circuit

In this test, the motor is prevented from rotating and rated current is applied to the three-phase stator winding with the help of an autotransformer. Under this condition, the readings of the wattmeter, voltmeter and ammeter are taken and since the core loss is very small (due to the application of very low voltage), it is neglected.

The windage and friction losses are also absent because the rotor of the motor is not rotating. With the help of this test, the following calculations are made,

$$\cos\theta_s = \frac{P_s}{\sqrt{3}V_s I_s} \quad (7.52)$$

$$R_{eq} = \frac{P_s}{3I_s^2} \quad (7.53)$$

$$Z_{eq} = \frac{V_s}{I_s} \tag{7.54}$$

$$X_{eq} = \sqrt{Z_{eq} - R_{eq}} \tag{7.55}$$

The test gives the total resistance and the total leakage reactance of the motor. Only stator resistance can be determined by giving a DC supply to the stator winding and measuring the current flowing through it. So the stator resistance can be obtained by the expression,

$$R_{stator} = \frac{V_{dc}}{I_{dc}} \tag{7.56}$$

7.20 CIRCLE DIAGRAM OF INDUCTION MOTOR

Circle diagram is a vector locus diagram of stator and rotor currents. It helps in obtaining the various characteristics of an induction motor. It is a graphical method of determining the performance of an induction motor.

7.20.1 Construction of Circle Diagram

Conduct no-load test and blocked rotor test on the induction motor and find out the per-phase values of no-load current I_0, short-circuit current I_{SC} and the corresponding phase angles φ_0 and φ_{SC}. Also find short-circuit current I_{SN} corresponding to normal supply voltage. With this data, the circle diagram shown in Fig. 7.16 can be drawn as follows.

1. With suitable scale, draw vector OA with length corresponding to I_0 at an angle φ_0 from the vertical axis. Draw a horizontal line AB.
2. Draw OS equal to ISN at an angle φ_{SC} and join AS.
3. Draw the perpendicular bisector to AS to meet the horizontal line AB at C.
4. With C as centre, draw a portion of circle passing through A and S. This forms the circle diagram, which is the locus of the input current.
5. From point S, draw a vertical line SL to meet the line AB.
6. Divide SL at point K so that SK : KL = Rotor resistance : Stator resistance.
7. For a given operating point P, draw a vertical line PEFGD as shown where, PE = Output power, EF = Rotor copper loss, FG = Stator copper loss, GD = Constant loss (iron loss + mechanical loss)
8. To find the operating points corresponding to maximum power and maximum torque, draw tangents to the circle diagram parallel to the output line and torque line respectively. The points at which these tangents touch the circle are, respectively, the maximum power point and maximum torque point.

Efficiency line

1. The output line AS is extended backwards to meet the X-axis at O′.
2. From any convenient point on the extended output line, draw a horizontal line QT so as to meet the vertical from O′. Divide the line QT into 100 equal parts.
3. To find the efficiency corresponding to any operating point P, draw a line from O′ to the efficiency line through P to meet the efficiency line at T_1. Now QT_1 is the efficiency.

Slip line

1. Draw line QR parallel to the torque line, meeting the vertical through A at R. Divide RQ into 100 equal parts.
2. To find the slip corresponding to any operating point P, draw a line from A to the slip line through P to meet the slip line at R_1. Now RR_1 is the slip

Power factor curve

1. Draw a quadrant of a circle with O as centre and any convenient radius. Divide OC_m into 100 equal parts.
2. To find power factor corresponding to P, extend the line OP to meet the power factor curve at C′. Draw a horizontal line $C'C_1$ to meet the vertical axis at C_1. Now OC_1 represents power factor.

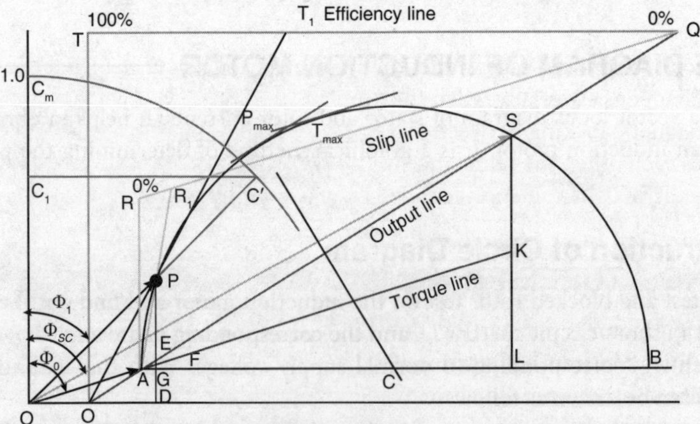

FIG. 7.16: Construction of circle diagram

7.21 CRAWLING AND COGGING OF INDUCTION MOTOR

Crawling

Sometimes, squirrel-cage induction motors exhibit a tendency to run at very slow speeds (as low as one-seventh of their synchronous speed). This phenomenon is called as crawling of an induction motor.

This action is due to the fact that the flux wave produced by a stator winding is not purely sine wave. Instead, it is a complex wave consisting of a fundamental wave and odd harmonics like 3rd, 5th, 7th etc. The fundamental wave revolves synchronously at synchronous speed N_s, whereas 3rd, 5th, 7th harmonics may rotate in forward or backward direction at $N_s/3$, $N_s/5$, $N_s/7$ speeds respectively. Hence, harmonic torques are also developed in addition to the fundamental torque. In a balanced three-phase system, 3rd harmonics are absent. Hence, 3rd harmonics do not produce rotating field and torque. The total motor torque now consist of three components as:

i) The fundamental torque with synchronous speed N_s.
ii) 5th harmonic torque with synchronous speed $N_s/5$.
iii) 7th harmonic torque with synchronous speed $N_s/7$ (provided that higher harmonics are neglected).

Now, 5th harmonic currents will have phase difference of $5\times120=600°=2\times360-120=-120°$. Hence, the revolving speed set up will be in reverse direction with speed $N_s/5$. The small amount of 5th harmonic torque produces breaking action and can be neglected.

The 7th harmonic currents will have phase difference of $7\times120=840°=2\times360+120=+120°$. Hence, they will set up a rotating field in the forward direction with synchronous speed equal to $N_s/7$. If we neglect all the higher harmonics, the resultant torque will be equal to sum of fundamental torque and 7th harmonic torque. Seventh harmonic torque reaches its maximum positive value just before1/7th of N_s. If the mechanical load on the shaft involves constant load torque, the torque developed by the motor may fall below this load torque. In this case, motor will not accelerate up to its normal speed, but it will run at a speed which is nearly 1/7th of its normal speed. This phenomenon is called as crawling in induction motors.

Cogging

Sometimes, the rotor of a squirrel-cage induction motor refuses to start at all, particularly if the supply voltage is low. This happens especially when the number of rotor teeth is equal to the number of stator teeth, because of magnetic locking between the stator teeth and the rotor teeth. When the rotor teeth and stator teeth face each other, the reluctance of the magnetic path is minimum; that is why, the rotor tends to remain fixed. This phenomenon is called cogging or magnetic locking of induction motor.

7.22 STARTING A THREE-PHASE INDUCTION MOTOR

Three phase induction motor is self starting, but when at standstill, it takes a very high current (usually 5 to 9 times the full load current) at starting if started at full voltage. This large starting current will produce large voltage drop in line, which may affect the operation of other devices connected in the line. Special methods are therefore required to reduce the starting current while starting the three phase induction motor.

7.22.1 Starting of Squirrel-cage Induction motor

There are several methods of starting squirrel-cage induction motors, which include:

i) Direct on-line starting.
ii) Autotransformer starting
iii) Star–delta starting

7.22.1.1 Direct on-line (DOL) starter

DOL starter is used only for motors with a rating of less than 5KW. It is the simplest and the most inexpensive of all starting methods and is usually used for squirrel-cage induction motors. The circuit of DOL starter is shown in Fig. 7.17. STOP switch is a normally closed switch while START switch is a normally open switch. When the START switch is pressed the circuit ACDE gets completed. This energises the control circuit CO which connects the MAIN CONTACT of the DOL starter to the supply, there by connecting the motor with the supply.

During overload, the OVERLOAD contact opens and the motor gets disconnected from the supply. Again when the STOP push button is pressed, the circuit ACDE gets disconnected and the control circuit CO will get no supply; so the MAIN CONTACT opens up, thereby disconnecting the motor from the mains. Fuses provide protection from over current.

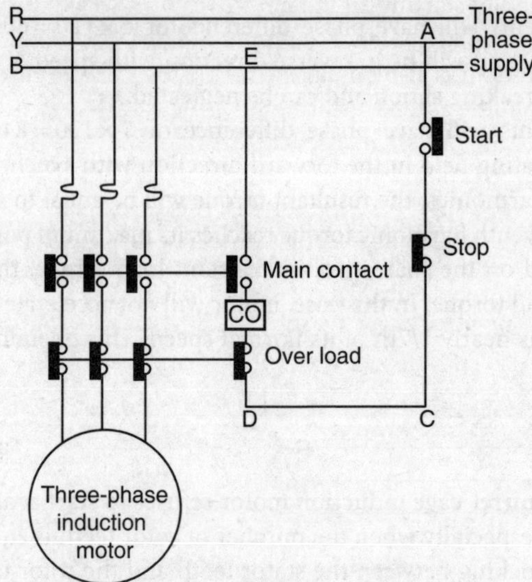

FIG. 7.17: DOL starter

7.22.1.2 Autotransformer starter

In autotransformer method of starting an induction motor, a reduced voltage is applied across the stator terminals of a three-phase induction motor with the help of an autotransformer. This, in the process, limits the starting current for the motor. The circuit diagram for the autotransformer starter is shown in Fig. 7.18

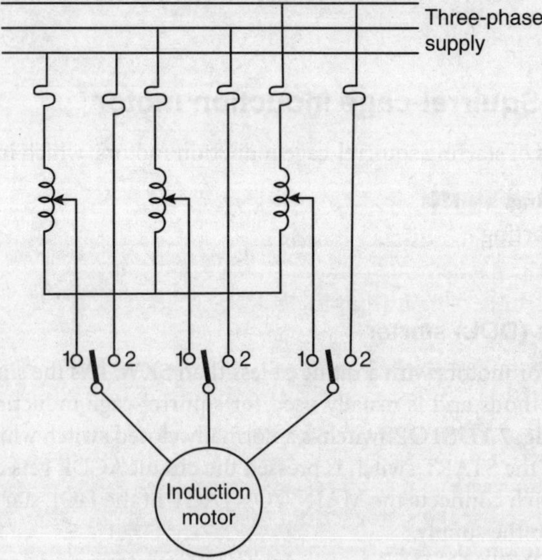

FIG. 7.18: Autotransformer starter

Initially while starting the motor, the switch is in Position 1. When the motor starts rotating, the switch is moved to Position 2, thereby disconnecting the autotransformer from the motor circuit.

The autotransformer method of starting an induction motor is very expensive and complex in operation. It also makes the system bulkier. Hence, autotransformer starter can be used for both star- or delta-connected induction motor. It is also possible to adjust the starting current and starting torque by taking out proper tappings from the auto transformer.

7.22.1.3 Star–delta starter

It is the most common method of starting an induction motor. The connection diagram of star–delta starter is shown in Fig. 7.19. First, the stator windings are connected in star (when the switch is in Position A). After the motor picks up a desired speed, the winding arrangement is changed to delta by changing the position of the switch (Position B). The line current in the case of star is one-third of the current in delta connection. Also star–delta reduces the starting torque to one-third of that obtainable by direct delta starting.

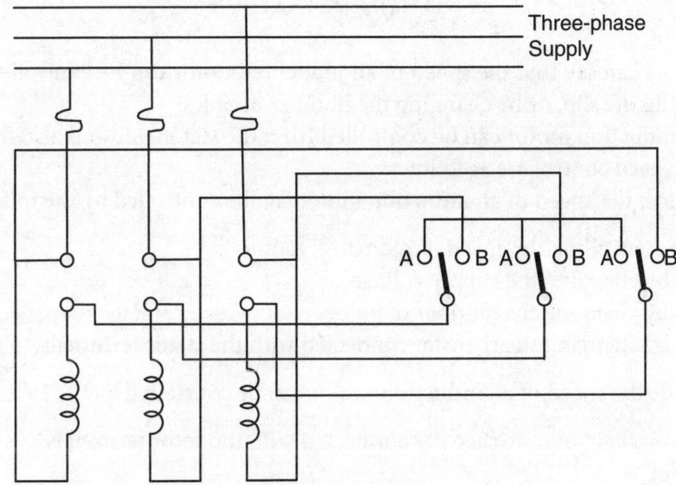

Fig. 7.19: Star–Delta starter

7.22.1.4 Rotor resistance starter

In this method of starting, external resistance is connected with the rotor through slip-rings and brushes. At starting, the rotor resistance is set to maximum and is then gradually decreased as the motor speed increases, before slowly disconnecting it from the motor circuit. This method has a very high maintenance cost. Also, a considerable amount of heat is generated through the resistors when heavy current passes through it. In this method of starting an induction motor, the motor can also be started on load.

7.22.2 Slip-ring Induction Motor Starter

In slip-ring induction, motor variable resistances are connected across each phase of the rotor circuit. Full-load voltage is applied across the stator terminals of the induction motor. At starting the variable resistances are kept at maximum position. This not only reduces the starting current but also increases the starting torque. As the motor picks up speed, the external resistance is slowly cut out in steps and when the motor attains the normal speed the external resistance is completely cut off from the rotor circuit.

In slip-ring induction motor, at starting, high starting torque is developed because of the external resistance in the rotor circuit, and the rotor current also stays within limit. Hence, this motor can be started under load condition also.

7.23 SPEED CONTROL OF INDUCTION MOTOR

The synchronous speed of an induction motor is given by the expression

$$N_s = \frac{120f}{P} \tag{7.57}$$

Also, the speed of rotation of the rotor is given by the expression

$$N = (1-s)N_s = (1-s)\frac{120f}{P} \tag{7.58}$$

From Eq. (7.58), we can say that the speed of an induction motor can be controlled by changing the frequency, by changing the slip, or by changing the number of poles.

The speed on an induction motor can be controlled from the stator side and also from the rotor side. Various methods of speed control are as follows:

From the stator side, the speed of an induction motor can be controlled by the following methods:

i) *V/f* method of controlling the speed of induction motor
ii) Speed control by changing the supply voltage.
iii) Speed control by changing the number of poles.
iv) Speed control by changing the rheostat connected with the stator terminals.

From the rotor side the speed of an induction motor can be controlled by the following methods:

i) Speed control by changing the rheostat connected with the rotor terminals.
ii) Cascade control

7.23.1 *V/f* Method of Controlling the Speed of Induction Motor

When a three-phase supply is given to the stator of an induction motor, it produces a rotating magnetic field, which rotates at a speed called the synchronous speed (N_s). The expression of synchronous speed in given by Eq (7.57). Also, the e.m.f. equation of a three-phase induction motor is given by the expression,

$$E \text{ or } V = 4.44 f \varphi_m N$$

$$\text{or } \varphi_m = \frac{V}{4.44 fN} \tag{7.59}$$

Where N is the number of turns per phase and f is frequency. From Eq. (7.57), we can say that by change frequency, it is possible to change the synchronous speed of the motor

But again from Eq. (7.59), flux φ_m also depends on frequency f. If the frequency decreases then the flux φ_m will increase. This may cause saturation of rotor and stator cores, which will increase the no-load current of the motor. So to maintain flux φ as constant, we have to change the voltage. On decreasing the frequency, flux increases. However, if with frequency, voltage is also decreased simultaneously then the flux will not change and hence it remains constant. In other word V/f ratio must be kept constant. So this method is also called V/ f method of speed control of induction motor. Variable voltage and frequency can be easily obtained by using converter and inverter set.

7.23.2 Speed Control by Changing the Supply Voltage

From Eq. (7.44), torque is directly proportional to the square of the voltage. So by decreasing the supply voltage, it is possible to decrease the torque. But in this method, a small change in speed requires large reduction of voltage. Hence, the current drawn by the motor will increase, which will cause heating of the motor.

7.23.3 Speed Control by Changing the Number of Poles

From Eq. (7.57), it can be easily said that the speed of an induction motor can be controlled by changing the number of poles. There are two methods of speed control:

i) Multiple stator winding method.
ii) Pole amplitude modulation method.

Multiple stator winding method

In this method the stator of an induction motor will be having two separate windings, electrically isolated from each other. Each winding is divided into coils to which, pole changing with consequent poles facility is provided. Using switching arrangement, supply is given to only one winding at a time and hence speed control is possible. The disadvantages of this method are that in this arrangement smooth speed control is not achieved. This method is more costly and less efficient as two different stator windings are required. This method of speed control can only be applied for squirrel-cage motor.

Pole amplitude modulation method (PAM)

It is the modulation of the original sinusoidal m.m.f. wave by another sinusoidal m.m.f. wave having different number of poles.

Let $f_1(\theta)$ be the original m.m.f. wave of induction motor whose speed is to be controlled, $f_2(\theta)$ be the modulation m.m.f. wave, P_1 be the number of poles of induction motor whose speed is to be controlled, P_2 be the number of poles of modulation wave, θ be the mechanical angle

Then,

$$f_1(\theta) = F_1 \sin\left(\frac{P_1\theta}{2}\right)$$

$$f_2(\theta) = F_2 \sin\left(\frac{P_2\theta}{2}\right)$$

The resultant m.m.f. after modulation we get,

$$F_{\mathrm{m}} = F_1 F_2 \sin\left(\frac{P_1\theta}{2}\right)\sin\left(\frac{P_2\theta}{2}\right)$$

$$= \frac{F_1F_2}{2}\left[\cos\left(\frac{P_1 - P_2}{2}\right)\theta - \cos\left(\frac{P_1 + P_2}{2}\right)\theta\right]$$

Therefore the resultant m.m.f. wave will have two different numbers of poles

$$P_{11} = P_1 + P_2 \text{ and } P_{12} = P_1 - P_2$$

Therefore, by changing the number of poles we can easily change the speed of the three-phase induction motor.

7.23.4 Speed Control by Connecting Rheostat at the Stator Terminals

In this method of speed control, a variable rheostat is connected in each phase of the stator winding. By changing the stator resistances, the voltage can be varied which in turn changes the torque of the induction motor.

7.23.5 Speed Control by Connecting Rheostat at the Rotor Terminals

From Eq.(7.35), the torque is inversely proportional to rotor resistance. So by connecting a variable resistance with the rotor winding, it is possible to obtain speed control over a limited range. But this method of speed control reduces the efficiency of the motor.

7.23.6 Cascade Control

Two motors, mounted on the same shaft, are used for the speed control of induction motor by cascade control. One motor is fed from a three-phase AC supply and the other motor is fed from the induced e.m.f of the first motor via slip-rings.

Let P_1 be the number of poles of motor M_1 and P_2 be the number of poles of motor M_2. Let f_1 and f_2 be the stator frequency for the motor M_1 and M_2 respectively. If N_{s1} and N_{s2} are the synchronous speed of the motor M_1 and M_2 respectively, then,

$$N_{s1} = \frac{120 f_1}{P_1} \text{ and } N_{s2} = \frac{120 f_2}{P_2}$$

If s_1 and s_2 are the slip of the motor M_1 and M_2 respectively, then the rotor speed of the two motors will be,

$$N_{r1} = N_{s1}(1 - s_1) = \frac{120 f_1}{P_1}(1 - s_1) \text{ for motor } M_1 \tag{7.60}$$

$$N_{r2} = N_{s2}(1 - s_2) = \frac{120 f_2}{P_2}(1 - s_2) \text{ for motor } M_2 \tag{7.61}$$

Since the two motors are mechanically coupled, we can write, $f_2 = s_1 f_1$ and the speed of the motors will also be the same. So,

$$\frac{120 f_1}{P_1}(1 - s_1) = \frac{120 s_1 f_1}{P_2}(1 - s_2) \tag{7.62}$$

During normal operating condition the slip-rings of motor M_2 are shorted, so the slip s_2 will be almost equal to zero. So, from Eq. (7.62), we get,

$$s_1 = \frac{P_2}{P_1 + P_2} \tag{7.63}$$

Substituting Eq. (7.63) in Eq. (7.60), we get,

$$N_{r1} = \frac{120 f_1}{P_1}\left(1 - \frac{P_2}{P_1 + P_2}\right) = \frac{120 f_1}{P_1 + P_2} \tag{7.64}$$

So, with this method, four different speeds can be obtained

i) When only motor M_1 works, corresponding speed $= N_{s1} = \dfrac{120 f_1}{P_1}$

ii) When only motor M_2 works, corresponding speed $= N_{s2} = \dfrac{120 f_2}{P_2}$

iii) When commutative cascade is done, speed $= N = \dfrac{120 f_1}{P_1 + P_2}$

iv) When differential cascade is done, speed $= N = \dfrac{120 f_1}{P_1 - P_2}$

7.24 BRAKING OF INDUCTION MOTOR

Braking is a rapid method of stopping a motor from its running condition. The motor will be stopped electrically when the direction of developed torque is opposite to that of rotation. The braking methods for induction motor are as follows:

i) Regenerative braking of induction motor
ii) Braking by plugging
iii) Dynamic braking

7.24.1 Regenerative Braking of Induction Motor

Regenerative braking is a technique of slowing down a machine by converting its kinetic energy into another form of energy.

Let θ be the phase angle between stator supply voltage and stator current. In an induction motor, the motoring action takes place for $\theta < 90^0$ and braking operation takes place for $\theta > 90^0$. If the speed of rotation of the rotor exceeds synchronous speed, the relative speed between them reverses and hence, the phase angle becomes more than 90^0. This will cause the power to flow in the reverse direction and thus regenerative braking takes place.

If the supply frequency is constant then for regenerative braking, the rotor speed must be greater than synchronous speed. But for a variable frequency source, it is possible to have regenerative braking below the synchronous speed.

The problem with regenerative braking is that as the load decelerates, energy recovery decreases along with it and thus the braking force also diminishes. So to completely stop the motor, the backup brakes have to be applied.

7.24.2 Braking by Plugging

In this method of braking, the phase sequence of the motor is reversed by interchanging any two phases of the stator terminals. Under this condition let the slip be s_p. It can be shown that the slip is actually equal to $(2 - s)$.

$$s_p = \frac{-N_s - N_r}{-N_s} = \frac{N_s + (1-s)N_s}{N_s} = 2 - s \quad (7.65)$$

Thus by interchanging the two phases of the stator, it is possible to reverse the rotation of the magnetic field of the motor. Under this condition, the slip becomes greater than unity $(2 - s)$ and so the developed torque tries to rotate the motor in the opposite direction. This results in braking of the motor by plugging.

If the motor is needed to be stopped completely, then it should be disconnected from the supply at near-zero speed. In this method of braking, the heat produced in the rotor is approximately three times the initial energy of the drive system. So if this method is not properly used, then it may cause thermal damage to the rotor.

7.24.3 Dynamic Braking

There are four methods of dynamic braking by which an induction motor can be stopped.

i) AC dynamic braking
ii) DC dynamic braking
iii) Zero-sequence braking
iv) Self-excited braking

AC dynamic braking

In an AC dynamic braking, a three-phase induction motor is made to run on a single phase supply by disconnecting any one phase from the supply. In a single-phase operation, the motor is fed by positive and negative sequence; so the net torque will be the sum of torques due to positive and negative sequence voltage. At high resistance, this net torque becomes negative and braking takes place.

Self-excited braking

In this method of braking, capacitors are connected across the stator terminals. When the motor is disconnected from the source, these capacitors deliver enough reactive current to make the motor work as a generator. This will produce a torque in the opposite direction and the braking action takes place.

DC dynamic braking

In this method of braking of induction motor, the running induction motor is connected to a DC supply. Because of this DC supply, a stationary magnetic field is generated. Since the rotor is rotating in this field, a magnetic field will also be induced in the rotor circuit. This makes the motor work as a generator and the dynamic barking of induction motor takes place.

Zero-sequence braking

In this method of braking of induction motor, all the three phases of stator are connected in series with a single-phase AC or DC supply. This is called zero-sequence connection.

If a single-phase AC supply is connected to the series stator connection, the resultant field will be stationary in space but pulsates at the frequency of supply.

If DC supply is given, the resultant field will be stationary at constant magnitude. The main advantage of this method of braking is that it does not require large rotor resistance as in other braking methods.

7.25 DEEP BAR TYPE INDUCTION MOTOR

Deep bar rotors are used in induction motors to increase the starting torque. In this type of induction motor, the rotor has two different layers of bars, (Fig. 7.20). The outer layer has bars with small cross-sectional area. So, the outer bar has relatively large resistance. The bars are shorted at the both ends. The flux linkage is thus very less. Hence, the inductance is very low. Therefore, the resistance to inductive reactance ratio is high.

The inner layer has bars with large cross-sectional area as compared with the inner bar. So the resistance is very less, but the flux linkage is very high. As flux linkage is high, the inductance is also very high. Therefore, the resistance to inductive reactance ratio is poor.

At the standstill condition, the inner and outer bars get induced with voltage and current with the same frequency as that of the supply. The inductive reactance is more in the inner bars due to skin effect of the alternating quantity. Hence the current tries to flow through the outer rotor bars.

The outer rotor offers more resistance but poor inductive reactance. The ultimate resistance is somewhat higher than the resistance of the single bar rotor. The higher valued rotor resistance results in more torque to be developed at start-up.

When the speed increases, the frequency of the induced e.m.f. and current in the rotor get gradually decreased. Hence, the inductive reactance in the inner side gets decreased and the current faces less inductive reactance and less resistance, as a whole.

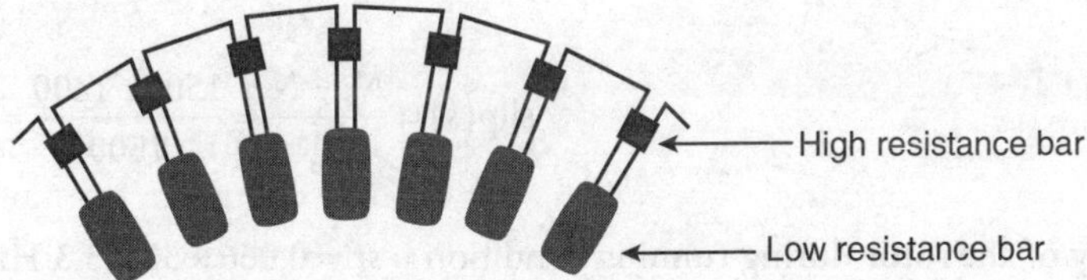

FIG. 7.20: Deep bar type induction motor

7.26 DOUBLE CAGE TYPE INDUCTION MOTOR

The problem of poor starting torque, which was the drawback of a squirrel-cage induction motor, has been addressed in double-cage induction motor construction. The stator of a double-cage induction motor is same as that of a squirrel-cage induction motor. But in case of a double-cage induction motor, the rotor has two layers of bar called cages which are short-circuited by end rings. The outer bar has high resistance and low reactance and is made up of material like brass or aluminium. The inner bar has low resistance and high reactance and is made up of material like copper. The two cages are separated from each other by a narrow slit.

During start-up, due to the large slip of the motor, the frequency of induced e.m.f. is high. Under this condition, the reactance of inner cage will be very high, thereby making the total impedance of the inner cage high. So when the motor is started, the current will flow through the outer cage, since the total impedance of the outer cage will be lower than the inner cage at start-up. Hence the starting torque will be large due to large resistance of the outer cage.

As the speed of the motor increases, its slip decreases; this will cause the rotor frequency to decrease. So under this condition, the reactance of the inner cage will be low and this will cause most of the current to flow through the inner cage because of its low resistance. This will increase the efficiency of the motor.

SOLVED PROBLEMS

Ex.7.1 The ratio of stator to rotor of a three-phase, 4-pole, 50 Hz induction motor is 4. When connected to 440 V supply, the speed of the motor at a certain load was observed as 1400 r.p.m. Calculate the following:

(i) Frequency of the rotor during running condition
(ii) Frequency of the rotor at standstill
(iii) Induced e.m.f. at standstill
(iv) Induced e.m.f. at running condition

Solution

$$K = \frac{1}{4} = 0.25$$

$$\text{Synchronous speed } (N_s) = \frac{120f}{P} = \frac{120 \times 50}{4} = 1500 \text{ r.p.m.}$$

$$\text{Speed of the rotor } (N) = 1400 \text{ r.p.m.}$$

$$\text{slip}(s) = \frac{N_s - N}{N_s} = \frac{1500 - 1400}{1500} = 0.066$$

$$\text{Frequency of the rotor during running condition} = sf = 0.066 \times 50 = 3.3 \text{ Hz}$$

$$\text{Frequency of the rotor at standstill} = f = 50 \text{ Hz}$$

At standstill,

$$\frac{E_{rotor}}{E_{stator}} = K = 0.25$$

Given,

$$E_{stator} = 440 \text{ V (Line voltage)}$$

Then,

$$E_{rotor} = 440 \times 0.25 = 110 \text{ V (Line voltage)}$$

During running condition,

$$E_{rotor\_run} = s \times 110 = 0.066 \times 110 = 7.26 \text{ V (Line voltage)}$$

Ex.7.2 A 50 Hz, three-phase, 8 pole, induction motor has full-load slip and no-load slip of 4% and 2 % respectively. Calculate the following:

(i) Synchronous speed
(ii) Rotor speed at no-load
(iii) Rotor speed at full-load
(iv) Frequency of the rotor at full-load
(v) Frequency of the rotor at standstill.

Solution

$$\text{Synchronous speed}\left(N_s\right)=\frac{120f}{p}=\frac{120\times 50}{8}=750 \text{ r.p.m.}$$

$$\text{Rotor speed at no load}\left(N_0\right)=N_s\left(1-\text{slip at no load}\right)=750\left(1-0.02\right)=735 \text{ r.p.m.}$$

$$\text{Rotor speed at full load}\left(N_{fl}\right)=N_s\left(1-\text{slip at full load}\right)=750\left(1-0.04\right)=720 \text{ r.p.m.}$$

$$\text{Frequency of the rotor at full load}\left(f_{fl}\right)=f\times\left(\text{slip at full load}\right)=50\times 0.04=2 \text{ Hz}$$

$$\text{Frequency of the rotor at standstill}\left(f_{st}\right)=f=50 \text{ Hz}$$

Ex.7.3 The rotor resistance and standstill reactance of a 10-pole, 50 Hz, three-phase induction motor are 0.02Ω and 0.08Ω respectively. If the motor runs at full load with slip of 5%, calculate the following:

(i) The ratio of full-load torque to maximum torque.
(ii) Speed at which full-load torque takes place.

Solution

$$\text{Synchronous speed }\left(N_s\right)=\frac{120f}{P}=\frac{120\times 50}{10}=600 \text{ r.p.m}$$

$$s_m=\frac{R_2}{X_2}=\frac{0.02}{0.08}=0.25$$

$$\frac{T_{fl}}{T_m}=\frac{2\times 0.05\times 0.25}{0.05^2+0.25^2}=0.384$$

$$\text{Speed of the rotor at full load}\left(N_r\right)=N_s\left(1-0.25\right)=600\left(1-0.25\right)=450 \text{ r.p.m.}$$

Ex.7.4 A three-phase, 50 Hz, 12-pole, 30KW, 440V slip-ring induction motor has rated output at rated voltage and frequency, when the slip rings are short-circuited. The maximum torque occurs at 0.06, which is two times the full-load torque. Calculate the following:

(i) Full-load slip.

(ii) Rotor speed at full-load
(iii) Rotor resistive loss at full-load
(iv) Full-load torque
(v) Starting torque.

Solution

$$\text{Synchronous speed } (N_s) = \frac{120 \times 50}{12} = 500 \text{ r.p.m.}$$

$$\frac{T_{fl}}{T_m} = \frac{2 \times 0.06 \times s_{fl}}{0.06^2 + s_{fl}^2}$$

$$\text{or } \frac{1}{2} = \frac{2 \times 0.06 \times s_{fl}}{0.06^2 + s_{fl}^2}$$

$$\text{or } s_{fl}^2 - 0.24 s_{fl} + 0.0036 = 0$$

Solving this quadratic equation, we get,

$$s_{fl} = 0.2239 \text{ and } 0.0161$$

Considering the smaller value of the slip, we get,

$$s_{fl} = 0.0161$$

$$\text{Speed of the rotor at full load} (N_{fl}) = 500(1 - 0.0161) = 492 \text{ r.p.m.}$$

$$P_g = \frac{30 \times 10^3}{1 - 0.0161} = 30490.91 \text{ W}$$

$$\text{Rotor resistive loss at full load} = s_{fl} \times P_g = 0.0161 \times 30490.91 = 490.91 \text{ W}$$

$$\text{Full load torque} (T_{fl}) = \frac{P_g}{2\pi\left(\frac{N_s}{60}\right)} = \frac{490.91}{2\pi\left(\frac{500}{60}\right)} = 582.33 \text{Nm}$$

$$\text{Starting torque } (T_{st}) = \frac{2 \times 2 \times T_{fl}}{0.06 + \frac{1}{0.06}} = \frac{2 \times 2 \times 582.33}{0.06 + \frac{1}{0.06}} = 139.26 \text{ Nm}$$

Ex.7.5 A star connected three-phase, 8-pole, 50 Hz, 440V induction motor has the following equivalent circuit parameters:

$$R_1 = 0.5\,\Omega,\ X_1 = 1\,\Omega,\ R_2' = 0.8\,\Omega,\ X_2' = 0.9\Omega,\ R_0 = 125\Omega,\ X_0 = 30\Omega$$

If the full-load slip is 0.06 and the stator loss is 2 KW, find the following:

(i) Synchronous speed
(ii) Rotor speed at full-load
(iii) Stator current at no-load
(iv) Stator current at full-load
(v) Motor input
(vi) Rotor input
(vii) Rotor copper loss
(viii) Motor output
(ix) Efficiency of the motor

Solution

$$\text{Synchronous speed}\left(N_s\right) = \frac{120 \times 50}{8} = 750 \text{ r.p.m.}$$

$$\text{Rotor speed at full load}\left(N_r\right) = 750(1-0.06) = 705 \text{ r.p.m.}$$

$$\text{Stator voltage}\left(V_{Line}\right) = 440 \text{ V}$$

$$\text{Stator voltage}\left(V_{ph}\right) = \frac{440}{\sqrt{3}} = 254.03\text{V}$$

$$\text{No load current, } I_0 = \frac{V_{ph}}{R_0 + jX_0} = \frac{254.03}{125 + j30} = 1.922 - j0.46A = 1.977\angle -13.46 \text{ A}$$

$$I'_2 = \frac{V_{ph}}{R_1 + jX_1 + \dfrac{R'_2}{s} + jX'_2} = \frac{254.0.3}{0.5 + j1 + \dfrac{0.8}{0.06} + j0.9} = 18.02 - j2.48A = 18.189\angle -7.84 \text{ A}$$

$$\text{Full load stator current} = I_0 + I'_2 = 1.922 - j0.46 + 18.02 - j2.48 = 19.94 - j2.94 \text{ A}$$

$$= 20.156\angle -8.39 \text{ A}$$

$$\text{Motor input} = \sqrt{3} \times 440 \times 18.189 \times \cos -7.84 = 13.73 \text{ KW}$$

$$\text{Rotor input} = 13.73 - 2 = 11.73 \text{ KW}$$

$$\text{Rotor copper loss} = 0.06 \times 11.73 = 0.7038 \text{ KW}$$

$$\text{Motor output} = 11.73 - 2 - 0.7038 = 9.0262 \text{ KW}$$

$$\text{Efficiency of the motor} = \frac{9.0262}{11.73} \times 100 = 76.94\%$$

Ex.7.6 A three-phase, 440V, star connected induction motor takes full-load current at 50V when the rotor is blocked. Calculate the tappings on the three-phase autotransformer to limit the starting current to five times the full-load current. Take slip to be 5%.

Solution

Let,

I_b be the block rotor current

I_{fl} be the full-load current

I_s be the starting current

T_s be the starting torque

T_{fl} be the full-load torque

Then,

$$\frac{I_s}{I_{fl}} = 5$$

$$\frac{I_b}{I_{fl}} = \frac{440}{50} = 8.8$$

$$\text{tapping}(k) = \frac{I_s}{I_b} \times 100 = \frac{I_s/I_{fl}}{I_b/I_{fl}} \times 100 = \frac{5}{8.8} \times 100 = 57\%$$

Ex.7.7 The no-load and the block rotor test result, conducted on a three-phase, star-connected induction motor are as follows:

No load test result:

Line voltage = 440 V

Current = 20 A

Power = 1000 W

Block rotor test result:

Line voltage = 50 V

Current = 60 A

Power= 3000 W

Determine the parameters of the equivalent circuit of the induction motor.

Solution

From no-load test data

$$V_0 = \frac{440}{\sqrt{3}} = 254.03\,\text{V}$$

$$P_0 = \frac{1000}{3} = 333.33\text{ W}$$

$$I_0 = 20\text{ A}$$

$$R_0 = \frac{254.03^2}{333.33} = 193.595\ \Omega$$

$$X_0 = \frac{254.03^2}{\sqrt{(254.03^2)\times(20^2) - 333.33^2}} = 12.73\ \Omega$$

From block rotor test data

$$V_s = \frac{50}{\sqrt{3}} = 28.87\ \Omega$$

$$I_s = 60\ \text{A}$$

$$P_0 = \frac{3000}{3} = 1000\ \text{W}$$

$$R_e = \frac{1000}{60^2} = 0.278\ \Omega$$

$$X_e = \frac{\sqrt{(28.87^2)\times(60^2) - (1000^2)}}{60^2} = 0.92\ \Omega$$

Ex.7.8 Consider a 6-pole, three-phase, 50 Hz, 400V induction motor. When running at 900 r.p.m., it takes a power input of 50 KW. If the stator loss is 1 KW and friction and windage losses are 1.5 KW, calculate the efficiency of the motor.

Solution

$$\text{Synchronous speed}(N_s) = \frac{120\times f}{P} = \frac{120\times 50}{6} = 1000\text{r.p.m.}$$

$$\text{slip} = \frac{1000-900}{1000} = 0.1$$

$$\text{Copper loss} = 0.1\times(50-1) = 4.9\text{KW}$$

$$\text{Efficiency}(\bullet) = \frac{50-(1+1.5+4.9)}{50}\times 100 = 85.2\%$$

Ex.7.9 A delta connected 30 HP, 230V, 6-pole, 50 Hz, three-phase squirrel-cage induction motor has the following circuit parameters:
$R_1 = 0.08\ \Omega$, $R_2 = 0.06\ \Omega$, $X_1 = 0.4\ \Omega$, $X_2 = 0.3\ \Omega$, $X_m = 12\ \Omega$.

(i) Calculate the starting current and the torque for a supply voltage of 230V

(ii) If the stator connection is connected in star for starting and then converted into delta for normal operation. Then,

(a) Calculate the parameters of the equivalent circuit of the motor during star connection of the stator windings.
(b) Calculate the starting current and the torque for a supply voltage of 230V, when the stator windings are connected in star.

Solution

$$V_{ph} = \frac{230}{\sqrt{3}} = 132.8 \text{ V}$$

$$R_{eq} + jX_{eq} = \frac{(R_1 + jX_1) \times jX_m}{R_1 + jX_1 + jX_m} = \frac{(0.08 + j0.4) \times j12}{0.08 + j0.4 + j12} = (0.075 + j0.387)\Omega$$

$$I_2 = \frac{V_{ph}}{\sqrt{\left(R_{eq} + \frac{R_2}{s}\right)^2 + (X_{eq} + X_2)^2}} = \frac{132.8}{\sqrt{\left(0.075 + \frac{0.06}{1}\right)^2 + (0.387 + 0.3)^2}} = 189.68 \text{ A}$$

$$\text{Torque}(T) = \frac{3I_2^2\left(\frac{R_2}{s}\right)}{\frac{2\pi}{60} \times \frac{120 \times 50}{6}} = \frac{3 \times 189.68^2\left(\frac{0.06}{1}\right)}{\frac{2\pi}{60} \times \frac{120 \times 50}{6}} = 61.84 \text{ Nm}$$

When the stator is connected in star,

$$R_1 = 3 \times 0.08 = 0.24\ \Omega$$

$$R_2 = 3 \times 0.06 = 0.18$$

$$X_1 = 3 \times 0.4 = 1.2$$

$$X_2 = 3 \times 0.3 = 0.9$$

$$X_m = 3 \times 12 = 36$$

Then,

$$R_{eq} + jX_{eq} = \frac{(R_1 + jX_1) \times jX_m}{R_1 + jX_1 + jX_m} = \frac{(0.24 + j1.2) \times j36}{0.24 + j1.2 + j36} = (0.225 + j1.163)\Omega$$

$$I_2 = \frac{V_{ph}}{\sqrt{\left(R_{eq} + \frac{R_2}{s}\right)^2 + (X_{eq} + X_2)^2}} = \frac{132.8}{\sqrt{\left(0.225 + \frac{0.18}{1}\right)^2 + (1.163 + 0.9)^2}} = 87.49 \text{ A}$$

$$\text{Torque}(T) = \frac{3I_2^2\left(\frac{R_2}{s}\right)}{\frac{2\pi}{60}\times\frac{120\times 50}{6}} = \frac{3\times 63.17^2\left(\frac{0.18}{1}\right)}{\frac{2\pi}{60}\times\frac{120\times 50}{6}} = 39.47 \text{ Nm}$$

Ex.7.10 A dc test is performed on a 460-V, delta-connected 100 HP induction motor. The readings of the voltmeter and ammeter are 21 V and 72 A respectively. Obtain the value of the stator resistance.

Solution

Let the stator resistance be R_1. Then,

$$\frac{V_{DC}}{I_{DC}} = \frac{R_1(R_1 + R_1)}{3R_1} = \frac{2}{3}R_1$$

$$\text{or, } R_1 = \frac{3}{2}\times\frac{V_{DC}}{I_{DC}} = \frac{3}{2}\times\frac{21}{72} = 0.438\ \Omega$$

Ex.7.11 A three-phase squirrel-cage type delta-connected induction motor, when connected across a 500 V, 50 Hz supply, takes a starting current of 100 A. Calculate the following:

(i) Line current for direct on-line starting
(ii) Line and phase current for star–delta starting
(iii) Line and phase current for 70 % tapping on autotransformer starting.

Solution

(i) $I_L(DOL) = \sqrt{3}\times 100 = 173.2$ A

(ii) $I_L(Star - Delta) = \frac{100}{\sqrt{3}} = 57.7\,A$

(iii) 500 V produces 100 A in phase winding, so (0.7×500) V will produce $(0.7\times 100) = 70$ A current.

$$\text{Motor line current} = \sqrt{3}\times 70 = 121.2\,\text{A}$$

$$\text{So supply line current} = 0.7\times 121.2 = 84.8\,\text{A}$$

Ex.7.12 Consider a three-phase induction motor with the following rating:
Rated output power = 75 kW, Rated voltage for two possible connections = 220/380 V, Rated power factor = 0.85, Rated efficiency = 0.92, Rated speed = 975 r.p.m., Rated mechanical loss = 375 kW, Stator resistance per phase = 0.033 Ω.
Calculate the following:

(i) Stator current for star- and delta-connected stator winding
(ii) Power drawn by the stator
(iii) Active and reactive power
(iv) Nominal torque

Solution

$$\text{Stator current for star connection} = \frac{75\times 10^3}{\sqrt{3}\times 380\times 0.92\times 0.85} = 145.7\,\text{A}$$

$$\text{Stator current for delta connection} = \frac{75\times10^3}{\sqrt{3}\times220\times0.92\times0.85} = 251.7\,A$$

$$\text{Power} = \frac{75}{0.92\times0.85} = 95.9\text{ KVA}$$

$$\text{Active power} = \frac{75}{0.92} = 81.52\text{ kW}$$

$$\text{Reactive power} = 95.9\times\left(\sqrt{1-0.85}\right) = 50.5\text{ KVAR}$$

$$\text{Torque} = 9.55\times\frac{75\times10^3}{975} = 734.6\text{ Nm}$$

Ex.7.13 A three-phase, 4-pole,3 HP, 440 V, 60 Hz induction motor is known to have a per phase stator resistance of 2.57 Ω. The following test data have been recorded for the motor:

No-load	**Blocked rotor**
$V_L = 440$ V	$V_L = 55$ V
$I_1 = 2.25$ A	$I_1 = 5.92$ A
$P_T = 220$ W	$P_T = 522$ W
$f = 60$ Hz	$f = 10$ Hz
$n_m = 1796$	

Determine the per-phase equivalent circuit parameters for this motor.

Solution

$$R_{eq} = \frac{P_{br}}{I_{br}^2} = \frac{552/3}{(5.92)^2} = 5.25\ \Omega$$

$$Z_{br} = \frac{V_{br}}{I_{br}} = \frac{55/\sqrt{3}}{5.92} = 5.364\ \Omega$$

$$X_{br} = \left[Z_{br}^2 - R_{eq}^2\right]^{1/2} = \left[(5.364)^2 - (5.25)^2\right]^{1/2} = 0.9456\ \Omega$$

$$R_2' = R_{eq} - R_1 = 5.25 - 2.57 = 2.68\ \Omega$$

$$X_{eq} = \frac{f}{f_{br}}X_{br} = \frac{60}{10}(0.9456) = 5.67\ \Omega$$

$$X_1 = 0.4X_{eq} = 0.4(5.67) = 2.268\ \Omega$$

$$X_2' = 0.6X_{eq} = 0.6(5.67) = 3.402\ \Omega$$

$$\theta_{n\ell} = \cos^{-1}\left[\frac{P_{n\ell}}{V_{n\ell}I_{n\ell}}\right] = \cos^{-1}\left[\frac{552/3}{\left(440/\sqrt{3}\right)(2.25)}\right] = 71.22°$$

$$\overline{E}_1 = V_1\angle 0° - I_{n\ell}\angle -\theta_{n\ell}\left(R_1 + jX_1\right)$$

$$\overline{E}_1 = \frac{440}{\sqrt{3}}\angle 0° - 2.25\angle -71.22°(2.57 + j2.268) = 247.37\angle 0.78°\ \text{V}$$

$$s_{n\ell} = \frac{n_s - n_{n\ell}}{n_s} = \frac{1800 - 1796}{1800} = 0.00222$$

$$I_2' = \frac{E_1}{\left[\left(\frac{R_2'}{s_{n\ell}}\right)^2 + (X_2')^2\right]^{1/2}} = \frac{247.37}{\left[\left(\frac{2.68}{0.00222}\right)^2 + (3.402)^2\right]^{1/2}} = 0.205\ \text{A}$$

$$P_c = P_{n\ell} - I_1^2R_1 - (I_2')^2\frac{R_2'}{s_{n\ell}}$$

$$P_c = \frac{220}{3} - (2.25)^2(2.57) - (0.205)^2\left(\frac{2.68}{0.00222}\right) = 22.60\ \text{W}$$

$$R_c = \frac{E_1^2}{P_c} = \frac{(247.37)^2}{22.60} = 2707.6\ \Omega$$

$$Q_m = V_{n\ell}I_{n\ell}\sin(\theta_{n\ell}) = I_{n\ell}^2X_1 - (I_2')^2X_2'$$

$$Q_m = \frac{440}{\sqrt{3}}(2.25)\sin(71.22°) - (2.25)^2(2.268) - (0.205)^2(3.402)$$

$$Q_m = 529.54\ \text{VARs}$$

$$X_m = \frac{E_1^2}{Q_m} = \frac{(247.37)^2}{529.54} = 115.6\ \Omega$$

Ex.7.14 A three-phase,. 4-pole, 600 V, 60 Hz induction motor is modelled by $\mathbf{Z}_{Th} = 0.6933 + j1.933\ \Omega$, $R_2' = 4.5\ \Omega$, and $X_2' = 2\ \Omega$. Find the shaft speed at which maximum torque occurs if the motor is absorbing power from the three-phase lines at rated frequency.

Solution

$$s_{\max+} = \frac{R_2'}{\sqrt{R_{Th}^2 + (X_{Th} + X_2')^2}} = \frac{4.5}{\sqrt{(0.6933)^2 + (3.933)^2}} = 1.127$$

$$n_s = \frac{120f}{p} = \frac{120(60)}{4} = 1800 \text{ rpm}$$

$$n_m = (1-s)n_s = (1-1.127)1800 = -228.2 \text{ rpm}$$

This motor is operating in the braking or plugging mode.

Ex.7.15 The three-phase, wound-rotor induction motor of Fig. Q7.15 has balanced three-phase voltage applied. It is operating in the steady-state when the three-pole switch is suddenly moved from Position 1 to Position 2. Will the motor reverse the direction of rotation?

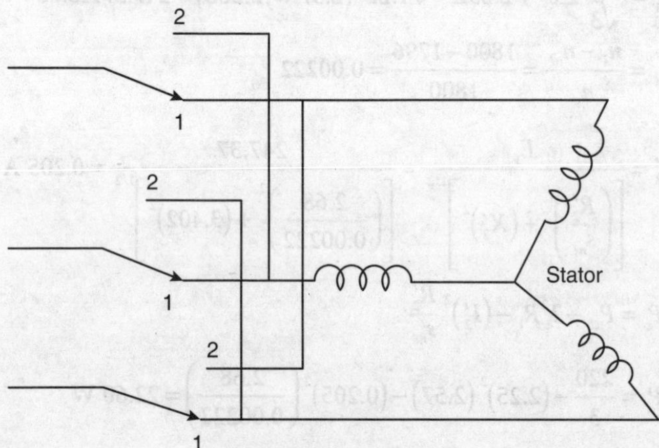

Fig. E7.15:

Solution

Study of the switch arrangement shows that after the switch moves, the impressed voltages are still connected in $a \cdot b \cdot c$ phase sequence in a counter-clockwise direction of the stator winding. The stator-produced m.m.f. wave does not change direction of rotation; thus, the rotor does not change the direction of rotation.

Ex.7.16 A three-phase, 50 Hz, 4-pole slip-ring induction motor has a star connected rotor. The full-load speed of the motor is 1460 rpm. The rotor resistance and standstill reactance per phase are 0.1 Ω and 1.5 Ω respectively. The open-circuit voltage on open circuit between the slip-rings is 90 volts. Calculate the following:

(i) Percentage slip
(ii) Induced e.m.f. in rotor per phase
(iii) Rotor reactance per phase at full-load
(iv) Rotor current and full-load power factor.

Solution

$$\text{Synchronous speed} = \frac{120 \times 50}{4} = 1500 \text{ r.p.m.}$$

$$\text{percentage slip} = \frac{1500-1460}{1500} \times 100 = 2.67$$

$$\text{Induced e.m.f.} = \frac{90}{\sqrt{3}} = 51.96 \text{ V per phase}$$

$$\text{Rotor induced e.m.f.} = 0.0267 \times 51.96 = 1.38 \text{ V}$$

$$\text{Rotor reactance at full load} = 0.0267 \times 1.5 = 0.04\ \Omega \text{ per phase}$$

$$\text{Rotor impedance} = \sqrt{0.1^2 + 0.04^2} = 0.1077\ \Omega \text{ per phase}$$

$$\text{rotor current} = \frac{1.38}{0.1077} = 12.81 \text{ A per phase}$$

$$\text{Power factor at full load} = \frac{0.1}{0.1077} = 0.928$$

Ex.7.17 If the induced e.m.f. in the stator of an 8-pole induction motor has a frequency of 50 Hz and that in the rotor is 1.5 Hz, at what speed is the motor running and what is the slip?

Solution

$$\text{slip} = \frac{1.5}{50} = 0.03$$

$$N_s = \frac{120 \times 50}{8} = 750 \text{ r.p.m.}$$

$$\text{speed of rotor} = (1 - 0.03) \times 750 = 728 \text{ r.p.m.}$$

Ex.7.18 A three-phase, 12-pole, salient pole alternator is coupled to a diesel engine running at 500 rpm. It supplies an induction motor which has a full-load speed of 1440 rpm. Find the percentage slip and number of poles of the induction motor.

Solution

$$\text{Frequency} = \frac{12 \times 500}{120} = 50 \text{ Hz}$$

$$\text{Number of poles} = \frac{120 \times 50}{1440} = 4.16 \cong 4$$

$$N_s = \frac{120 \times 50}{4} = 1500 \text{ r.p.m.}$$

$$\text{percentage slip} = \frac{1500 - 1440}{1500} \times 100 = 4\%$$

Ex.7.19 A three-phase, six-pole, star connected 400 V, 50 Hz induction motor has an output power of 80 KW. Its rated power factor and efficiency are 0.8 and 95 % respectively. The resistance of the stator winding is 0.05Ω and its rated speed and mechanical loss are 950 rpm and 0.5 % of output power respectively. Calculate the following:

(i) Stator current
(ii) Power drawn by the stator
(iii) Active power
(iv) Reactive power
(v) Torque
(vi) Slip

Solution

$$I = \frac{80 \times 10^3}{\sqrt{3} \times 400 \times 0.95 \times 0.8} = 151.93 \text{ A}$$

$$S = \frac{80}{0.95 \times 0.8} = 105.26 \text{ KVA}$$

$$P = \frac{80}{0.95} = 84.21 \text{ KW}$$

$$Q = 105.26 \times 0.6 = 63.16 \text{ KVAr}$$

$$T = 9.55 \frac{80 \times 10^3}{950} = 804.21 \text{ Nm}$$

$$N_s = \frac{120 \times 50}{6} = 1000 \text{ r.p.m.}$$

$$\%\text{slip} = \frac{1000 - 950}{1000} = 5\%$$

Ex.7.20 A 50 Hz, three-phase, 6-pole, induction motor has full-load slip and no-load slip of 5.2% and 3.8 % respectively. Calculate the following:

(i) Synchronous speed
(ii) Rotor speed at no-load
(iii) Rotor speed at full-load
(iv) Frequency of the rotor at full-load
(v) Frequency of the rotor at standstill.

Solution

$$\text{Synchronous speed}(N_s) = \frac{120f}{p} = \frac{120 \times 50}{8} = 1000 \text{ r.p.m.}$$

$$\text{Rotor speed at no-load}\left(N_0\right) = N_s\left(1 - \text{slip at no load}\right) = 1000\left(1 - 0.038\right) = 962 \text{ r.p.m.}$$

$$\text{Rotor speed at full-load}\left(N_{fl}\right) = N_s\left(1 - \text{slip at full load}\right) = 1000\left(1 - 0.052\right) = 948 \text{ r.p.m.}$$

$$\text{Frequency of the rotor at full load } (f_{fl}) = f \times (\text{slip at full load}) = 50 \times 0.052 = 2.6 \text{ Hz}$$

$$\text{Frequency of the rotor at standstill } (f_{st}) = f = 50 \text{ Hz}$$

Ex.7.21 In a three-phase induction motor, the input power is 60KW and corresponding stator loss is 3KW. Calculate the following:

(i) Total mechanical power developed
(ii) Rotor copper loss
(iii) Output of the motor if the frictional and windage loss is 1KW
(iv) Efficiency of the motor

Take slip to be 3%

Solution

$$\text{Power input} = 60 - 3 = 57 \text{ KW}$$

$$\text{Rotor copper loss} = 0.03 \times 57 = 1.71 \text{ KW}$$

$$\text{Mechanical power} = 57 - 1.71 = 55.29 \text{ KW}$$

$$\text{Power} = 55.29 - 1 = 54.29 \text{ KW}$$

$$\text{Efficiency}(\eta) = \frac{54.29}{60} \times 100 = 90.48\%$$

Ex.7.22 The stator and the rotor frequency of a 4-pole induction motor are 50 Hz and 1.5 Hz respectively. What is the rotor speed?

Solution

$$N_s = \frac{120 \times 50}{4} = 1500 \text{ r.p.m.}$$

$$s = \frac{1.5}{50} = 0.03$$

$$N_r = 1500(1 - 0.03) = 1455 \text{ r.p.m.}$$

EXERCISES

1. Explain why induction motor is the most widely used motor for applications.
2. Give the classification of induction motor.
3. Explain in detail the construction of induction motor.
4. Explain the construction of squirrel-cage induction motor.
5. Explain the construction of wound-type induction motor.

6. Differentiate between slip-ring and squirrel-cage induction motor.
7. Explain how rotating magnetic field is produced, when a three-phase supply is given to the stator of a three-phase induction motor.
8. Show that the magnitude of the rotating magnetic field in the stator of a three-phase induction motor is constant.
9. Explain the working of three-phase induction motor.
10. Define slip.
11. How does slip affect the rotor parameters of an induction motor?
12. Explain in detail the effect of slip on the following rotor parameters:
    (i) Frequency
    (ii) Induced e.m.f.
    (iii) Resistance, Reactance and Impedance
    (iv) Current.
13. Compare induction motor and transformer.
14. Derive the torque equation of a three phase induction motor.
15. On what factor does the torque of an induction motor depend?
16. Derive the torque equation of a three-phase induction motor under running condition.
17. Obtain the condition for maximum running torque.
18. Obtain the condition for maximum starting torque.
19. Draw and explain the torque–slip characteristic of a three-phase induction motor.
20. Draw and explain the torque–speed characteristic of a three phase induction motor.
21. Derive the relationship between full-load torque and maximum torque.
22. Derive the relationship between starting torque and maximum torque.
23. Explain the effect of supply voltage on the starting torque of an induction motor.
24. Explain the effect of rotor resistance on the torque–slip characteristic of an induction motor.
25. Derive the power output of a three-phase induction motor.
26. Step by step, obtain the equivalent circuit of a three-phase induction motor.
27. Explain no-load test on induction motor.
28. Explain block-rotor test on induction motor.
29. Explain how the parameters of the equivalent circuit on an induction motor can be obtained from no-load test and the block-rotor test.
30. How is a circle diagram constructed?
31. What information can be obtained from the circle diagram?
32. What is crawling and cogging of induction motor?
33. What are the different methods of starting a squirrel-cage induction motor
34. Draw and explain the operation of direct on-line starting.
35. Draw and explain the operation of autotransformer starting
36. Explain in detail the operation of star–delta starting.
37. Explain the operation of rotor resistance starter.
38. Explain the starting method of a slip-ring induction motor.
39. List the methods of speed control of induction motor from the stator side.
40. List the methods of speed control of induction motor from the rotor side.
41. Explain *v*/*f* method of controlling the speed of an induction motor
42. How can the speed of an induction motor be controlled by changing the supply voltage?
43. How can the speed of an induction motor be controlled by changing the number of poles?
44. Explain speed control of induction motor by connecting rheostat at the stator terminals
45. Explain Speed control of induction motor by connecting rheostat at the rotor terminals

46. Explain cascade control.
47. What is braking?
48. Explain regenerative braking of induction motor.
49. Explain braking by plugging.
50. Explain dynamic braking.
51. What are the different methods of dynamic braking?
52. Explain the flowing dynamic braking techniques:
    (i) AC dynamic braking
    (ii) DC dynamic braking
    (iii) Zero-sequence braking
    (iv) Self-excited braking
53. The ratio of stator to rotor of a three-phase, 6-pole, 50 Hz induction motor is 3. When connected to 410 V supply, the speed of the motor at a certain load was observed as 1420 r.p.m. Calculate the following:
    (i) Rotor frequency during running condition
    (ii) Rotor frequency at standstill
    (iii) E.m.f. at standstill
    (iv) E.m.f. at running condition
54. A 50 Hz, three-phase, 6-pole, induction motor has full-load slip and no-load slip of 3% and 1.7 % respectively. Calculate the following:
    (i) Speed of rotation of the stator field
    (ii) Rotor speed at no-load
    (iii) Rotor speed at full-load
    (iv) Rotor frequency at full-load
    (v) Rotor frequency at standstill.
55. The rotor resistance and standstill reactance of an 8-pole, 50 Hz, three-phase induction motor are 0.05Ω and 0.1Ω respectively. If the motor runs at full load with slip of 4%, calculate the ratio of full-load torque to maximum torque and the speed at which full-load torque takes place.
56. A three-phase, 50 Hz, 10-pole, 38KW, 440V slip-ring induction motor has rated output at rated voltage and frequency, when the slip-rings are short-circuited. The maximum torque occurs at 0.08, which is 3 times the full-load torque. Calculate the following:
    (i) Full-load slip.
    (ii) Rotor speed at full-load
    (iii) Rotor resistive loss at full-load
    (iv) Full-load torque
    (v) Starting torque.
57. A star-connected three-phase, 6-pole, 50 Hz, 410V induction motor has the following equivalent circuit parameters:

$$R_1 = 0.8\Omega, X_1 = 1.3\Omega, R'_2 = 0.9\Omega, X'_2 = 1.5\Omega, R_0 = 130\Omega, X_0 = 45\Omega$$

If the full-load slip is 0.08 and the stator loss is 2.5 KW, find the following:
    (i) Synchronous speed
    (ii) Rotor speed at full-load
    (iii) Stator current at no-load
    (iv) Stator current at full-load
    (v) Motor input

(vi) Rotor input
(vii) Rotor copper loss
(viii) Motor output
(ix) Efficiency of the motor

58. A three-phase, 410V, star-connected induction motor takes full-load current at 70V when the rotor is blocked. Calculate the tappings on the three-phase autotransformer to limit the starting current to 4 times the full-load current. Take slip to be 8%.

59. The no-load and the block-rotor test result conducted on a three-phase, star-connected induction motor are as follows:

No load test result
Line voltage = 400 V
Current = 21 A
Power= 1120 W

Block rotor test result
Line voltage = 45 V
Current = 50 A
Power = 3020 W

Determine the parameters of the equivalent circuit of the induction motor.

60. Consider a 12-pole, three-phase, 50 Hz, 420V induction motor. When running at 930 r.p.m., it takes a power input of 70 KW. If the stator loss is 1.2 KW and friction and windage losses are 1.2 KW, calculate the efficiency of the motor.

CHAPTER 8

SINGLE-PHASE INDUCTION MOTOR

8.1 INTRODUCTION

A single-phase induction motor is structurally similar to a three-phase induction motor; the difference is only in the stator winding arrangements. The stator windings of a single-phase induction motor are distributed, pitched and skewed to produce a sinusoidal m.m.f. in space. They are simple in construction but more difficult to analyse as compared with three-phase induction motors. Efficiency of a single-phase induction motor is less than the efficiency of a three-phase induction motor. The single-phase induction motor operates at low power factor.

8.2 CONSTRUCTION

The construction of a single-phase induction motor is similar to that of a three-phase induction motor. The single-phase motor stator has a laminated iron core with two windings arranged perpendicularly – one is the main and the other is the auxiliary winding or starting winding. The motor uses a squirrel-cage rotor, which has a laminated iron core with slots in it. Aluminium bars are moulded on the slots and short-circuited at both ends with a ring. Single-phase induction motor is not self-starting and requires some mechanism to assist it in the starting process.

8.3 DOUBLE-FIELD REVOLVING THEORY

According to this theory, a pulsating field of a single-phase motor can be resolved into two rotating field of half its amplitude rotating in opposite direction at synchronous speed.

When the stator of a single-phase induction motor is connected to a single-phase supply, a sinusoidal flux (φ_s) will be produced in the stator. The value of this sinusoidal flux will change from $+\varphi_{sm}$ to $-\varphi_{sm}$ with respect to time for each cycle, φ_{sm} being the maximum value of the flux.

This flux can be broken down into two components, φ_{sf} and φ_{sb}. The magnitude of these two components of flux will be $\frac{1}{2}\varphi_{sm}$ but they will be rotating in the opposite direction with respect to time.

Figure 8.1 shows the rotation of φ_{sf} and φ_{sb} with respect to time t_1, t_2, t_3, t_4 and t_5. A 45^0 time interval is taken between them.

From Fig. 8.1 (a), at time $t = t_1$, $\varphi_s = 0$ (φ_{fs} and φ_{bs} are equal in magnitude but opposite in direction and hence they cancel out.)

From Fig. 8.1 (b), at time $t = t_2$, $\varphi_s = 0.707\varphi_{sm}$

From Fig. 8.1 (c), at time $t = t_3$, $\varphi_s = \varphi_{sm}$ (maximum value)

From Fig. 8.1 (d), at time $t = t_4$, $\varphi_s = 0.707\varphi_{sm}$

From Fig. 8.1 (e), at time $t = t_5$, $\varphi_s = 0$ (φ_{fs} and φ_{bs} are equal in magnitude but opposite in direction and hence they cancel out.)

So we can say that in a half-cycle, φ_s changes from 0 to maximum value to 0. φ_{fs} which is rotating in clockwise direction is called the forward field and φ_{bs} which is rotating in anti-clockwise direction is called the backward field of φ_s.

The torque slip curve due to φ_{fs} and φ_{bs} is shown in Fig. 8.2.

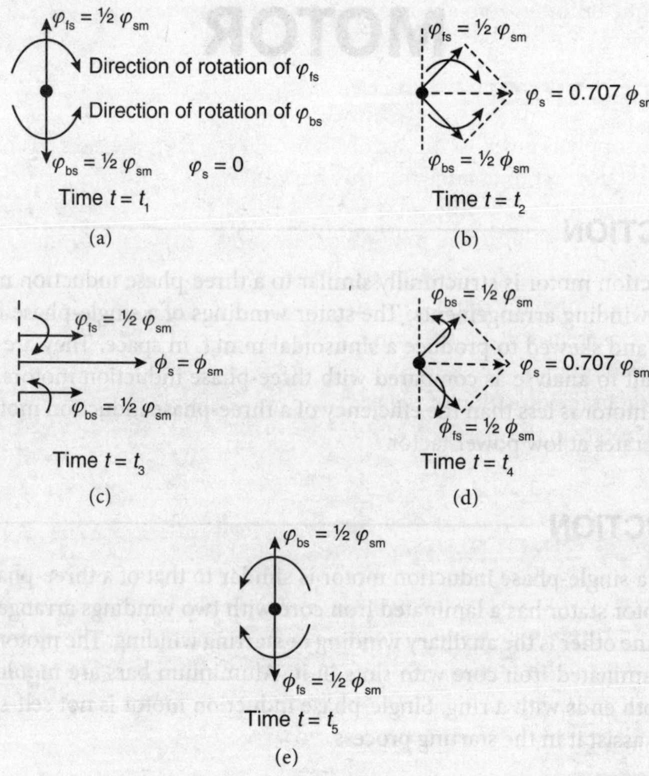

Fig. 8.1: Rotation of flux with respect to time

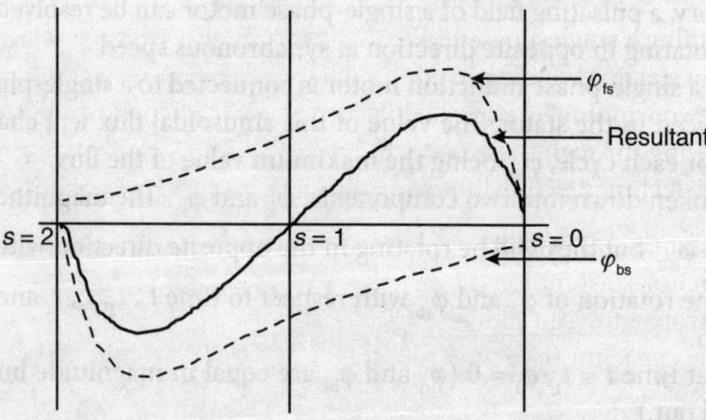

Fig. 8.2: Torque-slip curve

From the resultant torque-slip curve, it can be concluded that

- At slip $s = 1$, the torque is zero; so there is no starting torque. So a single-phase induction motor is not self starting.
- The maximum torque and breakdown torque are less when compared with a three-phase induction motor.
- In a single-phase induction motor, increase in rotor resistance increases the backward field, thus reducing the breakdown torque and the efficiency of the motor. It also increases the slip at which the maximum torque will take place.
- The average torque becomes zero at some value of speed below the synchronous speed.

8.4 CROSS FIELD THEORY

When a single-phase supply is given to the stator of a single-phase induction motor, an alternating flux φ_s, is produced in the stator. Let us assume that this flux is along the horizontal axis. This alternating flux φ_s, will produce an induced e.m.f. in the rotor windings, which is initially at standstill. No torque will be produced in the rotor since the axis of both the stator and the rotor fields lie in the same direction.

Now, if the rotor is rotated by giving some external force, an e.m.f. will be induced in the rotor and thus a current will flow through the rotor. This current will be displaced by 90^0 (electrical) from the stator axis. That is, the rotor current will lag the stator induced e.m.f. by 90^0 (electrical). Hence, there will be a 90^0 phase-shift between the rotor field φ_r and the stator field φ_s. Now, these two fluxes will produce a revolving field rotating in the direction of the rotor movement. Once this rotating field is produced, the rotor will continue to rotate in the direction of the rotating field, as in the case of a three-phase induction motor.

8.5 STARTING OF SINGLE-PHASE INDUCTION MOTOR

Single-phase induction motor is not self-starting. Therefore, some additional methods are required to make the motor self-starting. In fact, single-phase induction motors are classified based on the starting methods used for it. Some of the starting methods are,

i) Split-phase method
ii) Shaded-pole method
iii) Reluctance-start method

8.5.1 Split-phase Method

It is a technique of running a single-phase induction motor as a two-phase induction motor. This can be obtained by providing a starting winding in additional to main single-phase winding. The resistance and the reactance of these two windings are selected in such a way that when connected across a single-phase supply, the current flowing through these two windings will have a phase difference between them. This is called the split-phase technique. The different split-phase techniques are:

i) Split-phase resistance-start motor.
ii) Split-phase capacitor-start motor.
iii) Capacitor-start–capacitor-run induction motor
iv) Permanent single capacitor induction motor.

8.5.1.1 Split-phase resistance-start motor

The schematic diagram of a split-phase resistance-start induction motor is shown in Fig. 8.3. In this type of motor, the auxiliary winding is placed at 90⁰ with respect to the main winding. The auxiliary winding is connected in parallel with the stator main winding. It is selected in such a way that the R/X ratio of the auxiliary winding is higher than that of the main winding.

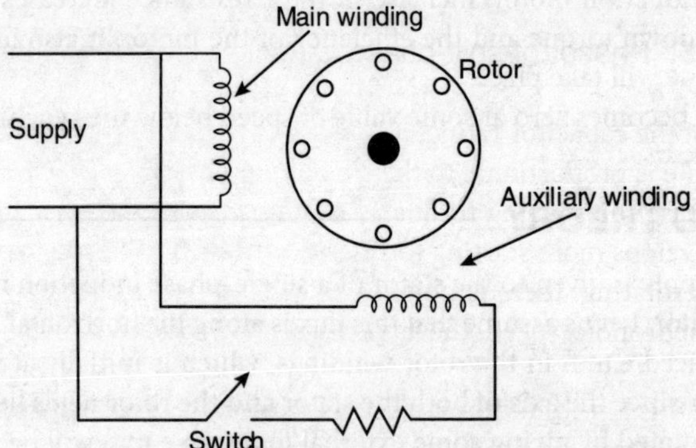

Fig. 8.3: Split-phase resistance-start induction motor

Let the current through the auxiliary winding be I_{au} and that through the main winding be I_m. Since the R/X ratio of the auxiliary winding is more than that of the main winding, current I_{au} will lag the voltage V by a smaller angle as compared with I_m. The phasor is shown in Fig. 8.4.

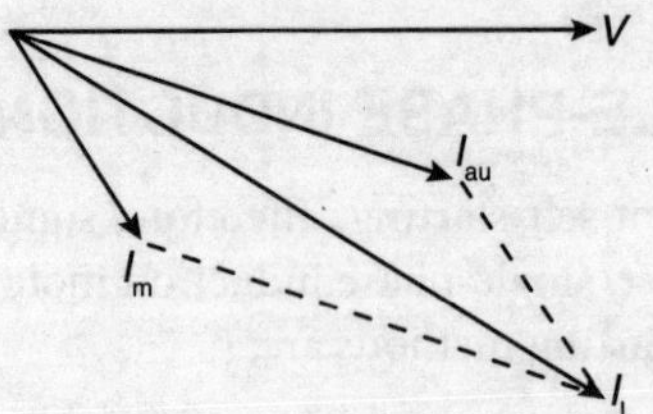

Fig. 8.4: Phasor of split-phase resistance-start induction motor

Since the current in the two windings are different in magnitude and phase, it will produce a rotating magnetic field. This rotating magnetic field will produce a torque on the rotor, which will cause the rotor to rotate.

Once the rotor starts rotating, there is no need of auxiliary winding. So when the motor picks up about 75% to 80% of synchronous speed, the auxiliary winding is cut out from the circuit with the help of a centrifugal switch.

The direction of rotation of the motor can be reversed by interchanging the connection of either the auxiliary winding or the main winding.

This method of starting is mostly used in motor with low inertia load or continuous operating load.

8.5.1.2 Split-phase capacitor-start motor

In this type of motor, an electrolytic capacitor is connected in series with the auxiliary winding. Due to the presence of the capacitor, the auxiliary winding current will now lead the applied voltage and the main winding current will lag the applied voltage, as shown in the phasor diagram, Fig. 8.5.

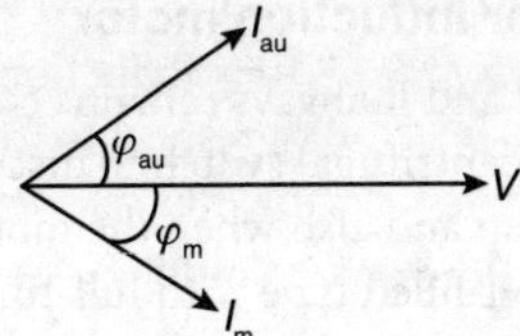

Fig. 8.5: Phasor of split-phase capacitor-start induction motor

By careful selection of the capacitor rating, it is possible to have 90° phase shift between the currents I_{au} and I_m. Since the torque is proportional to the sine of the phase angle between I_{au} and I_m, an increase in phase angle will increase the starting torque as compared with the resistance start method. So, the capacitor-start motor develops more starting torque as compared to resistance-start motor.

Once the rotor starts rotating, there is no need of auxiliary winding. So, when the motor picks up about 75% to 80% of synchronous speed, the auxiliary winding is cut out from the circuit with the help of a centrifugal switch.

The capacitor used in this type of motor is electrolytic type designed for short-duty service. This type of motor is used in fans, blowers, air conditioners pumps, etc.

8.5.1.3 Capacitor-start–capacitor-run induction motor

These motors are used for such applications where large starting torque and quiet operations are required. These motors produce constant torque and have better efficiency and power factor. In this starting method, two capacitors, C_{st} and C_{rn} are connected with the auxiliary winding in the manner as shown in Fig. 8.6.

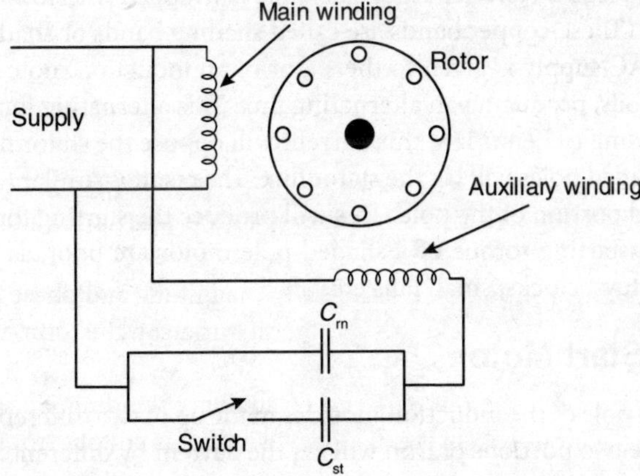

Fig. 8.6: Capacitor-start–capacitor-run motor

Capacitor C_{st} is a short-duty electrolytic capacitor and C_{rn} is an oil-impregnated paper continuous rating capacitor. When the motor is started, both the capacitors remain connected with the auxiliary winding. However, when the motor picks up about 75% to 80% of synchronous speed, the capacitor C_{st} is cut out from the circuit with the help of a centrifugal switch. Direction of rotation of the motor can be reversed by interchanging the connection of either the main winding or the auxiliary winding.

8.5.1.4 Permanent single-capacitor induction motor

In this motor, only one capacitor is used and it always remains connected with the auxiliary winding of the motor. The reason is that there is no centrifugal switch to disconnect it. So this capacitor will remain connected with the supply during start-up and also when the motor is in running condition.

The capacitor used in this case is an oil-filled type with full-time rating. The starting torque obtained in this case will be better with improved power factor. The rotating field produced in this case will be uniform and the motor will be less noisy. The efficiency of the motor will also be high.

8.5.2 Shaded Pole Method

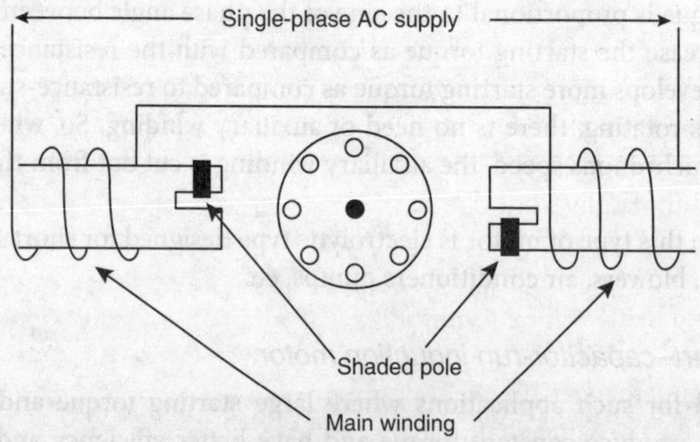

Fig. 8.7: Shaded pole motor

The shaded-pole motor is shown in Fig. 8.7. This method of starting is employed for small motors that require small starting torque. Here, a part of each pole is wrapped with low resistance copper bands, which form a closed loop (These copper bands are called shading bands or shaded poles).

When a single-phase AC supply is given to the stator of an induction motor, an alternating current flows through the stator coils, producing an alternating flux. This alternating flux will set up a current in the shading bands. According to Lenz's law, this current will oppose the stator flux.

Thus, the flux in the shaded poles will lag the stator flux. The result is similar to a rotating field moving from un-shaded to shaded portion of the pole. This will produce the starting torque.

The efficiency and the starting torque of a shaded pole motor are poor, as also is its power factor. These motors are used in toys, clocks, small fans etc.

8.5.3 Reluctance Start Motor

In this type of motor, each pole of the induction motor is made up of two different amounts of iron. Thus, the flux produced by these two portions of iron will lag the current by different angles. This will create a rotating magnetic field and enable the rotor to start rotating. Reluctance motor has limited applications due to its low torque.

8.6 EQUIVALENT CIRCUIT OF A SINGLE-PHASE INDUCTION MOTOR

At standstill condition, a single phase induction motor works as a single-phase transformer. So, the equivalent circuit of the motor will be similar to the equivalent circuit of a single-phase transformer with its secondary short-circuited. The equivalent circuit is shown in Fig. 8.8.

Where,

r_1 and x_1 are the stator resistance and stator reactance.

R_0 and X_0 are the iron loss component and magnetising component respectively of the no-load current.

r_2' and x_2' are the rotor resistance and rotor reactance referred to stator side.

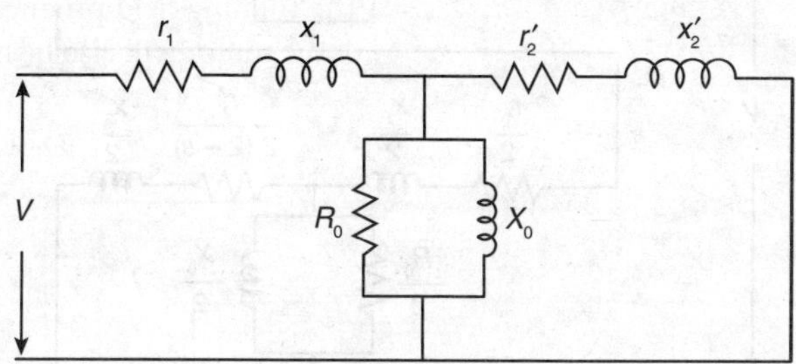

FIG. 8.8: Equivalent circuit of single-phase induction motor

According to double-field revolving theory, due to positive and negative sequence magnetising current, two opposite rotating fields will be produced which rotates at synchronous speed with respect to the stator.

Due to these two rotating magnetising fields, a single-phase induction motor is treated as two polyphase induction motors on the same shaft, having equal and opposite torque at standstill. Thus, the equivalent circuit of an induction motor at standstill based on double-field revolving theory will be as shown in Fig. 8.9.

When the rotor starts rotating at speed N, let s be the slip with respect to the forward rotating field and $(2 - s)$ be the slip with respect to the backward rotating field. Then the rotor resistance because of forward rotating field will be $\frac{r_2'}{2s}$ and the rotor resistance because of backward rotating field will be $\frac{r_2'}{2(2-s)}$.

So, the equivalent circuit of single-phase induction motor at any slip s will be as shown in Fig. 8.10.

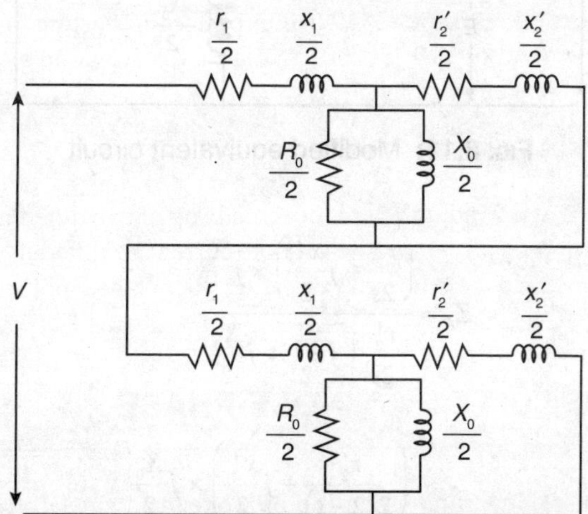

FIG. 8.9: Equivalent circuit at standstill based on double-field revolving theory

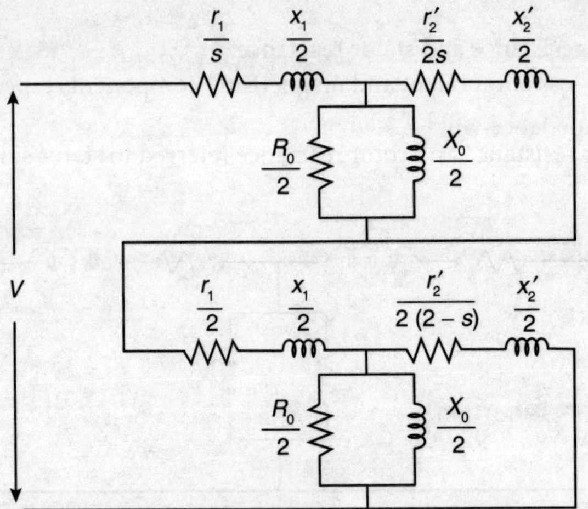

Fig. 8.10: Equivalent circuit at any slip

If the core loss is neglected, then the modified equivalent circuit is as shown in Fig. 8.11

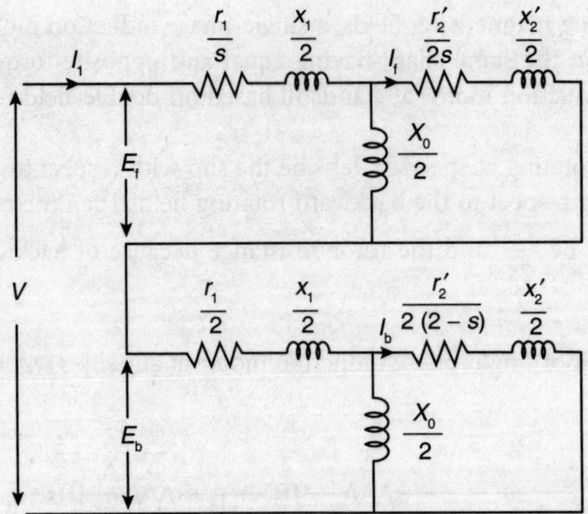

Fig. 8.11: Modified equivalent circuit

Let,

$$Z_f = \frac{\left(\frac{r_2'}{2s} + j\frac{x_2'}{2}\right) \times j\frac{X_0}{2}}{\frac{r_2'}{2s} + j\frac{x_2'}{2} + j\frac{X_0}{2}} \tag{8.1}$$

$$Z_b = \frac{\left(\frac{r_2'}{2(2-s)} + j\frac{x_2'}{2}\right) \times j\frac{X_0}{2}}{\frac{r_2'}{2(2-s)} + j\frac{x_2'}{2} + j\frac{X_0}{2}} \tag{8.2}$$

And

$$Z_1 = r_1 + jx_1 \tag{8.3}$$

Then the equivalent impedance will be,

$$Z_{eq} = Z_1 + Z_f + Z_b \tag{8.4}$$

The current,

$$I_1 = \frac{V}{Z_{eq}} \tag{8.5}$$

Again from the circuit we can write,

$$E_f = I_1 Z_f \text{ and } E_b = I_1 Z_b$$

And

$$I_f = \frac{E_f}{\frac{r'_2}{2s} + j\frac{x'_2}{2}} \tag{8.6}$$

$$I_b = \frac{E_b}{\frac{r'_2}{2(2-s)} + j\frac{x'_2}{2}} \tag{8.7}$$

Then the power input to the rotor due to forward field is

$$P_f = I_f^2 \times \frac{r'_2}{2s} \tag{8.8}$$

And the power input to the rotor due to backward field is

$$P_b = I_b^2 \times \frac{r'_2}{2(2-s)} \tag{8.9}$$

Mechanical power output due to forward field is

$$P_{mf} = (1-s)P_f \tag{8.10}$$

Mechanical power output due to backward field is

$$P_{mb} = (1-(2-s))P_b \tag{8.11}$$

Net mechanical power output is

$$P_{net} = P_{mf} + P_{mb} \tag{8.12}$$

The torque developed due to forward field is

$$T_f = \frac{P_f}{(2\pi N_s)/60} \text{ Nm} \tag{8.13}$$

The torque developed due to backward field is

$$T_{\mathrm{b}} = \frac{P_{\mathrm{b}}}{(2\pi N_{\mathrm{s}})/60}\ \mathrm{Nm} \tag{8.14}$$

Net torque is

$$T_{\mathrm{net}} = T_{\mathrm{f}} - T_{\mathrm{b}} \tag{8.15}$$

$$\text{Rotor copper loss due to forward field} = sP_{\mathrm{f}} \tag{8.16}$$

$$\text{Rotor copper loss due to backward field} = (2-s)P_{\mathrm{b}} \tag{8.17}$$

$$\text{Total copper loss} = sP_{\mathrm{f}} + (2-s)P_{\mathrm{b}} \tag{8.18}$$

Net power output = P_{net} – Friction and windage loss – Copper loss.

$$\text{Efficiency}\ (\eta) = \frac{\text{Output}}{\text{Input}} \times 100 \tag{8.19}$$

8.7 TESTS TO DETERMINE THE EQUIVALENT CIRCUIT PARAMETERS

It is possible to conduct simple tests on induction motor to determine the circuit parameters of its equivalent circuit. These tests are discussed in this section.

8.7.1 DC Resistance Test

In this test, DC supply is given to the stator of a single-phase induction motor. The reading of the voltmeter and the ammeter connected in the circuit is noted. The voltmeter gives the value of the supply voltage and the ammeter gives the current flowing through the stator winding. Then, the DC resistance of the stator is calculated as

$$R_{\mathrm{DC}} = \frac{V_{\mathrm{DC}}}{I_{\mathrm{DC}}} \tag{8.20}$$

And the AC resistance is obtained by

$$R_{\mathrm{AC}} = 1.15\, R_{\mathrm{DC}} \tag{8.21}$$

8.7.2 Block Rotor Test

As the name suggests, in this test the rotor of the single-phase induction motor is prevented from rotating (that is, it is blocked). Since the rotor is blocked, the slip due to forward rotating field (s_{f}) and the slip due to backward rotating field (s_{b}) will be 1. That is, $s_{\mathrm{f}} = s_{\mathrm{b}} = 1$.

Under this condition, the value of the rotor impedance becomes very less as compared to the value of the magnetising branch. Since these two branches are connected in parallel, the magnetising branch can be neglected. So the equivalent circuit of Fig. 8.10 can be redrawn as shown in Fig. 8.12.

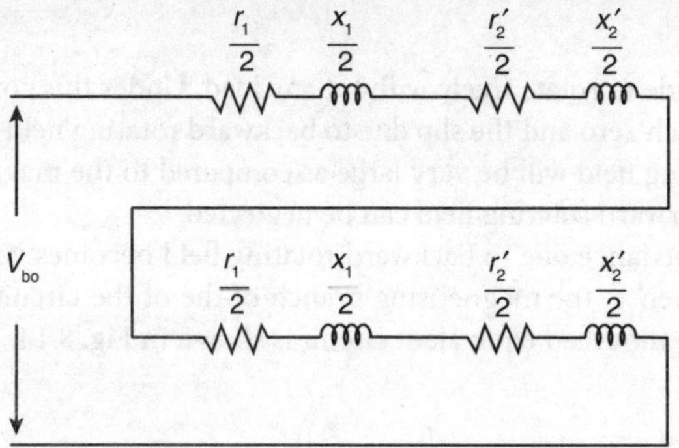

FIG. 8.12: Equivalent circuit

The circuit can be further modified as shown in Fig. 8.13

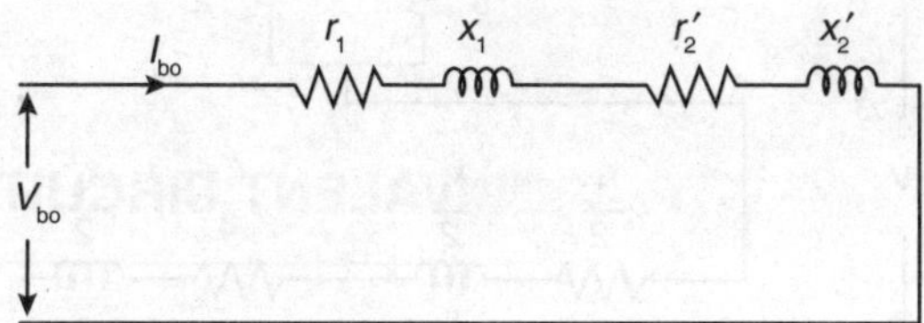

FIG. 8.13: Modified equivalent circuit

By connecting the voltmeter, ammeter and wattmeter at appropriate places, the following readings can be obtained for the phase voltage (V_{bo}), the phase current (I_{bo}) and the power (P_{bo}).

We know that,

$$P_{bo} = I_{bo}^2 R_{eq} \tag{8.22}$$

Then,

$$R_{eq} = \frac{P_{bo}}{I_{bo}^2} \tag{8.23}$$

Again,

$$Z_{eq} = \frac{V_{bo}}{I_{bo}} \tag{8.24}$$

Then,

$$X_{eq} = \sqrt{Z_{eq}^2 - R_{eq}^2} \tag{8.25}$$

So,

$$x_1 = x'_2 = \frac{X_{eq}}{2} \tag{8.26}$$

$$R_2' = R_{eq} - R_1 \tag{8.27}$$

8.7.3 No-load Test

In this test, the rotor is made to rotate freely without any load. Under this condition, the slip due to forward rotating field will reach zero and the slip due to backward rotating field will be 2. The rotor impedance due to forward rotating field will be very large as compared to the magnetising branch; hence, the rotor impedance due to forward rotating field can be neglected.

Similarly, the rotor impedance due to backward rotating field becomes very small as compared with the magnetising branch; hence, the magnetising branch of the of the circuit due to backward rotating field can be neglected. The modified equivalent circuit is shown in Fig. 8.14.

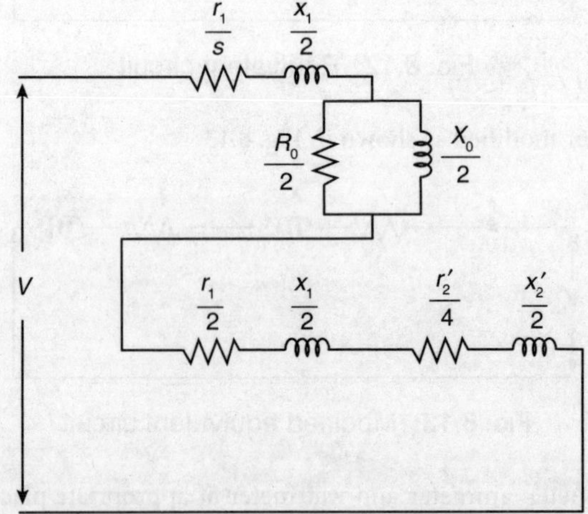

FIG. 8.14: Modified equivalent circuit for no-load test

The circuit can be further simplified with some assumptions; the modified circuit is called the approximate equivalent circuit, as shown in Fig. 8.15

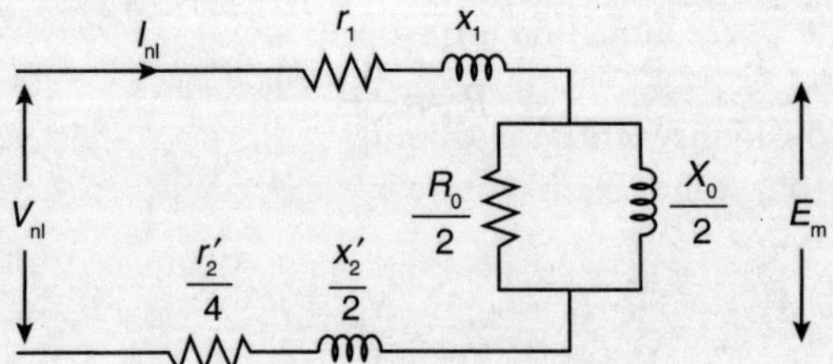

FIG. 8.15: Simplified circuit

By connecting the voltmeter, ammeter and wattmeter at appropriate places, the following readings can be obtained for the phase voltage (V_{nl}), the phase current (I_{nl}) and the power (P_{nl}). After getting the readings, we can calculate the following:

$$\text{Mechanical Power}(P_m) = P_{nl} - I_{nl}^2\left(R_1 + \frac{R_2'}{4}\right) \tag{8.28}$$

$$E_m = V_{nl} - I_{nl}\left[\left(R_1 + \frac{R_2'}{4}\right) + j\left(x_1 + \frac{x_2'}{2}\right)\right] \tag{8.29}$$

$$\cos\theta = \frac{P_{nl}}{V_{nl}I_{nl}} \tag{8.30}$$

$$R_0 = 2\frac{E_m^2}{P_m} \tag{8.31}$$

$$I_w = \frac{2E_m}{R_0} \tag{8.32}$$

$$I_m = \sqrt{I_{nl}^2 - I_w^2} \tag{8.33}$$

$$X_m = \frac{2E_m}{I_m} \tag{8.34}$$

8.8 COMPARISON BETWEEN SINGLE-PHASE AND THREE-PHASE INDUCTION MOTOR

1. Single-phase induction motors are simple in construction, reliable and economical for small power rating as compared to three-phase induction motors.
2. The electrical power factor of single-phase induction motors is low as compared to three-phase induction motors.
3. For the same size, single-phase induction motors develop about 50% of the output as that of three-phase induction motors.
4. The starting torque is low for asynchronous motors.
5. The efficiency of single-phase induction motors is less as compared to that of three-phase induction motors.

SOLVED PROBLEMS

Ex.8.1 A 2-pole single-phase induction motor has the equivalent circuit parameters as follows: $R_1 = 4\Omega$, $R_2 = 3.5\Omega$, $X_1 = 5\Omega$, $X_2 = 6\Omega$, $X_0 = 95\Omega$. When supplied with 120V, 50 Hz supply it runs at 2500 r.p.m., calculate the following:

(i) The supply current
(ii) Power factor

(iii) Input power
(iv) Output power
(v) Efficiency.

Solution

$$\text{Synchronous speed}\left(N_s\right) = \frac{120f}{P} = \frac{120 \times 50}{2} = 3000 \text{ r.p.m.}$$

$$\text{Slip}\left(s\right) = \frac{3000 - 2500}{3000} = 0.16$$

$$Z_1 = r_1 + jx_1 = \left(4 + j5\right)\Omega$$

$$Z_f = \frac{\left(\frac{r'_2}{2s} + j\frac{x'_2}{2}\right) \times j\frac{X_0}{2}}{\frac{r'_2}{2s} + j\frac{x'_2}{2} + j\frac{X_0}{2}} = \frac{\left(\frac{3.5}{2 \times 0.16} + j\frac{6}{2}\right) \times j\frac{95}{2}}{\frac{3.5}{2 \times 0.16} + j\frac{6}{2} + j\frac{95}{2}} = \frac{(10.93 + j3) \times j47.5}{10.93 + j3 + j47.5}$$

$$= \left(8.9 + j4.67\right)\Omega$$

$$Z_b = \frac{\left(\frac{r'_2}{2(2-s)} + j\frac{x'_2}{2}\right) \times j\frac{X_0}{2}}{\frac{r'_2}{2(2-s)} + j\frac{x'_2}{2} + j\frac{X_0}{2}} = \frac{\left(\frac{3.5}{2 \times (2 - 0.16)} + j\frac{6}{2}\right) \times j\frac{95}{2}}{\frac{3.5}{2 \times (2 - 0.16)} + j\frac{6}{2} + j\frac{95}{2}} = \frac{(0.95 + j3) \times j47.5}{0.95 + j3 + j47.5}$$

$$= \frac{-142.5 + j45.13}{0.95 + j50.5} = \left(0.298 + j2.88\right)\Omega$$

$$Z_{eq} = Z_1 + Z_f + Z_b = 4 + j5 + 8.9 + j4.67 + 0.298 + j2.88 = \left(13.19 + j12.55\right)\Omega$$

The supply current,

$$I_1 = \frac{V}{Z_{eq}} = \frac{120}{13.19 + j12.55} = 6.58\angle -43.58 \text{ A} = \left(4.77 - j4.54\right)\Omega$$

$$E_f = I_1 Z_f = \left(4.77 - j4.54\right) \times \left(8.9 + j4.67\right) = \left(63.65 - j18.13\right)\text{V}$$

$$E_b = I_1 Z_b = \left(4.77 - j4.54\right) \times \left(0.298 + j2.88\right) = \left(14.49 + j12.38\right)\text{V}$$

$$I_f = \frac{E_f}{\frac{r'_2}{2s} + j\frac{x'_2}{2}} = \frac{63.65 - j18.13}{\frac{3.5}{2 \times 0.16} + j\frac{6}{2}} = \frac{63.65 - j18.13}{10.93 + j3} = 8.4940 - j2.4152 \text{ A} = 8.83\angle -15.87 \text{ A}$$

$$I_b = \frac{E_b}{\frac{r'_2}{2(2-s)} + j\frac{x'_2}{2}} = \frac{14.49 + j12.38}{\frac{3.5}{2\times(2-0.16)} + j\frac{6}{2}} = \frac{14.49 + j12.38}{0.95 + j3} = -7.0948 - j6.0663\text{ A}$$

$$= 9.33\angle -139.47\text{ A}$$

$$P_f = I_f^2 \times \frac{r'_2}{2s} = (8.83)^2 \frac{3.5}{2\times 0.16} = 818.51\text{ W}$$

$$P_b = I_b^2 \times \frac{r'_2}{2(2-s)} = (9.33)^2 \times \frac{3.5}{2\times(2-0.16)} = 83.09\text{ W}$$

$$P_{mf} = (1-s)P_f = (1-0.16)\times 371.75 = 687.55\text{ W}$$

$$P_{mb} = (1-(2-s))P_b = (1-(2-0.16))\times 33.78 = -69.79\text{ W}$$

$$P_{net} = P_{mf} + P_{mb} = 687.55 - 69.79 = 617.75\text{ W}$$

$$T_f = \frac{P_f}{(2\pi N_s)/60} = \frac{818.51}{\left(2\times\pi\times 3000/60\right)} = 2.61\text{ Nm}$$

$$T_b = \frac{P_b}{(2\pi N_s)/60} = \frac{83.09}{\left(2\times\pi\times 3000/60\right)} = 0.264\text{ Nm}$$

$$T_{net} = T_f - T_b = 2.346\text{ Nm}$$

Rotor copper loss due to forward field $= sP_f = 0.16\times 818.51 = 130.96$ W

Rotor copper loss due to backward field $= (2-s)P_b = (2-0.16)\times 83.09 = 152.89$ W

Total copper loss $= sP_f + (2-s)P_b = 130.96 + 152.89 = 283.85$ W

Net power output $= P_{net}$ – Friction and windage loss – Copper loss.

Net power output $= 617.75 - 283.85 = 333.9$ W

Input power $= 120\times 6.58\times \cos 43.58 = 571.99$ W

$$\text{Efficiency }(\eta) = \frac{\text{Output}}{\text{Input}}\times 100 = \frac{333.9}{571.99}\times 100 = 58.38\ \%$$

Ex.8.2 A single-phase, split-phase induction motor runs at 1700 r.p.m. when supplied from a 115V, 50 Hz supply. The block rotor test results are as follows:

Main winding: 23 V 4 A 60 W

Auxiliary winding: 23 V 1.5 A 30 W

Find the current drawn by the motor.

Solution

Main winding:

$$\cos\theta_{\mathrm{m}} = \frac{60}{23\times 4} = 0.65$$

$$\theta_{\mathrm{m}} = cos^{-1}0.65 = 49.6^{0}$$

Auxiliary winding:

$$\cos\theta_{\mathrm{A}} = \frac{30}{23\times 1.5} = 0.87$$

$$\theta_{\mathrm{m}} = \cos^{-1}0.87 = 29.6^{0}$$

$$\text{Total current at 23 V} = 4\angle 49.6 + 1.5\angle 29.6 = 5.43\angle 44.23 \text{ A}$$

$$\text{Current at 115 V} = 5.43\times\frac{115}{23} = 27.15 \text{ A}$$

Ex.8.3 A 3KW, 130V, 50 Hz, capacitor-start single-phase induction motor has the following main and auxiliary winding impedance (at starting).

$$Z_{\mathrm{m}} = 5 + j4\,\Omega$$

$$Z_{\mathrm{a}} = 10 + j4\,\Omega$$

Find the value of the starting capacitor that will place the main and auxiliary winding current in quadrature at starting.

Solution

Let X_c be the capacitor reactance that will place the main and auxiliary winding current in quadrature at starting.

$$\varphi_{\mathrm{m}} = \tan^{-1}\frac{4}{5} = 38.66^{0}$$

$$\varphi = 90 - 38.66 = 51.34^{0}$$

$$Z_{\mathrm{tot}} = Z_a + jX_c = 10 + j4 + jX_c$$

Then,

$$\tan^{-1}\frac{4+X_c}{10} = 51.34$$

$$\text{or, } \tan 51.34 = \frac{4 + X_c}{10}$$

$$\text{or, } 1.24 = \frac{4 + X_c}{10}$$

$$\text{or, } X_c = 16.4\ \Omega$$

$$\text{Again, } X_c = \frac{1}{2\pi fC}$$

$$\text{Then, } \frac{1}{2\pi fC} = 16.4$$

$$\text{or, } C = \frac{1}{2 \times \pi \times 50 \times 16.4} = 0.194\ \text{mF}$$

Ex.8.4 A single-phase 4-pole 50 Hz induction motor has a supply voltage of 200V, input current 4A, output of 500W, power factor of 0.8 and speed 1400 r.p.m. Calculate the efficiency and the slip.

Solution

$$\text{Input power} = 200 \times 4 \times 0.8 = 640\ \text{W}$$

$$\text{Efficiency } (\eta) = \frac{500}{640} \times 100 = 78.13\ \%$$

$$N_s = \frac{120 \times 50}{4} = 1500\ \text{r.p.m.}$$

$$\text{Slip } (s) = \frac{1500 - 1400}{1500} \times 100 = 6.67\ \%$$

Ex.8.5 Consider a 230V, 50 Hz, single-phase capacitor-start induction motor that has the following parameters at standstill:

Main winding parameter:	100 V	2 A	40 W
Auxiliary winding parameter:	80 V	1 A	50 W

Calculate the value of the capacitor for the maximum starting torque. Assume

$$R_1 = R_2' \text{ and } X_1 = X_2'$$

Solution

From main winding:

$$Z_m = \frac{100}{2} = 50\ \Omega$$

$$R_m = \frac{40}{2^2} = 10\ \Omega$$

$$X_m = \sqrt{Z_m^2 - R_m^2} = \sqrt{50^2 - 10^2} = 48.99\,\Omega$$

$$X_1 = X_2' = \frac{48.99}{2} = 24.5\,\Omega$$

$$R_1 = R_2' = \frac{10}{2} = 5\,\Omega$$

$$\theta_m = \tan^{-1}\frac{24.5}{5} = 78.46^0$$

From auxiliary winding:

$$Z_{eq} = \frac{80}{1} = 80\,\Omega$$

$$R_{eq} = \frac{50}{1^2} = 50\,\Omega$$

$$X_{eq} = \sqrt{Z_{eq}^2 - R_{eq}^2} = \sqrt{80^2 - 50^2} = 62.45\,\Omega$$

$$R_a = 50 - 5 = 45\,\Omega$$

$$X_a = 62.45 - 24.5 = 37.95\,\Omega$$

Let X_c be the value of the capacitance, then,

$$\tan(90 - 78.46) = \frac{37.95 - X_c}{45}$$

$$\text{or } X_c = 28.76\,\Omega$$

Ex.8.6 A single-phase, split-phase induction motor runs at 1000 r.p.m. when supplied from a 200V, 50 Hz supply. The block rotor test results are as follows:
Main winding: 28 V 5.2 A 68 W
Auxiliary winding: 28 V 3 A 34 W
Find the current drawn by the motor.

Solution
Main winding:

$$\cos\theta_m = \frac{68}{28 \times 5.4} = 0.45$$

$$\theta_m = \cos^{-1}0.45 = 63.3^0$$

Auxiliary winding:

$$\cos\theta_A = \frac{34}{28 \times 3} = 0.405$$

$$\theta_m = \cos^{-1} 0.405 = 66.1^0$$

$$\text{Total current at 28 V} = 5.4\angle 63.3 + 3\angle 66.2 = 8.39\angle 64.29 \text{ A}$$

$$\text{Current at 200 V} = 8.39 \times \frac{200}{28} = 60 \text{ A}$$

Ex.8.7 A single-phase 6-pole 50 Hz induction motor has a supply voltage of 200V, input current of 4A, output of 500W, power factor of 0.8 and speed 965 r.p.m. Calculate the efficiency and the slip.

Solution

$$\text{Input power} = 200 \times 4 \times 0.8 = 640 \text{ W}$$

$$\text{Efficiency } (\eta) = \frac{500}{640} \times 100 = 78.13\,\%$$

$$N_s = \frac{120 \times 50}{9} = 1000 \text{ r.p.m.}$$

$$\text{Slip } (s) = \frac{1000 - 965}{1000} \times 100 = 3.5\,\%$$

Ex.8.8 A single-phase induction motor has main and auxiliary winding impedance as (5 + j4) Ω and (9.5 + j3.5) Ω respectively. Find the value of the capacitor to place the main and auxiliary winding current in quadrature during starting.

Solution

$$\theta_m = \tan^{-1}\left(\frac{4}{5}\right) = 38.66^0$$

$$\theta_a = 38.66 - 90 = -51.34^0$$

Again,
Let X_c be the capacitive reactance. Then,

$$\tan\theta_a = \frac{3.5 - X_c}{9.5}$$

$$\text{or, } -1.25 = \frac{3.5 - X_c}{9.5}$$

$$X_c = 15.38\ \Omega$$

Then,

$$C = \frac{1}{2\pi f X_c} = \frac{1}{2\times\pi\times 50\times 15.38} = 0.21\,\text{mF}$$

Ex.8.9 A 200W, 100V, 50 Hz, 4-pole, single-phase induction motor rotating at rated voltage and frequency has $R_1 = 2\,\Omega, R'_2 = 4\,\Omega, X_m = 60\,\Omega, X_1 = X'_2 = 2\,\Omega$ and slip is 0.05. Find the input current.

Solution

$$Z_b = \frac{1}{2}\times\frac{j60\times\left(\frac{2}{2-0.05}+j2\right)}{\frac{2}{2-0.05}+j2+j60} = (0.96+j1.95)\,\Omega$$

$$Z_f = \frac{\frac{j60}{2}\times\left(\frac{4/2}{0.05}+\frac{j2}{2}\right)}{\frac{4/2}{0.05}+\frac{j60}{2}+\frac{j2}{2}} = (14.06+j19.11)\,\Omega$$

$$Z_t = (2+j2)+(0.96+j1.95)+(14.06+j19.11) = (17.02+j23.06)\,\Omega = 28.66\angle 53.57\,\Omega$$

$$I = \frac{100}{28.66} = 3.49\text{ A}$$

Ex.8.10 For the motor of Ex.8.9, find the resultant torque.

Solution

$$N_s = \frac{120\times 50}{4} = 1500\text{ r.p.m.}$$

$$\text{Torque} = \frac{3.49^2\times 14.06 - 3.49^2\times 0.96}{1500} = 0.106\text{ Nm}$$

EXERCISES

1. Explain the construction of a single-phase induction motor
2. Why is a single-phase induction motor not self-starting?
3. With a suitable phasor diagram, explain the double-field revolving theory.
4. Explain the cross field theory.
5. What are the different methods of starting of a single-phase induction motor?
6. Explain the split-phase method of starting a single-phase induction motor.
7. How does shaded-pole method help in starting an induction motor?
8. Explain the reluctance start method of starting an induction motor.

9. Explain the following starting methods for a single-phase induction motor:
   (i) Split-phase resistance-start motor.
   (ii) Split-phase capacitor-start motor.
   (iii) Capacitor start – capacitor run induction motor
   (iv) Permanent single capacitor induction motor.
10. Explain the shaded pole method.
11. Write a note on reluctance-start motor.
12. Obtain the equivalent circuit of a single-phase induction motor
13. From the equivalent circuit of a single-phase induction motor obtain an expression for:
   (i) Mechanical power output.
   (ii) Torque
   (iii) Copper loss
   (iv) Efficiency
14. Explain DC resistance test.
15. What parameters are obtained from DC resistance test?
16. Explain the block rotor test on a single-phase induction motor.
17. Explain the no-load test on a single-phase induction motor.
18. How can the parameters of the equivalent circuit be obtained from block-rotor test and no-load test?
19. Compare single-phase and three-phase induction motors.
20. If *s* is the slip of the forward flux, prove that the slip of backward flux is 2*s*.
21. An 800W, 4-pole, 50 Hz, 240 V, single-phase induction motor has the following parameters:
   $R_1 = 2.5\ \Omega, R_2 = 4\ \Omega, X_1 = 3\ \Omega, X_2 = 3\ \Omega, X_m = 60\ \Omega$
   The friction and windage loss is 20 W
   Calculate the performance of the motor at 4% slip.
22. What are the inherent characteristics of a plain 1-phase induction motor?
23. A 220V, single-phase induction motor has the following parameters:
   $R_1 = R'_2 = 9\,\Omega$
   $X_1 = X'_2 = 15\,\Omega$
   $X_m = 210\,\Omega$
   Calculate the following:
   (i) Input current
   (ii) Input power
   (iii) Torque
   Take slip to be 5% and motor speed to be 1700 r.p.m.
24. A 4-pole, 100V, 50 Hz single-phase induction motor has the following parameters:
   $R'_2 = 8\,\Omega$
   $X_1 = X'_2 = 10\,\Omega$
   $X_m = 200\,\Omega$
   If the core loss is 15 W and the friction and windage loss is 8 W, calculate the efficiency of the motor at 8% slip.
25. The no-load and the block rotor test of a single-phase induction motor is as follows:
   No-load test result: 100 V 4 A 50 W
   Block rotor test result: 50 V 7 A
   If the friction and the windage loss is 5W and stator resistance is 3 Ω. Assume that $X_1 = X'_2$, determine the parameters of the double-field revolving field equivalent circuit.

CHAPTER 9

SYNCHRONOUS GENERATOR

9.1 INTRODUCTION

Synchronous generator is one of the primary sources of electrical energy and is one of the largest energy converters. Synchronous generator is a device which converts mechanical energy into electrical energy. It works on the principle of electromagnetic induction. It is also called an alternator. The difference between a DC generator and an alternator is that in the former, the armature rotates while the field system is stationary; whereas in the latter, the armature (called as stator) remains stationarywhile the field system (called as rotor) rotates.

9.2 CLASSIFICATION OF SYNCHRONOUS MACHINES

A synchronous machine can be classified under three categories:

i) Classification based on the arrangement of field and armature winding.
ii) Classification based on the shape of the shape of the field
iii) Classification based of the type of prime mover used

We shall now discuss these classifications in detail.

9.2.1 Classification Based on the Arrangement of Field and Armature Winding

Under this classification a synchronous machine can be categorised as:

a) **Rotating armature-type machine**:This type of machine has rotating armature and stationary field. It is usually of low voltage rating of small KVA. The power is brought to the load through slip-rings.
b) **Rotating field-type machine**:In this type of machine, the armature is stationary and the field is rotating. To the filed the current is supplied via slip rings from the exciter and the stationary armature is directly connected to the load.

9.2.2 Classification Based on the Shape of the Shape of the Field

Synchronous machines can be categorised as

a) **Non-salient pole machine (Cylindrical rotor)** :These machines have cylindrical smooth rotor construction with distributed field winding in slots. This type of rotor construction is employed for machines driven by steam turbines. They are operated at high speeds and are usually two-pole machines. They have small diameter-to-length ratio to avoid excessive mechanical stress on the rotor due to large centrifugal forces.

A cylindrical rotor synchronous generator does not have any projection coming out from the surface of the rotor; rather,the central polar area is provided with slots for housing the field windings as we can see from the diagram above. The field coils are so arranged around these poles that flux density is maximum on the polar central line and falling away gradually as we move out towards the periphery. The cylindrical rotor type machine gives better balance and quieteroperation along with lesser windage losses.

b) **Salient pole machine**:This type of machine has salient pole or projecting poles with concentrated field windings on the bodies of such poles. This type of construction is for machines that are driven by hydraulic turbines or diesel engines. They have low-speed operation due to their large diameter-to-length ratio and large number of poles.

9.2.3 Classification based on the type of prime movers used

The different categories of induction machines are:

a) Generator driven by hydraulic turbine called the hydro-generator.
b) Generator driven by steam turbine called the turbo-generator.
c) Generator driven by IC

9.3 ADVANTAGES OF ROTATING FIELD AND STATIONARY ARMATURE

1. Since high rating generators produce large amount of current corresponding to voltages ranging from 11kV to 33 kV, a large number of conductors are required to obtain the current and insulations are required to handle such large current. It is easier to insulate stationary winding of alternators that are usually designed for high voltages.This is because stationary windings are not subjected to centrifugal forces and also sinceextra space is available due to the stationary arrangement of the armature.
2. It is comparatively easier to collect current from stationary parts of the generator rather than from rotating parts with the help of slip rings and brushes.
3. When a rotating field is used, slip-rings are connected to around 110 to 220 V DC supply, through which the field is excited. However in the case of rotating armature, the slip-rings and brushes are connected across a generated voltage ranging from 11kV to 33 kV.
4. If armature is used as rotor, the high currents produced may cause sparking in the slip-rings and damage the generator setup; so slip rings are preferred only for the low-voltage circuits.
5. Since field circuit is smaller than armature circuit both in weight and size, the inertia of field circuit is lower compared to that of the latter. This facilitates high-speed rotation and produces output with enhanced efficiency.
6. If armature is rotated, then centrifugal forces arise and high currents are produced.Hence, this setup is avoided so as to keep the generator free from the interaction of mechanical and electrical forces which may otherwise damage the machine.

9.4 CONSTRUCTION

The stator is a stationary member. It is a hollow cylinder inside which the rotor rotates. Sufficient air gap is provided between the stator and the rotor. The armature winding, made up of silicon steel material,is housed in the slots cut in the stator. Stator stampings consist of annular rings with slots on the inner periphery, within which the windings are placed. Usually, enamelled copper wire is used as winding material. The stator carries three-phase armature windings, physically displaced from each other by 120 degrees

The rotors of a synchronous motor are of two types, salient pole and cylindrical rotor, as shown in Fig 9.1.

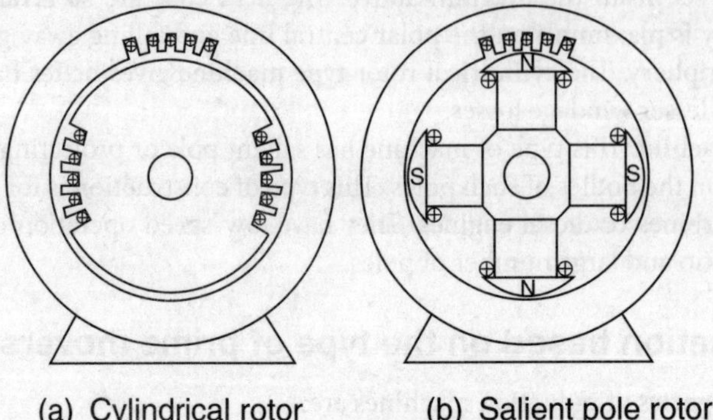

(a) Cylindrical rotor (b) Salient pole rotor

Fig. 9.1: Rotors of synchronous motor

The cylindrical pole rotor has the DC field winding embedded in it. The cylindrical rotor provides greater mechanical strength and permits more accurate dynamic balancing. It is particularly used in high-speed turbo generators.

The salient pole rotor has projecting poles in it. These projecting poles lessen its mechanical strength. This type of rotor construction is used for low-speed applications such as hydroelectric generators. Large number of poles in the rotor makes the rotor larger in diameter and smaller in length.

The rotor can be made of solid steel forging or stacked by laminations. Windings are placed on the slots that are provided on the outer surface of the rotor. After inserting the windings in these slots, they are generally closed by manganese bronze or steel wedges. The two ends of the winding are connected to the slip-rings mounted on the shaft. Two sets of brushes touch the slip-ring by spring action.

By passing DC current through the brush and slip-ring to the field winding, alternating North- and South-pole magnetic fields are set up in the rotor.

9.5 OPERATION OF SYNCHRONOUS GENERATOR

The operating principle of synchronous machine is similar to that of the DC machine. The only difference is that in a synchronous machine, there is no need to rectify the varying sinusoidal e.m.f. of the armature winding.Thus, a synchronous machine does not require commutator.

The field winding which is present on the rotor of a synchronous machine is supplied with DC current. The DC current will set up a magnetic field on the rotor. The rotor is then rotated at synchronous speed with the help of a prime mover. The magnetic field of the rotor conductor will cut the stationary stator coils and an e.m.f. will be induced in it (Faraday`s law). This e.m.f. will be displaced in phase by 120^0. The voltage waveform of all the three phases (R-Y-B) of the stator winding is shown in Fig. 9.2.

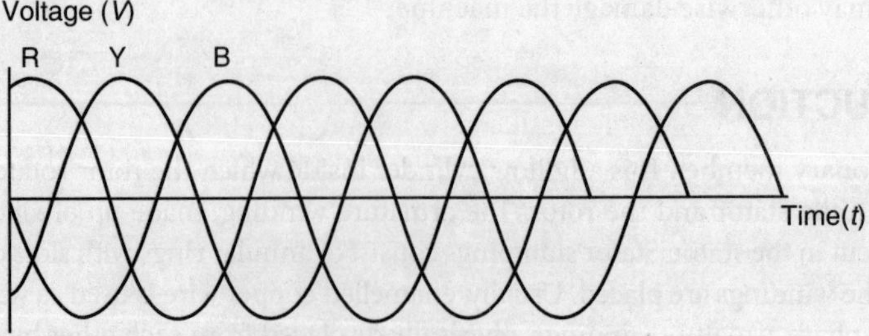

Fig. 9.2: Voltage waveform

Where

$$V_R = V_m \sin \omega t \tag{9.1}$$

$$V_Y = V_m \sin(\omega t - 120) \tag{9.2}$$

$$V_B = V_m \sin(\omega t + 120) \tag{9.3}$$

If the number of poles on the rotor is P, then in one revolution $P/2$ cycles of e.m.f. is completed. Then the frequency of induced e.m.f. will be

$$f = \frac{P}{2} n \tag{9.4}$$

Where n is the number of revolution per second of the rotor.

$$f = \frac{P}{2} \times \frac{N}{60} = \frac{PN}{120} \tag{9.5}$$

Where N is the number of revolution per minute of the rotor.

9.6 WINDINGS IN ALTERNATORS

In case of three-phase alternators, the following types of windings are employed. (i) Lap winding, (ii) wave winding and (iii) mesh winding.

Based on pitch of the coil, the types of windings are (i) full-pitched (ii) short-pitched windings

Based on number of layers, the types are (i) Single-layer (ii) Double-layer

9.7 TERMINOLOGIES

- **Pole pitch**: it is the ratio of distance between two adjacent opposite main poles to the number of armature conductors.
- **Coil span**: It is the distance between two coils coil sides of a coil. If the coil span is equal to one pole pitch, it is called a full pitch winding. (180 electrical degree). If the coil span is less than one pole pitch, it is called a short pitch or fractional chorded winding. (less than 180 electrical degree).
- **Slot angle(β)**: The distance between any two conductors is called slot angle (β).
- **Pitch factor (or) chording factor (or) coil span factor (K_c)**: It is the ratio of the vector sum of e.m.fs induced in the two sides of the coil to their arithmetic sum.

$$Kc = \frac{\text{Vector sum of induced e.m.f.s per coil}}{\text{Arithmetic sum of induced emf per coil}} \tag{9.6}$$

$$= \frac{\text{Voltage induced in short-pitch winding}}{\text{Voltage induced in full-pitch winding}} \tag{9.7}$$

- **Distribution factor (or) breadth factor (K_d)**:When the coils comprising a phase of the windings are distributed in two or more slots per pole, the e.m.fs in the adjacent coils will be out of phase with respect to one another and their resultant will be less than their algebraic sum. The ratio of the vector sum of the e.m.fs induced in all the coils distributed in a number of slots under one pole to the arithmetic sum of the e.m.fs induced (or to the resultant of the e.m.fs induced in all the coils concentrated in one slot under one pole) is known as distribution factor K_d. It is the ratio of voltage induced in a distribution winding to the voltage induced in the concentric winding.

9.8 E.M.F. EQUATION OF AN ALTERNATOR

Let

Z be the number of conductors of coil side in series per phase
φ be the flux per pole in Wb
P be the number of poles in the rotor
N be the speed of the rotor in rpm
f be the frequency in Hz
T be the number of coils or turns per phase.

$$\text{The average induced e.m.f. in one conductor} = \frac{P\varphi N}{60} \tag{9.8}$$

$$\text{The average induced e.m.f. per phase} = \frac{P\varphi NZ}{60} \tag{9.9}$$

$$\text{Speed}(N) = \frac{120f}{P} \tag{9.10}$$

Substituting Eq. (9.10) in Eq. (9.9) we get,

$$\text{Average induced e.m.f.} = \frac{P\varphi Z}{60} \times \frac{120f}{P} = 2\varphi Zf \tag{9.11}$$

Again, Z=2T
Substituting in Eq. (9.11) we get,

$$\text{Average induced e.m.f.} = 4\varphi Tf \tag{9.12}$$

Form factor of sine wave = 1.11.

r.m.s. value of induced e.m.f. = Form factor × Average induced e.m.f.

$$= 1.11 \times 4\varphi Tf$$

$$= 4.44\varphi Tf \tag{9.13}$$

In a synchronous generator, the turns are distributed over armature periphery and short-pitched. To compensate this, the e.m.f. equation is multiplied by distribution factor (K_d) and pitch factor (K_c). So Eq. (9.13) becomes,

$$\text{r.m.s. value of induced e.m.f.} = K_d K_c 4.44\varphi Tf \tag{9.14}$$

9.9 ARMATURE REACTION IN SYNCHRONOUS GENERATOR

A rotating electrical machine works based on Faraday's law. Every electrical machine requires a magnetic field and a coil (known as armature) with a relative motion between them. In case of an alternator, we supply electricity to pole to produce magnetic field and output power is taken from the armature. Due to relative motion between field and armature, the conductor of the armature cuts the flux of magnetic field and hence, there would be changing flux linkage with the armature conductor. According to Faraday's law of electromagnetic induction, there would be an e.m.f. induced in the armature. Thus, as soon as the load is connected with armature terminals, there is a current flowing in the armature coil. When thecurrent starts flowing through the armature conductor, there is a reverse effect of this current on the main field flux of the alternator (or synchronous generator). This reverse effect is referred as armature reaction in alternator or synchronous generator. We already know that a current-carrying conductor produces its own magnetic field, and this magnetic field affects the main magnetic field of the alternator. This has two undesirable effects: either it distorts the main field or it reduces the main field flux or both. These effects deteriorate the performance of the machine. When the field gets distorted, it is known as cross-magnetising effect and when the field flux gets reduced, it is known as demagnetising effect. The electromechanical energy conversion takes place through magnetic field as a medium. Due to relative motion between armature conductors and the main field, an e.m.f. is induced in the armature windings whose magnitude depends upon the relative speed and as well as the magnetic flux. Due to armature reaction, flux is reduced or distorted, the net e.m.f. induced is also affected and hence, there is degradation in the performance of the machine.

The armature reaction of alternator or synchronous generator, depends upon the phase angle between stator armature current and induced voltage across the armature winding of alternator.

The phase difference between these two quantities, i.e. armature current and voltage may vary from – 90° to + 90°. Let this angle be θ.

To understand the actual effect of this angle on armature reaction of alternator, we will consider three cases,

1) When θ = 0
2) When θ = 90°
3) When θ = –90°

Case1: Armature reaction of alternator when $\theta = 0$

At unity power factor, the angle between armature current and induced e.m.f. is zero. The e.m.f. induced in the armature is due to changing of main field flux, linked with the armature conductor.

Since the field of excitation is DC, the main field flux is constant with respect to field magnets, and it will alternate with respect to armature as there is a relative motion between field and armature. The main field flux can then be expressed as,

$$\varphi_{\text{mff}} = \varphi_{\text{m}} \sin \omega t \tag{9.15}$$

The induced e.m.f. will be,

$$\text{Induced e.m.f.} = \frac{d\varphi_{\text{mff}}}{dt} = -\omega\varphi_{\text{m}} \cos \omega t \tag{9.16}$$

Again, the armature flux φ_{a} is proportional to armature current and also they are in phase with each other. At unity power factor armature current and induced e.m.f. are is phase, so at unity power factor armature flux φ_a is phase with induced e.m.f. So at unity power factor armature flux is in phase with induced e.m.f. and the field flux is in quadrature with the induced e.m.f.

As these two fluxes are perpendicular to each other, the armature reaction of alternator at unity power factor is purely distorting or cross-magnetising type.

Case2: Armature reaction of alternator at lagging zero power factor

In this case, the armature current lags the induced e.m.f. by 90⁰. Again, the induced e.m.f. leads the main field flux by 90°. From equation (9.15) and equation (9.16),

When,

ωt= 0: Induced e.m.f. is maximum and φ_{ffm} is zero.

ωt = 90°: Induced e.m.f is zero and φ_{ffm} has maximum value.

ωt= 180°: Induced e.m.f. is maximum and φ_{ffm} is zero.

ωt = 270°: Induced e.m.f. is zero and φ_{ffm} has negative maximum value.

The armature flux φ_a is proportional to armature current. Since the armature current lags induced e.m.f. by 90⁰,we can say the armature flux φ_a lags induced e.m.f. by 90⁰. Similarly, the field flux φ_f will lead induced e.m.f. by 90⁰.

Therefore, armature flux and field flux act directly opposite to each other. Thus, armature reaction of alternator at lagging power factor is purely demagnetising type.

Case3: Armature reaction of alternator at leading power factor

Here, the armature current will lead the induced e.m.f. by an angle of 90°. Since the armature flux φ_a is proportional to armature current, it will also lead the induced e.m.f. by an angle of 90°. We know that field flux φ_f leads the induced e.m.f. by 90°.

So, both the armature flux and field flux will lead induced e.m.f. by 90°. Hence, the resultant flux is simply the arithmetic sum of field flux and armature flux. Thus, the armature reaction of alternator due to leading power factor is magnetising type.

9.10 SYNCHRONOUS REACTANCE AND SYNCHRONOUS IMPEDANCE

Synchronous reactance is the combined effect of the armature leakage reactance and the armature reaction. When the synchronous reactance is combined vectorially with the armature resistance, the quantity is called the synchronous impedance. The vector is shown in Fig. 9.3.

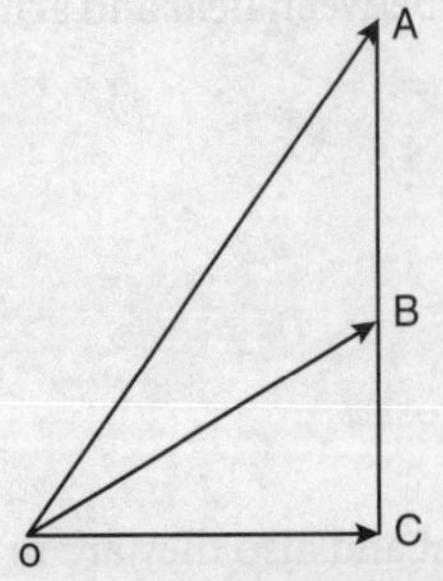

OC...Armature Resistance

CB...Leakage Reactance

AB...Reactance of Armature Reaction (equivalent)

AC...Synchronous Reactance

OA...Synchronous Impedance

FIG. 9.3: Vector of synchronous reactance

The sum of leakage reactance (X_l) and the equivalent reactance of armature reaction (X_a) is called synchronous reactance (X_s). Under this condition, impedance of the armature winding is called the synchronous impedance (Z_s).

Thus, synchronous reactance $X_s = X_l + X_a$ per phase

and synchronous impedance $Z_s = R_a + jX_s$ per phase

Where, R_a is the armature resistance.

As armature reaction reactance is dependent on armature current,the synchronous reactance and hence synchronous impedance will also depend on armature current or load current.

9.11 EQUIVALENT CIRCUIT OF SYNCHRONOUS GENERATOR

Let

V_t be the terminal voltage.

I_a be the armature current.

X_l be the leakage reactance.

X_a be the reactance of the armature.

E_{ph0} be the induced e.m.f. per phase on no load.

E_{phl} be the induced e.m.f. per phase on load.

E_{pl0} and E_{phl} are different because of armature reaction.

The equivalent circuit of the generator per phase is shown in Fig. 9.4.

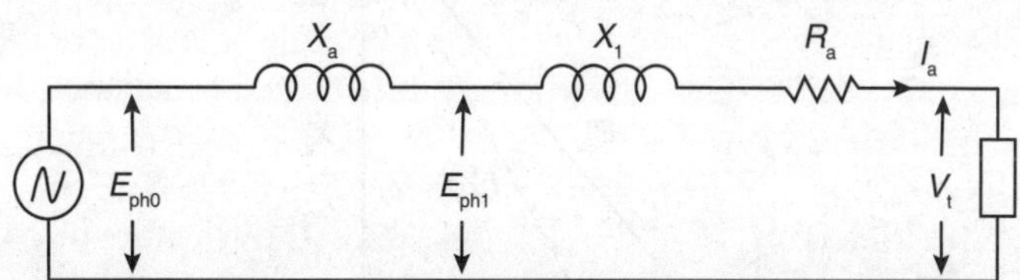

Fig. 9.4: Equivalent circuit of generator per phase

Applying KVL in the circuit shown in Fig. 9.4, we get,

$$V_t = E_{phl} - I_a R_a - I_a X_l$$

$$= E_{phl} - I_a \left(R_a + X_l\right) \quad (9.17)$$

Let,

$$X_s = \text{Synchronous reactance} = X_r + X_l$$

Then, the modified equivalent circuit is shown in Fig. 9.5

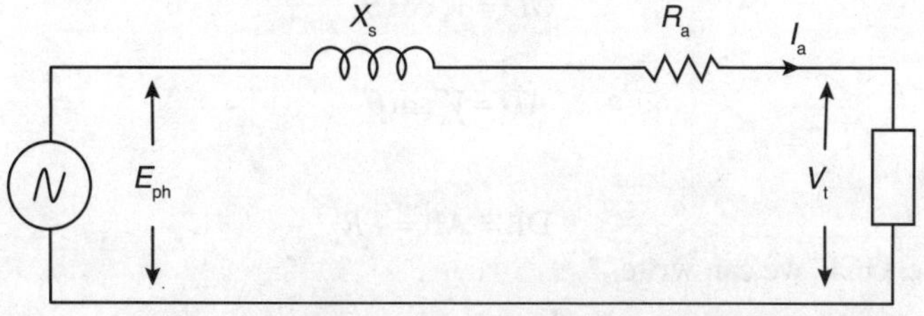

Fig. 9.5: Modified equivalent circuit

Applying KVL in the circuit shown in Fig. 9.5 we get,

$$V_t = E_{ph} - I_a R_a - I_a X_s$$

$$= E_{ph} - I_a(R_a + X_s)$$

$$= E_{ph} - I_a Z_s \tag{9.18}$$

Where, Z_s is called the synchronous impedance.
Equation (9.18) is called the voltage equation of an alternator.

9.12 PHASOR DIAGRAM OF AN ALTERNATOR AT LAGGING LOAD

At lagging load, the armature current I_a will lag the voltage V_t by an angle of θ. The complete phasor diagram is shown in Fig. 9.6.

Let,
OA= V_t; AB = $I_a R_a$; BC = $I_a X_s$; OC = E_{ph}; AC = $I_a Z_s$;

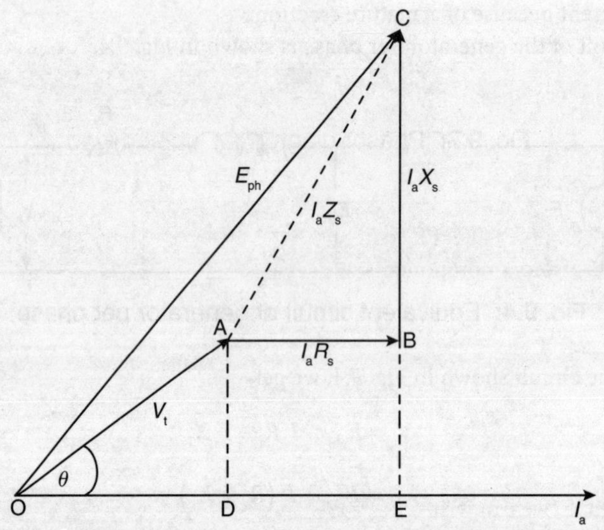

Fig. 9.6: Phasor diagram at lagging load

Resolving V_t into its X and Y components, we get,

$$OD = V_t \cos\theta$$

$$AD = V_t \sin\theta$$

Also,

$$\text{DE} = \text{AB} = I_a R_a.$$

From the triangle, OCE, we can write,

$$\text{OC}^2 = \text{OE}^2 + \text{EC}^2$$

$$OC^2 = (OD + DE)^2 + (EB + BC)^2$$

$$= (V_t \cos\theta + I_a R_a)^2 + (V_t \sin\theta + I_a X_s)^2$$

$$E_{ph} = \sqrt{(V_t \cos\theta + I_a R_a)^2 + (V_t \sin\theta + I_a X_s)^2} \quad (9.19)$$

9.13 PHASOR DIAGRAM OF AN ALTERNATOR AT LEADING LOAD

At lagging load, the armature current I_a will lead the voltage V_t by an angle of θ. The complete phasor diagram is shown in Fig. 9.7.

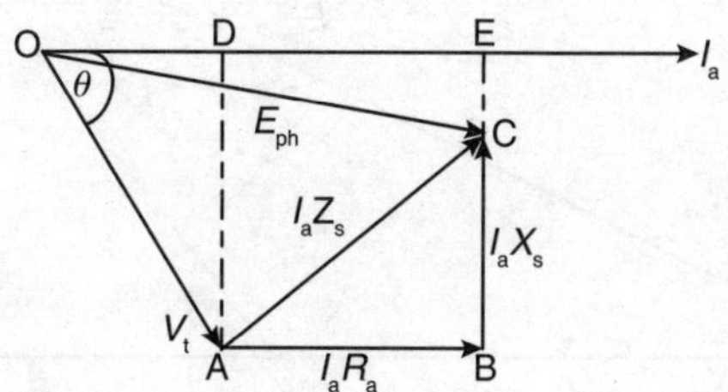

Fig. 9.7: Phasor diagram at leading load

Let,
OA= V_t; AB = I_aR_a; BC = I_aX_s; OC = E_{ph} ; AC = I_aZ_s;
Resolving V_t, into its X and Y components, we get,

$$OD = V_t \cos\theta$$

$$AD = V_t \sin\theta = BE$$

Also,

$$DE = AB = I_aR_a.$$

From the triangle OCE, we can write,

$$OC^2 = OE^2 + EC^2$$

$$OC^2 = (OD + DE)^2 + (EB - BC)^2$$

$$= (V_t \cos\theta + I_a R_a)^2 + (V_t \sin\theta - I_a X_s)^2$$

$$E_{ph} = \sqrt{(V_t \cos\theta + I_a R_a)^2 + (V_t \sin\theta - I_a X_s)^2} \quad (9.20)$$

From Eq.(9.19) and Eq. (9.20), we can write a combined equation,

$$E_{ph} = \sqrt{(V_t \cos\theta + I_a R_a)^2 + (V_t \sin\theta \pm I_a X_s)^2} \quad (9.21)$$

Where,
+ indicates lagging load
– indicates leading load

9.14 PHASOR DIAGRAM OF AN ALTERNATOR AT UNITY POWER FACTOR

At unity power factor, the armature current I_a and the voltage V_t will be in phase. The complete phasor diagram is shown in Fig. 9.8.

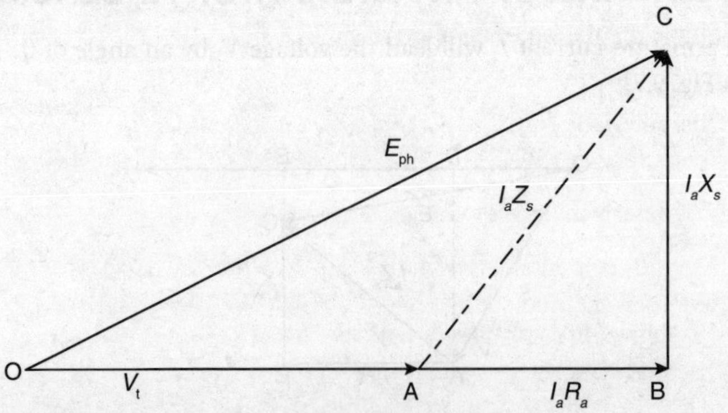

Fig. 9.8: Phasor diagram at unity power factor

From the triangle OBC, we can write,

$$OC^2 = OB^2 + BC^2$$

$$OC^2 = (OA + AB)^2 + (BC)^2$$

$$= (V_t + I_aR_a)^2 + (I_aX_s)^2$$

$$E_{ph} = \sqrt{(V_t + I_aR_a)^2 + (I_aX_s)^2} \tag{9.22}$$

9.15 VOLTAGE REGULATION OF AN ALTERNATOR

In an alternator, the terminal voltage changes with change in the load on the generator. So in loaded condition, the terminal voltage of an alternator is not the same as the induced e.m.f. The voltage regulation of an alternator is the change of voltage from full load to no load, expressed as a percentage of full-load volts, when the speed and dc field current are held constant.

That is,

$$\text{Voltage regulation} = \frac{\text{Terminal voltage under no-load} - \text{Terminal voltage under full-load}}{\text{Terminal voltage under full-load}}$$

$$= \frac{E_f - V_t}{V_t} \times 100 \quad (9.23)$$

Where,
E_f is the no-load voltage,
V_t is the full-load voltage.
At lagging condition, power factor $E_f > V_t$ and so the voltage regulation is positive.
At leading condition, power factor $E_f < V_t$ and so the voltage regulation is negative.

9.15.1 Methods to Measure Voltage Regulation of an Alternator

To determine the voltage regulation of an alternator, the following results needs to be obtained,

- Armature resistance (R_a)
- Open-circuit Characteristic (OCC)
- Short-circuit Characteristic (SCC)

Various methods of determination of voltage regulation are,

- E.M.F. or synchronous impedance method
- M.M.F. or ampere turn method
- Zero Power Factor Method (ZPF) or Potier Method

9.15.1.1 Determination of armature resistance (R_a)

The armature resistance of a synchronous generator is obtained from DC tests method. This test is conducted to measurewinding resistance of a synchronous generator when it is at rest and the field winding is open. The circuit diagram for the test is shown in Fig. 9.9. Here, the armature is assumed to be star connected.

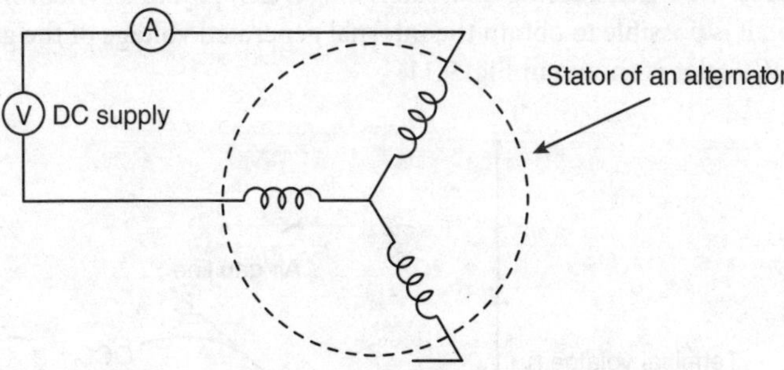

Fig. 9.9: Circuit to determine armature resistance

A low-voltage DC supply is used to measure the DC resistance. DC supply is used because the AC method would include coupled losses in the field pole structure and its surrounding iron, yielding spurious values. The DC resistance per phase is given by

$$R_{dc} = \frac{1}{2} \times \frac{V}{A}$$

The AC resistance is obtained by multiplying R_{dc} by a factor between 1.2 to 1.5. This factor depends on the frequency, quality of insulation, size, capacity etc. Here, we will use the factor 1.5. So,

$$R_{ac} = 1.5\, R_{dc}$$

9.15.1.2 Open-circuit characteristic (OCC)

It is the plot of open-circuit terminal voltage at varying values of excitation current, when the machine is running at synchronous speed. The circuit diagram for this characteristic is shown in Fig. 9.10.

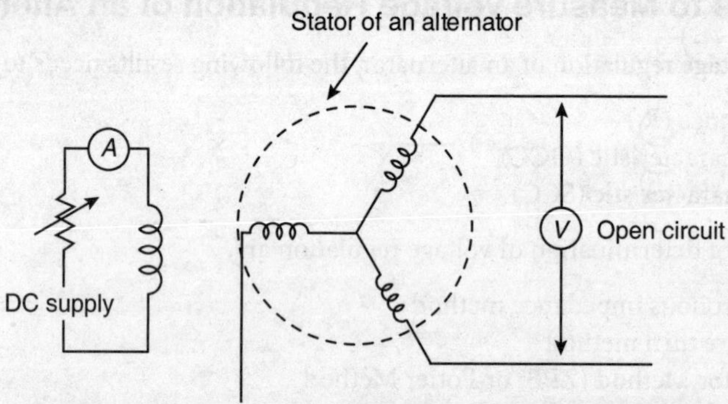

FIG. 9.10: Circuit for determination of OCC characteristic

In this test, the generator is rotated at rated speed. The load terminal is kept open; that is, no load is connected at the terminals. The field current is increased in steps from zero to maximum value and the corresponding terminal voltage is tabulated.

At no load, armature current, $I_a = 0$, so $E_{ph} = V_t$. A plot is obtained between E_{ph} or V_t versus field current, I_f. This plot is called the Open Circuit Characteristic (OCC) plot of a synchronous generator. From this characteristic plot, it is possible to obtain the internal generated voltage of the generator for any field current. The nature OCC plot is shown in Fig. 9.11.

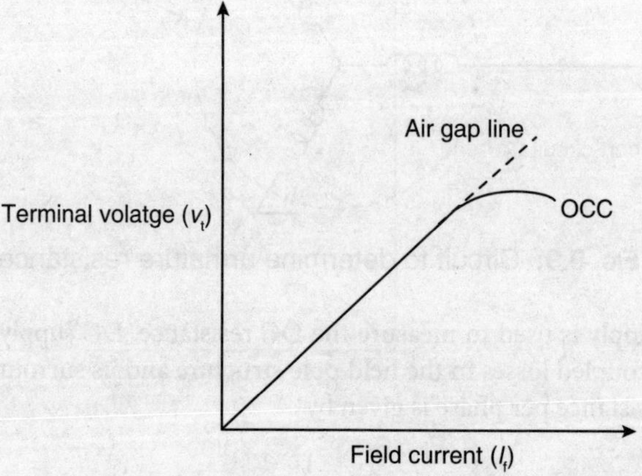

FIG. 9.11: OCC plot

Generally, Open Circuit Characteristic (OCC) is a straight line as long as the magnetic saturation is not reached. After magnetic saturation, the characteristic becomes non-linear.

9.15.1.3 Short-circuit characteristic (SCC)

The circuit diagram for determination of short-circuit characteristic is shown in Fig. 9.12

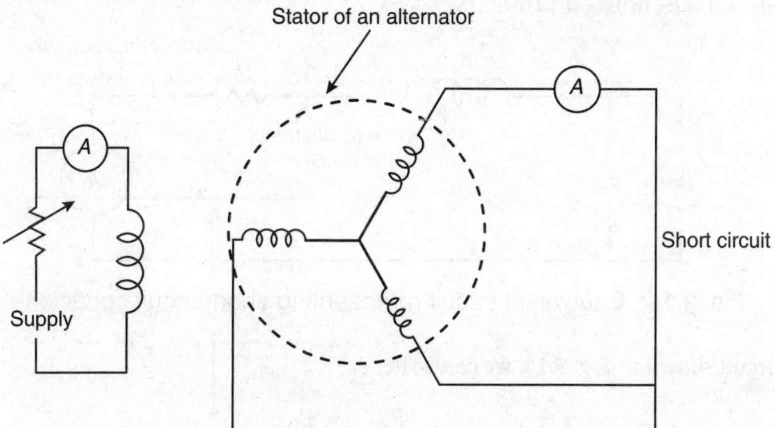

Fig. 9.12: Circuit for short-circuit characteristic

All the three phases of the load terminal are short-circuited and an ammeter is connected to any one phase for the current measurement. The machine is run at rated speed and the excitation is gradually increased in steps and the corresponding short-circuit current is noted. This excitation is increased in steps until the rated armature current flows under short-circuit condition. The plot between the field current (I_f) and the short-circuit current (I_{sc}) is called the short-circuit characteristic (SCC) of the generator. This plot is shown in Fig. 9.13.

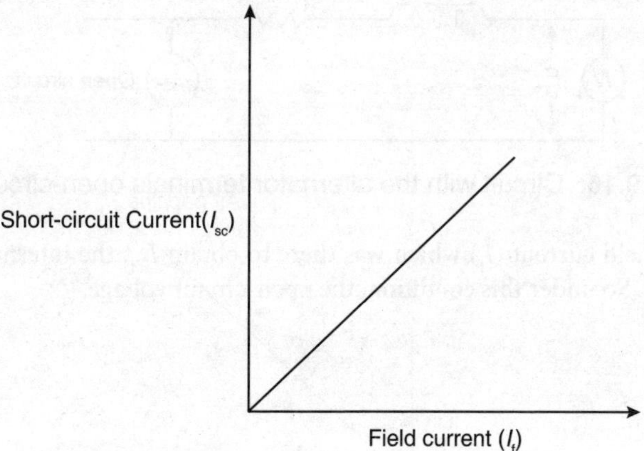

Fig. 9.13: SCC plot

9.15.1.4 E.M.F. or synchronous impedance method of determination of voltage regulation

The synchronous impedance (Z_s) of an alternator depends on the load on the generator. The equivalent circuit of generator shown in Fig. 9.5 represents the condition of synchronous generator on load. This equivalent circuit can be redrawn to represent short-circuited synchronous generator condition. The modified circuit is shown in Fig. 9.14.

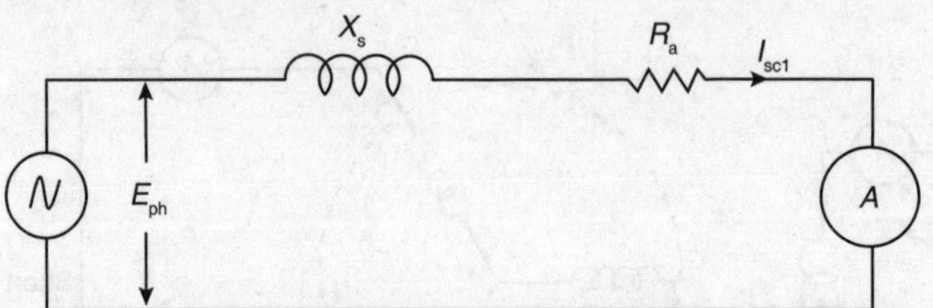

Fig. 9.14: Equivalent circuit representing short-circuit condition

From the circuit shown in Fig. 9.14, we can write,

$$I_{scl} = \frac{E_{ph}}{R_a + jX_s} = \frac{E_{ph}}{Z_s} \tag{9.24}$$

$$Z_s = \frac{E_{ph}}{I_{scl}} \tag{9.25}$$

The value of I_{scl} can be obtained from the ammeter reading. But to determine the value of E_{ph}, the alternator terminals are open-circuited, as shown in Fig. 9.15.

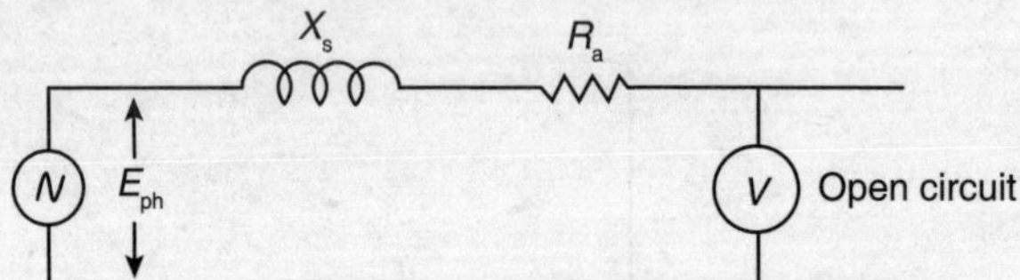

Fig. 9.15: Circuit with the alternator terminals open-circuited

Now for the same field current (I_f) which was there to obtain I_{scl}, the internally induced e.m.f. will remain the same as E_{ph}. So under this condition, the open-circuit voltage,

$$V_{oc} = E_{ph}$$

So,

$$Z_s = \frac{E_{ph}}{I_{scl}} = \left.\frac{V_{oc}}{I_{scl}}\right|_{\text{same field current}} \tag{9.26}$$

The value of Z_s can also be obtained from OCC and SCC graph of a synchronous generator. Both the OCC and SCC characteristic graph are plotted on the same field current scale as shown in Fig. 9.16.

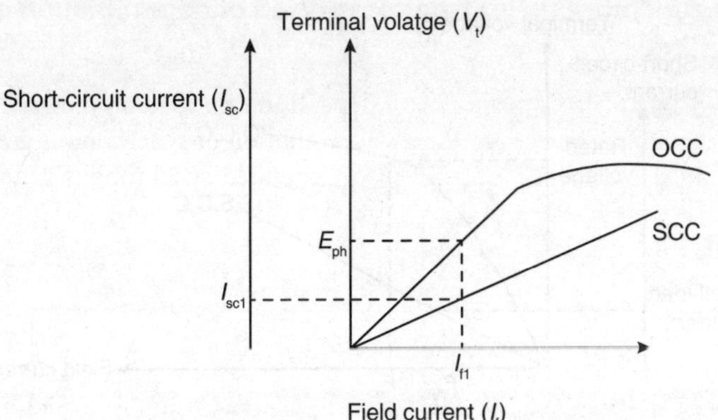

Fig. 9.16: OCC and SCC characteristic

Let us obtain the value of Z_s at full load. Let the full load current be $I_{sc1.}$

The field current is obtained from the SCC characteristic corresponding to I_{sc1}.Let it beI_{f1}.

Now for this field current I_{f1}, the corresponding open-circuit voltage is obtained from the OCC characteristic.Let it be E_{ph}.

So, the value of Z_s at full load will be,

$$Z_s = \frac{E_{ph}}{I_{sc1}} \tag{9.27}$$

The value of Z_s will be different for different values of field current, I_f, as the OCC graph is non- linear in nature.

After obtaining the values of armature resistance (R_a) and synchronous impedance (Z_s), we can get the value of synchronous reactance (X_s).

$$X_s = \sqrt{(Z_s)^2 - (R_a)^2} \tag{9.28}$$

So after obtaining the value of R_a and X_s, we can calculate the value of E_{ph} from Eq.(9.21) and voltage regulation from Eq. (9.23).

9.15.1.4.1 ***Advantages of Synchronous Impedance Method*****:** The main advantage of this method is that the value of synchronous impedance Z_s for any load condition can be calculated. Hence, regulation of the alternator at any load condition and load power factor can be determined. Actual load need not be connected to the alternator and hence, this method can be used for very high capacity alternators.

9.15.1.4.2 ***Limitation of Synchronous Impedance Method*****:** The main limitation of this method is that the method gives large values of synchronous reactance. This leads to high values of percentage regulation than the actual results.

9.15.1.5 M.M.F. or ampere-turn method

This method is also called Rothert's M.M.F. method. In short-circuit test, field m.m.f. circulates the full-load current balancing the armature reaction effect, which is wholly demagnetising in nature. The value of ampere-turns required to circulate full-load current can be obtained from short-circuit characteristics. This is denoted as F_{AR}.The field m.m.f. required to induce the rated terminal voltage on open circuit can be obtained from open-circuit test results and open-circuit characteristics. This is denoted as F_O. The two values,F_0 and F_{AR}, is shown in OCC and SCC as in Fig. 9.17.

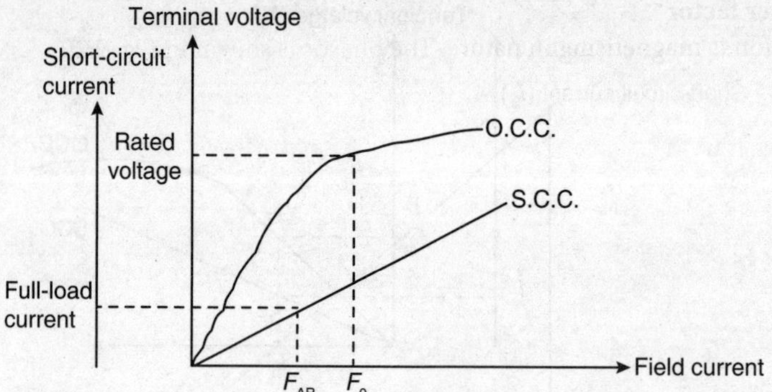

Fig. 9.17: Determination of F_0 and F_{AR} from OCC and SCC

Total field m.m.f. is the vector sum of its two components, F_O and F_{AR}, which also depends on the power factor of the load supplied by the alternator. Let the resultant of the two components be F_R. We will consider three cases to obtain the resultant of the two m.m.f. component.

Case-1: Zero lagging power factor

The armature reaction is demagnetising in nature. The phasor is shown in Fig. 9.18.

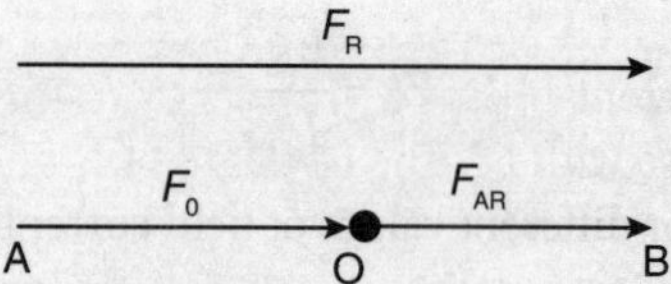

Fig. 9.18: Phasor for zero lagging power factor

From the phasor of Fig. 9.18,

$$OA = F_{AR} \text{ (Demagnetising)}$$
$$OB = F_0$$
$$AB = \text{Resultant } (F_R)$$
$$F_R = F_{AR} + F_0$$

Case-2: Zero leading power factor

The armature reaction is magnetising in nature. The phasor is shown in Fig. 9.19

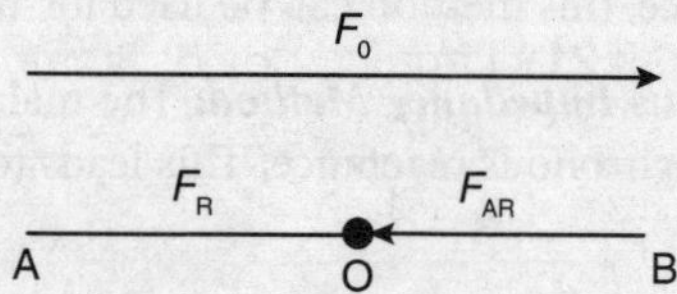

Fig. 9.19: Phasor for zero leading power factor

From the phasor of Fig. 9.19

$$AB = F_0$$
$$OB = F_{AR} \text{ (Magnetising)}$$
$$AO = \text{Resultant } (F_R) = AB - OB = F_0 - F_{AR}$$

Case-3: Unity power factor

The armature reaction is magnetising in nature. The phasor is shown in Fig. 9.20.

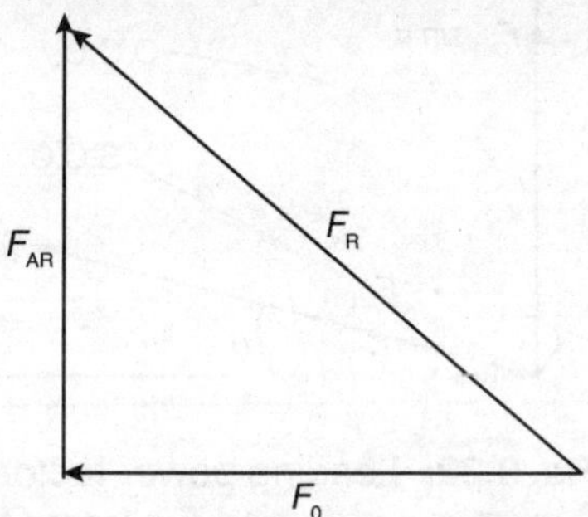

Fig. 9.20: Phasor for unity power factor

From the phasor of Fig. 9.20,

F_{AR} (Cross-magnetising)

$$\text{Resultant}\left(F_R\right) = \sqrt{F_{AR}^2 - F_0^2}$$

Case-4: Power factor lagging (cos *θ*)

In this case, $\cos\theta$ is lagging. The current I_{ph} lags V_{ph} by angle θ. The component F_0 is at right angle to V_{ph} and the component F_{AR} is in phase with I_{ph}. The resultant of F_0 and F_{AR} is F_R. The complete phasor is shown Fig. 9.21.

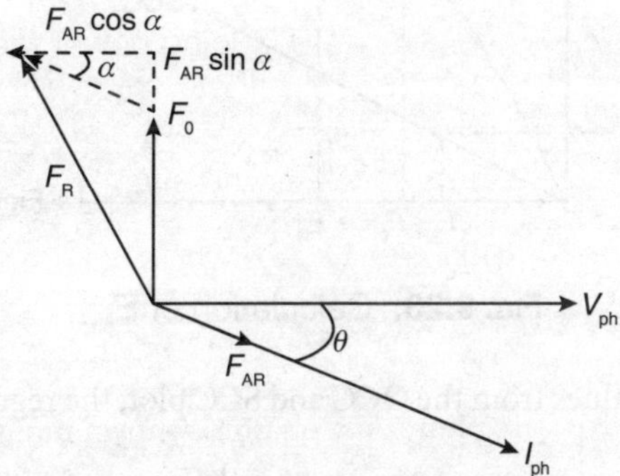

Fig. 9.21: Lagging power factor

From the phasor of Fig. 9.21,

$$F_R = \sqrt{\left(F_0 + F_{AR}\sin\alpha\right)^2 + \left(F_{AR}\cos\alpha\right)^2}$$

Case-5: Power factor leading (cos*θ*)

In this case, $\cos\theta$ is leading. The current I_{ph} leads V_{ph} by an angle θ. The component F_0 is at right angle to V_{ph} and the component F_{AR} is in phase with I_{ph}. The resultant of F_0 and F_{AR} is F_R. The complete phasor is shown Fig. 9.21.

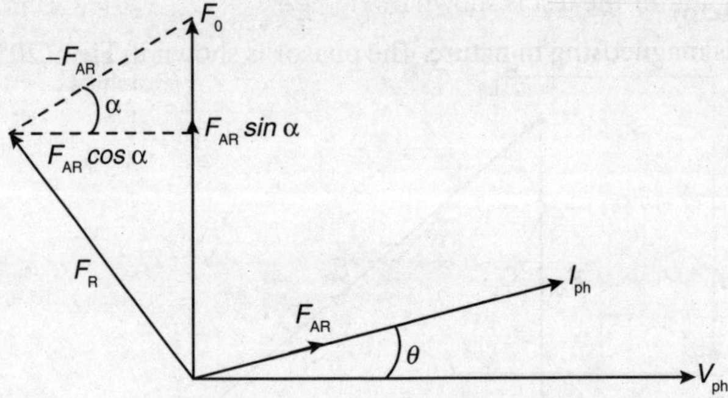

Fig. 9.22: Leading power factor

From the phasor of Fig. 9.22.

$$F_R = \sqrt{(F_0 - F_{AR}\sin\alpha)^2 + (F_{AR}\cos\alpha)^2}$$

Once we have calculated the value of F_R, we can get the value of E_{ph} from the OCC and SCC characteristic as shown in Fig. 9.23.

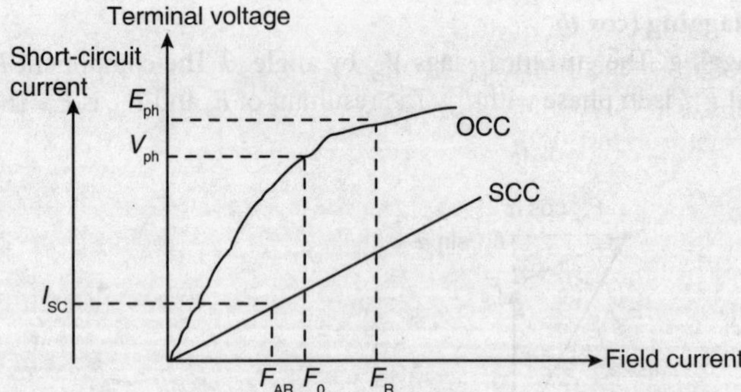

Fig. 9.23: Calculation of E_{ph}

One we have obtained the values from the OCC and SCC plot, the regulation is calculated as follows,

$$\%\text{regulation} = \frac{E_{ph} - V_{ph}}{V_{ph}} \times 100$$

9.15.1.6 Zero power factor method (ZPF) or Potier method

This method is also called Potier method and it is based on the separation of armature leakage reactance and armature reaction effects. There are two tests on alternators for obtaining the voltage regulation by this method. The two tests are:

1. Open-circuit test
2. Zero power factor test

Open-circuit test: The circuit for the test is shown in Fig. 9.24.

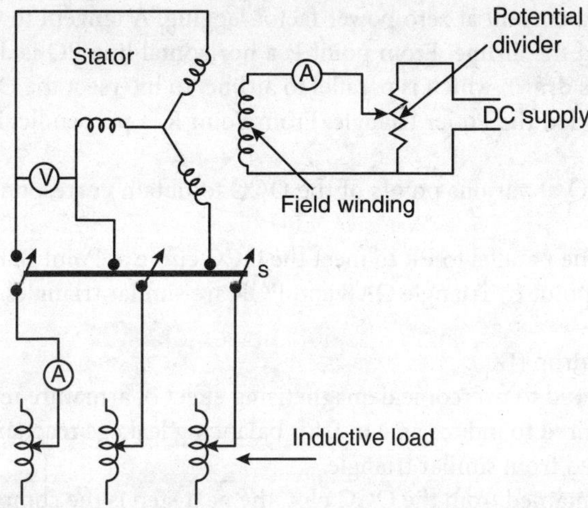

Fig. 9.24: Open-circuit test

The switch (*S*) is kept open. The alternator is brought to its rated synchronous speed with the help of a prime mover. The excitation is varied from zero to its rated value in steps with the help of a potential divider and the open-circuit voltage is measured with the help of voltmeter (V_{oc}) and the corresponding excitation current with the help of ammeter (I_f). A graph of I_f against V_{oc} is plotted. This is the open-circuit characteristic of the alternator in test.

Zero power factor test: Now the switch *S* is closed so that a purely inductive load gets connected to the alternator. The load current delivered by the alternator is maintained constant at its rated full-load value by varying excitation and by adjusting variable inductance of the inductive load. Only two values are noted from this test. One set of values is the zero terminal voltage (short-circuit condition) and the field current required to deliver full-load short circuit armature current. The other set of values is the field current required to obtain rated terminal voltage while delivering rated full-load armature current. From these two sets of data, the zero p.f. saturation curve is plotted. The OCC and zero p.f. full-load plot is shown in Fig. 9.25.

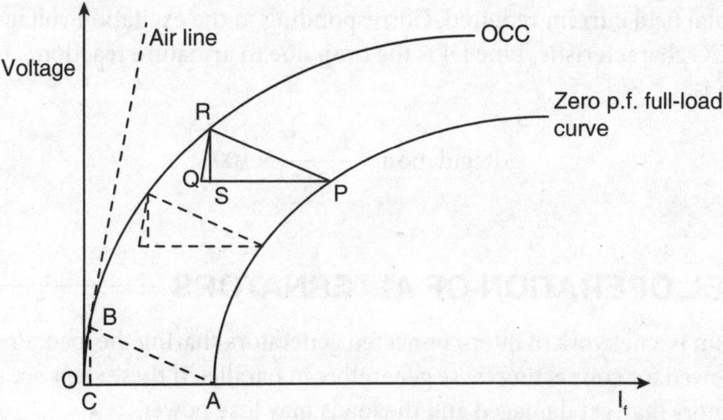

Fig. 9.25: OCC and zero power factor plot

Consider the plot of Fig. 9.25. In the OCC, Point A is the short-circuit full-load zero power factor armature current at zero terminal voltage. Point P is the point which is the rated voltage when the alternator is delivering full-load current at zero power factor lagging. A tangent to OCC is drawn from the origin O. This line is called the airline. From point P, a horizontal line PQ is drawn, which is equal to OA. From Point Q, a line is drawn which is parallel to airline, to intersect the OCC line at point R. Join PR. The triangle PRQ is called the Potier triangle. From Point R, a perpendicular is drawn to meet the line PQ line at S.

Impose the triangle PRQ at various points of the OCC to obtain corresponding points on the zero power factor curve.

From Point A, draw a line parallel to PR to meet the OCC curve at Point B. From B, a perpendicular is drawn to meet x-axis at point C. Triangle OAB and PQR are similar triangles.

In the plot of Fig. 9.25,

RS= Leakage reactance drop (IX_L).

PS = Field current required to overcome demagnetising effect of armature reaction at full load.

SQ = Field current required to induce an e.m.f. for balancing leakage reactance drop.

These values are obtained from similar triangle.

Once required data is obtained from the OCC plot, the next step is the complete phasor diagram.

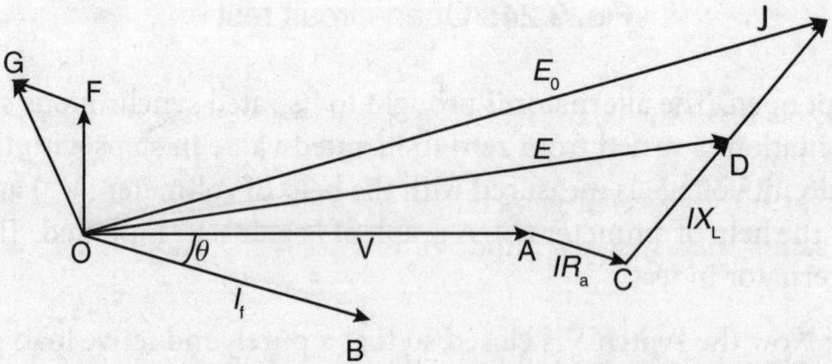

Fig. 9.26: The complete phasor

In the phasor of Fig. 9.26, line OA is the terminal voltage (V) and line OB is the full-load current(I_f) at a given power factor. Line AC is the resistance drop (IR_a), which is parallel to OB. Line CD is the reactance drop (IX_L),which is perpendicular to OB. Line OD is the generated e.m.f.(E). The field current corresponding to this generated e.m.f. is obtained from the OCC and is drawn perpendicular to OD. This has been indicated by the line OF. The line FG is the load current drawn parallel to the line OB. Then the line OG is the total field current required. Corresponding to the excitation voltage OG, the e.m.f. E_0 is obtained from OCC characteristic. Line DJ is the drop due to armature reaction.

So the regulation is,

$$\%\text{Regulation} = \frac{E_0 - V}{V} \times 100$$

9.16 PARALLEL OPERATION OF ALTERNATORS

Modern power system is a network of inter-connected generators sharing the load. Proper layered-down steps need to be followed for connecting these generators in parallel. If these steps are not followed properly, then the generators may get damaged and the loads may lose power.

There are many advantages of operating the alternators in parallel.

i) It helps in dealing with bigger load demands.
ii) Presence of many generators in the system increases the reliability of the system. The failure of one generator will not cause total black out.
iii) With more than one generator in the system, it becomes easy to remove the faulty generator from the system for the purpose of maintenance.
iv) With many generators operating in parallel, depending on the load condition, it is possible to run only a fraction of them, so that it is operated at near to full load, thus increasing the efficiency of the system.

9.17 SYNCHRONISING OF ALTERNATORS

The process of connecting an alternator in parallel with an existing alternator or with a common bus bar is called synchronising.

9.17.1 Condition for Parallel Operation of Alternators

The alternators which are already connected with the system and sharing the load, are called the running machines. The alternators which is to be connected to the existing system, is called the incoming machine. Certain conditions are to be followed to connect a new alternator to the system.

i) The line voltage (R.M.S.) of the incoming alternator must be same as that of the running alternator or of the bus bars connecting them.
ii) All the alternators, running as well as incoming, must have the same phase sequence. If the phase sequence is not same, then the magnitude of the current will be too high, thus damaging the machines.
iii) The frequency of the new generator to be connected with the system must be same as that of the existing generators. If this frequency is not maintained the same, then large power transients will take place until the generator stabilises at a common frequency.
iv) The generators' output phase angles must be the same.

9.18 METHOD OF SYNCHRONISING A SINGLE-PHASE ALTERNATOR

Consider a single-phase system as shown in Fig. 9.27.

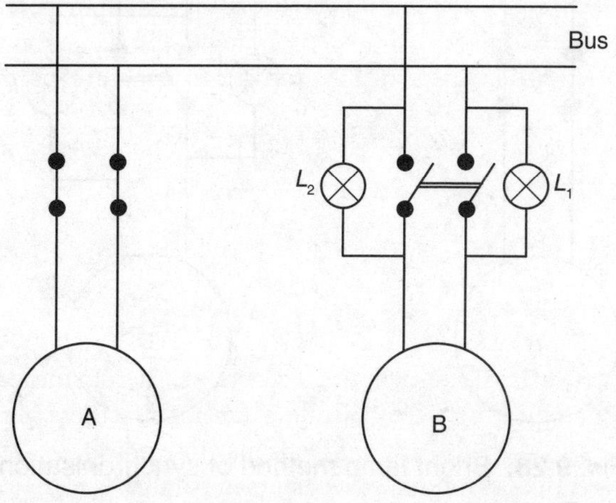

Fig. 9.27: Dark lamp method of synchronisation

In the circuit shown in Fig. 9.27, Alternator A is already connected with the system and Alternator B is the new incoming machine. To connect Alternator B with the system, all the four conditions discussed in Section 9.17.1 need to be satisfied. The bus bar voltage can be easily obtained with the help of a voltmeter. Once the bus bar voltage is known, the incoming alternator is started and its speed is brought close to its rated speed. The alternator is then excited and its voltage is made equal to the bus bar voltage by increasing the excitation.

Now the synchronising lamps L_1 and L_2 are connected across the incoming alternator in parallel as shown in Fig. 9.27. If the frequency of the incoming Alternator B and the frequency of the running machine are same, then their terminal voltages are in exact phase opposition with respect to the circuit, and no voltage will appear across the lamps L_1 and L_2.Thus, the lamps will not glow.

On the other hand if the frequencies of the Alternators A and B are not same, then the phase angle between their voltages will continue to change and current will flow through the lamps. Hence, the lamps will flicker. The frequency of flickering is proportional to the difference of frequencies of the two alternators. Greater the difference in frequency, faster will be the flickering; and smaller the difference in frequency, lesser will be the flickering.

In the dark period (when the lamps are not glowing), the voltages of the two alternators (A and B) will be in phase opposition with respect to the circuit. Synchronisation is done in this period. So this method is called the dark lamp method of synchronisation.

At the time of synchronising, the speed of the incoming alternator is adjusted until the lamps turn ON and OFF very slowly.

The terminal voltage of the incoming alternator is made equal to the bus bar voltage by adjusting the excitation of the incoming alternator.

When both these situations are met, the switch of the incoming alternator is closed during the dark period of the lamps, thereby connecting the incoming alternator to the grid.

After the incoming alternator has been properly synchronised, the next step is to transfer load from the running alternator to the newly connected alternator.

This transfer of load is done by increasing the mechanical power input to the prime mover of the incoming alternator and at the same time decreasing the mechanical power input to the prime mover of the existing alternator.

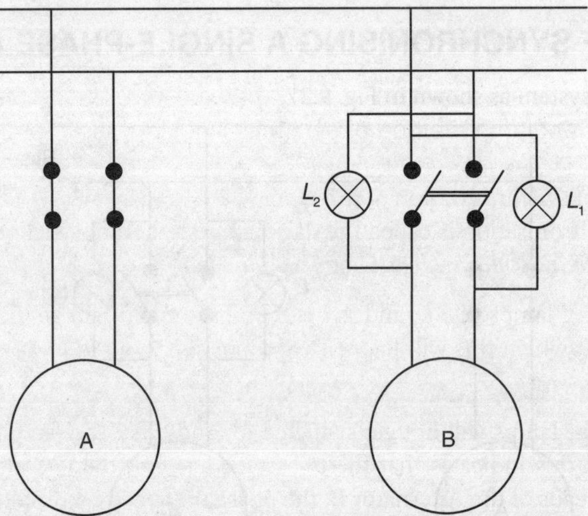

Fig. 9.28: Bright lamp method of synchronisation

It is also possible to synchronise the two alternators (A and B) in the middle of the bright period of the lamps rather than in the dark period of the lamps. This method of synchronisation is called the bright lamp method of synchronisation. The circuit for this method is shown in Fig. 9.28. All the steps of synchronisation are same as that of dark lamp method.

9.19 SYNCHRONISATION OF THREE-PHASE ALTERNATOR

For synchronisation of three-phase alternators, three lamps are used. There are two methods of synchronisation of three-phase alternator.

i) Dark lamp method
ii) Two-bright–one-dark lamp method

The conditions to be satisfied for the synchronisation of three-phase alternators are the same as those of single-phase alternator. Here, we have to take care of the phase sequence also. That is, the phase sequence of both the alternators must be same.

Consider two three-phase alternators, A and B, as shown in Fig. 9.29. Alternator A is already connected with the bus and is supplying power and Alternator B is the new incoming alternator.

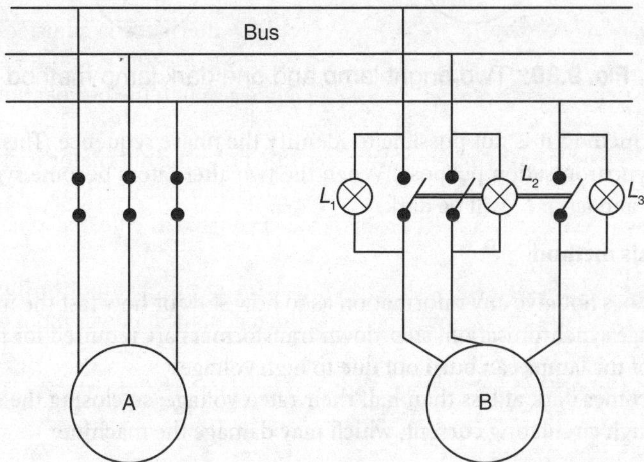

Fig. 9.29: Dark lamp method of synchronisation of three-phase alternator

Dark lamp method of synchronization of three phase alternator

This method of synchronisation is called the dark lamp method of synchronisation. When both the machines are ON, one of the following situations will be seen:

Situation 1: All the three lamps (L_1, L_2 and L_3) will be bright and will go dark at the same time, that is, in unison. The rate at which this will happen is proportional to the frequency difference of the two alternators.

Situation 2: All the three lamps will not get switched on and off at the same time. Since the lamps are not working in unison, this indicates that the phase sequence of the two alternators is not same. By interchanging any two leads of the Alternator B, the phase sequence can be altered. Now, the lamps will go bright and dark in unison.

The speed of the incoming alternator is adjusted until the lamps go bright and dark very slowly. This is done by varying the frequency of the two alternators until they become equal.

The same procedure is followed for synchronisation by the two bright lamp and one dark lamp method. The circuit for this method is shown in Fig. 9.30.

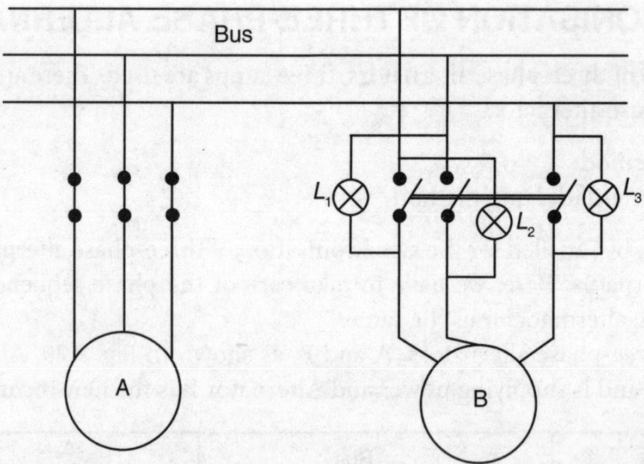

Fig. 9.30: Two bright lamp and one dark lamp method

However, by this method it is not possible to identify the phase sequence. This method is only used for the purpose of synchronisation purpose. When the two alternators become synchronised, lamps L_1 and L_2 will be bright and lamp L_3 will be dark.

Disadvantages of this method

i) This method does not give any information as to how slow or how fast the machines are.
ii) For high-voltage synchronisation, step-down transformers are required for the lamps.
iii) The filament of the lamps can burn out due to high voltage.
iv) The lamps becomes dark at less than half their rated voltage; so closing the synchronising switch will result in high circulating current, which may damage the machine.

9.20 BLONDEL'S TWO-REACTION THEORY

It is known that air gap is uniform in case of non-salient pole type alternators. The field flux as well as armature flux vary sinusoidallywithin the uniform air gap. In non-salient rotor alternators, air gap length is constant and reactance is also constant. Due to this, the m.m.f. of the armature and the field act upon the same magnetic circuit all the time.Hence, these two quantities can be added vectorially. But in salient-pole type alternators, the length of the air gap varies and the reluctance also varies. Hence, the armature flux and field flux cannot vary sinusoidally in the air gap. The reluctances of the magnetic circuits on which m.m.fs act, are different in case of salient-pole alternators. Hence the armature and field m.m.f.s cannot be treated in a simple way as they can be in non-salient pole alternators.

The theory which gives the method of analysis of the distributing effects caused by salient-pole construction is called two-reaction theory. Professor Andre Blondel has put forward the two-reaction theory. According to this theory, the armature m.m.f. can be divided into two components

i) Components acting along the pole axis called **direct axis**
ii) Component acting at right angles to the pole axis called **quadrature axis.**

The component acting along direct axis can be magnetising or demagnetising. The component acting along quadrature axis is cross-magnetising. According to this theory, the disturbing effects are caused these components of armature m.m.f.

9.21 DETERMINATION OF DIRECT- AND QUADRATURE-AXIS SYNCHRONOUS REACTANCE USING SLIP TEST

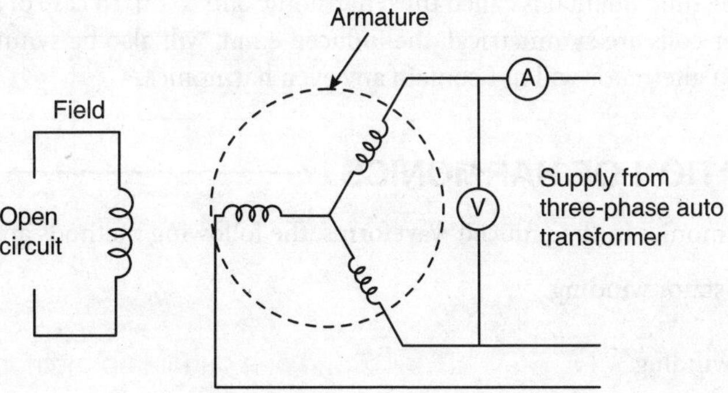

Fig. 9.31: Slip test circuit

The circuit diagram for the test is shown in Fig. 9.31.

In slip test, with the help of a three-phase auto transformer, a three-phase reduced voltage is applied to the armature with the field winding open. Due to the three-phase supply, three-phase current flows through the armature winding thereby producing a rotating magnetic flux. This will make the rotor rotate at a speed close to its synchronous speed.

If N_s is the synchronous speed and N is the actual speed of the rotor, then the armature m.m.f., which rotates at synchronous speed, will cut the field poles at a slip speed ($N_s - N$). This will induce an e.m.f. in the field circuit. When the stator m.m.f. aligns itself with the d-axis of field poles, the flux φ_d per pole is set up.Let the effective reactance offered by the alternator beX_d. Similarly, when the stator m.m.f. aligns itself with the q-axis of field poles, the flux, φ_q, per pole is set up.Let the effective reactance offered by the alternator beX_q.

Since the air gap is not uniform throughout, the reactance offered will also not be same. Thus, the current drawn by the armature will vary. This current will be minimum when machine reactance is X_d and will be maximum when machine reactance is X_q.

Similarly, the terminal voltage will be maximum when current and various drops are minimum and it will be minimum when current and various drops are maximum. The voltage in the field winding is maximum when the alternator reactance is X_q and it is minimum whenthe alternator reactance is X_d.

Thus,

$$X_d = \frac{\text{Maximum voltage}}{\text{Minimum current}}$$

$$X_q = \frac{\text{Minimum voltage}}{\text{Maximum current}}$$

9.22 HARMONICS

When the uniformly sinusoidally distributed air gap flux iscut by either the stationary or rotating armature, a sinusoidal e.m.f. is induced in the alternator. Hence, the nature of the waveform of induced e.m.f. and current is sinusoidal. However when the alternator is loaded, the waveform will not be sinusoidal. Such non-sinusoidal wave form is called a complex wave form. By using Fourier series representation, it is possible to represent complex non-sinusoidal waveform in terms of a series of sinusoidal components called harmonics, whose frequencies are integral multiples of the fundamental wave. The fundamental wave form is one which is having the frequency same as that of complex wave. The waveform, which is of a frequency twice that of the fundamental, is called second harmonic. The one which has a frequency three times that of the fundamental is called third harmonic and so on. In case of alternators, as the field system and the stator coils are symmetrical, the induced e.m.f. will also be symmetrical and hence the generated e.m.f. in an alternator will not contain any even harmonics.

9.23 MINIMISATION OF HARMONICS

To minimise the harmonics in the induced waveforms, the following methods are employed:

1. Distribution of stator winding.
2. Short chording
3. Fractional slot winding
4. Skewing
5. Larger air gap length.

SOLVED PROBLEMS

Ex.9.1 A 13.8 KV, 50 Hz, 3000 r.p.m. synchronous generator has a synchronous reactance of 15 Ω. It is operated at rated voltage and speed. The excitation voltage is 11.5 KV and the torque angle is 15^0. Calculate

(i) Current of the stator
(ii) Power factor
(iii) Output power

Solution

$$\text{Current}(I) = \frac{(11.5\times10^3)\angle 15 - \dfrac{13.8\times10^3}{\sqrt{3}}}{15\angle 90} = \frac{(11.5\times10^3)\angle 15 - 7.97\times10^3}{15\angle 90}$$

$$= \frac{3138.15 + j2976.42}{j15}(= -198.43 + j209.21)A = 288.35\angle 133.49\,\text{A}$$

$$\text{Power factor} = \cos 133.49 = -0.688$$

$$\text{Power} = \sqrt{3}\times 13.8\times10^3\times 288.35\times 0.688 = 4741.85\text{KW}$$

Ex.9.2 A 25 KV, 500 MVA, 50 Hz, three-phase alternator is operated at rated voltage and frequency. It has a power factor of 0.8 lag. The leakage reactance and synchronous reactance are 0.15Ω and 0.8Ω respectively. If the excitation voltage is 15 KV, find the following:

(i) Torque angle
(ii) Output power
(iii) Line current
(iv) E.m.f.

Solution

$$\text{Torque angle}(\delta) = \cos^{-1}\left[\frac{\frac{25\times10^3}{\sqrt{3}}\times0.8}{15\times10^3}\right] - \cos^{-1}(0.8) = 39.66 - 36.87 = 2.79^0$$

$$\text{Power}(P) = \frac{3\times15\times10^3\times\frac{25\times10^3}{\sqrt{3}}}{0.8}\times\sin 2.79 = 39.52\text{ MW}$$

$$\text{Current}(I) = \frac{39.52\times10^6}{\sqrt{3}\times25\times10^3\times0.8} = 1140.84\text{ A}$$

$$\text{e.m.f.} = \frac{25\times10^3}{\sqrt{3}} + j0.15\times1140.84\angle-36.87 = 14537.05\angle0.52\text{ V}$$

Ex.9.3 A 250 MVA, 24 KV, 50 Hz, three-phase alternator is operating at its rated values. If the excitation is 20 KV and power factor is 0.8 lag, calculate the torque angle and synchronous reactance.

Solution

$$\theta = \cos^{-1}0.8 = 36.87^0$$

$$\text{Torque angle }(\delta) = \cos^{-1}\left[\frac{\frac{24\times10^3}{\sqrt{3}}\times0.8}{20\times10^3}\right] - \cos^{-1}(0.8) = 19.47^0$$

$$\text{Current}(I) = \frac{250\times10^6}{\sqrt{3}\times24\times10^3}\angle-36.87 = 6014.2\angle-36.87A$$

$$\text{Synchronous reactance}(X) = \frac{20\times10^3\angle19.47 - \frac{24\times10^3}{\sqrt{3}}}{j(6014.2\angle-36.87)} = 1.38'$$

Ex.9.4 Consider a 12 KVA, 230V, three-phase star-connected round rotor synchronous generator with synchronous reactance of 1Ω per phase and armature resistance of 0.6 Ω per phase. Calculate the voltage regulation at full load, 0.8 power factor lag.

Solution

$$V_p = \frac{230}{\sqrt{3}} = 132.8\,\text{V}$$

$$I_p = \frac{\dfrac{\left(12\times10^3\right)}{3}}{132.8} = 30.12\,\text{A}$$

$$V_0 = \sqrt{\left(132.8\times0.8+30.12\times0.6\right)^2+\left(132.8\times0.6+30.12\times1\right)^2} = 165.86\,\text{V}$$

$$\%\ \text{regulation} = \frac{165.86-132.8}{132.8}\times100 = 24.89\%$$

Ex.9.5 A 6 KVA, 230 V three-phase, star-connected salient pole alternator delivers full-load current at unity power factor. Calculate the armature current, load angle and excitation voltage if the direct and quadrature axis reactances are 10Ω and 5Ω respectively.

Solution

$$\cos\theta = 1 \text{ and} \sin\theta = 0$$

$$\text{Armature current}(I) = \frac{6\times10^3}{\sqrt{3}\times230\times1} = 15.06\,\text{A}$$

$$\delta = \tan^{-1}\left[\frac{15.06\times5\times1}{\dfrac{230}{\sqrt{3}}+15.06\times5}\right] = 19.89^0$$

$$\beta = 0+\delta = 0+19.89 = 19.89$$

$$I_d = I\sin\beta = 15.06\times0.34 = 5.12\,\text{A}$$

$$\text{Excitation voltage} = \frac{230}{\sqrt{3}}\times\cos19.89+5.12\times10 = 176.07\,\text{V}$$

Ex.9.6 The test data of a three-phase, star-connected, 8 KVA, 440 V 50 Hz, 4-pole alternator is as follows:

Excitation current (A)	1	1.5	2	2.5	3	3.5	4	4.5	5	6	7
Open-circuit voltage (V)	80	145	175	210	230	250	260	265	273	279	285

Excitation current (A)	1	2	3
Short-circuit line current (A)	4	7.6	11.2

Draw the OCC and calculate the synchronous reactance per phase if the armature resistance is 3Ω.

Solution

The OCC plot is shown in Fig. E9.6.

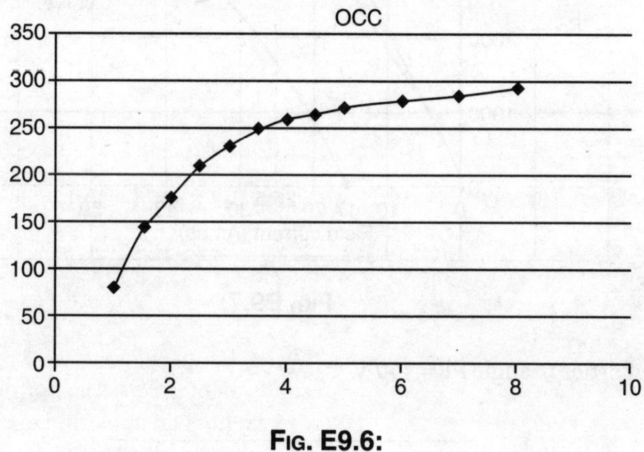

FIG. E9.6:

For excitation current of 1 A, the open-circuit voltage is 80V and short-circuit current is 4A.

$$\text{Synchronous impedance}(Z) = \frac{80}{4} = 20\ \Omega$$

$$\text{Synchronous reactance}(X) = \sqrt{20^2 - 3^2} = 19.77\Omega$$

Ex.9.7 A three-phase star-connected alternator has the following test data

Field current (A)	14	18	23	30	43
Terminal voltage (V)	4000	5000	6000	7000	8000

At full-load current with armature short-circuited, the field current is 17A.At full load of 20,000 KVA and unity power factor, the field current is 42.5 A.The terminal voltage is 6000 V. Calculate the field current when the machine is supplying full load at 0.8 power factor lagging.

Solution

The terminal voltage after converting into phase voltage we get,

Terminal voltage (V)	4000	5000	6000	7000	8000
Terminal voltage, phase voltage (V)	2309	2887	3464	4041	4619

The full-load zero power factor curve has two points (17,0) and (42.5, 3464).
The Portier triangle is shown in Fig. E9.7.

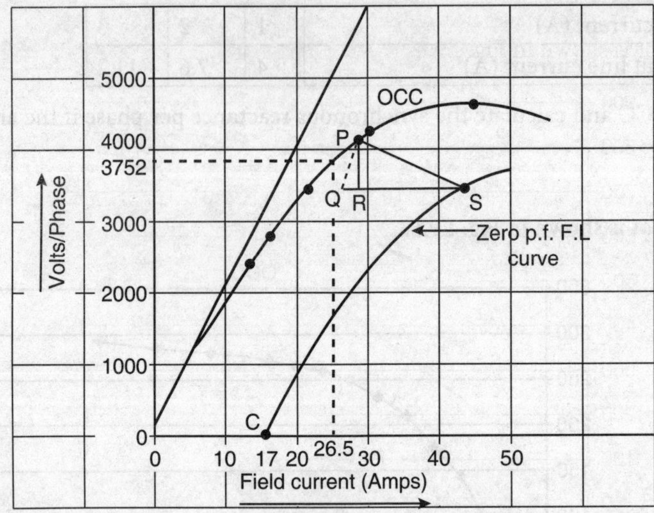

Fig. E9.7:

From the Portier triangle PR= 450 V
So,

$$E = \sqrt{(3464 \times 0.8)^2 + (3464 \times 0.6 + 450)^2} = 3752\,\text{V}$$

From the plot shown in Fig.E9.7,
The field current corresponding to 3752 V is 26.5 A
The field current required to overcome armature reaction = RS = 14.5 A

$$\text{Total field current} = \sqrt{26.5^2 + 14.5^2 - 2 \times 26.5 \times 14.5 \times \cos(90 + 36.8)} = 37.05\,\text{A}$$

Ex.9.8 Consider a 1600 MW, 0.8 p.f., 13KV, 50 Hz, 1800 r.p.m. three-phase round rotor synchronous motor. The resistance per phase is 2Ω. The magnetising curve data is as follows:

Field current, I (A)	200	300	400	500	600	720	780	800
Armature voltage (L–L) KV	4	6	8	10	12	15	17	19

The short-circuit armature-current test gives a straight line through the origin and through-rated
armature current at 700A field current.
Draw the OCC curve of the same.

(i) Determine the unsaturated synchronous impedance per phase.
(ii) Determine the saturated synchronous impedance per phase.
(iii) Draw a phasor diagram and determine the voltage regulation for the condition of rated load and lagging power factor of 0.8.

Solution

The OCC plot is shown in Fig. E9.8.

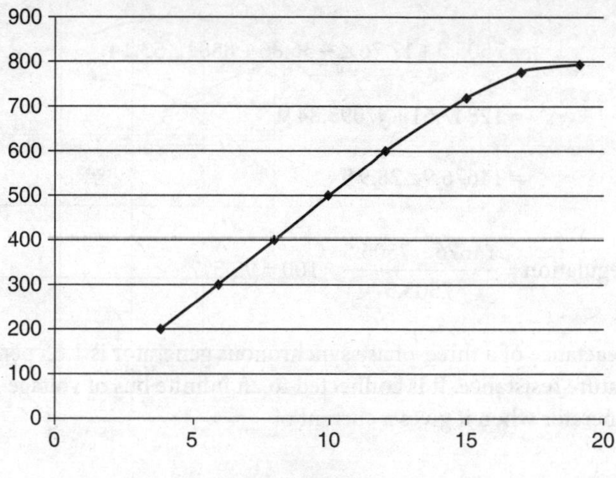

FIG. E9.8:

$$\text{Rated current}\left(I_a\right) = \frac{\left(160\times 10^6\right)/3}{\frac{13\times 10^3}{\sqrt{3}}\times 0.8} = 8882\,\text{A}$$

From the plot,

$$Z_{usat} = \frac{13/\sqrt{3}}{7.4} = 1\Omega$$

$$Z_{sat} = \frac{13/\sqrt{3}}{8.1} = 0.9\Omega$$

$$\theta = cos^{-1}0.8 = 36.86^0$$

The phasor diagram is shown in Fig. E9.8 (a)

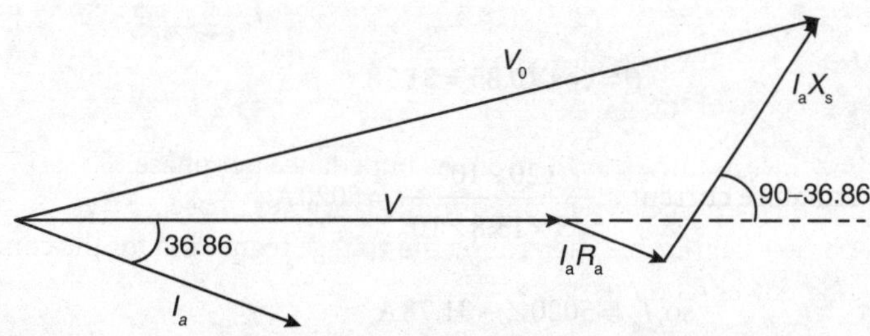

FIG. E9.8 (a):

$$V_0 = \frac{13\times10^3}{\sqrt{3}} + 8882\times2\times10^{-3}\angle-36.86 + 8882\times1\angle53.14$$

$$= 7505.5 + 17.76\angle-36.86 + 8882\angle53.14$$

$$= 12847.61 + j7095.84\,\text{V}$$

$$= 14676.9\angle28.9\,\text{V}$$

$$\%\ \text{regulation} = \frac{14676-7505.5}{7505.5}\times100 = 95.5\%$$

Ex.9.9 Synchronous reactance of a three-phase synchronous generator is 1.65 per unit (p.u.) and it has negligible armature resistance. It is connected to an infinite bus of voltage 1.0 p.u. Calculate the e.m.f. of the generator when it gives a current of

(i) $1.030\,\text{p.u.}$
(ii) $1.0\angle0\,\text{p.u.}$
(iii) $1.0\angle-30\,\text{p.u.}$

Solution

$$E = 1\angle0 + 1\angle30\times1.65\angle90 = 1.44\angle83\,\text{p.u.}$$

$$E = 1\angle0 + 1\angle0\times1.65\angle90 = 1.93\angle58.8\,\text{p.u.}$$

$$E = 1\angle0 + 1\angle-30\times1.65\angle90 = 2.32\angle38\,\text{p.u.}$$

Ex.9.10 A three-phase 120 MVA, 13.8 kV, 0.85 star-connected synchronous generator is rated 120 MVA, 13.8 kV, 0.85 PF lag, and 60 Hz. Its synchronous reactance is 0.7 Ω, and its resistance may be ignored.

(a) What is its voltage regulation?
(b) What would the voltage and apparent power rating of this generator be if it were operated at 50 Hz with the same armature and field losses as it had at 60 Hz?
(c) What would the voltage regulation of the generator be at 50 Hz?

Solution

$$\theta = \cos^{-1}0.85 = 31.78$$

$$\text{Rated armature current} = \frac{120\times10^6}{\sqrt{3}\times13.8\times10^3} = 5020\,\text{A}$$

$$\text{so}, I_a = 5020\angle-31.78\,\text{A}$$

$$\text{Generated voltage} = \frac{13.8\times10^3}{\sqrt{3}} + \left(\left(0.7\angle90\right)\times\left(5020\angle-31.78\right)\right)$$

$$= 10260\angle 16.9\,\text{V}$$

$$\text{Voltage regulation} = \frac{10260 - \dfrac{13.8\times 10^3}{\sqrt{3}}}{\dfrac{13.8\times 10^3}{\sqrt{3}}} \times 100 = 28.8\%$$

At 50% frequency

$$V = \frac{5}{6}\times 13.8 = 11.5\,\text{kV}$$

$$S = \frac{5}{6}\times 120 = 100\,\text{MVA}$$

$$X = \frac{5}{6}\times 0.7 = 0.583'$$

$$\text{Rated armature current} = \frac{100\times 10^6}{\sqrt{3}\times 11.5\times 10^3} = 5020\,\text{A}$$

$$\text{so}, I_a = 5020\angle -31.78\text{A}$$

$$\text{Generated voltage} = \frac{11.5\times 10^3}{\sqrt{3}} + \left((0.583\angle 90)\times(5020\angle -31.78)\right)$$

$$= 8550\angle 16.9\,\text{V}$$

$$\text{Voltage regulation} = \frac{8550 - \dfrac{11.5\times 10^3}{\sqrt{3}}}{\dfrac{11.5\times 10^3}{\sqrt{3}}} \times 100 = 28.8\%$$

Ex.9.11 One of the phases of a three-phase synchronous generator is shown in Fig.E9.11. It is operated at zero p.f. leading with a capacitive load of -10Ω. Calculate the ratio of V/V_n

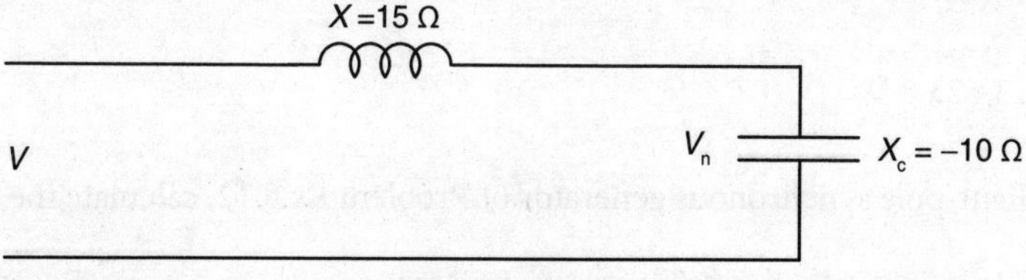

Fig. E.9.11:

Solution

$$V_n = \frac{jX_c}{j(X_c + X)} V = \frac{-j10}{j15 - j10} V = -2V$$

$$\text{or } \frac{V}{V_n} = -\frac{1}{2}$$

Ex.9.12 A 100 MVA, 13.8 kV, 60 Hz, salient-pole, synchronous generator with 8 poles has and $X_q = 1.1\ \Omega$. The stator coil resistance is negligible. If the generator supplies $X_d = 1.9\ \Omega$ rated apparent power at a power factor of 0.866 lagging,determine the value of excitation voltage E_f and torque angle δ.

Solution

$$\bar{I}_a = \frac{S_T}{\sqrt{3}V_L} \angle -\cos^{-1}(PF) = \frac{100 \times 10^6}{\sqrt{3}(13{,}800)} \angle \cos^{-1}(0.866) = 4183.8\angle -30° \text{ A}$$

$$\bar{E}_f' = \bar{V}_{an} + jX_q\bar{I}_a = \frac{13{,}800}{\sqrt{3}} \angle 0° + (1.1\angle 90°)(4183.8\angle -30°)$$

$$\bar{E}_f' = 11{,}015\angle 21.21° \text{ V}$$

$$\delta = \angle \bar{E}_f' = 21.21°$$

$$\psi = (\theta + \delta) = (30 + 21.21) = 51.21°$$

$$I_d = I_a \sin\psi = 4183.8 \sin(51.21°) = 3261 \text{ A}$$

$$\bar{I}_d = I_d\angle\delta - 90° = 3261\angle -68.79° \text{ A}$$

$$\bar{E}_f = \bar{E}_f + j(X_d - X_q)\bar{I}_d$$

$$\bar{E}_f = 11{,}015\angle 21.21° + (0.8\angle 90°)(3261\angle -68.79°)$$

$$\bar{E}_f = 13{,}623.8\angle 21.21° \text{ V}$$

$$E_f = 13{,}623.8 \text{ V}$$

Ex.9.13 For the salient-pole synchronous generator of Problem Ex.9.12, calculate the developed torque.

Solution

$$\omega_s = \frac{2}{p}\omega = \frac{2}{8}(2\pi 60) = 30\pi \text{ rad/s}$$

$$3T_d = \frac{3V_{an}E_f}{\omega_s X_d}\sin\delta = \frac{3V_{an}^2\left(X_d - X_q\right)}{2\omega_s X_d X_q}\sin 2\delta$$

$$3T_d = \frac{3\left(\frac{13{,}800}{\sqrt{3}}\right)(13{,}623.8)}{30\pi(1.9)}\sin(21.21°) + \frac{3\left(\frac{13{,}800}{\sqrt{3}}\right)^2(1.9-1.1)}{2(30\pi)(1.9)(1.1)}\sin(2\times 21.21°)$$

$$3T_d = 918.8 \text{ kN}\cdot\text{m}$$

Ex.9.14 Determine the maximum power that the salient-pole synchronous generator of Problem Ex.9.12 can convert if the field circuit were in open circuit.

Solution

$$3P_{d\max} = \frac{3V_{an}^2\left(X_d - X_q\right)}{2X_d X_q}$$

$$3P_{d\max} = \frac{3\left(\frac{13{,}800}{\sqrt{3}}\right)^2(1.9-1.1)}{2(1.9)(1.1)} = 36.45 \text{ MW}$$

Ex.9.15 A 15 KV, synchronous generator has a synchronous reactance of 20 Ω. It is operated at rated voltage and speed. The excitation voltage is13 KV and the torque angle is 20^0. Calculate the output power.

Solution

$$\text{Current}(I) = \frac{\left(13\times 10^3\right)\angle 20 - \frac{15\times 10^3}{\sqrt{3}}}{20\angle 90} = \frac{\left(13\times 10^3\right)\angle 20 - 8660.26}{20\angle 90}$$

$$= \frac{3555.74 + j4446.26}{j20}(= -222.31 + j177.79)\text{A} = 284.66\angle 141.35\text{A}$$

$$\cos 141.35 = -0.781$$

$$\text{Power} = \sqrt{3}\times 15\times 10^3 \times 284.66\times 0.781 = 5776.03\text{KW}$$

Ex.9.16 A 30 KV, three-phase alternator is operated at rated voltage and frequency. It has a power factor of 0.8 lag. The leakage reactance and synchronous reactance are 0.2 Ω and 1 Ω respectively. If the excitation voltage is 20 KV, find the e.m.f. of the alternator

Solution

$$\text{Torque angle}(\delta) = \cos^{-1}\left[\frac{\frac{30\times10^3}{\sqrt{3}}\times0.8}{20\times10^3}\right] - \cos^{-1}(0.8) = 46.15 - 36.87 = 9.28^0$$

$$\text{Power}(P) = \frac{3\times20\times10^3\times\frac{30\times10^3}{\sqrt{3}}}{0.8}\times\sin 9.28 = 209.48\text{ MW}$$

$$\text{Current}(I) = \frac{209.48\times10^6}{\sqrt{3}\times30\times10^3\times0.8} = 5039.31\text{A}$$

$$\text{e.m.f.} = \frac{30\times10^3}{\sqrt{3}} + j0.2\times5039.31\angle-36.87 = 17943.35\angle2.57\text{V}$$

Ex.9.17 A 200 MVA, 24 (30) KV, 50 Hz, three phase alternator is operating at its rated values. If the excitation is 20 (22) KV and power factor is 0.8 lag, calculate the torque angle and synchronous reactance.

Solution

$$\theta = \cos^{-1}0.8 = 36.87^0$$

$$\text{Torque angle}(\delta) = \cos^{-1}\left[\frac{\frac{30\times10^3}{\sqrt{3}}\times0.8}{22\times10^3}\right] - \cos^{-1}(0.8) = 14.09^0$$

$$\text{Current}(I) = \frac{200\times10^6}{\sqrt{3}\times30\times10^3}\angle-36.87 = 3849\angle-36.87\text{A}$$

$$\text{Synchronous reactance}(X) = \frac{22\times10^3\angle14.09 - \frac{30\times10^3}{\sqrt{3}}}{j(3849\angle-36.87)53.13} = 1.74\Omega$$

Ex.9.18 A three-phase star-connected, 8-pole synchronous generator is rotating at 900 r.p.m. The stator has 100 slots and 10 conductors per slot. What will be the voltage generated by the machine if the flux per pole is 0.07 Wb and the winding factor is 0.9.

Solution

$$\text{Frequency}(f) = \frac{8 \times 900}{120} = 60\,\text{Hz}$$

$$\text{Total number of conductors per phase} = \frac{100 \times 10}{3} = 333$$

$$\text{Generated voltage} = 2.22 \times 0.9 \times 60 \times 0.07 \times 333 = 2794.4\,V \text{ per phase}$$

Ex.9.19 Find the voltage regulation and the torque angle of a three-phase 15 KVA, 450V, 50 Hz synchronous generator supplying a rated load at 0.8 power factor lag. Take the armature and the field resistance to be 1Ω and 15Ω respectively.

Solution

$$\text{Impedance}(Z) = 1 + j15 = 15\angle 86.1\Omega$$

$$\text{Phase voltage}(V_P) = \frac{450}{\sqrt{3}} = 259.81\text{V}$$

$$\cos^{-1} 0.8 = 36.89$$

$$I_L = I_P = \frac{15 \times 10^3}{\sqrt{3} \times 450} = 19.25A = 19.25\angle -36.89A = 15.39 - j11.56\text{A}$$

$$E_P = 259.81 + (15.39 - j11.56) \times (1 + j15) = 448.6 + j219.29\,\text{V} = 499.32\angle 26.05\,\text{V}$$

So the torque angle is 26.05 degree.

$$\%\text{regulation} = \frac{499.32 - 259.81}{259.81} \times 100 = 92.19\%$$

Ex.9.20 A star-connected, three-phase alternator having resistance and reactance of 1Ω and 15Ω respectively, delivers 300 A at 0.8 power factor lag to a 11 KV constant frequency bus bar. Find the percentage change in the induced e.m.f. which is necessary to raise the power factor to unity.

Solution

$$V_P = \frac{11000}{\sqrt{3}} = 6350.85\,\text{V}$$

$$\theta = \cos^{-1} 0.8 = 36.88$$

$$I_L = I_P = 300\angle -36.87\text{A} = 240 - j180\text{A}$$

$$Z = 1 + j15\ \Omega$$

$$E = 6350.85 + (1 + j15) \times (240 - j180) = 9290.85 + j3420\text{V} = 9900.32\angle 20.21\text{V}$$

For the power factor to be unity,

$$I_2 = \frac{300 \times 0.8}{1} = 240\text{ A}$$

Again,

$$E_1 = 6350.85 + (1 + j15) \times (240 - j0) = 6590.85 + j3600V = 7509.9\angle 28.64\text{V}$$

$$\text{Change in excitation} = \frac{9900.32 - 7509.9}{9900.32} \times 100 = 24.14\%$$

EXERCISES

1. Give a detailed classification of synchronous machines.
2. Differentiate between salient pole and non-salient pole machine.
3. Give the advantages of rotating field and stationary armatures.
4. Explain the construction of a synchronous generator.
5. What are the different types of windings that are used in the alternator?
6. Explain the following:
   (i) Pole pitch
   (ii) Coil span
   (iii) Slot angle
   (iv) Pitch factor (or) chording factor (or) coil span factor
   (v) Distributions factor (or) breadth factor
7. Derive the e.m.f. equation for an alternator.
8. Write a note on armature reaction in synchronous generator.
9. Explain the armature reaction in synchronous generator for lagging and leading power factor.
10. Obtain the equivalent circuit of a synchronous generator.
11. Draw the phasor diagram of a alternator for:
    (i) Lagging load
    (ii) Leading load
    (iii) Unity power factor
12. Define voltage regulation. What the different methods to measure voltage regulation of an alternator?
13. How can you determine the armature resistance of an alternator?
14. Explain the OCC and SCC characteristic of an alternator. How are these characteristics obtained?
15. Draw and explain the synchronous impedance method of determination of voltage regulation.
16. What are the advantages of synchronous impedance method?
17. What are the limitations of synchronous impedance method?
18. Draw and explain the MMF method of determining the voltage regulation.
19. Draw and explain the ZPF method of determining the voltage regulation.
20. What are the advantages of operating the alternators in parallel?
21. What are the conditions for parallel operation of alternators?

22. Draw and explain the dark lamp method of synchronisation of single-phase alternator.

23. Draw and explain the bright lamp method of synchronisation of single-phase alternator.

24. Draw and explain the dark lamp method of synchronising a three-phase alternator

25. Draw and explain the two-bright – one-dark lamp method synchronising a three-phase alternator.

26. Consider a 12 KVA, 230V, three-phase star-connected round rotor synchronous generator with synchronous reactance of 1Ω per phase and armature resistance of 0.6 Ω per phase. Calculate the voltage regulation at full load and 0.8 power factor lead.

27. A 270 MVA, 25 KV, 50 Hz, three-phase alternator is operating at its rated values. If the excitation is 21 KV and power factor is 0.8 lag, calculate the torque angle and synchronous reactance.

28. A 35 KV, 510 MVA, 50 Hz, three-phase alternator is operated at rated voltage and frequency. It has a power factor of 0.8 lag. The leakage reactance and synchronous reactance are 0.18Ω and 0.8Ω respectively. If the excitation voltage is 10 KV, find the following:

(i) Torque angle
(ii) Output power
(iii) Line current
(iv) E.m.f.

CHAPTER 10

SYNCHRONOUS MOTOR

10.1 INTRODUCTION

Synchronous motors are widely used in the industry for high-precision applications. This motor runs at constant speed and it does not depend on the torque acting on it. So it has a constant-speed torque characteristic. The efficiency of synchronous motor is around 90%–93%. It is a doubly fed motor; three-phase power is given to the stator while the rotor is fed from a DC source for excitation of the field winding. The field excitation can be changed, which in turn will change the power factor of the motor. The synchronous motor can operate both on lagging and leading power factor.

10.2 CONSTRUCTION

There is no constructional difference between synchronous motor and synchronous generator (or alternator). Synchronous motor construction is similar to that of an alternator with rotating field. It has mainly two parts,namely, the stator and the rotor (Fig. 10.1).

The stator has a laminated core with slots to hold the three-phase windings. These windings can be either star- or delta-connected. It is excited by a three-phase AC supply.

Rotor holds the field winding. The rotor can be of salient-pole type or cylindrical type. Mostly synchronous motors are of salient-pole type. The field windings is generally excited by a separate DC supply through slip rings.

Synchronous motor is likely to hunt (Oscillation of the rotor about its final equilibrium position) and so damper windings are also provided in the rotor poles.

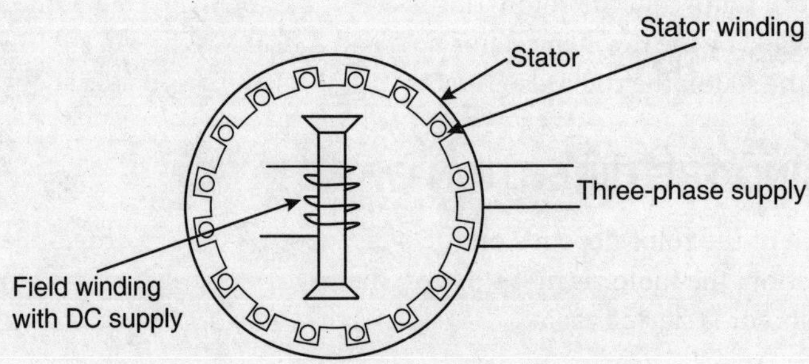

FIG. 10.1: Parts of synchronous motor

10.3 PRINCIPLE OF OPERATION OF SYNCHRONOUS MOTOR

In a synchronous motor, the stator is wound on the same number of poles as that of the rotor. Synchronous motor is a doubly excited machine, that is, two electrical inputs are provided to it. The stator winding which houses three-phase winding is fed from a three-phase supply and the rotor is fed with a DC supply.

When a three phase-supply is given to the stator of the synchronous motor, it produces a rotating magnetic flux of constant magnitude, rotating at synchronous speed. The direction of rotation depends on the phase sequence of the supply. Similarly, a DC supply on the rotor will also produce a flux of constant magnitude.

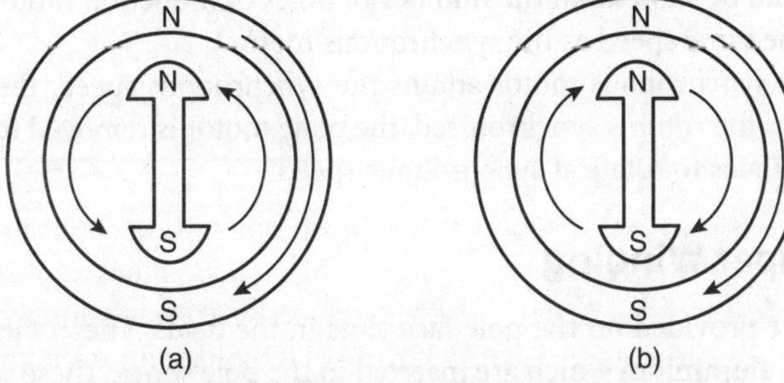

Fig. 10.2: Two-pole synchronous motor

Consider a two-pole synchronous machine as shown in Fig. 10.2. Here, it has been assumed that the stator flux is rotating in clockwise direction at synchronous speed.

Now, let us consider a situation when the N-pole of the rotor is near to the N-pole of the stator (Fig. 10.2(a)). They will repel each other and hence the torque produced will be anti-clockwise. Since the stator flux is rotating at synchronous speed, their positions will interchange very fast. However, the rotor will not be able to rotate at the same angle because of its inertia and hence,its position will be as shown in Fig. 10.2 (b), will reach. That is, the N-pole of the rotor will be near to the S-pole of the stator. In this case, they will attract each other and hence the torque produced will be clockwise.

As this process is repeated continuously, the rotor is unable to rotate due to the continuous attraction and repulsion force acting on it because of stator rotating flux. Thus the rotor will remain in standstill position. Thus a three phase synchronous motor is not self-starting.

If the rotor of the synchronous motor is rotated by some external means at the start so that it also reverses its polarity at the same speed as that of the stator flux, then there will be a continuous force of attraction between the stator and the rotor. This is called magnetic locking. Once this stage is reached, the rotor pole is dragged by the revolving stator field and thus the rotor will continue to rotate. Since the rotor is dragged by the stator, the rotor also rotates with the same speed as that of the stator.

10.4 REVERSING THE DIRECTION OF ROTATION

Direction of rotation of the rotor depends on the phase sequence of the three-phase supply. To reverse the direction of rotation, the motor is first stopped; then, the phase sequence of the three-phase supply is changed and the motor is started again.

10.5 STARTING METHODS FOR SYNCHRONOUS MOTOR

Three-phase synchronous motor is not self-starting and does not have a self-starting torque. To start the synchronous motor, its rotor must be rotated by some external force. The different methods that are generally followed to start the synchronous motor are:

i) By using a pony motor (Small induction motor)
ii) By using a damper winding
iii) By using DC motor
iv) Starting as an induction motor

10.5.1 Using a Pony Motor (Small Induction Motor)

In this method, the rotor of the synchronous motor is brought to its synchronous speed with the help of an external induction motor. This external motor is called the pony motor. The number of poles of the synchronous motor should be more than the number of poles of induction motor, to enable the induction motor to rotate at the same speed as the synchronous motor.

Once the rotor of the synchronous motor attains the synchronous speed, the DC excitation to the rotor is switched on. Once the rotor is synchronised, the pony motor is removed from the circuit and the synchronous motor continues to rotate at synchronous speed.

10.5.2 Using Damper Winding

The damper windings are provided on the pole face slots in the fields. These winding are usually solid bars of copper, bronze or aluminium which are inserted in the pole shoes. These windings are short-circuited at both ends with the help of end rings, thus forming a squirrel-cage system. Now, when a three-phase supply is given to the stator of a synchronous motor, it will start as a three-phase induction motor, which will rotate at a speed near to synchronous speed. When a DC excitation is given to the rotor, the rotor will pull into synchronism.

Since the rotor is rotating at synchronous speed, the relative speed between the damper winding and the rotor magnetic field will be zero; thus, there will be no induced e.m.f. in the damper winding. This shows that the damper winding comes into play only during the starting of the synchronous motor; after that, it is not in use.

Since the damper windings are short-circuited at both the ends, it will draw large amount of current from the supply at starting. So in this case, induction motor starter is used to start the synchronous motor.

10.5.3 Using a DC Motor

In this method of starting, the synchronous motor is brought to its synchronous speed with the help of a DC motor coupled to it.Once the rotor of the synchronous motor attains synchronous speed, the DC excitation to the rotor is switched on.

10.5.4 As an Induction Motor

In the method of starting of synchronous motor by damper winding, the synchronous motor is started as a squirrel-cage induction motor. However when the synchronous motor is started as a slip-ring induction motor,the three ends of the windingsare connected to an external resistance in series through slip-rings. The resistance is gradually reduced as the motor picks up speed and at synchronous speed, the DC excitation is provided to the rotor.

The motor gets synchronised and starts to rotate at synchronous speed. The damper windings are shorted by shorting the slip-rings.

10.6 POWER ANGLE AND DEVELOPED TORQUE OF SYNCHRONOUS MOTOR

The synchronous motor rotates at synchronous speed. But increase in shaft load causes the rotor magnet to change its angular position with respect to the rotating flux of the stator by an electrical angle δ. This angle is called the power angle or load angle or torque angle.

The average speed of rotation of synchronous motor remains the same from no-load to peak-load. When the load on the motor is increased, the motor first slows down a bit to allow the rotor to shift its angular position with respect to the stator rotating fluxand then it again regains its synchronous speed. When the load on the motor is decreased, the motor speeds up to allow the rotor to decrease its angular position with respect to the stator rotating flux and then it again regains its synchronous speed.

If the load exceeds the peak load the motor pulls out of synchronism and its stops.

10.7 BEHAVIOUR OF SYNCHRONOUS MOTOR ON LOAD

In a synchronous motor, the rotor rotates at synchronous speed and the stationary stator conductors cuts the rotor flux. Due to this, as per Lenz's law, an e.m.f. will be induced in the stator called the back e.m.f. (E_b). Since speed does not change in a synchronous motor, the frequency remains constant. So the magnitude of back e.m.f. can only be changed by changing the flux φ as produced by the rotor.

LetR_a and X_s be the resistance and synchronous reactance per phase. Then,

$$Z_s = R_a + jX_s \tag{10.1}$$

So the voltage equation of synchronous motor will be,

$$V_P = E_b + I_a Z_s \tag{10.2}$$

$$\text{or,} I_a = \frac{V_P - E_b}{Z_s} \tag{10.3}$$

Where, V_p is the supply voltage.

10.7.1 Synchronous Motor at No-load

Let us first consider an ideal synchronous motor, that is, a motor having zero losses. It is possible to make the magnitude of V_p equal to E_p by controlling the flux of the rotor. That is,

$$|V_P| = |E_P| \tag{10.4}$$

Under no-load condition, the magnetic axis of the stator and the rotor will coincide with each other. Although magnitude-wise $|V_P| = |E_P|$, they will be in phase opposition as shown in Fig. 10.3.

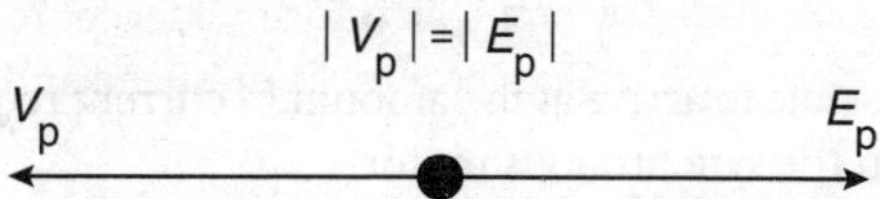

Fig. 10.3: No-load condition

Since the magnitude of V_P and E_P is same and they are in phase opposition as seen from Equation (10.3), we can say, I_a=0. Practically, this condition is not possible.

Not let us see the performance of synchronous motor at no-load after considering the losses in the motor. For instance, when the magnetic locking between the stator and the rotor takes place, the magnetic axis of the stator and the rotor do not coincide. There will be some angular difference (δ) between the stator and the rotor magnetic axis. This angle δ is called the load angle or power angle or torque angle.

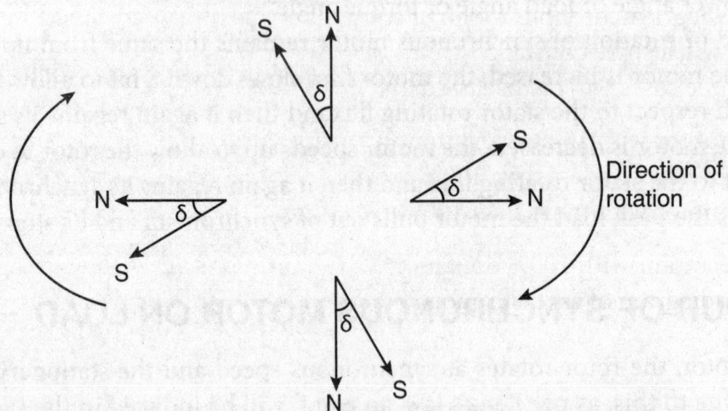

FIG. 10.4: Phasor diagram showing the load angle

The phasor notation, shown in Fig. 10.4, shows that the stator axis leads the rotor axis by an angle δ. Although there is an angular difference between the stator and the rotor magnetic axis, the magnetic locking between them will remain intact and thus the rotor of the synchronous motor will rotate at synchronous speed.

Although $|V_P| = |E_P|$, but now there will now be a phase difference between them, as shown in Fig. 10.5.

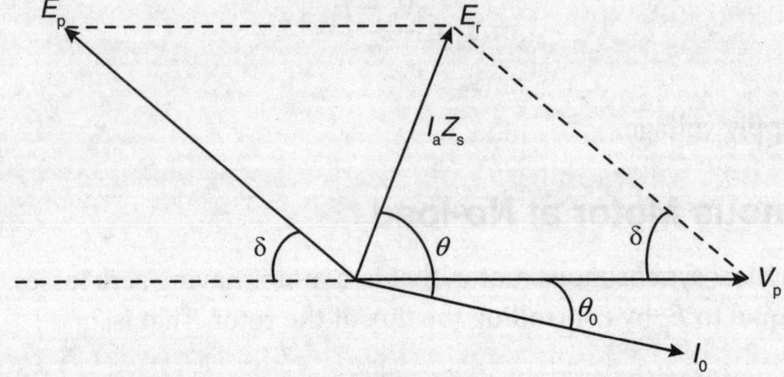

FIG. 10.5: Phase difference between $|V_p|$ and $|E_p|$

From the phase of Fig. 10.5, we can write,

$$V_P - E_P = I_a Z_s = E_r \tag{10.4}$$

From the expression, it is possible to arrive at the amount of current required to produce the required torque to overcome the losses in the synchronous motor.

10.7.2 Synchronous Motor on Load

As the load on the synchronous motor increases, its speed does not change; but its load angle, δ, changes according to the load condition. As the load increases, the angle δ increases and hence the magnitude of the current drawn by the motor increases. Similarly as the load decreases, the angle δ decreases and hence the magnitude of the current drawn by the motor decreases.

10.8 EFFECT OF CHANGE IN EXCITATION ON SYNCHRONOUS MOTOR

Let us consider a loss-less synchronous motor at constant mechanical load. Since the load is constant, the output of the motor will also be constant.

In this section, three cases will be discussed.

i) Synchronous motor with 100% excitation
ii) Synchronous motor with less than 100% excitation
iii) Synchronous motor with more than 100% excitation

Case 1: Synchronous motor with 100% excitation

Consider the phasor diagram of a synchronous motor on load, as shown in Fig. 10.6

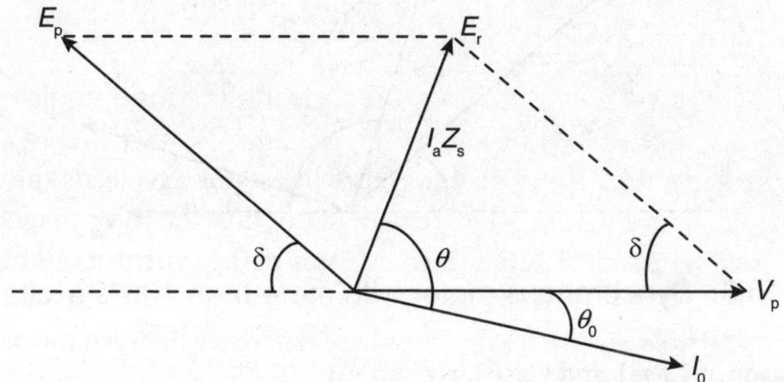

Fig. 10.6: Synchronous motor at 100% excitation

At 100% excitation $|V_P| = |E_P|$ and as indicated in the phasor diagram, current I_0 will lag the voltage V_P by some angle θ_0. Also, I_0 will lag voltage E_r by θ.

Again we can write,

$$\tan\theta = \frac{X_s}{R_a} \tag{10.5}$$

Case 2: Synchronous motor with less than 100% excitation

At excitation less than 100%, $|V_P| > |E_P|$. The phasor is shown in Fig. 10.7.

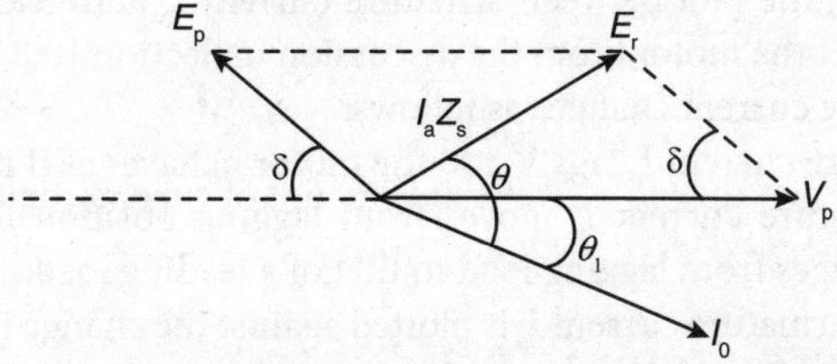

Fig. 10.7: Synchronous motor with less than 100% excitation

The current I_0 will lag the voltage V_P by some angle θ_1. Also, I_0 will lag voltage E_r by θ.
Comparing the phasor of Case1 and Case2, we can say,

- As the magnitude of E_b becomes less than V_P, the resultant phasor, E_r, rotates in the clockwise direction.
- The armature current I_0, will also rotate in the clockwise direction to the same extent, as the angle θ between E_r and I_0. So, the angle θ between E_r and I_0 will not change.
- The phase angle between V_P and I_0 will increase,changing from θ_0 to θ_1,where $\theta_1 > \theta_0$
- Since $\theta_0$increases to θ_1, the power factor will decrease.

Case 3: Synchronous motor with more than 100% excitation
At excitation more than 100%, $|V_P| < |E_P|$. The phasor is shown in Fig. 10.8.

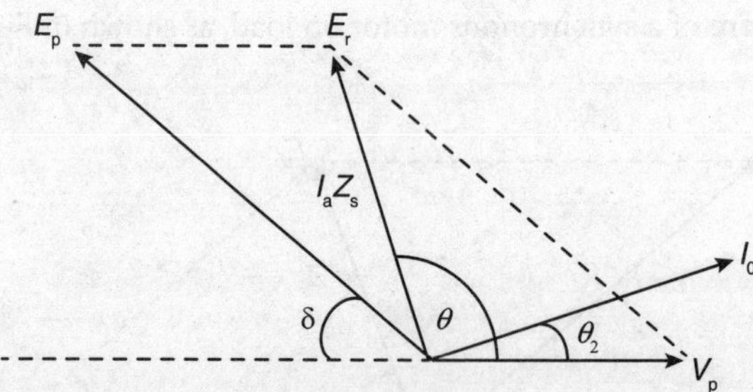

Fig. 10.8: Synchronous motor with more than 100% excitation

Comparing the phasor of Case1 and Case3, we can say,

- As the magnitude of E_b becomes more than V_P, the resultant phasor, E_r, rotates in anti-clockwise direction.
- The armature current I_0 will also rotate in the anti-clockwise direction to the same extent. So, the angle θ between E_r and I_0 will not change.
- The phase angle between V_P and I_0 will now be θ_2.
- In the phasor of Fig. 10.8., it has been shown that current I_0 is leading V_P by θ_2. It is also possible that for a particular value of excitation, V_p and I_0 will be in phase (unity power factor). At this stage, the current drawn by the motor will be minimum.

10.9 V-CURVE AND INVERTED V-CURVE

In the case of constant load, the plot between armature current I_0 and field current I_f of synchronous motor is called the V-curve of the motor.From the discussion in Section 10.8, it is clear that as the excitation is changed, the armature current changes as follows:

When $E_b < V_P$, the armature current I_0 lags V_P; so the motor behaves as if it has an inductive load.

AsE_b increases, the armature current I_a moves from lagging position to leading position; so, the behaviour of the motor changes from lagging load to that of a leading load.

When this magnitude of armature current I_a is plotted against the change in field current I_f at constant load, we get a plot as shown in Fig. 10.9.

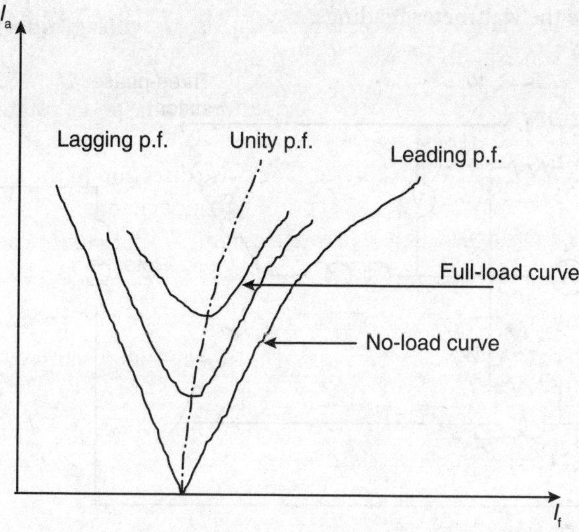

Fig. 10.9: V-curve

Since this plot takes the shape of the alphabet 'V', the plot is called the V-curve.Similarly, if power factor is plotted with change in field current I_f, we get a plot as shownin Fig. 10.10. Since the plot takes the shape of an inverted 'V', it is called the inverted V-curve. This curve is obtained by plotting power factor against the field current I_f under various load conditions.

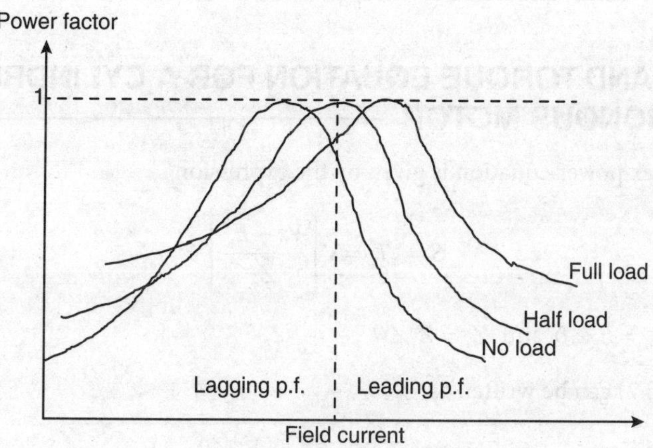

Fig. 10.10: Inverted V-curve

The circuit to obtain the V-curve and the inverted V-curve is shown in Fig. 10.11.

Here, the synchronous motor is assumed to have a three-phase balanced load. The motor is tested by varying the DC excitation in steps. The readings of the two wattmeters, two ammeters and voltmeter with varying DC excitation are tabulated.

The power factor is obtained by the expression

$$\cos\theta = \cos\left[\tan^{-1}\left\{\frac{\sqrt{3}(W_1 - W_2)}{W_1 + W_2}\right\}\right] \tag{10.6}$$

Where, W_1 and W_2 are the wattmeter readings.

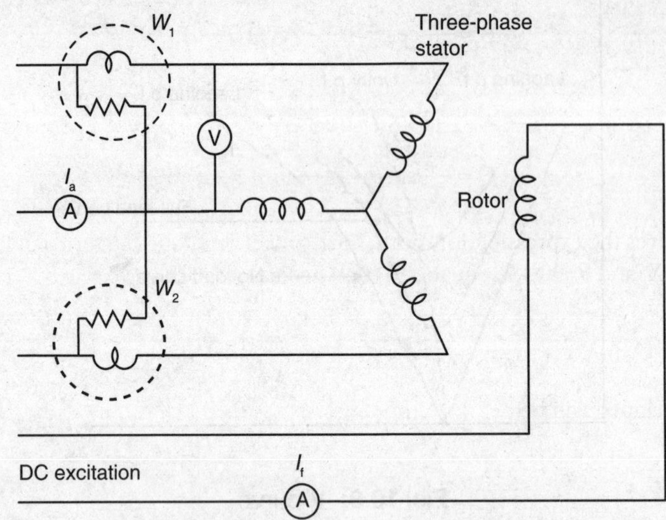

Fig. 10.11: Circuit connection for V-curve and inverted V-curve

Of the two ammeters connected in the circuit, one ammeter reads the armature current I_a and the other gives the field current I_f. The plot between the two currents gives the V-Curve. And the plot between Cos θ (power factor) and field current I_f gives the inverted V-curve.

10.10 POWER AND TORQUE EQUATION FOR A CYLINDRICAL ROTOR SYNCHRONOUS MOTOR

The per-phase complex power equation is given by the expression

$$S = VI_a^* = V\left[\frac{V_P - E_b}{Z_s}\right]^* \tag{10.7}$$

Let, $V_p = V_p\angle 0$ $E_b = E_b\angle\delta$ and $Z_s = Z_s\angle\theta$

Then Equation (10.7) can be written as,

$$S = V_p\angle 0\left[\frac{V_P\angle 0 - E_b\angle -\delta}{Z_s\angle -\theta}\right] = \frac{V_p^2\angle 0}{Z_s\angle -\theta} - \frac{V_P E_b\angle -\delta}{Z_s\angle -\theta}$$

$$= \frac{V_P^2}{Z_s}\angle\theta - \frac{V_P E_b}{Z_s}\angle\theta - \delta \tag{10.8}$$

Again, the complex power S per phase can be expressed as,

$$S = P + jQ$$

So from Equation (10.8), we can write,

$$P = \frac{V_P^2}{Z_s}\cos\theta - \frac{V_P E_b}{Z_s}\cos(\theta - \delta) \tag{10.9}$$

And,

$$Q = \frac{V_P^2}{Z_s}\sin\theta - \frac{V_P E_b}{Z_s}\sin(\theta - \delta)$$

Equation (10.9) gives the expression for mechanical power developed in terms of load angle δ. If R_a is neglected, then $Z_s = X_s$ and θ=90°. So equation (10.9) becomes,

$$P = \frac{V_P E_b}{X_s}\sin\delta \tag{10.10}$$

So the torque developed will be,

$$\text{Torque } (T) = \frac{P}{\omega_s} = \frac{V_p E_b}{\omega_s X_s}\sin\delta \tag{10.11}$$

$$\text{or, Torque } (T) = T_m \sin\delta \tag{10.12}$$

Where,

$$T_m = \frac{V_p E_b}{\omega_s X_s} \tag{10.13}$$

From equation (10.12), we can say that the torque produced by a synchronous motor depends on the load angle δ. As this angle increases, the magnetic locking force between the stator and the rotor decreases. At δ=90°, sin δ = 1, the magnetic locking between the stator and the rotor breaks and the motor loses its synchronism and hence it stops.

So, $T = T_m$, is called the maximum torque or the pull-out torque, beyond which the synchronous motor will stop. The plot between angle, δ and torque T is shown in Fig. 10.12.

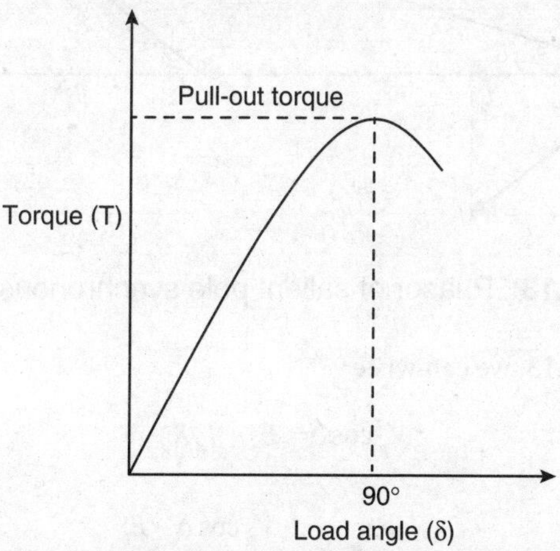

Fig. 10.12: Plot of load angle and torque

Condition for maximum power can be obtained by differentiating equation (10.9) with respect to δ and equating it to zero.

$$\frac{dP}{d\delta} = 0$$

$$\text{or, } \frac{V_p E_b}{Z_s}\sin(\theta - \delta) = 0$$

$$\text{or, } \sin(\theta - \delta) = 0$$

$$\text{or, } \theta = \delta \tag{10.14}$$

Substituting equation (10.14), in equation (10.9) we get,

$$P_{max} = \frac{V_p^2}{Z_s}\cos\delta - \frac{V_p E_b}{Z_s} \tag{10.15}$$

10.11 POWER AND TORQUE EQUATION FOR A SALIENT POLE SYNCHRONOUS MOTOR

A salient pole synchronous motor has a field pole axis, called the direct axis or d-axis and quadrature axis or q-axis. The armature current I_a can also be resolved into its d-axis and q-axis. Let it be I_d and I_q respectively.

The phasor diagram of a salient pole synchronous motor is shown in Fig. 10.13.

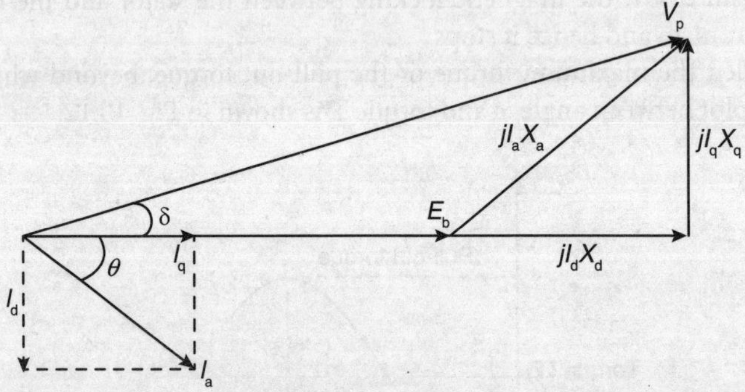

Fig. 10.13: Phasor of salient pole synchronous motor

From the phasor of Fig. 10.13, we can write

$$V_p \cos\delta = E_b + I_d X_d$$

$$\text{or, } I_d = \frac{V_p\cos\delta - E_b}{X_d} \tag{10.16}$$

$$V_p \sin\delta = I_q X_q$$

$$\text{or, } I_q = \frac{V_P \sin\delta}{X_q} \tag{10.17}$$

The power input per phase is given by the expression,

$$P = V_d I_d + V_q I_q \tag{10.18}$$

Where,

$$V_d = -V_P \sin\delta \left(\text{Since } \delta \text{ is negative for a motor}\right) \tag{10.19}$$

$$\text{and } V_q = V_P \cos\delta \tag{10.20}$$

Substituting Equation (10.16), (10.17), (10.19) and (10.20) in Equation (10.18), we get,

$$P = -V_P \sin\delta \times \frac{V_P \cos\delta - E_b}{X_d} + V_P \cos\delta \times \frac{V_P \sin\delta}{X_q}$$

$$= \frac{V_P E_b \sin\delta}{X_d} - \frac{V_P^2 \sin\delta\cos\delta}{X_d} + \frac{V_P^2 \sin\delta\cos\delta}{X_q}$$

$$= \frac{V_P E_b \sin\delta}{X_d} + V_P^2 \sin\delta\cos\delta\left[\frac{1}{X_q} - \frac{1}{X_d}\right]$$

$$= \frac{V_P E_b \sin\delta}{X_d} + V_P^2 \frac{\sin 2\delta}{2}\left[\frac{1}{X_q} - \frac{1}{X_d}\right] \quad (\because \sin 2\delta = 2\sin\delta\cos\delta)$$

$$= \frac{V_P E_b \sin\delta}{X_d} + V_p^2 \frac{\sin 2\delta}{2}\left[\frac{X_d - X_q}{X_q X_d}\right]$$

So the total power will be,

$$P_{Tot} = \frac{3V_P E_b \sin\delta}{X_d} + 3V_P^2 \frac{\sin 2\delta}{2}\left[\frac{X_d - X_q}{X_q X_d}\right] \tag{10.21}$$

The component $3V_P^2 \frac{\sin 2\delta}{2}\left[\frac{X_d - X_q}{X_q X_d}\right]$ is called the reluctance torque and is independent of excitation.

$$\text{Torque}(T) = \frac{P_{tot}}{\omega_s} \tag{10.22}$$

10.12 HUNTING IN SYNCHRONOUS MOTOR

We know that as the load on the synchronous motor changes, its speed does not change, but the load angle δ (angle between the stator and the rotor magnetic axis) changes. At no-load $\delta = 0$, that is the magnetic axis of stator and rotor coincides. But when the motor is loaded, the rotor axis lags the stator axis by an angle δ. If the load is now suddenly changed, then the rotor will try to achieve its new equilibrium position according to the new changed load.

But due to the inertia of the rotor it will not achieve this position at one go. Instead, it will pass beyond its actual position. This will produce more torque than required. So now the rotor will try to reduce the load angle and swing in the opposite direction. Thus, there will be periodic swing of the rotor on either side before finally coming to a stop at the equilibrium position. This oscillation of the rotor due to change in load is called hunting of rotor.

10.12.1 Causes of Hunting in Synchronous Motor

Some of the causes of hunting in synchronous motor are as follows:

1. Sudden change in load
2. Sudden change in field current
3. A load containing harmonic torque
4. Fault in supply system.

10.12.2 Effects of Hunting in Synchronous Motor

The effects of hunting are:

1. It may lead to loss of synchronism.
2. It produces mechanical stresses.
3. Increases machine loss and causes temperature rise.
4. Causes greater surges in current and power flow.

10.12.3 Reduction of Hunting in Synchronous Motor

Two techniques can be used to reduce hunting.

i) *By using damper winding:* Damper winding damps out hunting by producing torque opposite to slip of rotor. The magnitude of damping torque is proportional to the slip speed.
ii) *By using flywheels:* By providing large and heavy flywheel to the prime mover, its inertia can be increased, which in turn, helps in maintaining the rotor speed constant.

10.13 APPLICATION OF SYNCHRONOUS MOTOR

1. Synchronous motor having no load connected to its shaft is used for power factor improvement.
2. Synchronous motor finds application where operating speed is less and high power is required.
3. As synchronous motor is capable of operating under either leading or lagging power factor, it can be used for power factor improvement. A synchronous motor under no-load with leading power factor is connected in a power system where static capacitors cannot be used.
4. It is used where high power at low speed is requiredsuch as rolling mills, chippers, mixers, pumps, pumps, compressors etc.

10.14 DIFFERENCE BETWEEN SYNCHRONOUS AND INDUCTION MOTOR

1. Synchronous motors operate at synchronous speed, while induction motors operate at less than synchronous speed. Slip is nearly zero at zero-load torque and increases as load torque increases.
2. Synchronous motors require DC excitation to be supplied to the rotor windings, induction motors do not require DC excitation.
3. Synchronous motors require a DC power source for rotor excitation.
4. Synchronous motors require slip-rings and brushes to supply rotor excitation. Induction motors do not require slip-rings, but some induction motors have them for soft starting or speed control.
5. Synchronous motors require rotor windings while induction motors are most often constructed with conduction bars in the rotor that are shorted together at the ends to form a "squirrel cage."
6. Synchronous motors require a starting mechanism in addition to the mode of operation that is in effect,on reaching synchronous speed. Three phase induction motors can start by simply applying power, but single phase motors require an additional starting circuit.
7. The power factor of a synchronous motor can be adjusted to be lagging, unity or leading while induction motors must always operate with a lagging power factor.
8. Synchronous motors are generally more efficient than induction motors.
9. Synchronous motors can be constructed with permanent magnets in the rotor eliminating slip-rings, rotor windings, DC excitation system and power factor adjustability.
10. Synchronous motors are usually built only in sizes larger than about 1000 HP (750 kW) because of their cost and complexity. However, permanent magnet synchronous motors and electronically controlled permanent synchronous motors called brushless DC motors are available in smaller sizes.

SOLVED PROBLEMS

Ex.10.1 The terminal voltage and the current of a three-phase, star-connected, 50 Hz, synchronous motor is 440 V, 110 Amps respectively. The power factor is lagging by 0.9. If the synchronous reactance is 2 Ω, calculate the following:

(i) Generated voltage
(ii) Input power
(iii) Current required for unity power factor

Solution

$$V_p = \frac{440}{\sqrt{3}} = 254.03\text{V}$$

$$\cos\theta = 0.9$$

$$\theta = \cos^{-1}0.9 = 25.84^0$$

$$\therefore I_{\text{line}} = 110\angle -25.84\text{A} = 99 - j48\text{A}$$

$$\therefore \text{Generated voltage}\left(E_g\right) = V_p - jXI_{line}$$

$$= 254.03 - j2 \times \left(99 - j48\right)$$

$$= 158.03 - j198\,\text{V} = 253.33\angle -51.41\,\text{V}$$

$$\text{Three-phase power input}\left(P_{in}\right) = 3V_P I_{line} \cos\theta$$

$$= 3 \times 254.03 \times 110 \times 0.9 = 75.45\,\text{KW}$$

$$\text{Current for unity power factor}\left(I\right) = \frac{75.45 \times 10^3}{3V_P} = \frac{75.45 \times 10^3}{\sqrt{3} \times 254.03} = 99\,\text{A}$$

Ex.10.2 Consider a four-pole, 50 Hz, 440 V, synchronous motor. At full load the current is 45 A, with a power factor of unity. Calculate the following:

(i) Input power
(ii) Synchronous speed
(iii) Output torque
(iv) Field current at 0.8 power factor

Solution

$$\text{Synchronous speed} = \left(N_s\right) = \frac{120f}{P} = \frac{120 \times 50}{4} = 1500\,\text{r.p.m}$$

$$\text{Input power}\left(P_{in}\right) = \sqrt{3}V_L I_L \cos\theta = \sqrt{3} \times 440 \times 45 \times 1 = 34.29\,\text{KW}$$

$$\text{Output power}\left(P_{out}\right) = \frac{34.29}{1500 \times \dfrac{2\pi}{60}} = 218.3\,\text{Nm}$$

$$\text{Field current at 0.8 power factor} = \frac{34.29 \times 10^3}{\sqrt{3} \times 440 \times 0.8} = 56.24\,\text{A}$$

Ex.10.3 Consider a 500 V, 50 Hz, 300 KW, 0.9 power factor, 6-pole, delta-connected synchronous motor. It has synchronous reactance of 2 Ω. Calculate the following:

(i) e.m.f.
(ii) Torque
(iii) Torque angle

Solution

$$\text{Synchronous speed}\left(N_s\right) = \frac{120f}{P} = \frac{120 \times 50}{6} = 1000\ \text{r.p.m.}$$

$$I_L = \frac{300 \times 10^3}{\sqrt{3} \times 500 \times 0.9} = 385\text{A}$$

$$I_P = \frac{385}{\sqrt{3}} = 222.28\,\text{A}$$

$$\theta = \cos^{-1} 0.9 = 25.8$$

$$\text{so}, I_p = 222.28\angle -25.8\,\text{A} = 200 - j96.7\,\text{A}$$

$$\text{Generated e.m.f.} = 500 - j2(200 - j96.7) = (306.6 - j400)\text{V} = 503.99\angle -52.53\,\text{V}$$

$$\text{T} = \frac{300 \times 10^3}{1000 \times \frac{2\pi}{60}} = 2864.79\,\text{Nm}$$

$$\text{Torque angle} = -52.54^0$$

Ex.10.4 Consider a 500 V, 50 Hz, 300 KW, 0.9 power factor, 6-pole, star-connected synchronous motor. It is having armature resistance and synchronous reactance of 0.09 Ω and 1 Ω respectively. Its efficiency at full load is 94%. Calculate the following:

(i) Synchronous speed
(ii) Torque
(iii) Input power
(iv) e.m.f.

Solution

$$\text{Synchronous speed}(N_s) = \frac{120f}{P} = \frac{120 \times 50}{6} = 1000\,\text{r.p.m.}$$

$$T = \frac{300 \times 10^3}{1000 \times \frac{2\pi}{60}} = 2864.79\,\text{Nm}$$

$$\text{Input power}(P_{\text{in}}) = \frac{300 \times 10^3}{0.94} = 319.15\text{KW}$$

$$V_P = \frac{500}{\sqrt{3}} = 288.67\,\text{A}$$

$$I_L = \frac{300 \times 10^3}{\sqrt{3} \times 500 \times 0.9} = 385\,\text{A}$$

$$\text{Generated e.m.f.} = 288.67 - 385 \times 0.09 - j385 \times 1$$

$$= 254.02 - j385\,\text{V} = 461.25\angle -56.58\,\text{V}$$

Ex.10.5 Consider a three-phase star-connected 400 V, synchronous motor with synchronous reactance of 2 Ω per phase. The power supplied is 100 KW and the torque angle is 20⁰. Obtain the generated voltage.

Solution

$$V_L = 400\,\text{V}$$

$$V_p = \frac{400}{\sqrt{3}} = 230.9\,\text{V}$$

$$\text{Generated e.m.f.} = \frac{2 \times 100 \times 10^3}{3 \times 230.9 \times \sin 20} = 844\,\text{V}$$

$$\text{Armature current}(I) = \frac{230.9\angle 0 - 844\angle -20}{j2} = (144 + j281)\,\text{A} = 315.74\angle 62.87\,\text{A}$$

Ex.10.6 A 1492 kW,unity power factor,three-phase,star-connected, 2300 V, 50 Hz, synchronous motor has a synchronous reactance of 1.95 Ω/phase. Compute the maximum torque in N-m, which this motor can deliver if it is supplied from a constant frequency source and if the field excitation is constant at the value which would result in unity power factor at rated load. Assume that the motor is of cylindrical rotor type. Neglect all losses.

Solution

$$\text{Rated KVA per phase} = \frac{1492}{3} = 497.33\,\text{KVA}$$

$$\text{Rated voltage per phase} = \frac{2300}{\sqrt{3}} = 1327.9\,\text{V}$$

$$\text{Rated current per phase} = \frac{497.33 \times 10^3}{1327.9} = 374.52\,\text{A}$$

$$E = \sqrt{497.33^2 + (374.52 \times 1.95)^2} = 1515.4\text{V}$$

$$P_m = \frac{1515.4 \times 1327.9}{1.95} = 1032 \text{ kW per phase}$$

$$\text{Maximum torque} = \frac{1032 \times 10^3}{2\pi f} = \frac{1032 \times 10^3}{314.15} = 3285 \text{ Nm per phase}$$

Ex.10.7 A three-phase, star-connected 6.6 KV synchronous motor, with direct and qudrature axis synchronous reactance of 10 is running in parallel with an infinite bus. Calculate the maximum load, armature current and maximum power when the field current is reduced to zero.

Solution

$$P_{max} = 3 \times \left(\frac{1}{2} \times \left(\frac{6.6 \times 10^3}{\sqrt{3}} \right)^2 \times \left(\frac{1}{5} - \frac{1}{10} \right) \right) = 2178\,\text{KW}$$

$$I_d = \frac{6.6 \times 10^3}{\sqrt{3}} \times \frac{\cos 45}{10} = 269.45\,\text{A}$$

$$I_q = \frac{6.6 \times 10^3}{\sqrt{3}} \times \frac{\sin 45}{10} = 538.9\,\text{A}$$

$$\text{Armature current}(I) = \sqrt{269.45^2 + 538.9^2} = 602.5\,\text{A}$$

Ex.10.8 A 100-HP, 440V, 0.8-PFleading, Δ-connected synchronous motor has an armature resistance of 0.22 Ω and a synchronous reactance of 3.0 Ω. Its efficiency at full load is 89 %.

(a) What is the input power to the motor at rated conditions?
(b) What is the line current of the motor at rated conditions? What is the phase current ofthe motor at rated conditions?
(c) What is the reactive power consumed by or supplied by the motor at rated conditions?
(d) What is the generated voltage of this motor at rated conditions?
(e) What are the stator copper losses in the motor at rated conditions?
(f) What is P_{conv} at rated conditions?
(g) If generated voltage is decreased by 10 percent, how much reactive power will be consumed by orsupplied by the motor?

Solution

$$\cos\theta = 0.8$$

$$\theta = \cos^{-1} 0.8 = 36.86$$

$$P_{in} = \frac{100 \times 746}{0.89} = 83.8\,\text{KW}$$

$$I_L = \frac{83.8}{\sqrt{3} \times 440 \times 0.8} = 137\,\text{A}$$

$$I_p = \frac{I_L}{\sqrt{3}} = \frac{137}{\sqrt{3}} = 79.4\,\text{A}$$

$$\text{Reactivepower} = 3 \times 440 \times 79.4 \times \sin 36.87) = 62.9\text{KVAR}$$

$$\text{Generatedvoltage} = 440 - 0.22 \times 79.4\angle 36.87 - j3 \times 79.4\angle 36.87 = 603\angle -19.5\text{V}$$

$$\text{Copperloss} = 3 \times 79.4^2 \times 0.22 = 4.16\text{KW}$$

$$P_{conv} = 83.8 - 4.16 = 79.6\text{KW}$$

$$\text{Modifiedgeneratedvoltage} = 0.9 \times 603 = 543\text{V}$$

Ex.10.9 A three-phase, 60 Hz 1200 HP, 2300 V, synchronous machine has $R_s = 0.2\ \Omega$ and $X_s = 5.6\ \Omega$. When operated with the mechanical load disconnected with rated voltage and frequency, the field current is adjusted until line current has a minimum value. At this point, $I_a = 22.1$ A, the measured input power is $P_T = 17.5$ kW, and the measured field circuit values are $V_f = 276$ V and $I_f = 53.2$ A. Determine the following:

(a) The rotational losses.
(b) The power factor, line current, and efficiency, if the field current, frequency, and impressed terminal voltage are unchanged, but a mechanical load of600 HP is attached.

Solution

$$P_{FW} = P_T - 3 \times I_a^2 \times R_s = 17500 - 3 \times 22.1 \times 22.1 \times 0.2 = 17206.95\text{W}$$

$$\bar{E}_f = \bar{V}_{an} - \bar{I}_a\left(R_s + jX_s\right) = \frac{2300}{\sqrt{3}} - 22.1(0.2 + j5.6) = 1329.28\angle -5.34°\ \text{V}$$

$$\delta = \sin^{-1}\left[\frac{\left(P_s + P_{FW}\right)X_s}{3V_{an}E_f}\right] = \sin^{-1}\left[\frac{(600 \times 746 + 17{,}793.05)(5.6)}{3\left(2300/\sqrt{3}\right)(1329.28)}\right] = 31.84°$$

$$\bar{I}_a = \frac{\bar{V}_{an} - \bar{E}_f}{R_s + jX_s} = \frac{\frac{2300}{\sqrt{3}}\angle 0° - 1329.28\angle -31.84°}{0.2 + j5.6} = 130.07\angle -13.77°\ \text{A}$$

$$PF_{in} = \cos\left(\angle\bar{V}_{an} - \angle\bar{I}_a\right) = \cos(11.35°) = 0.97\ \text{lagging}$$

$$\text{Losses} = P_{FW} + 3I_a^2R_a + V_fI_f$$

$$\text{Losses} = 17{,}793.05 + 3(130.07)^2(0.2) + (276)(53.2) = 42{,}627.17\ \text{W}$$

$$\eta = \frac{100P_s}{P_s + \text{Losses}} = \frac{100(600 \times 746)}{600 \times 746 + 42627.17} = 91.30\%$$

Ex.10.10 If the synchronous motor of Problem Ex.10.9 still operates the 600 HP load with terminal voltage and frequency unchanged, and the field current increased to 80 A, determine

(a) the input power factor
(b) the leading VARs supplied to the three-phase grid.

Assume that core losses are valid.

Solution

$$E_{\text{f}} = \frac{80}{53.2}(1329.28) = 1998.92 \text{ V}$$

$$\delta = \sin^{-1}\left[\frac{(P_{\text{s}} + P_{\text{FW}})X_{\text{s}}}{3V_{\text{an}}E_{\text{f}}}\right] = \sin^{-1}\left[\frac{(600\times 746 + 17{,}793.05)(5.6)}{3(2300/\sqrt{3})(1998.92)}\right] = 19.10°$$

$$\bar{I}_{\text{a}} = \frac{\bar{V}_{\text{an}} - \bar{E}_{\text{f}}}{R_{\text{s}} + jX_{\text{s}}} = \frac{\frac{2300}{\sqrt{3}}\angle 0° - 1998.92\angle -19.10°}{0.2 + j5.6} = 153.45\angle 42.86° \text{ A}$$

$$PF_{\text{in}} = \cos\left(\angle\bar{V}_{\text{an}} - \angle\bar{I}_{\text{a}}\right) = \cos(-42.86°) = 0.733 \text{ leading}$$

$$Q_{\text{T}} = \sqrt{3}V_{\text{L}}I_{\text{a}}\sin\theta = \sqrt{3}(2300)(153.45)\sin(-42.86°)$$

$$Q_{\text{T}} = 415.80 \text{ KVARs}$$

Ex.10.11 If the synchronous motor of Problem Ex.10.9 now supplies rated power to a coupledmechanical load with rated voltage and frequency applied to the stator and $I_{\text{f}} = 80$ A, determine:

(a) Line current
(b) Efficiency.

Assume that R_s can be neglected in torque angle.
Also assume that core losses have changed negligibly.

Solution

The excitation voltage for $I_{\text{f}} = 80$ A is determined in Problem Ex. 10.10 to be

$$\bar{I}_{\text{a}} = \frac{\bar{V}_{\text{an}} - \bar{E}_{\text{f}}}{R_{\text{s}} + jX_{\text{s}}} = \frac{\frac{2300}{\sqrt{3}}\angle 0° - 1998.02\angle -39.97°}{0.2 + j5.6} = 229.46\angle 11.75° \text{ A}$$

$$R_{\text{f}} = \frac{V_{\text{f}}}{I_{\text{f}}} = \frac{276}{53.2} = 5.19\ \Omega$$

$$\text{Losses} = P_{\text{FW}} + 3I_{\text{a}}^2R_{\text{s}} + I_{\text{f}}^2R_{\text{f}}$$

$$\text{Losses} = 17{,}793.05 + 3(229.46)^2(0.2) + (80)^2(5.19) = 82{,}600.18 \text{ W}$$

$$\eta = \frac{100P_s}{P_s + \text{losses}} = \frac{100(1200\times746)}{1200\times746 + 82{,}600.18} = 91.55\%$$

Ex.10.12 The synchronous machine of Problem Ex.10.9 is now operated as a 2300 V, 60 Hz alternator supplying 800 kW to a 0.8 PF lagging load. Field current $I_f = 53.2$ A. Determine the required input mechanical power to drive the rotor and coupled exciter if the exciter has an efficiency of 91% when supplying the field voltage and current given in Problem Ex.9.9.

Solution

$$\bar{I}_a = \frac{P_T}{\sqrt{3}V_L PF} = \frac{800{,}000}{\sqrt{3}(2300)(0.8)} = 251.02 \text{ A}$$

$$P_s = P_T + 3I_a^2R_s + P_{FW} + V_fI_f/\eta_{ex}$$

$$P_s = 800{,}000 + 3(251.02)^2(0.2) + 17{,}793.05 + (276)(53.2)/0.91$$

$$P_s = 871.73 \text{ kW}$$

Ex.10.13 A three-phase synchronous motor has a synchronous reactance of 10 Ω per phase at 1000V and 9 KW. Calculate the induced e.m.f. at full load if the efficiency of the motor is 90%. Assume the power factor to be unity.

Solution

$$\text{Input power} = \frac{9000}{0.9} = 10\text{KW}$$

$$\text{Current}(I) = \frac{10000}{\sqrt{3}\times1000\times1} = 5.77\text{A}$$

$$\text{e.m.f.} = \sqrt{\left(\frac{1000}{\sqrt{3}}\right)^2 + (5.77\times10)^2} = 580.23\text{V per phase}$$

Ex.10.14 A three-phase, star-connected 6.6 KV synchronous motor is running in parallel with an infinite bus. Its direct and quadrature axis synchronous reactances are 10 Ω and 5 Ω respectively. Find the maximum load that can be applied to the motor if the field current is made zero. Also obtain the armature current at maximum power.

Solution

$$P_m(\text{per phase}) = \left(\frac{6.6\times10^3}{\sqrt{3}}\right)^2 \times \frac{1}{2}\times\left(\frac{1}{5}-\frac{1}{10}\right) = 726\text{KW}$$

At maximum power, $\delta - = 45$

So,

$$I_d = \frac{6.6\times10^3\times\cos 45}{10\times\sqrt{3}} = 269.45\,\text{A}$$

$$I_q = \frac{6.6\times10^3\times\sin 45}{10\times\sqrt{3}} = 538.9\,\text{A}$$

$$\text{Current}(I) = \sqrt{269.45^2 + 538.9^2} = 602.5\text{A}$$

Ex.10.15 A synchronous motor takes 40 KW and it is connected in parallel with a load of 300KW with 0.85 p.f. lag. How much leading reactive KVA should be supplied by the synchronous motorif the combined load should have a lagging power factor of 0.9? What would be the power factor of the synchronous motor?

Solution

$$\text{Total power} = 300 + 40 = 340\ \text{KW}$$

$$\theta_1 = \cos^{-1}0.85 = 31.79$$

$$\theta_2 = \cos^{-1}0.9 = 25.84$$

$$\text{Initial reactive KVA} = 300\times\tan 31.79 = 185.92\ \text{KVA}$$

$$\text{KVA at 0.9 p.f.} = 340\times\tan 25.84 = 164.67\ \text{KVA}$$

$$\text{Thus the p.f.of the synchronous motor will be} = \cos\left(\tan^{-1}\left(\frac{185.92-164.67}{40}\right)\right) = 0.883$$

Ex.10.16 Determine the current drawn by a three-phase star-connected synchronous motor, when connected across a 500V supply with a net output of 17 KW, 0.9 p.f. lag, at full load. Take losses to be 1300W and armature resistance to be 0.8 Ω. What will be its efficiency at full load?

Solution

$$V_p = \frac{500}{\sqrt{3}} = 288.68\,\text{V}$$

$$P_m = 17000 + 1300 = 18300\,\text{W}$$

Again

$$3\times288.68\times I_a\times0.9 - 3\times I_a^2\times0.8 = 18300$$

Solving the quadratic equation we will get two values of I_a, the lower values of I_a has been considered. So

$$I_a = 25.48\text{A}$$

Then,

$$P_{inp} = 3 \times 288.68 \times 25.48 \times 0.9 = 19857\,\text{W}$$

$$\eta = \frac{17000}{19857} \times 100 = 85.6\%$$

Ex.10.17 A three-phase salient-pole synchronous motor, 150 KW, 2300 V, 50 Hz and 1000 r.p.m. has the direct axis and quadrature axis synchronous reactance as 32 Ω and 20 Ω per phase respectively. If the field excitation is adjusted in such a way that the back e.m.f. is twice the applied voltage, what will be the torque developed by the motor. Take $\alpha = 16$

Solution

$$V_p = \frac{2300}{\sqrt{3}} = 1328\,\text{V}$$

$$\text{and } E_b = 2 \times V_p = 2 \times 1328 = 2656\,\text{V}$$

$$P_e = \frac{2656 \times 1328}{32} \times \sin 16 = 30382\,\text{W}$$

$$P_r = \frac{1328^2 \times (32 - 20)}{2 \times 32 \times 20} \times \sin 2 \times 16 = 8760\,\text{W}$$

$$\text{Total power} = 30382 + 8760 = 117425\,\text{W}$$

$$\text{Torque} = 9.55 \times \frac{117425}{1000} = 1120\,\text{Nm}$$

Ex.10.18 A single-phase 500 V, synchronous motor has a mechanical power output of 7460 W at 0.9 p.f. lag. The iron and frictional losses are 500 W and 800 W respectively. What is the armature current and efficiency? Take the armature resistance to be 0.8 Ω

Solution

$$P_m = 7460 + 500 + 800 = 8760\text{W}$$

$$I = \frac{500 \times 0.9 \pm \sqrt{(500 \times 0.9)^2 - 4 \times 0.8 \times 8760}}{2 \times 0.8}$$

Solving we get,

$$\text{I} = 20.2 \text{ A}$$

$$\eta = \frac{7460}{500 \times 20.2 \times 0.9} \times 100 = 82.06\%$$

Ex.10.19 A synchronous motor supplies a load of 1000 KW at 0.5 power factor lag. Find the value of the synchronous condenser that is needed to be added to raise the power factor to 0.9.

Solution

Without the synchronous condenser

$$\theta_1 = \cos^{-1} 0.5 = 60^0$$

$$S_1 = \frac{1000}{0.5} = 2000\,\text{KVA}$$

$$Q_1 = 2000 \times \sin 60 = 1732\,\text{KVAR}$$

With synchronous condenser

$$\theta_2 = \cos^{-1} 0.9 = 25.84^0$$

$$S_2 = \frac{1000}{0.9} = 1111.11\,\text{KVA}$$

$$Q_1 = 1111.11 \times \sin 25.84 = 484.29\,\text{KVAR}$$

So the rating of the condenser will be = 1732 – 484.29 = 1247.7 KVAR

Ex.10.20 A 2000 KW, 1500 V, 20-pole, 50 Hz synchronous motor has reactance of 2 Ω per phase. Calculate the maximum torque under unity power factor at rated load.

Solution

$$V_p = \frac{1500}{\sqrt{3}} = 866.03\text{V}$$

$$I = \frac{2000 \times 10^3}{\sqrt{3} \times 1500 \times 1} = 769.8\text{A}$$

$$\text{e.m.f.}(E) = \sqrt{866.03^2 + (769.8 \times 2)^2} = 1766.46\,\text{V}$$

$$P_m = \frac{3 \times 866.03 \times 1766.46}{2} = 2294.71\,\text{KW}$$

$$T_m = \frac{2294.71 \times 10^3}{\frac{2 \times \pi \times 2 \times 50}{20}} = 73042.89\,\text{Nm}\,.$$

EXERCISES

1. Give the construction and principle of operation of a synchronous motor.
2. How can you change the direction of rotation of a synchronous motor?
3. List the different starting methods for synchronous motor.
4. Explain the following starting methods:
   (i) By using a pony motor (Small induction motor)
   (ii) By using a damper winding
   (iii) By using DC motor
   (iv) Starting as an induction motor
5. Define power angle and developed torque of a synchronous motor
6. With the help of phasor, explain the behaviour of a synchronous motor
   (i) On no-load
   (ii) On load
7. Explain the operation of a synchronous motor under the following conditions:
   (i) With 100% excitation
   (ii) With less than 100% excitation
   (iii) With more than 100% excitation
8. What is V-curve and inverted V-curve?
9. Draw and explain the operation of the circuit to obtain the V-curve and the inverted V-curve.
10. Derive the power and torque equation for a cylindrical rotor synchronous motor
11. Derive the power and torque equation for a salient-pole synchronous motor.
12. What is hunting in synchronous motor?
13. What are the causes and effect of hunting?
14. How can hunting be reduced in synchronous motor?
15. Give some application of synchronous motor.
16. Differentiate between synchronous and induction motor.
17. The terminal voltage and the current of a three-phase, star-connected, 50 Hz, synchronous motor is 400V, 100 Amps respectively. The power factor lags by 0.8. If the synchronous reactance is 2.5Ω, calculate the following:
   (i) Generated voltage
   (ii) Input power
   (iii) Current required for unity power factor
18. Consider a four-pole, 50 Hz, 400V, synchronous motor. At full load, the current is 35A, at 0.9 power factor. Calculate the following:
   (i) Input power
   (ii) Synchronous speed
   (iii) Output torque
   (iv) Field current at 0.8 power factor

CHAPTER 11

SPECIAL MACHINES

11.1 INTRODUCTION

Conventional electrical machines are generally used for bulk energy conversion. However, there are some electrical machines that are used for specific applications. These machines are increasingly used in position control systems, robotics and mechatronics, electric vehicles and high-speed transportation. This chapter gives an introduction to some of these special machines.

11.2 STEPPER MOTOR

This motor rotates at fixed angular steps in response to input current pulse. When a series of pulses are supplied, the motor rotates through a definite known angle. The angle through which the motor shaft rotates for each pulse input is called a step angle. It is denoted by the symbol θ. The value of the step angle θ is obtained by,

$$\theta = \frac{P_s - P_r}{P_s P_r} \times 360 \tag{11.1}$$

Or

$$\theta = \frac{360}{mP_r} \times 360 \tag{11.2}$$

Where,
P_s is the number of stator poles
P_r is the number of rotor teeth
m is the number of stator phases

11.2.1 Classification of Stepper Motor

Stepper motors are classified as follows:

i) Variable reluctance stepper motor
ii) Permanent magnet stepper motor
iii) Hybrid stepper motor

11.2.2 Variable Reluctance Stepper Motor

It has a salient pole stator with concentric winding, which forms the different phases of the stator. Its rotor has no winding nor is it a permanent magnet. It is simply a salient pole structure with teeth in it.

Variable reluctance stepper motor is of two types

i) Single-stack stepper motor
ii) Multiple-stack stepper motor

Single-stack stepper motor

The motor is constructed as a single unit. The stator have an even number of poles and the windings of opposite poles are connected in series to form one phase. The windings connected in this way forms different phases of the motor. When current flows through these windings it produces a magnetic flux that aids the windings.

Rotor is made up of laminated silicon steel or solid silicon steel. In general, the material used for stator or rotor must have high permeability. The number of stator poles and the rotor teeth should not be equal. This will make the motor self starting.

Consider a six-pole stator with four rotor teeth as shown in Fig.11.1.

The position of the rotor before the excitation of the stator windings is shown in Fig.11.1. The different phases of the stator are excited with the help of a switch as shown in Fig.11.2. Let the switch S_1 be first closed to energise the coil AA′ of the stator (Fig.11.4). This will magnetise the stator pole AA′ and pull the rotor teeth 1 and 3, thereby moving the rotor to a position of minimum reluctance. The new position of the rotor is shown in Fig.11.3

Now, when the switch S_2 is closed (S_1 open now), the stator pole BB′ gets energised (Fig.11.6). This will attract the rotor teeth 2 and 4 to a position of minimum reluctance. Hence, the rotor will move anti-clockwise as shown in Fig.11.5.

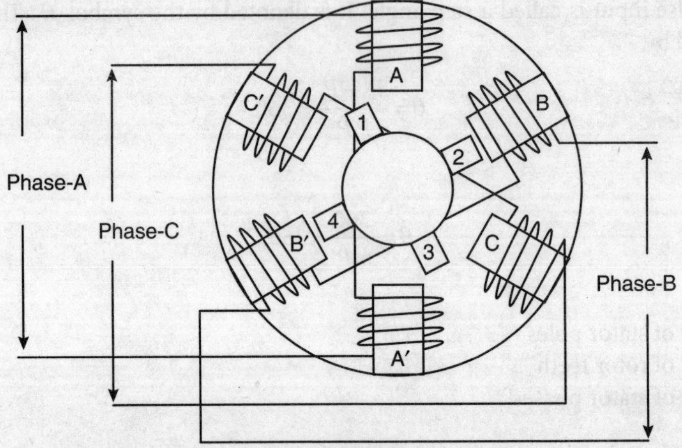

Fig. 11.1: Six-pole stator with four rotor teeth

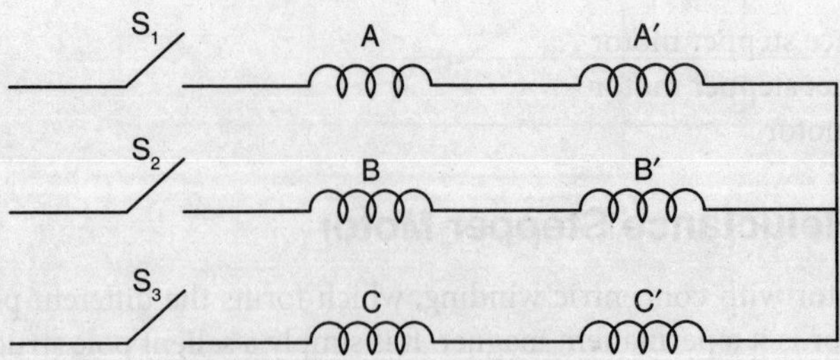

Fig.11.2: Switch connection

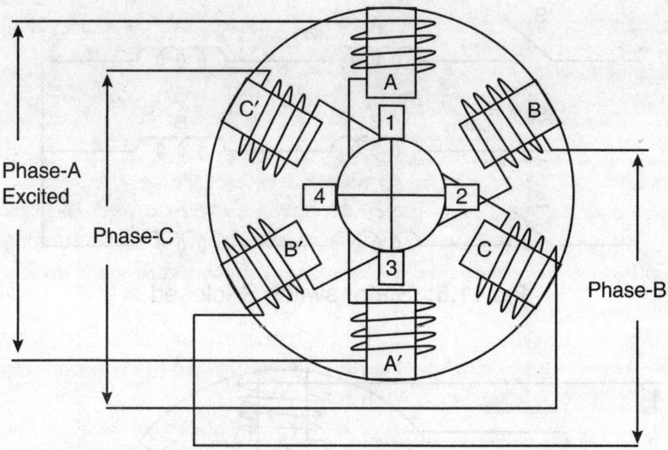

FIG. 11.3: Rotor position when switch S_1 is closed

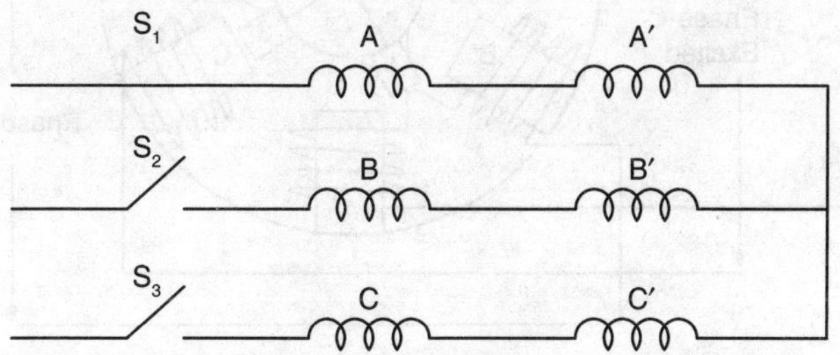

FIG.11.4: Stator switch S_1 closed

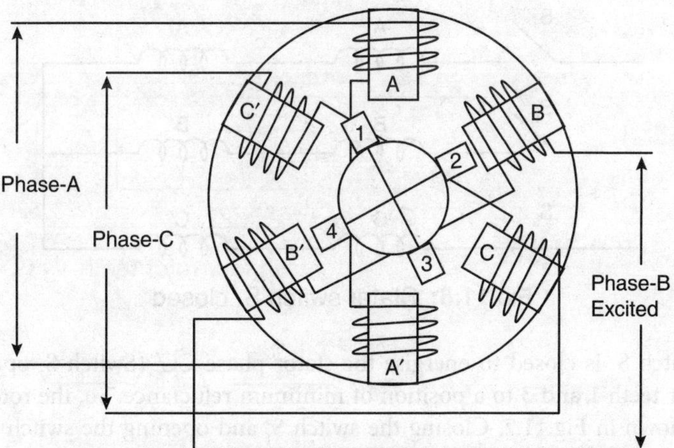

FIG.11.5: Rotor position when switch S_2 is closed

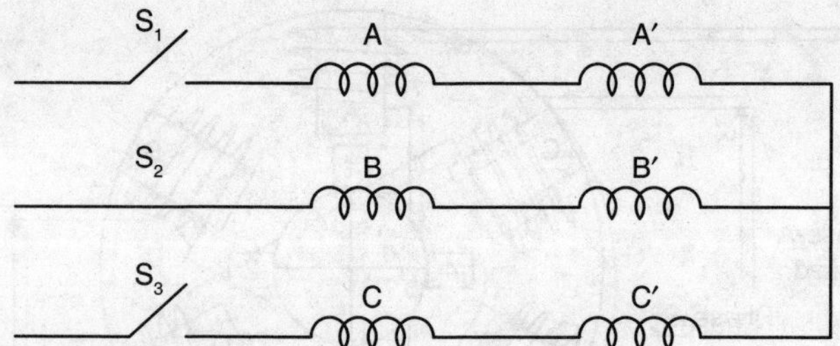

Fig.11.6: Stator switch S_2 closed

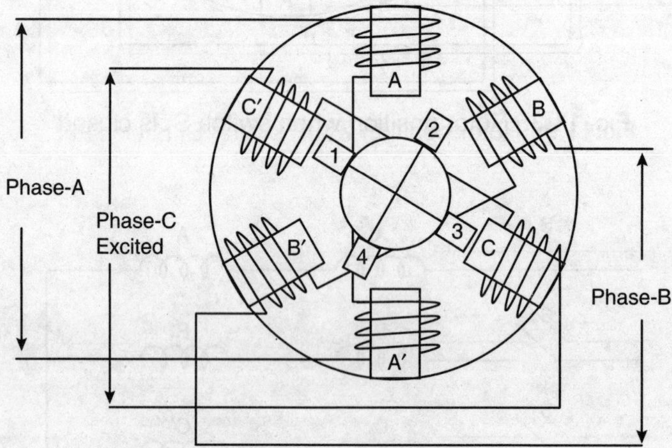

Fig.11.7: Rotor position when switch S_3 is closed

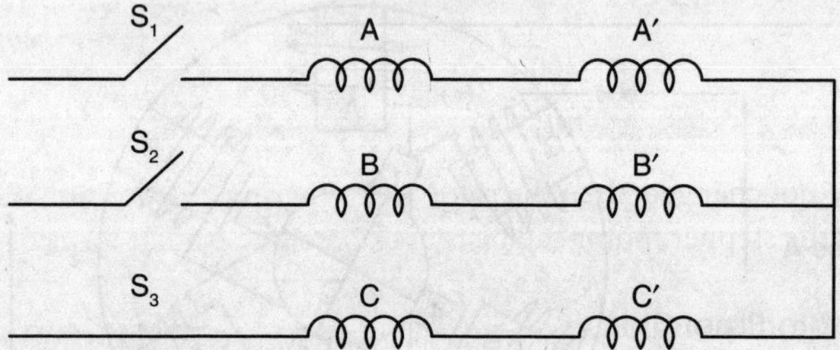

Fig.11.8: Stator switch S_3 closed

Again, when switch S_3 is closed to energise the stator phase CC′ (Switch S_2 open) (Fig.11.8), this will attract the rotor teeth 1 and 3 to a position of minimum reluctance. So, the rotor will again move anti-clockwise as shown in Fig.11.7. Closing the switch S_1 and opening the switch S_3 (Fig.11.10), will attract the rotor teeth 2 and 4 to a position of minimum reluctance, thus rotating the motor again in anti-clockwise direction (Fig.11.9).

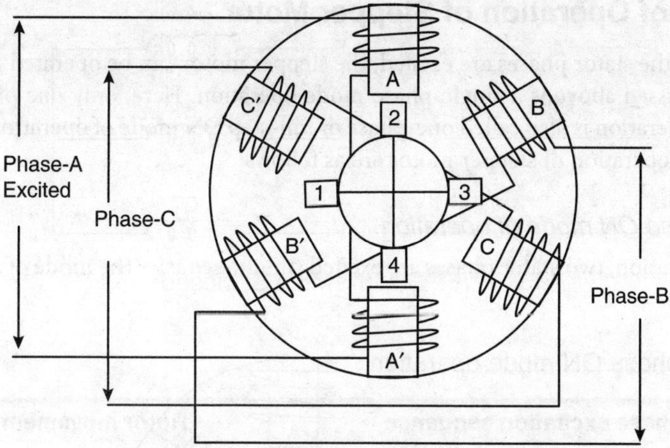

Fig.11.9: Rotor position when switch S_1 is closed

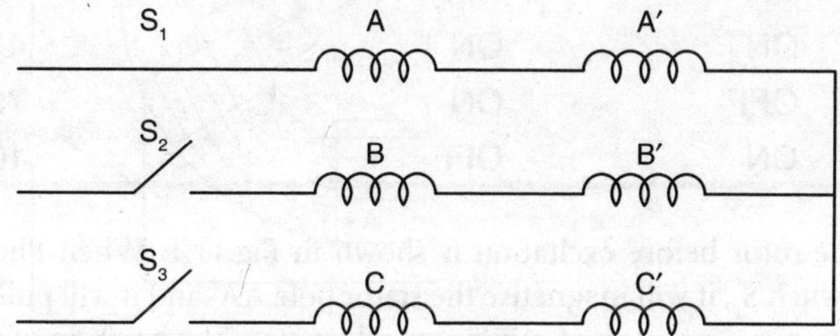

Fig.11.10: Stator switch S_1 closed

Hence, each excitation of the stator phases will rotate the stepper motor by some angle θ in the anti-clockwise direction. The value of θ can be obtained from the equation,

$$\theta = \frac{360}{mP_r} \times 360 = \frac{360}{3 \times 4} = 30^0 \tag{11.3}$$

If the stator phases are not changed, then the rotor will remain in the position of the last closed stator switch. This operation of the stepper motor can be represented in a tabular form, as shown in Table.11.1.

Table 11.1 Stepper motor operation

Stator phase excitation sequence			Rotor movement on degree
S_1	S_2	S_3	θ
ON	OFF	OFF	0
OFF	ON	OFF	30
OFF	OFF	ON	60
ON	OFF	OFF	90
OFF	ON	OFF	120

11.2.4 Modes of Operation of Stepper Motor

Depending on how the stator phases are excited, the stepper motor can be operated in different modes. The operation discussed above is a single-phase mode operation. Here, only one phase is excited at a time. This type of operation is also called one-phase or full-step ON mode of operation of stepper motor. The other modes of operation of stepper motor are as follows:

11.2.4.1 Two-phase ON mode of operation

In this mode of operation, two stator phases are excited simultaneously. The mode of operation is shown in Table.11.2

TABLE 11.2 Two-phase ON mode operation.

Stator phase excitation sequence			Rotor movement on degree
S_1	S_2	S_3	θ
ON	OFF	OFF	0
ON	ON	OFF	15
OFF	ON	ON	45
ON	OFF	ON	75
ON	ON	OFF	105

Initial position of the rotor before excitation is shown in Fig.11.1. When Phase A of the stator is excited by closing the Switch S_1, it will magnetise the stator pole AA′ and it will pull the rotor teeth 1 and 3. This will move the rotor to a position of minimum reluctance. The position of the rotor is shown in Fig. 11.3. The rotor movement is clockwise.

When the switch S_2 is also closed (S_1 is also closed), it causes excitation of stator phase BB′ and try to rotate the rotor anti-clockwise. However, phase AA′ will try to rotate the rotor clockwise. Due to these forces, the rotor will now rotate by 15^0 in anti-clockwise direction instead of 30^0 and take the position where the magnetic forces produced by the stator phase AA′ and BB′ are equal, as show in Fig.11.11.

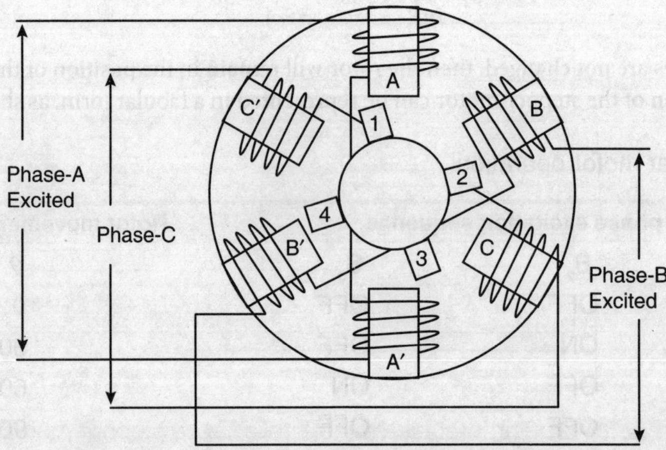

FIG.11.11: Two-phase ON mode of operation

In a similar manner, the other phases can be switched on in the sequence as indicated in Table.11.2, for the position of the rotor shown.

11.2.4.2 Half-step mode of operation

It is the combination of single-phase mode operation and two-phase ON mode of operation. The initial position of the rotor before excitation is shown in Fig.11.1. When switch S_1 is closed, it will excite the phase AA′ and the rotor teeth 1 and 3 will move to a position of minimum reluctance. Let it be the initial position of the rotor. Now when both the switches S_1 and S_2 are closed, it will rotate the rotor anti-clockwise by an angle of 15⁰. When S_1 is opened and S_2 remains closed, it will rotate the rotor anti-clockwise by 30⁰. With switch S_2 closed and switch S_3 also closed, the rotor will rotate by 45⁰. Similarly, when the switch S_3 remains closed and switch S_2 is opened, the rotor will rotate by 60⁰ and so on. The complete mode of operation is shown in Table 11.3.

TABLE 11.3 Half-step mode of operation

Stator phase excitation sequence			Rotor movement on degree
S_1	S_2	S_3	θ
ON	OFF	OFF	0
ON	ON	OFF	15
OFF	ON	OFF	30
OFF	ON	ON	45
OFF	OFF	ON	60

11.2.5 Multi-stack Variable Reluctance Stepper Motor

A multi-step variable reluctance stepper motor has the same number of poles on stator and rotor. In this type, the motor is divided along its axis into a number of stacks. Each stack is energised by a separate winding (phase) and magnetically isolated from each other.

Three-stack stepper motors are most common but motors with up to seven stacks and phases are also available.[1] The cross sectional view of a three-stack variable reluctance stepper motor is shown in Fig.11.12.

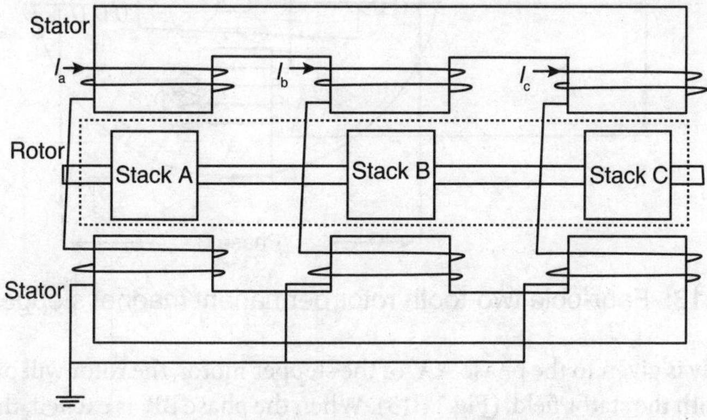

FIG.11.12: Three-stack variable reluctance stepper motor

[1] Sen PC, *Principles of Electric Machines and Power Electronics*, Second Edition, John Wiley & Sons, USA, 1997.

The step angle of a stepper motor with s number of stacks having N rotor teeth in each stack is calculated as follows,

$$\theta = \frac{360}{sN} \tag{11.4}$$

The tooth pitch is given by the expression,

$$\tau_p = \frac{360}{N} \tag{11.5}$$

11.2.6 Permanent Magnet Stepper Motor

In a permanent magnet stepper motor, the rotor is made up of permanent magnet and the stator is same as that of variable reluctance stepper motor. It operates on the reaction between a permanent magnet rotor and the electromagnetic field. The number of rotor teeth and the stator poles should not be same, so that there will be only a limited number of rotor teeth aligning themselves with the stator poles.

Some of the features of permanent magnet stepper motors are as follows:

i) It has higher inertia and lower acceleration/deceleration rates.
ii) Step pulse rate up to 300 pulses per second can be obtained as compared to 1200 pulses per second in variable reluctance stepper motors.
iii) In this motor, larger step sizes ranging from 30^0 to 90^0 can be obtained.
iv) It has a higher torque per ampere of stator currents.

Fig. 11.13 shows a four-pole two-tooth rotor permanent magnet stepper motor.

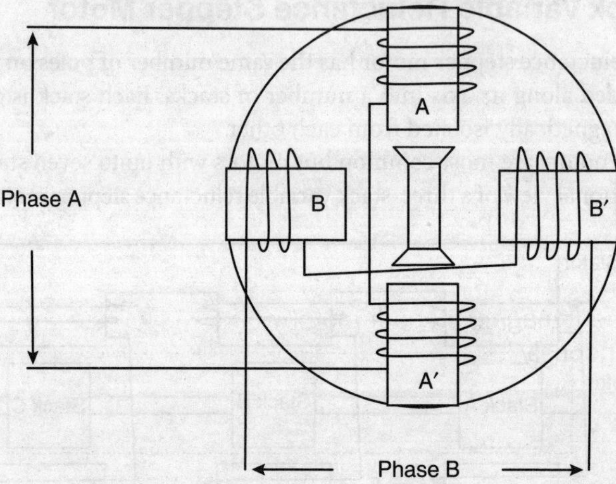

FIG. 11.13: Four-pole two-tooth rotor permanent magnet stepper motor

When a DC supply is given to the phase AA′ of the stepper motor, the rotor will overcome the residual torque and line up with the stator field. (Fig.11.13). When the phase BB′ is excited, the rotor will rotate by 90^0, either in clockwise direction or in anti-clockwise direction depending on the polarity of the supply (Fig.11.14).

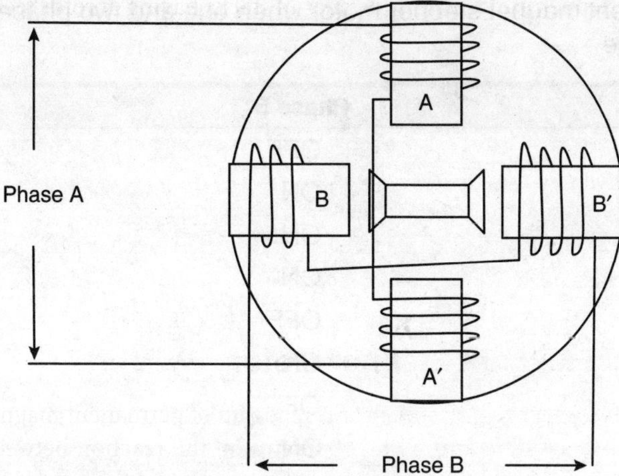

Fig.11.14: Anti-clockwise rotation

Table 11.4 gives the operation sequence of the motor when only one phase is energised at a time. Table.11.5 shows the operation sequence of the motor when two phases are energised simultaneously, while Table.11.6 shows the operation of the motor when one phase is energised and two phases are energised in sequence.

Table 11.4 Permanent magnet stepper motor when one phase is energised at a time

Phase A	Phase B	θ
ON	OFF	0
OFF	ON	90
ON	OFF	180
OFF	On	270
ON	OFF	360 = 0

Table 11.5 Permanent magnet stepper motor when two phases are energised simultaneously

Phase A	Phase B	θ
ON	ON	45
OFF	ON	135
ON	ON	225
ON	OFF	315
ON	ON	45

Table 11.6 Permanent magnet stepper motor when one and two phases are energised in sequence

Phase A	Phase B	θ
ON	OFF	0
ON	ON	45
OFF	ON	90
ON	ON	135
ON	OFF	180
ON	ON	255
OFF	OF	270
ON	ON	315
ON	OFF	0

11.2.7 Hybrid Stepper Motor

The hybrid stepper motor, as the name suggest is a motor designed to provide better efficiency by combining the pros of both the permanent magnet stepper motor and variable reluctance stepper motor. A hybrid stepper motor is more expensive than the individual types due to the following interesting features:

- Hybrid stepper motors have improved step resolution with increment angles ranging from 3.6 to 1 degree for 100 to 400 steps per revolution respectively.
- The system incorporates the advantages of both permanent magnet and variable reluctance types and utilises a special arrangement of permanent magnet poles over the entire concentric multi-toothed rotor/shaft assembly.
- The multi-toothed rotor assembly exhibits excellent coordination with magnetic fluxes generated by the stator windings, which guide the rotor movement in a well-controlled manner across the intermediate air gaps.

The above operations of the hybrid type produce more dynamic torque resolutions and high efficiency compared to the permanent magnet and variable reluctance types.

Construction of permanent magnet motors becomes very complex below 7.5 degrees step angles. Smaller step angles can be realised by combining the variable reluctance motor and the permanent magnet motor principles. Such motors are called hybrid motors (HB), which give much smaller step angles, as small as 0.9 degrees per step. A typical hybrid motor is shown in Fig.11.15.

The stator construction is similar to that in a permanent magnet motor. Likewise, the rotor is cylindrical and magnetised as in the permanent magnet motor with multiple teeth like in a variable reluctance motor. Each pole of a hybrid motor is covered with uniformly spaced teeth made of soft steel. The teeth on the two sections of each pole are misaligned with each other by a half-tooth pitch. Torque is created in the hybrid motor by the interaction of the magnetic field of the permanent magnet with the magnetic field produced by the stator. The teeth on the rotor provide a better path for the flux to flow through the preferred locations in the air gap. This increases the detent, holding, and dynamic torque characteristics of the motor compared to the other two types of motors. Hybrid motors have a smaller step angle compared to the permanent magnet motor, but they are very expensive.

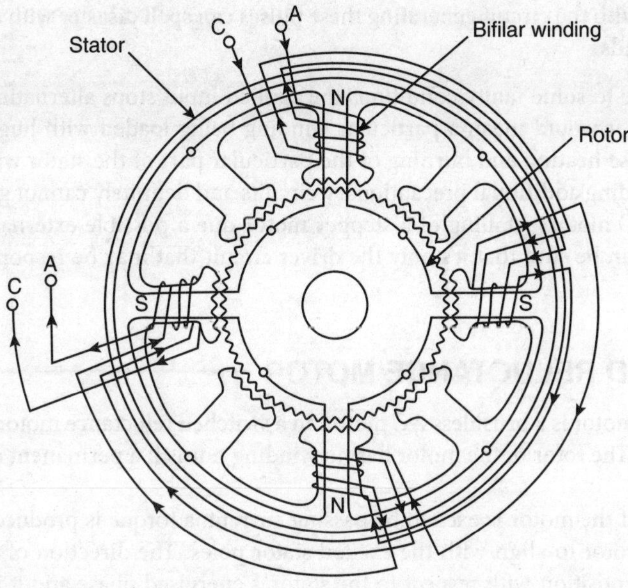

Fig.11.15: Hybrid motor

In low-cost applications, the step angle of a permanent magnet motor is divided into smaller angles using better control techniques. Permanent magnet motors and hybrid motors are more popular than the variable reluctance motor, and since the stator construction of these motors is very similar, a common control circuit can easily drive both types of motors.

11.2.8 Stepper Motor – Merits and Demerits

The following are their main advantages associated with stepper motors and the way they work:

- The rotation angle show perfect synchronisation with the applied electrical pulses.
- Even during minimum speeds or stationary periods the motor may exhibit full torque with the stator energised.
- These motors produce outstanding start-up, stop, and reversing response and adaptability.
- Being brushless means minimum wear and tear and therefore increased operating life; however, the bearing conditions can affect the performance.
- A stepper motor becomes comparatively easier to control than a DC motor due to its open loop operating feature, which enables accurate regulation of motor movements directly by adjusting the applied electrical pulses, thus avoiding the use of complex feedback circuitries. The feature also helps to enforce a wide range of rotational speeds.

However, there are a few disadvantages too with stepper motors, although most of them are quite negligible.

- With high speeds, controlling these motors becomes difficult and if the speed control goes unmonitored, it can sometimes give rise to resonance and a ringing effect in the motor.
- Another serious issue that may be associated with stepper motors is that the applied input electrical pulses need to be absolutely consistent and flawless. This is because the windings inside a stepper motor needs constant changeovers, and therefore fixed DC inputs can become highly undesirable.

- A malfunction with the circuit generating these pulses can spell disaster with the motor and burn it out within seconds.

For instance if due to some faulty conditions the applied input stops alternating and latches to produce direct currents, it would mean a particular winding being loaded with huge DC current, which would ultimately cause heating and burning of the particular part of the stator winding. However, this can be avoided by adding additional precautionary circuits and definitely cannot go unnoticed.

Unlike regular DC motors, stalling of a stepper motor due a possible external jamming will never burn it out, so one can be sure that it's only the driver circuit that may be responsible for the possible damages.

11.3 SWITCHED RELUCTANCE MOTOR

Switched reluctance motor is a brushless AC motor. In a switched reluctance motor only the stator of the motor has windings. The rotor of the motor has no winding nor is it a permanent magnet. It is made up of laminated steel.

When the stator of the motor is excited by passing current, a torque is produced in the rotor due to the tendency of the rotor to align with the excited stator poles. The direction of torque generated is a function of the rotor position with respect to the stator`s energised phase and it is independent of the direction of the current flowing through the stator. Continuous rotor torque can be produced by energising each stator phase in sequence. The number of stator poles and the rotor teeth must be different for switch reluctance motor. At least two stator phases are required to guarantee starting and at least three phases are required to insure starting direction.

Switched reluctance motor cannot run directly from DC supply or AC supply. It must always be electronically commutated.

Increasing the number of switched reluctance motor phases reduces the torque ripple but also with increase in number of stator phases, the requirement of electronic circuit also increases and becomes complex.

Fig.11.16 shows a six-stator pole and four-rotor teeth-switched reluctance motor.

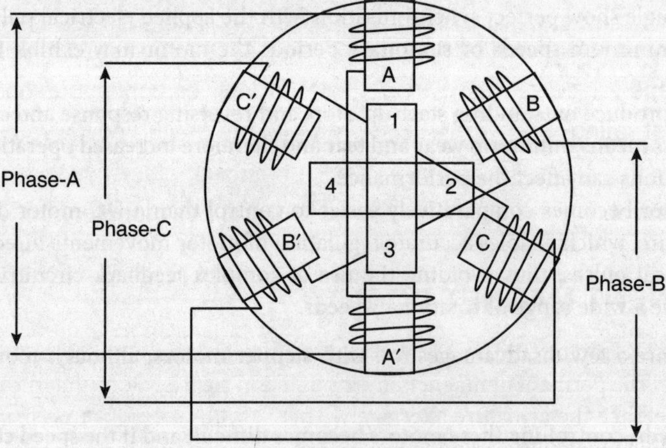

Fig.11.16: Switched reluctance motor

The stator coils are energised in sequence to rotate the rotor. The rotor will rotate in the clockwise direction when the stator phases are excited in the sequence, AA′, BB′, CC′, … and it will rotate in the anti-clockwise direction when the stator phases are excited in the sequence, AA′, CC′, BB′,…..

The nearest rotor poles are pulled into alignment with the energised stator poles and this makes the motor rotate.

11.3.1 Limitation of Switched Reluctance Motor

Some of the limitations of switched reluctance motor are

i) Switched reluctance motor cannot start directly from DC or AC supply.
ii) Both the stator and the rotor of the motor have salient pole configuration.
iii) Switched reluctance motor produces non linear magnetic characteristic.
iv) Control of switched reluctance motor is difficult.

11.3.2 Advantages of Switched Reluctance Motor

Some of the advantages of switched reluctance motor are

i) It has a high efficiency.
ii) It is robust in construction.
iii) It can run at high speed.
iv) Four-quadrant operation is possible.

11.3.3. Comparison between Variable Reluctance Stepper Motor and Switched Reluctance Motor

Table 11.7 gives the essential differences between a variable reluctance stepper motor and switched reluctance motor.

TABLE 11.7 Comparison of variable reluctance stepper motor and switched reluctance motor

Variable speed reluctance stepper motor	Switched reluctance motor
Rotates in steps	Gives continuous operation
Is of low power rating	Is of high power rating
Low efficiency	High efficiency
Open loop control	Closed loop control

11.4 PERMANENT MAGNET DC MOTOR

Permanent magnet DC motor (PMDC) is almost similar to DC shunt motor, but its field is made up of permanent magnet. The permanent magnets are mounted in such a way so that the north pole and south pole of each magnet face the armature alternately. There are three types of permanent magnet used in PMDC; they are, alnico magnets, ceramic magnets and rare earth magnets.

The armature construction is same as that of a DC motor and has armature slots for the windings, commutator segments and brushes. Armature is usually made up of laminated steel and the ends of the armature windings are connected to commutator segments where brushes are placed.

11.4.1 Working of Permanent Magnet DC Motor

As we said earlier, the working principle of PMDC motor is similar to the general working principle of a DC motor. That is, when a carrying conductor comes inside a magnetic field, a mechanical force will be experienced by the conductor and the direction of this force is governed by Fleming's left hand rule. As in a permanent magnet dc motor, the armature is placed inside the magnetic field of permanent magnet; the armature rotates in the direction of the generated force. Here each conductor of the armature experiences the mechanical force $F = B.I.L$ Newton where B is the magnetic field strength in Tesla (Weber / m^2), I is the current in Ampere flowing through that conductor and L is length of the conductor in meter comes under the magnetic field. Each conductor of the armature experiences a force and the compilation of those forces produces a torque, which tends to rotate the armature.

11.4.2 Applications of Permanent Magnet DC Motors

Permanent magnet dc motors are extensively used where smaller power ratings are required, such as in toys, small robots, computer disc drives etc.

11.4.3 Advantages of Permanent Magnet DC Motors

1. For smaller ratings, use of permanent magnets reduces manufacturing cost.
2. There is no need of field excitation winding; hence, construction is simpler.
3. There is no need of electrical supply for field excitation; hence, PMDC motor is relatively more efficient.
4. It is relatively smaller in size
5. It is cheap in cost

11.4.4 Disadvantages of Permanent Magnet DC Motors

1. Since the stator in PMDC motor consists of permanent magnets, it is not possible to add extra ampere-turns to reduce armature reaction. Thus, armature reaction is more in PMDC motors.
2. Stator side field control, for controlling speed of the motor, is not possible in permanent magnet dc motors.

11.5 BRUSHLESS DC MOTOR

Brushless DC (BLDC) motors have many similarities with AC induction motors and brushed DC motors in terms of construction and working principles respectively. Like all other motors, BLDC motors also have a rotor and a stator.

Similar to an induction AC motor, the BLDC motor stator is made out of laminated steel stacked up to carry the windings. Windings in a stator can be arranged in two patterns; i.e., a star pattern (Y) or delta pattern (Δ). The major difference between the two patterns is that the Y pattern gives high torque at low RPM while the Δ pattern gives low torque at low RPM. This is because, in the Δ configuration, half of the voltage is applied across the winding that is not driven, thus increasing losses and, in turn, efficiency and torque.

Steel laminations in the stator can be slotted or slot-less. A slot-less core has lower inductance; thus, it can run at very high speeds. Because of the absence of teeth in the lamination stack, requirements for the cogging torque also go down, thus making them an ideal fit for low speeds too. (Cogging torque develops when permanent magnets on the rotor and tooth on the stator align with each other and start interacting, causing ripples in speed). The main disadvantage of a slot-less core is higher cost because it requires more winding to compensate for the larger air gap.

Proper selection of the laminated steel and windings for the construction of stator are crucial to motor performance. An improper selection may lead to multiple problems during production, resulting in market delays and increased design costs. The rotor of a typical BLDC motor is made out of permanent magnets. Depending upon the application requirements, the number of poles in the rotor may vary. Increasing the number of poles does give better torque, but only at the cost of reducing the maximum possible speed.

Another rotor parameter that impacts the maximum torque is the material used for the construction of permanent magnet; the higher the flux density of the material, the higher the torque.

11.5.1 Working of BLDC

Motor operation is based on the attraction or repulsion between magnetic poles. Using the three-phase motor shown in Figure 11.1 the process starts when current flows through one of the three stator windings and generates a magnetic pole that attracts the closest permanent magnet of the opposite pole. The rotor will move if the current shifts to an adjacent winding. Sequentially charging each winding will cause the rotor to follow in a rotating field. The torque in this example depends on the current amplitude and the number of turns on the stator windings, the strength and the size of the permanent magnets, the air gap between the rotor and the windings, and the length of the rotating arm.

11.5.2 Brushless versus Brushed DC Motor

- Brushes require frequent replacement due to mechanical wear; hence, a brushed DC motor requires periodic maintenance. Also, as brushes transfer current to the commutator, sparking occurs. Brushes limit the maximum speed and number of poles the armature can have. These drawbacks are removed in a brushless DC motor. Electronic control circuit is required in a brushless DC motor for switching stator magnets to keep the motor running. This makes a BLDC motor potentially less rugged.
- Advantages of BLDC motor over brushed motors are: increased efficiency, reliability, longer lifetime, no sparking and less noise, more torque per weight etc.

11.5.3 Applications of BLDC

- For single-speed applications, induction motors are more suitable, but if the speed has to be maintained with the variation in load, then because of the flat speed-torque curve of BLDC motor, BLDC motors are a good fit for such applications.
- BLDC motors become a more suitable fit for applications where there is a variation in load because variable speed induction motors will also need an additional controller, thus adding to system cost. Brushed DC motors will also be a more expensive solution because of regular maintenance.
- Precise control is not required for applications such as induction cooker and because of low maintenance, BLDC motors are a winner here too. However, for such applications, BLDC motors use optical encoders, and complex controllers are required to monitor torque, speed, and position.
- Brushed DC motors are known for generating more EMI noise, so BLDC is a better fit. However, controlling requirements for BLDC motors also generate EMI and audible noise. This can, however, be addressed using field-oriented control (FOC) sinusoidal BLDC motor control.

11.6 LINEAR INDUCTION MOTOR

A Linear Induction motor (LIM) is a special type of induction motor which gives linear motion instead of rotational motion as in the case of conventional induction motor. It operates on the principle as that of a conventional induction motor.

The LIM operates on the same principle as that of a rotary squirrel-cage induction motor. The rotary induction motor becomes a LIM when the coils are laid out flat; the reaction plate in the LIM becomes the equivalent rotor. This is made from a non-magnetic highly conductive material. The induced field can be maximised by backing up the reaction plate with an iron plate (conducting sheet). The iron plate serves to amplify the magnetic field produced in the coil. The air gap between the stator and the reaction plate must typically be very small, much smaller than the allowable gap for the synchronous motor; otherwise, the amount of current required for the stator coils becomes unreasonable. When supplying an AC current to the coils, a rotating magnetic field is produced. Currents induced in the reaction plate by the rotating magnetic field create a secondary magnetic field. It is not necessary to keep the field of motion synchronised to the position of the reaction plate, since the second field is induced by the stator coil. A linear thrust is produced by the reaction between these two fields.

11.6.1 Advantanges of Linear Induction Motor

- Low maintenance cost because of absence of rotating parts
- Simplicity
- No limitation of tractive effort due to adhesion between wheel and the rail
- No limitation of maximum speed due to centrifugal forces
- Cost is less as we use less lamination and less winding.

11.6.2 Disadvantanges of Linear Induction Motor

- Poor utilisation of motor due to transverse effect and end effect.
- Larger air-gap and non-magnetic reaction rail need more magnetic current resulting in poor efficiency and low power factor
- Difficulties encountered in maintaining adequate clearances at points and crossing.

11.6.3 Applications of Linear Induction Motor

The use of linear induction motor is not as widespread as compared to that of a conventional motor, since it is less economical and less versatile. However, there are a few specialised operations for which the LIM is indeed necessary.

Few of the applications of a LIM have been listed below.

1. Automatic sliding doors in electric trains.
2. Mechanical handling equipment, such as in the propulsion of a train or tram along a certain route.
3. Metallic conveyor belts.

11.7 HYSTERESIS MOTOR

A hysteresis motor is a single-phase synchronous motor with no salient poles. Neither does it have DC excitation. It starts by virtue of the hysteresis loss induced in its hardened steel secondary member by the revolving field of the primary and continues to operate normally at synchronous speed, running on hysteresis torque because of the retentivity of the secondary core.

The stator of the hysteresis produces a synchronously revolving field from a single-phase supply. The stator can have main and auxiliary windings, in which case it will be called split-phase hysteresis motor; or it can also be shaded pole type, in which case it will be called shaded pole hysteresis motor.

The rotor of the hysteresis motor does not have any winding. It is made with magnetic material of high hysteresis losses (whose hysteresis loop area is very large).

When the stator is energised from a single-phase supply, it will produce a rotating magnetic field. The rotor, initially, starts to rotate due to eddy current torque and hysteresis torque developed on the rotor. Once the speed is near about the synchronous, the stator pulls rotor into synchronism. Due to the hysteresis effect, rotor pole axis lags behind the axis of the rotating magnetic field. Due to this, the rotor poles get attracted towards the moving stator poles. Thus, the rotor gets subjected to a torque called hysteresis torque. Due to principle of magnetic locking, the motor either rotates at synchronous speed or not at all. Only hysteresis torque is present, which keeps the rotor running at synchronous speed.

Advantages of Hysteresis Motor

- As rotor does not have teeth or winding, there are no mechanical vibrations.
- Due to absence of vibrations, the operation is quiet and noiseless.
- It is suitable for accelerating inertia loads.
- It can be used for multispeed operations by employing gear train.

Disadvantages of Hysteresis Motor

- The output is about one-quarter that of an induction motor of the same dimension.
- Low efficiency
- Low power factor
- Low torque
- It is available only in very small sizes

Applications of Hysteresis Motor

It is used in sound recording instruments, sound producing equipments, high-quality record players, electric clocks, printers, timing devices etc.

11.8 REPULSION–INDUCTION MOTOR

It is a combination of a repulsion motor and a squirrel-cage induction motor. Repulsion motors consist of a stator, rotor, commutator and brush. The stator of a repulsion motor is of non-salient pole type provided with slots. The rotor is connected to the commutator as in the case of the armature of a DC machine. The windings of the rotor are of distributed type while the commutator can be of axial type or vertical type. Carbon brushes are used to conduct current through the armature.

The difference between AC series motor and repulsion motor lies the way in which power is supplied to the armature. In AC series motor, the armature receives voltage by conduction through power supply. However in repulsion motors, the armature is supplied by induction from the stator windings.

A repulsion–induction motor is a single-phase motor with one stator winding and two rotor windings. At start, the repulsion winding is predominant; but as the motor speed increases, the squirrel-cage winding becomes predominant. Transition from repulsion to induction is smooth since no switching device is employed. This motor is ideally suited for applications where low voltage is a problem or high starting torque is required.

Various types of repulsion motors commonly used are: compensated repulsion motor, repulsion-start induction-run motor and repulsion induction motor.

The main disadvantages of repulsion motor are

- Its power factor is very poor
- It has very high no-load speed
- Its commutator and brushes wear our quickly

Because of excellent starting and accelerating characteristics, repulsion–induction motors are used in applications such as air compressors, laundry machines and farm motors.

11.9 UNIVERSAL MOTOR

It is a single-phase motor which can run on either DC or single-phase AC supply. The armature and field winding of this motor are connected in series and hence, it has a high starting torque. There are two types of universal motors – compensated type and uncompensated type.

The main parts of the universal motor are the frame assembly, the field and the armature. The frame is generally made up of rolled steel, aluminium, or cast-iron and is laminated. The field poles are generally held in the frame by means of bolts.

The armature of the universal motor is similar to that of a small DC motor with laminated core having either straight slots or skewed slots and a commutator to which the leads of the armature winding are connected. Both the core and commutator are pressed on the shaft.

A universal motor works on either DC or single-phase AC supply. When the universal motor is fed with DC supply, it works as a DC series motor. When current flows in the field winding, it produces an electromagnetic field. The same current also flows from the armature conductors. When a current carrying conductor is placed in an electromagnetic field, it experiences a mechanical force. Due to this mechanical force or torque, the rotor starts to rotate. The direction of this force is given by Fleming's left-hand rule.

When fed with AC supply, it still produces unidirectional torque. Since armature winding and field winding are connected in series, they are in same phase. Hence, as polarity of AC changes periodically, the direction of current in armature and field winding reverses at the same time. Thus, the direction of magnetic field and the direction of armature current reverse in such a way that the direction of force experienced by armature conductors remains same. Thus, regardless of AC or DC supply, universal motor works on the same principle that a DC series motor works on.

11.10 SERVO MOTOR

Servo motors, also called servos, are self-contained electric devices that rotate or push parts of a machine with great precision. It is used in toys, home electronics, aeroplanes and in many other places.

The servo motor is a small DC motor that runs on a battery supply and rotates at high speed but giving very low torque. With the help of gear mechanism, the speed of the motor is reduced and at the same time the torque is increased. These gears in a servo motor are generally made of plastic to keep it lighter and less costly.

There are basically three basic types of servo motors, namely, servos with positional rotation, continuous rotation and linear servo.

- **Positional rotation servo**: It is the most common type of servo motor. Here, the shaft rotates half-a-circle, or 180 degrees. It has physical stops placed in the gear mechanism to prevent turning beyond these limits to protect the rotational sensor.

- **Continuous rotation servo**: It is similar to positional rotation servo motor and it can turn in either direction indefinitely. Depending on the command signal given to the motor, the motor can rotate clockwise or anti-clockwise as desired, at varying speed.
- **Linear servo**: This is also like the positional rotation servo motor described above, but with additional gears (usually a **rack and pinion** mechanism) to change the output from circular to back-and-forth.

Servos take commands from a series of pulses sent from the computer or radio. In the battery device used in servos, "low" is considered to be ground or 0 volts and "high" is the battery voltage. Servos tend to work in a range of 4.5 to 6 volts.

The servo must be connected to a source of power (4.5 to 6 volts) and the control signal must come from a computer or other circuitry. Each servo's requirement may vary slightly, but a pulse of about 50 to 60 Hz works well for most models. The pulse width will vary from approximately 1 millisecond to 2 or 3 milliseconds. Any microcontroller can be programmed to generate these waveforms.

11.11 SCHRAGE MOTOR

Schrage motor is a variable-speed motor with wound rotor construction. The three-phase windings of the wound rotor are delta-connected. The end points of the delta-connected rotor are brought out and connected on to slip-rings. The three-phase supply is given to the motor through these slip-rings. The rotor has a second delta-connected winding with tappings in it, which are connected with the commutator. The three phases of the stator windings are connected to the commutator segment by two sets of brushes.

The two sets of brushes can be moved to connect to the same point on the commutator, or moved apart in either direction. When they are in alignment, the stator windings are shorted and the motor behaves like an induction motor with the rotor and stator swapped. As the brushes are separated, voltage from auxiliary rotor winding is coupled to the stator winding. The degree of separation and the direction of separation give the voltage and polarity of the voltage. The frequency of the voltage applied to the stator is dependent on the slip.

11.12 SINGLE-PHASE AC SERIES MOTOR

In a DC motor, if the direction of both field and armature current is reversed, the direction of torque does not change. When a DC series motor is connected to an AC supply, both field and armature current get reversed, thereby producing unidirectional torque. But it has certain limitations:

- The will be overheating due to eddy current losses in the yoke and field cores
- Operating power factor is very low.
- There will be sparking at brushes because of high voltage and current induced in the short circuited armature coils during the commutation period.

These limitations can be removed by some constructional modifications:

- Yoke and pole core are laminated.
- The power factor can be improved by reducing the magnitudes of field and armature reactance.
- To keep the torque same, it is necessary to increase the armature turns proportionately.

Now, to compensate for increased armature flux that produces severe armature reaction, it is necessary to use compensating winding. The flux produced by this winding is opposite to that produced by armature and it effectively neutralises the armature reaction.

If such a compensating winding is connected in series with the armature, the motor is said to be 'conductively compensated'. For motors that are operated on both DC and AC, the compensation should be conductive. If compensating winding is short-circuited, the motor is said to be 'inductively compensated'. In this, the compensating winding acts as a secondary of the transformer and armature, as its primary. The ampere turns produced by compensating winding neutralise the armature ampere turns.

The characteristics of such motor are similar to that of the DC series motor. The torque varies as square of the armature current and speed varies inversely as the armature current. The speed of such motor can be dangerously high at no-load condition and hence, it is always started with some load. Starting torque produced is high, being about 3 to 4 times the full-load torque.

11.13 ASYNCHRONOUS GENERATOR FOR WIND TURBINE

A generator converts mechanical energy into electrical energy. The electricity generated by the wind turbine is stepped up to a suitable voltage level with the help of a transformer so that it can be fed to the grid. Two types of generators are used with the wind turbine, namely, synchronous generator or asynchronous (induction) generator. Generally, induction generator is used with the wind turbine because of its low cost, ruggedness and brushless operation. Also, the speed of an induction generator changes according to the torque given to it. An important property of an induction generator is that it increases or decreases its speed slightly if the torque varies. Hence, the longevity of the gear box increases with the use of induction generator.

Induction generator is mechanically and electrically similar to a poly-phase induction motor. It has the ability to produce power at varying rotor speed and is not self-excited. In the case of wind power system connected to the power grid, the magnetising current produced by the rotating flux is supplied to the grid.

SOLVED PROBLEMS

Ex.11.1 A variable reluctance stepper motor has 10 poles, each having 6 teeth, and the rotor of the motor has 50 teeth. Calculate the step-angle of the motor.

Solution

$$P_s = 10 \times 6 = 60$$

$$P_r = 50$$

$$\text{Step angle}\left(\theta\right) = \frac{60-50}{60 \times 50} \times 360 = 1.2^\circ$$

Ex.11.2 A single-stack, three-phase variable reluctance motor has a step-angle of 30⁰. Calculate the number of stator and rotor poles.

Solution

From the equation

$$\theta = \frac{360}{\text{m}P_r}$$

$$\text{or } 30 = \frac{360}{3 \times P_r}$$

$$P_r = 4$$

Again,

$$\theta = \frac{P_s - P_r}{P_s P_r} \times 360$$

If $P_s > P_r$, then,

$$30 = \frac{P_s - 4}{4P_s} \times 360°.$$

$$\text{or, } P_s = 6.$$

If $P_r > P_s$, then,

$$30 = \frac{4 - P_s}{4P_s} \times 360$$

$$\text{or, } P_s = 3$$

Ex.11.3 A small 60-Hz hysteresis motor possesses 32 poles. In making one complete turn with respect to the revolution field, the hysteresis loss in the rotor amounts to 0.8 J. Calculate the following.

(i) Hysteresis torque
(ii) Maximum power output before the motor stall
(iii) Rotor losses when the motor is stalled
(iv) Rotor losses when the motor runs at synchronous speed.

Solution

$$\text{Hysteresis torque} = \frac{0.8}{2 \times \pi} = 0.127 \text{ Nm}$$

$$\text{Power output} = \frac{2\pi N_s}{60} \times 0.127 = \frac{2\pi \times \frac{120f}{P}}{60} \times 0.127 = \frac{2\pi \times \frac{120 \times 60}{32}}{60} \times 0.127 = 3 \text{ W}$$

$$\text{Energy loss} = 225 \times 0.8 = 180 \text{ J}$$

$$\text{Power dissipated} = \frac{180}{60} = 3 \text{ W}$$

There is no energy loss in the rotor when the motor runs at synchronous speed because the magnetic domains no longer reverse.

Ex.11.4 A 50 Hz hysteresis clock motor possesses 30 poles. In making one complete turn with respect to the revolving field, the hysteresis loss in the rotor amounts to 1 J. Calculate the following.

(i) Pull-in and pull-out torque
(ii) Synchronous speed
(iii) The maximum power output before the motor stalls
(iv) Rotor loss when the motor is stalled

Solution

$$\text{Pull-in torque} = \text{Pull-out torque} = \frac{1}{6.28} = 0.159 \text{ Nm}$$

$$\text{Synchronous speed} = \frac{120f}{p} = \frac{120 \times 50}{30} = 200 \text{ rpm}$$

$$\text{Maximum power} = \frac{200 \times 0.159}{9.55} = 3.3 \text{ W}$$

$$\text{Energy loss per min} = 200 \times 1 = 200 \text{ J}$$

$$\text{Power dissipated in the rotor} = \frac{200}{50} = 4 \text{ W}$$

Ex.11.5 The resistance and the inductance of a series universal motor are 40Ω and 0.1H respectively. When connected across 250V DC supply, it takes a load current of 1A at 1800 r.p.m. Calculate the value of back e.m.f. when there is (i) DC supply (ii) AC supply. Also find its speed when the motor is provided AC supply of the same load.

Solution

For DC supply:

$$N_{\text{dc}} = 1800 \text{ r.p.m.}$$

$$E_{\text{dc}} = 250 - 1 \times 40 = 210 \text{ V}$$

For AC supply

$$X_{\text{L}} = 2\pi fL = 2\pi \times 50 \times 0.1 = 31.425'$$

$$\text{Again, } V^2 = \left(I_{\text{a}}R_{\text{a}} + E_{\text{ac}}\right)^2 + \left(I_{\text{a}}X_{\text{L}}\right)^2$$

$$\text{or, } E_{\text{ac}} = \sqrt{V^2 - \left(I_{\text{a}}X_{\text{L}}\right)^2} - I_{\text{a}}R_{\text{a}}$$

$$= \sqrt{250^2 - \left(1 \times 31.4154\right)^2} - 1 \times 40$$

$$= 208.02 \text{ V}$$

$$N_{ac} = \frac{E_{ac}}{E_{dc}} \times N_{dc} = \frac{208.02}{210} \times 1800 = 1783 \text{ r.p.m.}$$

Ex.11.6 What is the step-angle of a 6-stack, 26-teeth variable reluctance stepper motor?

Solution

$$\text{Step-angle}(\alpha) = \frac{360}{6 \times 26} = 2.31 \text{ per slot}$$

Ex.11.7 What is the step-angle for a three-phase, 36-pole permanent magnet stepper motor?

Solution

$$\text{Step-angle}(\alpha) = \frac{360}{3 \times 36} = 3.33 \text{ per slot}$$

Ex.11.8 What is the step-angle of a 3-stack, 12-teeth variable reluctance stepper motor?

Solution

$$\text{Step-angle}(\alpha) = \frac{360}{3 \times 12} = 10 \text{ per slot}$$

Ex.11.9 A universal motor has a 2-pole armature with 100 conductors. At a particular load, the motor speed is 5500 r.p.m. with 5A armature current. The armature voltage is 150V and power is 300W, while the armature resistance is 4Ω. Calculate the maximum value of useful flux, the inductive reactance of the machine and the power factor.

Solution

$$\cos\theta = \left[\frac{300}{150 \times 5}\right] = 0.4$$

$$E_{dc} = 150 - 5 \times 4 = 130 \text{V}$$

$$E_{ac} = 150 \times 0.4 - 5 \times 4 = 40 \text{ V}$$

We know that

$$V^2 = \left(I_a R_a + E_{ac}\right)^2 + \left(I_a X_L\right)^2$$

$$X_L = \frac{1}{I_a} \times \sqrt{V^2 - \left(I_a R_a + E_{ac}\right)^2}$$

$$= \frac{1}{5} \times \sqrt{150^2 - \left(5 \times 4 + 40\right)^2} = 27.49 \ \Omega$$

$$\Phi = \frac{130 \times 60 \times 2}{5500 \times 2 \times 100} = 0.0142 \text{ Wb}$$

Ex.11.10 A single-phase AC motor, when fed from a 300V DC supply, runs at 3500 rpm and takes a current of 1.5 A. What will be its speed and power factor when connected across a 300V, 50Hz, AC supply, with the same load? Take resistance and inductance of the motor to be 50Ω and 0.3 H respectively.

Solution

$$E_{dc} = 300 - 1.5 \times 50 = 225 \text{ V}$$

$$N_{dc} = 3500 \text{ r.p.m.}$$

$$X_L = 2\pi f L = 2 \times \pi \times 50 \times 0.3 = 94.25\ \Omega$$

$$E_{ac} = \sqrt{V^2 - (I_a X_L)^2} - I_a R_a$$

$$= \sqrt{300^2 - (1.5 \times 94.25)^2} - 1.5 \times 50 = 189.6 \text{ V}$$

$$\frac{E_{ac}}{E_{dc}} = \frac{N_{ac}}{N_{dc}}$$

$$N_{ac} = \frac{E_{ac}}{E_{dc}} \times N_{dc} = \frac{189.6}{225} \times 3500 = 2949 \text{ r.p.m.}$$

Ex.11.11 A universal motor has a 4-pole armature with 100 conductors. At a particular load, the motor speed is 5500 r.p.m. with 5A armature current. The armature voltage is 150V and power is 300W, while the armature resistance is 4Ω. Calculate the maximum value of useful flux, the inductive reactance of the machine and the power factor.

Solution

$$\cos\theta = \left[\frac{300}{150 \times 5}\right] = 0.4$$

$$E_{dc} = 150 - 5 \times 4 = 130\text{V}$$

$$E_{ac} = 150 \times 0.4 - 5 \times 4 = 40 \text{ V}$$

We know that

$$V^2 = (I_a R_a + E_{ac})^2 + (I_a X_L)^2$$

$$X_L = \frac{1}{I_a} \times \sqrt{V^2 - (I_a R_a + E_{ac})^2}$$

$$= \frac{1}{5} \times \sqrt{150^2 - (5 \times 4 + 40)^2} = 27.49\,\Omega$$

$$\Phi = \frac{130 \times 60 \times 2}{5500 \times 4 \times 100} = 7.09\text{ MWb}$$

Ex.11.12 A single-stack, three-phase variable reluctance motor has a step-angle of 60^0. Calculate the number of stator and rotor poles.

Solution

From the equation,

$$\theta = \frac{360}{mP_r}$$

$$\text{or } 60 = \frac{360}{3 \times P_r}$$

$$P_r = 2$$

Again,

$$\theta = \frac{P_s - P_r}{P_s P_r} \times 360$$

If $P_s > P_r$ then,

$$60 = \frac{P_s - 4}{4P_s} \times 360$$

$$\text{or, } P_s = 12$$

If $P_r > P_s$, then,

$$60 = \frac{4 - P_s}{4P_s} \times 360$$

$$\text{or, } P_s = 3$$

EXERCISES

1. Give the classification of stepper motor.
2. Explain the single-phase mode operation of single-stack stepper motor.

3. Explain the two-phase ON mode operation of single-stack stepper motor.
4. Explain the half-step mode operation of single-stack stepper motor.
5. Write a note on multi-stack variable reluctance stepper motor.
6. What are the features of permanent magnet stepper motors?
7. Describe the operation of permanent magnet stepper motor under the following conditions:
   (i) When one phase is ON
   (ii) When two phases are simultaneously ON.
   (iii) When one and two phases are energised in sequence.
8. List the features of hybrid motor.
9. Explain the operation of hybrid motor.
10. What are the advantages and disadvantages of stepper motor?
11. Explain how a switched reluctance motor operates.
12. What are the advantages and disadvantages of switched reluctance motor?
13. What are the limitations of switched reluctance motor?
14. Differentiate between variable reluctance stepper motor and switched reluctance motor
15. Give the working of a permanent magnet DC motor
16. What are the merits and demerits of PMDC motor?
17. List some application of PMDC motor.
18. Explain the construction of BLDC motor.
19. Explain the working of BLDC motor
20. Differentiate between Brushless and Brushed DC Motor.
21. What are the application of BLDC motor
22. Explain the operation of linear induction motor.
23. What are the advantages, disadvantages and applications of linear induction motor?
24. Explain the operation of hysteresis motor.
25. What are the advantages, disadvantages and applications of hysteresis motor?

APPENDIX A

MULTIPLE CHOICE QUESTIONS

1. Star/star transformers work satisfactorily when
   (a) load is unbalanced only
   (b) load is balanced only
   (c) on balanced as well as unbalanced loads
   (d) None of the above.
2. Delta/star transformer works satisfactorily when
   (a) load is balanced only
   (b) load is unbalanced only
   (c) on balanced as well as unbalanced loads
   (d) None of the above.
3. Which type of winding is used in 3-phase shell-type transformer?
   (a) Circular type
   (b) Sandwich type
   (c) Cylindrical type
   (d) Rectangular type
4. Which of the following does not change in a transformer?
   (a) Current
   (b) Voltage
   (c) Frequency
   (d) All of the above.
5. In a transformer, the energy is transferred from primary to secondary
   (a) through a cooling coil
   (b) through air
   (c) by the flux
   (d) None of the above.
6. A transformer core is laminated to
   (a) reduce hysteresis loss
   (b) reduce eddy current loss
   (c) reduce copper loss
   (d) reduce all the above losses.
7. The no-load current drawn by a transformer is usually what percent of the full-load current?
   (a) 0.2 to 0.5 per cent
   (b) 2 to 5 per cent
   (c) 12 to 15 per cent
   (d) 20 to 30 per cent.
8. No-load test on a transformer is performed to determine
   (a) copper loss
   (b) magnetising current
   (c) eddy current loss
   (d) efficiency of the transformer.
9. The efficiency of a transformer will be maximum when
   (a) Copper loss = Hysteresis loss
   (b) Hysteresis loss = Eddy current loss
   (c) Eddy current loss = Copper losses
   (d) Copper loss = Iron loss
10. Which of the following is not part of a transformer?
   (a) Conservator
   (b) Breather
   (c) Buchholz relay
   (d) Exciter.
11. Short-circuit test on a transformer is generally performed on
   (a) high-voltage side
   (b) low voltage side
   (c) primary side
   (d) secondary side.
12. Which winding in a transformer has more number of turns?
   (a) Low-voltage winding
   (b) High-voltage winding

(c) Primary winding
(d) Secondary winding.

13. Efficiency of a power transformer is of the order of
(a) 100 per cent
(b) 98 per cent
(c) 50 per cent
(d) 25 per cent.

14. In a transformer, which loss is a constant loss?
(a) Friction and windage loss
(b) Copper loss
(c) Hysteresis and eddy current loss
(d) None of the above.

15. The function of a conservator is to
(a) provide fresh air for cooling the transformer
(b) supply cooling oil to transformer in time of need
(c) protect the transformer from damage when oil expends due to heating
(d) None of the above.

16. Power transformers are designed to have maximum efficiency at
(a) nearly full load
(b) 70% full load
(c) 50% full load
(d) no load.

17. Two transformers operating in parallel will share the load depending upon their
(a) leakage reactance
(b) per-unit impedance
(c) efficiencies
(d) ratings.

18. If R2 is the resistance of secondary winding of the transformer and K is the transformation ratio, then the equivalent secondary resistance referred to primary will be
(a) R2/2K
(b) R2/K^2
(c) $R2^2/K^2$
(d) R22/K. $R2^2$/K

19. In a transformer, the tappings are generally provided on
(a) primary side
(b) secondary side
(c) low-voltage side
(d) high-voltage side.

20. The chemical used in breather is
(a) asbestos fibre
(b) silica sand
(c) sodium chloride
(d) silica gel.

21. The transformer ratings are usually expressed in terms of
(a) volts
(b) amperes
(c) kW
(d) kVA.

22. The function of conservator in a transformer is
(a) to protect against internal fault
(b) to reduce copper as well as core loss
(c) to cool the transformer oil
(d) to take care of the expansion and contraction of transformer oil due to variation of temperature of surroundings.

23. In a transformer the resistance between its primary and secondary is
(a) zero
(b) 1 ohm
(c) 1000 ohms
(d) infinite.

24. A transformer oil must be free from
(a) sludge
(b) odour
(c) gases
(d) moisture.

25. Buchholz's relay gives warning and protection against
(a) electrical fault inside the transformer itself
(b) electrical fault outside the transformer in outgoing feeder
(c) both outside and inside faults
(d) None of the above

26. Which of the following does not change in an ordinary transformer?
(a) Frequency
(b) Voltage
(c) Current
(d) Any of the above

27. A transformer can have zero voltage regulation at
(a) leading power factor
(b) lagging power factor
(c) unity power factor
(d) zero power factor.

28. The full-load copper loss of a transformer is 1600 W. At half-load, the copper loss will be
(a) 6400 W
(b) 1600 W
(c) 800 W
(d) 400 W

29. The value of flux involved in the e.m.f. equation of a transformer is
(a) average value
(b) r.m.s. value
(c) maximum value
(d) instantaneous value.

30. Which of the following is the main advantage of an auto-transformer over a two-winding transformer?
(a) Hysteresis loss is reduced
(b) Saving in winding material
(c) Copper loss is negligible
(d) Eddy losses are totally eliminated.

31. During short-circuit test, iron loss is negligible because
(a) the current on secondary side is negligible
(b) the voltage on secondary side does not vary
(c) the voltage applied on primary side is low
(d) full-load current is not supplied to the transformer.

32. Two transformers are connected in parallel. These transformers do not have equal percentage impedance. This is likely to result in
(a) short-circuiting of the secondaries
(b) leading power factor for one of the transformers and lagging power factor for the other
(c) transformers having higher copper loss having negligible core loss
(d) loading of the transformers disproportionate to their kVA ratings.

33. An ideal transformer is one which has
(a) no loss or magnetic leakage
(b) interleaved primary and secondary windings
(c) a common core for its primary and secondary windings
(d) a core of stainless steel and winding of pure copper metal
(e) None of the above.

34. An ideal transformer will have maximum efficiency at a load such that
(a) Copper loss = Iron loss
(b) Copper loss < Iron loss
(c) Copper loss > Iron loss
(d) None of the above

35. If the supply frequency to the transformer is increased, the iron loss will
(a) not change
(b) decrease
(c) increase
(d) Any of the above.

36. Negative voltage regulation indicates that the load is
(a) capacitive only
(b) inductive only
(c) inductive or resistive
(d) None of the above.

37. During open-circuit test of a transformer,
(a) primary is supplied rated voltage
(b) primary is supplied full-load current
(c) primary is supplied current at reduced voltage
(d) primary is supplied rated kVA.

38. Open-circuit test on transformers is conducted to determine
(a) hysteresis loss
(b) copper loss
(c) core loss
(d) eddy current loss.

39. Short-circuit test on transformers is conducted to determine
(a) hysteresis loss
(b) copper loss
(c) core loss
(d) eddy current loss.

40. A good voltage regulation of a transformer means
(a) output voltage fluctuation from no-load to full-load is least
(b) output voltage fluctuation with power factor is least
(c) difference between primary and secondary voltage is least
(d) difference between primary and secondary voltage is maximum.

41. The efficiency of two identical transformers under load conditions can be determined by
(a) short-circuit test
(b) back-to-back test
(c) open circuit test
(d) Any of the above.

42. Which of the following loss in a transformer is zero even at full load?
(a) Core loss
(b) Friction loss
(c) Eddy current loss
(d) Hysteresis loss.

43. The resistance of armature winding of a DC generator depends on
(a) length
(b) cross-sectional area
(c) number of conductors
(d) All the above.

44. The commutator segments are connected to the armature conductors by
(a) copper lugs
(b) resistance wires
(c) insulated pads
(d) brazing.

45. In lap winding, the number of brushes will be
(a) half the number of poles
(b) double the number of poles
(c) equal to the number of poles
(d) always two

46. In a DC generator, the mechanical loss depends on
(a) voltage
(b) current
(c) speed
(d) All of these.

47. In a DC machine, the iron loss does not depends on
(a) voltage
(b) load
(c) speed
(d) speed and voltage.

48. Brushes are made up of
(a) copper
(b) aluminium
(c) soft copper
(d) carbon

49. The pole pitch for a four-pole, two-layer lap winding and sixteen coils will be
(a) 16
(b) 8
(c) 4
(d) 2

50. In an unsaturated DC machine, the armature reaction will be
(a) magnetising
(b) cross-magnetising
(c) demagnetising
(d) None of these

51. The series field-winding of a short-shunt DC generator is excited by
(a) load current
(b) armature current
(c) external current
(d) None of these.

52. DC generator operates on the principle of
(a) Biot–Savort law
(b) Ohm's law
(c) Lenz's law
(d) Faraday`s law of electromagnetic induction.

53. The critical resistance is the resistance of
(a) armature
(b) field
(c) load
(d) commutator.

54. If the resistance of the field winding is increased, then its output voltage will
(a) increase
(b) decrease
(c) remain the same
(d) None of these.

55. When stepper motor is operated at high speed, it is referred to as
(a) inching
(b) cogging
(c) crawling
(d) slewing.

56. Stepper motor can also be considered as
(a) AC-to-DC converter
(b) DC-to-AC converter
(c) analog-to-digital converter
(d) digital-to-analog converter.

57. In a synchronous motor, the breakdown torque
(a) is proportional to half of applied voltage
(b) is proportional to square of applied voltage
(c) is inversely proportional to applied voltage
(d) is directly proportional to applied voltage.

58. For maximum power on synchronous motor, the load angle is
(a) 90^0
(b) 60^0
(c) 30^0
(d) 0^0.

59. Exciter for synchronous machine is a
(a) DC series motor
(b) DC shunt motor
(c) DC compound motor
(d) Any motor can be used.

60. The duration of short-circuit test on a synchronous motor is about
(a) one hour
(b) one minute
(c) one second
(d) It has no such duration.

61. The speed of a universal motor is usually controlled by
(a) belts
(b) gears
(c) brakes
(d) None of these.

62. Which motor is used in tap-recorder?
(a) Universal motor
(b) Reluctance motor
(c) Capacitor-run motor
(d) Hysteresis motor.

63. Which motor is used in compressors?
(a) DC series motor
(b) Reluctance motor
(c) Capacitor-start–capacitor-run motor
(d) Nonc of thcsc.

64. Centrifugal switch disconnects the auxiliary winding of a single-phase motor at about
(a) 40% to 50 % of synchronous speed
(b) 50% to 60 % of synchronous speed
(c) 60% to 70 % of synchronous speed
(d) 70% to 80 % of synchronous speed.

65. Which motor is used in portable drills?
(a) DC series motor
(b) Reluctance motor
(c) Capacitor-start–capacitor-run motor
(d) Universal motor.

66. Which motor can run on both AC and DC?
(a) Universal motor
(b) Reluctance motor
(c) Capacitor-start–capacitor-run motor
(d) DC series motor.

67. The power factor of a single-phase induction motor is
(a) leading
(b) lagging
(c) unity
(d) It does not have any power factor.

68. The revolving field in a shaded-pole single-phase induction motor is produced by
(a) resistance
(b) inductance
(c) capacitor
(d) shaded coils.

69. Which motor is used in mixies?
(a) DC series motor
(b) Reluctance motor
(c) Capacitor-start–capacitor-run motor
(d) Universal motor.

70. In a capacitor-start–single-phase induction motor, the displacement between starting and running winding is about
(a) 90^0
(b) 60^0
(c) 30^0
(d) 0^0.

71. The rotor of a three-phase synchronous motor rotates at
(a) synchronous speed
(b) more than synchronous speed
(c) less than synchronous speed
(d) It has no relation with synchronous speed.

72. A three-phase, 6-pole, 440 V, 50 Hz induction motor runs at 960 r.p.m. The slip of the motor is
(a) 0.02
(b) 0.04
(c) 0.06
(d) It cannot be calculated.

73. The shape of torque-slip curve of a three-phase induction motor is a
(a) rectangular parabola
(b) parabola
(c) hyperbola
(d) straight line.

74. If N_s is the synchronous speed of a three-phase induction motor and N_r is the speed of rotation of the rotor, then the slip of the motor is given by
(a) $\% \text{ slip} = \frac{N_r - N_s}{N_s} \times 100$
(b) $\% \text{ slip} = \frac{N_r - N_s}{N_r} \times 100$
(c) $\% \text{ slip} = \frac{N_s - N_r}{N_s} \times 100$
(d) $\% \text{ slip} = \frac{N_s - N_r}{N_r} \times 100$

75. The number of slip-rings in a squirrel-cage three-phase induction motor is
(a) 0
(b) 2
(c) 4
(d) 6.

76. A synchronous motor with synchronous speed of 1500 r.p.m. has
(a) 2 poles
(b) 4 poles
(c) 6 poles
(d) 8 poles.

77. The noise produced by a transformer is termed as
(a) ringing
(b) zoom
(c) humming
(d) buzzing.

78. In a transformer, the resistance between primary winding and secondary winding is
(a) 0
(b) infinity
(c) very low
(d) None of these.

79. The efficiency of two identical transformers on load is determined by
(a) open-circuit test
(b) short-circuit test
(c) back-to-back test
(d) All of these.

80. Short-circuit test on a transformer gives
(a) eddy current loss
(b) core loss
(c) hysteresis loss
(d) copper loss.

81. Open-circuit test on a transformer gives
(a) eddy current loss
(b) core loss
(c) hysteresis loss
(d) copper loss.

82. At maximum efficiency of a transformer,
(a) Copper loss = Iron loss
(b) Copper loss > Iron loss
(c) Copper loss < Iron loss
(d) None of these.

83. If the full-load copper loss in a transformer is 1600 W, then at half load it will be
(a) 3200 W
(b) 800 W
(c) 400 W
(d) 200 W.

84. In the e.m.f. equation of the transformer, the flux is
(a) average value
(b) r.m.s. value
(c) maximum value
(d) instantaneous value.

85. Power transformer has maximum efficiency at near to
(a) no-load
(b) half-load
(c) full-load
(d) It does not depend on load.

86. Buchholz's relay protects against
(a) fault inside the transformer
(b) fault outside the transformer
(c) all kinds of faults.
(d) None of these.

87. Zero voltage regulation is possible in transformer at
(a) zero power factor
(b) unity power factor
(c) lagging power factor
(d) leading power factor.

88. The rating of the transformer is in
(a) Kilo Volt
(b) Ampere
(c) Kilo Watt
(d) Kilo Volt Ampere (KVA).

89. Harmonics in transformer is due to
(a) voltage
(b) saturation of the core
(c) vibrations
(d) power factor.

90. The maximum efficiency of a distribution transformer is at
(a) full-load
(b) 50 % of full-load
(c) 25 % of full load
(d) no-load

91. The tappings in transformer are provided on
(a) primary side
(b) secondary side
(c) low-voltage side
(d) high-voltage side.

92. Transformer breaths in when
(a) load decreases
(b) load increases
(c) load remains constant
(d) It does not depend on the load.

93. Sumpner's test on transformer gives
(a) all-day efficiency
(b) windage loss
(c) temperature
(d) All of these.

94. The iron loss in a DC motor occurs in
(a) the field
(b) the brush
(c) the commutator
(d) the armature.

95. The retardation test on a DC shunt motor gives
(a) eddy current loss
(b) windage loss
(c) stray loss
(d) copper loss.

96. Hopkinson's test on a DC machines is conducted at
(a) no-load
(b) half-load
(c) full-load
(d) It does not depend on load.

97. In elevators, the braking used is
(a) plugging
(b) rheostatic braking
(c) regenerative braking
(d) None of these.

98. If the speed of a DC motor is increased, its back e.m.f. will
(a) decrease
(b) increase
(c) remain constant
(d) become zero.

99. A DC generator can be considered as a
(a) rectifier
(b) prime mover
(c) pump
(d) rotating amplifier.

100. The e.m.f. of a DC generator is proportional to its
(a) speed
(b) flux
(c) number of poles
(d) All the above.

ANSWERS TO MULTIPLE CHOICE QUESTIONS

Question	Answer	Question	Answer
1	B	2	C
3	B	4	C
5	C	6	B
7	B	8	C
9	D	10	D
11	A	12	B
13	B	14	C
15	C	16	A
17	D	18	B
19	C	20	D
21	D	22	D
23	D	24	D
25	A	26	A
27	A	28	D
29	C	30	B
31	C	32	D
33	A	34	A
35	C	36	A
37	A	38	C
39	B	40	A
41	B	42	B
43	D	44	A
45	C	46	C
47	B	48	D
49	B	50	B
51	A	52	D
53	B	54	B
55	D	56	D
57	D	58	A
59	B	60	C
61	B	62	D
63	C	64	D
65	D	66	A
67	B	68	D

69	D	70	A
71	C	72	B
73	A	74	C
75	A	76	B
77	C	78	B
79	C	80	D
81	C	82	A
83	C	84	C
85	C	86	A
87	D	88	D
89	B	90	B
91	C	92	A
93	C	94	D
95	C	96	C
97	A	98	B
99	D	100	A

Appendix B

SHORT QUESTIONS AND ANSWERS

1. What will happen if a transformer is connected to a DC supply?
2. Give the application of equivalent circuit of a single-phase transformer.
3. Define synchronous speed.
4. Define torque.
5. How can eddy current loss in an electrical machine be reduced?
6. Explain, why in a DC generator, the series field winding has low resistance while the shunt field winding has high resistance.
7. What is the no-load power factor of a transformer?
8. What is the power factor of a transformer on load?
9. What is the phase angle between the voltage and the no-load current of a transformer?
10. How can you reduce the magnetic leakage in a transformer?
11. What is the turns ratio of a transformer?
12. What is the voltage ratio of the transformer?
13. Why is the L.T. winding of a transformer placed near to the core?
14. What is a power transformer?
15. What is a distribution transformer?
16. What does the size and the construction of bushings in a transformer depend upon?
17. What will happen if a DC motor is operated below its rated speed?
18. What will be the direction of rotation of a shaded-pole single-phase induction motor?
19. How can the synchronous motor be used as a synchronous condenser?
20. What happens when the field current of a synchronous motor is increased beyond the normal value at constant input?
21. At what load angle does power develop in a synchronous motor to become its maximum value?
22. What is the effect on speed if the load is increased on a three-phase synchronous motor?
23. What is circle diagram of an I M?
24. State the advantages of skewing.
25. What happens if the air gap flux density in an induction motor increases?
26. What would happen if a three-phase induction motor is switched on with one phase disconnected?
27. Why are centrifugal switches provided on many single-phase induction motors?

28. What is meant by infinite bus-bars?
29. How is synchronoscope is used for synchronising alternators?
30. Why are alternators rated in kVA and not in kW?
31. What is the relation between electrical degree and mechanical degree?
32. How does electrical degree differ from mechanical degree?
33. What is the function of commutator in a DC machine?
34. Name the various methods for predetermining the voltage regulation of a three-phase alternator
35. If P_s is the number of stator poles and P_r is the number of rotor teeth of a stepper motor, then what will be the expression for the step-angle?
36. What are the advantages of three-phase motor over single-phase motor?
37. What are turbo alternators?
38. What are the advantages and disadvantages of short-pitched winding?
39. What are the various methods of measuring slip?
40. What is the advantage of skewed stator slots in the rotor of induction motors?
41. What is meant by armature reaction?
42. Why should dry-type transformers never be overloaded?
43. State the difference between generator and alternator
44. What are the different methods for the starting of a synchronous motor?
45. Name the types of motors used in vacuum cleaners, phonographic appliances, vending machines, refrigerators, rolling mills, lathes, power factor improvement and cranes.
46. Do commutators convert AC to DC or DC to AC?
47. Why are armature teeth of a DC machine skewed?
48. Which motor has high starting torque and starting current? DC motor, Induction motor or Synchronous motor?
49. Define stepper motor. What is the use of stepper motor?
50. What is a two-phase motor?

ANSWERS

1. If DC supply is given to the transformer, the current will not change due to constant supply. So, mutual induction is not possible and thus the transformer will not work. The primary winding resistance is very small. For DC supply, the primary winding inductive reactance is zero. Hence, the primary will draw very high current for DC supply, and as a result, extra heat is generated. This heat damages the transformer. Therefore, DC supply should not be applied to the transformer.
2. The equivalent circuit is the electrical model of the transformer. Once the equivalent circuit parameters are obtained, the regulation and efficiency of the transformer at any power factor and load conditions can be obtained without actually loading the transformer.
3. For synchronous machines, there exists a fixed relationship between the number of poles P, frequency f and the speed of the machine. The speed of the synchronous machine for the given number of poles and the rated frequency is called the synchronous speed (N_s). The expression for synchronous speed is

$$N_s = \frac{120f}{P}$$

where

f = frequency

P = No of poles of the machine

4. A turning or twisting force about an axis is defined as torque.
5. By using laminated core construction, the eddy current loss can be reduced. The laminated core divides the solid iron core into thin laminations. The path of eddy currents is broken due to the insulating material sheets between the laminations. Thus, the eddy currents and losses generated by them can be minimised.
6. Series field winding is always connected in series with the armature. Hence, it has to carry the armature current which directly gets decided by the load. So, the current passing though series field winding is of high level. The voltage drop across series field winding gets added to the voltage drop across armature winding while deciding the back e.m.f. This voltage drop must be very small. Hence, as the current through series field winding is high, its resistance is kept very low in order to restrict the voltage drop across it to a small value. The shunt field winding is directly connected across the rated supply voltage. Hence, to limit the current through it, resistance is very high.
7. Very low and lagging
8. The power factor on load is equal to the power factor of the load the transformer is carrying.
9. The no-load current lags the voltage by 80^0 to 85^0
10. By sectionalising and interleaving the primary and secondary winding.
11. It is the ratio of the number of turns of primary to the number of turns of secondary
12. It is the ratio of the voltage between the line terminals of one winding to that between the line terminals of another winding at no load.
13. To reduce the dielectric strength of the insulating material of the winding.
14. The transformers that are used in transmission lines for transmission of large amount of power.
15. It is a transformer used for distribution of power from the transmission lines. Its secondary is directly connected with the consumer load.
16. Voltage and the size of the tank.
17. The motor will overheat because more field current is required to maintain the rated voltage.
18. The motor rotates in the direction specified from unshaded to shaded region in the pole phase
19. Synchronous motor is operated on over-excitation so as to draw leading reactive current and power from the supply lines. This compensates the lagging current and power requirement of the load, making the system power factor become unity. The motor does the job of capacitors and hence, it is called as synchronous condenser.
20. Increase in e.m.f. causes the motor to have reactive current in the leading direction. The additional leading reactive current decreases the magnitude of line current, accompanied by the decrease in power factor.
21. When load angle become equal to impedance angle.
22. The speed of operation remains constant from no-load to maximum load in the motor operating at constant frequency bus bars.
23. When an I M operates on constant voltage and constant frequency source, the loci of stator current phasor is found to fall on a circle. This circle diagram is used to predict the performance of the machine at different loading conditions as well as mode of operation.
24. It reduces humming and hence, quiet running of the motor is achieved. It reduces magnetic locking of the stator and rotor.

25. The increase in air gap flux increases iron loss and hence, efficiency decreases.
26. The motor is likely to burn.
27. Centrifugal switches are provided on many single-phase induction motors to disconnect the starting / auxiliary winding from the supply when the motor reaches about 70% of its synchronous speed.
28. The source or supply lines with non-variable voltage and frequency are called infinite bus-bars. The source lines are said to have zero source impedance and infinite rotational inertia.
29. Synchronoscope can be used for permanently connected alternators where the correctness of phase sequence is already checked by other means. Synchronoscope is capable of rotating in both directions. The rate of rotation of the pointer indicates the extent of frequency difference between the alternators. The direction of rotation indicates whether incoming alternator frequency is higher or lower than the existing alternator. The TPST switch is closed to synchronise the incoming alternator when the pointer faces the top thick line marking.
30. The continuous power rating of any machine is generally defined as the power that can be delivered by the machine or apparatus for a continuous period so that the loss incurred in the machine gives rise to a steady temperature rise not exceeding the limit prescribed by the insulation class. Apart from the constant loss incurred in alternators, there is also copper loss, which occurs in the three-phase winding and depends on I^2R, the square of the current delivered by the generator. As the current is directly related to the apparent power delivered by the generator , alternators have only their apparent power in VA/kVA/MVA as their power rating.
31. $\theta_e = \frac{P}{2}\theta_m$
32. Mechanical degree is the unit for accounting the angle between two points based on their mechanical or physical placement. Electrical degree is used to account the angle between two points in rotating electrical machines. Since all electrical machines operate with the help of magnetic fields, the electrical degree is accounted with reference to the magnetic field. The angle between adjacent North and South poles is accounted as 180 electrical degrees.
33. A DC machine is an alternating current machine. The e.m.f. generated in the conductor is alternating in nature. With the help of commutator segments, this alternating e.m.f. is converted into a direct one.
34. E.M.F. method, MMF method and ZPF method.
35. The expression for the step-angle is

$$\theta = \frac{P_s - P_r}{P_s P_r} \times 360$$

36. Three-phase motors have the following advantages over single-phase motor:
    (i) Higher starting torques
    (ii) Improved speed regulation
    (iii) Less vibration
    (iv) Quieter operation
37. Turbo alternators are high-speed alternators. Because of high speed of rotation, the rotor diameter is reduced and the axial length is increased. Two or four poles are generally used and steam turbines are used as prime movers.

38. The advantages are:
1. They save copper for end connections
2. They improve the waveform of the generated E.M.F.
3. The generated E.M.F. can be made to approximate to a sine wave more easily and the distorting harmonics can be reduced.

The disadvantages are:
1. The total voltage around the coils is somewhat reduced because the voltage induced in the two coil sides is slightly out of phase.

39. 1. By actual measurement of rotor speed
2. By measurement of rotor frequency
3. Stroboscopic method

40. In the induction motor design, the rotor slots are designed with a slight skew arrangement. It will not be parallel to the shaft. This is for reducing the magnetic locking or magnetic attraction between stator and rotor teeth. Moreover, this arrangement will help to reduce the magnetic hum and noise.

41. The effect of armature flux on the main flux is called armature reaction. The armature flux may either support or oppose the main flux.

42. Overloading of a transformer results in excessive temperature. This excessive temperature causes overheating, which will result in rapid deterioration of the insulation and cause complete failure of the transformer coils.

43. Generator and alternator are two devices, which convert mechanical energy into electrical energy. Both work on the principle of electromagnetic induction, the only difference being in their mode of construction. The generator has a stationary magnetic field and rotating conductor which rolls on the armature with slip-rings and brushes riding against each other. Hence, it converts the induced e.m.f. into dc current for external load. In the case of an alternator, it has a stationary armature and rotating magnetic field for high voltages but for low voltage output, rotating armature and stationary magnetic field is used.

44. Synchronous motor can be started by the following two methods:
By means of an auxiliary motor: The rotor of a synchronous motor is rotated by auxiliary motor. Then, rotor poles are excited due to which the rotor field is locked with the stator-revolving field and continuous rotation is obtained.
By providing damper winding: Here, bar conductors are embedded in the outer periphery of the rotor poles and are short-circuited with the short-circuiting rings at both sides. The machine is started as a squirrel-cage induction motor first. When it picks up speed, the rotor is excited and it starts rotating continuously as the rotor field is locked with the stator revolving field.

45. The following motors are used:
Vacuum cleaners – Universal motor.
Phonographic appliances – Hysteresis motor.
Vending machines – Shaded pole motor.
Refrigerators – Capacitor split-phase motors.
Rolling mills – Cumulative motors.
Lathes – DC shunt motors.
Power factor improvement – Synchronous motors.

46. Commutator in a DC machine converts AC to DC current and DC to AC current. Consider a DC generator. The current produced in the armature due to electro-magnetic induction is AC. However, the commutator converts bi-directional current to uni-directional current and supplies it to the external load circuit. In DC motor, the voltage we apply at the terminals is DC. This DC current is converted to AC and supplied to the armature circuit so as to produce the rotational torque. Hence, commutator converts both AC to DC variables and DC to AC variables.

47. The air gap length changes between the slot and teeth of the armature when the armature is rotating under constant magnetic field. This causes a small variation in the torque, which can be observed as small vibrations in the frame of the machine. So, in order to avoid this vibration effect, the armature teeth is skewed.

48. DC series motor has high starting torque. We can start the induction motor and synchronous motors on load, but cannot start the DC series motor without load.

49. The motor which works or acts on an applied input pulse is called a stepper motor. It comes under the category of synchronous motor and often does not fully depend on the complete cycle. It works in either direction, in steps. Hence, it is mainly used in automation parts.

50. A two-phase motor is a motor in which the starting winding and the running winding have a phase split. The AC servo motor is an example for this, where the auxiliary winding and the control winding have a phase split of 90 degrees.

APPENDIX C

MATLAB CODES

C.1 MATLAB CODE FOR EX.1.1

```
% Solution to Ex.1.1
clc;clear;
B= input('Enter the value of magnetic field in Tesla');
r= input('Enter the value of radius in metre');
I= input('Enter the value of current in ampere');
l= input('Enter the value of length in metre');
T=-2*B*r*I*l;
fprintf('Torque (T)= %f sin(alfa)\n',T)
```

C.2 MATLAB CODE FOR EX.1.2

```
% Solution to Ex.1.2
clc;clear;
N= input('Enter the value of N in number of turns');
g= input('Enter the value of g in millimetre');
d= input('Enter the value of d in metre');
x= input('Enter the value of x in metre');
l= input('Enter the value of l in metre');
i= input('Enter the value of i in amp');
L=((4*pi*10^-7)*N*N*l*(d-x))/(2*g*10^-3);
W=0.5*L*i*i;
F=((4*pi*10^-7)*l*N*N*i*i)/(4*g*10^-3);
fprintf('Inductance(L)= %f H \n',L)
fprintf('Magnetic stored energy(W)= %f J \n',W)
fprintf('Force acting on the plunger (F)= %f N \n',F)
```

C.3 MATLAB CODE FOR EX.2.1

```
% Solution to Ex.2.1
clc;clear;
P= input('Enter the number of poles');
N= input('Enter the speed of rotation in RPM');
Z= input('Enter the number of conductors');
```

```
phi= input('Enter the flux per pole in Wb');
display('FOR LAB WINDING')
A=P;
el=(phi*Z*N*P)/(60*A);
fprintf('\n Induced e.m.f.= %f Volt \n',el)
display('FOR WAVE WINDING')
A=2;
ew=(phi*Z*N*P)/(60*A);
fprintf('\n Induced e.m.f.= %f Volt \n',ew)
```

C.4 MATLAB CODE FOR EX.2.2

```
% Solution to Ex.2.2
clc;clear;
P= input('Enter the number of poles');
N= input('Enter the speed of rotation in RPM');
Z= input('Enter the number of conductors');
e= input('Enter the generated e.m.f. in volts');
display('FOR WAVE WINDING')
A=2;
phi=(60*A*e)/(Z*N*P);
fprintf('\n Flux per pole= %f wb \n',phi)
```

C.5 MATLAB CODE FOR EX.2.3

```
% Solution to Ex.2.3
clc;clear;
P= input('Enter the number of poles');
N= input('Enter the speed of rotation in RPM');
S= input('Enter the number of slots');
C=input('Enter the number of conductors per slot');
phi= input('Enter the flux per pole in Wb');
Z=S*C;
display('FOR LAB WINDING')
A=P;
el=(phi*Z*N*P)/(60*A);
fprintf('\n Induced e.m.f.= %f Volt \n',el)
display('FOR WAVE WINDING')
A=2;
ew=(phi*Z*N*P)/(60*A);
fprintf('\n Induced e.m.f.= %f Volt \n',ew)
```

C.6 MATLAB CODE FOR EX.2.4

```
% Solution to Ex.2.4
clc;clear;
P= input('Enter the number of poles');
```

```
e= input('Enter the induced e.m.f. in volts');
S= input('Enter the number of slots');
C=input('Enter the number of conductors per slot');
phi= input('Enter the flux per pole in Wb');
Z=S*C;
display('FOR LAB WINDING')
A=P;
N1=(60*A*e)/(phi*Z*P);
fprintf('\n Speed of rotation= %f r.p.m. \n',N1)
display('FOR WAVE WINDING')
A=2;
N2=(60*A*e)/(phi*Z*P);
fprintf('\n Induced e.m.f.= %f Volt \n',N2)
```

C.7 MATLAB CODE FOR EX.2.5

```
% Solution to Ex.2.5
clc;clear;
P= input('Enter the number of poles');
N= input('Enter the rotation in r.p.m.');
S= input('Enter the number of turns');
C=input('Enter the number of coils');
l=input('Enter the length of the armature in metre');
d=input('Enter the diameter of the armature in metre');
b=input('Enter the flux density in Wb/sq metre');
a= input('Enter the angle in degree');
display ('WAVE WINDING');
A=2;
Z=C*S*A;
r=d/2;
ag=(2*pi*r*l*a)/(P*180);
phi=b*ag;
e=(phi*Z*N*P)/(60*A);
fprintf('\n Induced e.m.f.= %f volts \n',e)
```

C.8 MATLAB CODE FOR EX.2.6

```
% Solution to Ex.2.6
clc;clear;
P=input('Enter the number of poles');
N=input('Enter the speed of rotation in r.p.m.');
e=input('Enter the e.m.f. in volts');
Z=input('Enter the number of conductors');
d=input('Enter the diameter of the pole shoe');
l=input('Enter the length of the pole shoe');
r=input('Enter the ratio of pole arc to pole pitch');
```

```
pp=(pi*d)/P;
A=P;
pa=r*pp;
ap=pa*l;
phi=(60*A*e)/(Z*N*P);
B=phi/ap;
display('Lap winding')
fprintf('\n Flux per pole = %f Wb \n',phi)
fprintf('\n Flux density = %f Wb/m^2 \n',B)
```

C.9 MATLAB CODE FOR EX.2.8

```
% Solution to Ex.2.8
clc;clear;
P=input('Enter the number of poles');
Z=input('Enter the number of conductor');
I=input('Enter the current in ampere');
B1=input('Enter the brush shift in electrical degree');
A=2;
display('WAVE WINDING');
DA=(Z*I*B1)/(180*A*P);
CA=((Z*I)/(A*P))*((180-(2*B1))/180);
display('When the brush shift is given in  electrical degree')
fprintf('\n Demagnetising ampere turns = %f AT/pole \n',DA)
fprintf('\n Cross-magnetising ampere turns = %f AT/pole \n',CA)
display('When the brush shift is given in  mechanical degree')
B2=input('Enter the brush shift in mechanical degree');
bd=(P/2)*B2;
DA=(Z*I*bd)/(180*A*P);
CA=((Z*I)/(A*P))*((180-(2*bd))/180);
display('When the brush shift is given in  mechanical degree')
fprintf('\n Demagnetising ampere turns = %f AT/pole \n',DA)
fprintf('\n Cross-magnetising ampere turns = %f AT/pole \n',CA)
display('When the brush is in GNP')
B3=0;
DA=(Z*I*B3)/(180*A*P);
CA=((Z*I)/(A*P));
display('When the brush is in GNP')
fprintf('\n Demagnetising ampere turns = %f AT/pole \n',DA)
fprintf('\n Cross-magnetising ampere turns = %f AT/pole \n',CA)
fprintf('\n \n \n');
A=P;
display('LAP WINDING');
DA=(Z*I*B1)/(180*A*P);
CA=((Z*I)/(A*P))*((180-(2*B1))/180);
```

```
display('When the brush shift is given in  electrical degree')
fprintf('\n Demagnetising ampere turns = %f AT/pole \n',DA)
fprintf('\n Cross-magnetising ampere turns = %f AT/pole \n',CA)
display('When the brush shift is given in  mechanical degree')
B2=input('Enter the brush shift in mechanical degree');
bd=(P/2)*B2;
DA=(Z*I*bd)/(180*A*P);
CA=((Z*I)/(A*P))*((180-(2*bd))/180);
display('When the brush shift is given in mechanical degree')
fprintf('\n Demagnetising ampere turns = %f AT/pole \n',DA)
fprintf('\n Cross-magnetising ampere turns = %f AT/pole \n',CA)
display('When the brush is in GNP')
B3=0;
DA=(Z*I*B3)/(180*A*P);
CA=((Z*I)/(A*P));
display('When the brush is in GNP')
fprintf('\n Demagnetising ampere turns = %f AT/pole \n',DA)
fprintf('\n Cross-magnetising ampere turns = %f AT/pole \n',CA)
```

C.10 MATLAB CODE FOR EX.2.12

```
% Solution to Ex.2.12
clc;clear;
P=input('Enter the power of each lamp');
C=input('Enter the number of lamps');
V=input('Enter the terminal voltage');
ra=input('Enter the armature resistance');
rf=input('Enter the field resistance');
bv=input('Enter the brush voltage');
Pole=input('Enter the number of poles');
iff=V/rf;
ia=(P*C)/V;
i=ia+iff;
display('WAVE WINDING')
A=2;
iaa=ia/A;
e=V+ia*ra+2*bv;
fprintf('Current in each armature conductor = %f  A \n', iaa)
fprintf('Generated e.m.f. = %f  V \n', e)
display('LAP WINDING')
A=Pole;
iaa=ia/A;
e=V+ia*ra+2*bv;
fprintf('Current in each armature conductor = %f  A \n', iaa)
fprintf('Generated e.m.f. = %f  V \n', e)
```

C.11 MATLAB CODE FOR EX.2.13

```
% Solution to Ex.2.13
clc;clear;
ra=input('Enter the value of armature resistance');
rsh=input('Enter the value of shunt field resistance');
rse=input('Enter the value of series field resistance');
il=input('Enter the value of load current');
v=input('Enter the value of voltage');
display('LONG SHUNT')
iff=v/rsh;
ia=il+iff;
e=ia*ra+ia*rse+v;
fprintf('\n Generated e.m.f. = %f volt \n',e)
display('SHORT SHUNT')
vsh=v+il*rse;
iff=vsh/rsh;
ia=il+iff;
e1=ia*ra+iff*rsh+v;
fprintf('\n Generated e.m.f. = %f volt \n',e)
```

C.12 MATLAB CODE FOR EX.3.2

```
% Solution to Ex.3.2
clc;clear;
P=input('Enter the number of poles');
C=input('Enter the number of conductors');
B=input('Enter the flux density in Tesla');
lp=input('Enter the length of pole shoe in metres');
la=input('Enter the length of armature in metres');
N=input('Enter the speed in r.p.m.');
i=input('Enter the current in ampere');
A=2; % wave winding
phi=(pi*la*lp*B)/P;
e=(phi*C*N*P)/(60*A);
pm=e*i;
t=(pm*60)/(2*pi*N);
display('WAVE WINDING')
fprintf('\n E.M.F. = %f volts \n',e)
fprintf('\n Mechanical Power = %f W \n',pm)
fprintf('\n Torque = %f Nm \n',t)
```

C.13 MATLAB CODE FOR EX.3.3

```
% Solution to Ex.3.3
clc;clear;
ra=input('Enter the armature resistance in ohm');
```

```
rf=input('Enter the field resistance in ohm');
N=input('Enter the speed in r.p.m');
i=input('Enter the current in ampere');
v=input('Enter the voltage in volts');
ex=input('Enter the external resistance in ohm');
e1=v-i*ra-i*rf;
k_phi1=e1/N;
e2=v-i*ra-12.5*rf;
k_phi_2=0.8*k_phi1;
n1=e2/k_phi_2;
fprintf('\n SPEED = %f r.p.m. \n',n1)
```

C.14 MATLAB CODE FOR EX.3.4

```
% Solution to Ex.3.4
clc;clear;
v=input('Enter the voltage in volt');
i=input('Enter the current in ampere');
N=input('Enter the speed in r.p.m.');
ex=input('Enter the excitation in %');
ra=input('Enter the armature resistance');
eb=v-i*ra
i1=i/(ex/100)
eb1=v-i1*ra
r=(eb/eb1)*(ex/100)
n1=N/r;
fprintf('\n SPEED = %f r.p.m. \n',n1)
```

C.15 MATLAB CODE FOR EX.3.6

```
% Solution to Ex.3.6
clc;clear;
v=input('Enter the voltage in volt');
N=input('Enter the speed in r.p.m.');
ra=input('Enter the armature resistance in ohm');
i=input('Enter the current in ampere');
sr=[2 1.5 1 0.5 ra];
p(1)=sr(1)+sr(2)+sr(3)+sr(4)+sr(5);
p(2)=sr(2)+sr(3)+sr(4)+sr(5);
p(3)=sr(3)+sr(4)+sr(5);
p(4)=sr(4)+sr(5);
p(5)=sr(5);
e=0;
 e1=v-i*ra;
for a= 1:length(sr)
   ist=(v-e)/p(a);
   fprintf('\n Position = %f \n',a)
```

```
   fprintf('\n Starting current = %f amp \n',ist)
   e(a+1)=v-i*p(a);
   n1=(e(a+1)/e1)*N;
   fprintf('\n Speed = %f r.p.m. \n',n1)
   e=e(a+1);
end
```

C.16 MATLAB CODE FOR EX.3.7

```
% Solution to Ex.3.7
clc;clear;
rf=input('Enter the field resistance in ohm');
ra=input('Enter the armature resistance in ohm');
v=input('Enter the voltage in volts');
il=input('Enter the load current in ampere');
ie=input('Enter the current need to calculate the efficiency');
p=v*il;
iff=v/rf;
ia=il-iff;
culoss=ia*ia*ra;
constloss=p-culoss;
ia1=30-iff;
culoss1=ia1*ia1*ra;
tloss=constloss+culoss1;
eff=(((v*ie)-tloss)/(v*ie))*100;
fprintf('\n  Efficiency in percentage = %f \n ',eff)
```

C.17 MATLAB CODE FOR EX.3.9

```
% Solution to Ex.3.9
clc;clear;
v=input('Enter the supply voltage in volt');
i=input('Enter the line current in amp');
iam=input('Enter the armature current of the motor in amp');
ifm=input('Enter the field current of the motor in amp');
ifg=input('Enter the field current of the generator in amp');
ram=input('Enter the armature resistance of the motor in ohm');
rag=input('Enter the armature resistance of the generator in ohm');
vb=input('Enter the brush voltage drop per brush');
ip=v*i;
% Motor
cla=iam*iam*ram;
bl=iam*(vb*2);
clf=v*ifm;
tlm=cla+bl+clf;
% Generator
cla1=(iam-i+ifm+ifg)*(iam-i+ifm+ifg)*rag;
```

```
bl1=(iam-i+ifm+ifg)*(vb*2);
clf1=v*ifg;
tlg=cla1+bl1+clf1;
 T=tlm+tlg;
sl=ip-T;
 % Efficiency of motor
slpm=sl/2;
tlm1=tlm+slpm;
mip=v*(iam+ifm);
eff=((mip-tlm1)/mip)*100;
fprintf('\n Motor efficiency in percentage = %f \n', eff)
 % Efficiency of generator
tlg1=tlg+slpm;
gip=v*(iam-i+ifm);
effg=(gip/(gip+tlg1))*100;
fprintf('\n Generator efficiency in percentage = %f \n', effg)
```

C.18 MATLAB CODE FOR EX.3.12

```
ar=input('Enter the value of the armature resistance in ohm');
fr=input('Enter the value of field resistance in ohm');
v=input('Enter the value of the supply voltage in volts');
nlc=input('Enter the value of the no load current in amp');
sp=input('Enter the value of speed in RPM');
flc=input('Enter the value of the full load current in amp');
iff=v/fr
io=nlc-iff
eo=v-io*ar
irf=flc-iff
eof=v-irf*ar
nf=(eof/eo)*sp
lnl=v*nlc-iff*iff*fr-io*io*ar
mo=v*flc-iff*iff*fr-irf*irf*ar-lnl
t=(60*mo)/(2*pi*nf)
```

C.19 MATLAB CODE FOR EX.4.1 AND EX.4.2

```
clear;clc;
kva=input('Enter the KVA rating of the transformer');
v1=input('Enter the primary side voltage');
v2=input('Enter the secondary side voltage');
f=input('Enter the frequency');
emf=input('Enter the emf per turn');
bm=input('Enter the flux denisty');
phim=emf/(4.44*f)
ca=phim/bm
n1=v1/(4.44*f*phim)
```

```
n2=v2/(4.44*f*phim )
i1=kva/v1
i2=kva/v2
```

C.20 MATLAB CODE FOR EX.4.4

```
clear;clc;
v1=input('Enter the primary side voltage');
v2=input('Enter the secondary side voltage');
f=input('Enter the frequency of supply');
io=input('Enter the no load current');
pf=input('Enter the power factor');
n1=input('Enter the number of turns on the primary');
theta=acos(pf)
im=io*sin(theta)
ic=io*cos(theta)
loss=v1*io*pf
n2=(v2/v1)*n1
phim=v1/(4.44*f*n1)
```

C.21 MATLAB CODE FOR EX.4.5

```
clear;clc;
n1=input('Enter the primary side turns');
n2=input('Enter the secondary side turns');
io=input('Enter the no load current');
pfo=input('Enter the no load power factor');
pf2=input('Enter the power factor on load');
i2=input('Enter the load current');
thetao=acos(pfo)
theta2=acos(pf2)
i2d=i2*(n2/n1)
im=io*sin(thetao)
ic=io*cos(thetao)
ioc=ic+j*im
i2c=i2d*cos(theta2)+j*i2d*sin(theta2)
i1=ioc+i2c
```

C.22 MATLAB CODE FOR EX.4.8

```
clear;clc;
z1=input('Enter the impedance of the H.V. side in a+bi form');
z2=input('Enter the impedance of the L.V. side in a+bi form');
ro=input('Enter the valur of Ro');
xo=input('Enter the valur of Xo');
v1=input('Enter the voltage on the H.V. side');
v2=input('Enter the voltage on the L.V. side');
```

```
display('EQUIVALENT CIRCUIT REFERRED TO H.V.SIDE')
r1=real(z1)
x1=imag(z1)
r2=real(z2);
x2=imag(z2);
k=v2/v1;
r2d=r2/(k*k)
x2d=x2/(k*k)
display('EQUIVALENT CIRCUIT REFERRED TO L.V.SIDE')
r1d=r1*(k*k)
x1d=x1*(k*k)
rod=ro*(k*k)
xod=xo*(k*k)
```

C.23 MATLAB CODE FOR EX.4.9

```
clear;clc;
kva=input('Enter the KVArating of the transformer')
pfc=input('Enter the power factor at which eff to be calculated')
display('Open Circuit test Data');
vo=input('Enter the voltage value');
io=input('Enter the current value');
wo=input('Enter the power');
display('Short Circuit test Data');
vs=input('Enter the voltage value');
is=input('Enter the current value');
ws=input('Enter the power');
display('From OC test data')
pfo=wo/(vo*io)
thetao=acos(pfo)
ic=io*cos(thetao)
im=io*sin(thetao)
ro=vo/ic
xo=vo/im
display('From SC test data')
pfs=ws/(vs*is)
ze=vs/is
re=ws/(is*is)
xe=sqrt((ze*ze)-(re*re));
loss=wo+ws
eff=(kva*pfc)/((kva*pfc)+loss)
```

C.24 MATLAB CODE FOR EX.4.14

```
clear;clc;
il=input('Enter the iron loss of the transformer');
cl=input('Enter the copper loss of the transformer');
```

```
kva=input('Enter the KVA rating of the transformer');
a=[7 410 0.8;9 298 0.79;5 150 0.75;3 50 0.8];
l=length(a)
tcll=0;
op=0;
for i=1:l
   lo(i)=a(i,2)/a(i,3);
   lm(i)=((lo(i)/kva)^2)*cl;
   tcll=tcll+(lm(i)*a(i,1));
   op=op+(a(i,1)*a(i,2));
end
til=il*24;
eff=(op/(op+tcll+til))*100
```

C.25 MATLAB CODE FOR EX.5.5

```
clear;clc;
vp=input('Enter the primary voltage');
vs=input('Enter the secondary voltage');
kva=input('Enter the KVA rating of the transformer');
lr=input('Enter the leakage reactance per transformer');
lo=input('Enter the load per phase');
display('STAR-STAR CONNECTION')
vlp=sqrt(3)*vp
vls=sqrt(3)*vs
rld=lo/((vls/vlp)^2)
zld=rld+lr*j
display('DELTA-STAR CONNECTION')
vlp=vp
vls=sqrt(3)*vs
rld=lo/((vls/vlp)^2)
zld=rld+lr*j
display('STAR-DELTA CONNECTION')
vlp=sqrt(3)*vp
vls=vs
rld=lo/((vls/vlp)^2)
zld=rld+lr*j
display('DELTA-DELTA CONNECTION')
vlp=vp
vls=vs
rld=lo/((vls/vlp)^2)
zld=rld+lr*j
```

C.26 MATLAB CODE FOR EX.6.2

```
clear;clc;
va = input('Enter the VA rating of the CT');
```

```
ip = input('Enter the primary current');
is = input('Enter the secondary current');
r = input('Enter the ratio of resistance to reactance');
il = input('Enter the iron loss');
im = input('Enter the magnetising current');
k = ip/is
del=atan(1/r)
cs = cos(del)
ss = sin(del)
ep = va/ip
ic = il/ep
ar = k+((ic*cs+im*ss)/is)
rr = ((k-ar)/ar)*100
pr = (180/pi)*((im*cs-ic*ss)/(k*is))
```

C.27 MATLAB CODE FOR EX.6.5

```
clear;clc;
vp = input('Enter the primary voltage');
vs = input('Enter the secondary voltage');
nvs = input('Enter the terminal voltage of secondary');
bd = input('Enter the durden');
r = input('Enter the resistance referred to secondary');
x = input('Enter the inductance referred to secondary');
k=vp/vs
z=sqrt(((real(bd))^2)+((imag(bd))^2))
del = atan(imag(bd)/real(bd))
r=nvs+(nvs*(r*cos(del)+x*sin(del)))/z
rr = ((nvs-r)/r)*100
```

C.28 MATLAB CODE FOR EX.7.5

```
clear;clc;
p = input('Enter the number of poles');
f=input('Enter the supply frequency');
v=input('Enter the supply voltage');
r1 = input('Enter the value of R1');
x1=input('Enter the value of x1');
r2d=input('Enter the value of R2 dash');
x2d=input('Enter the value of X2 dash');
ro=input('Enter the value of R0');
xo=input('Enter the value of X0');
sfl=input('Enter the value of slip atn full load');
sl=input('Enter the stator loss');
ns=(120*f)/p
nr=ns*(1-sfl)
vp=v/sqrt(3)
```

```
io=vp/(ro+xo*j)
[th0,mo]=cart2pol(real(io),imag(io))
i2d=vp/((r1+x1*j)+((r2d/sfl)+x2d*j))
[thi2d,moi2d]=cart2pol(real(i2d),imag(i2d))
ifl=io+i2d
mi=sqrt(3)*v*moi2d*cos(thi2d)
ri=mi-sl
rc=sfl*ri
mo=ri-sl-rc
eff=(mo/ri)*100
```

C.29 MATLAB CODE FOR EX.7.7

```
clear;clc;
display('Enter the NO LOAD test data')
vo=input('Enter the voltage');
io=input('Enter the current');
po=input('Enter the power');
display('Enter the BLOCK ROTOR  test data')
vb=input('Enter the voltage');
ib=input('Enter the current');
pb=input('Enter the power');
display('FROM NO LOAD TEST')
vop=vo/sqrt(3)
pop=po/3
ro=(vop*vop)/pop
xo=(vop*vop)/(sqrt(vop*vop*io*io)-(pop*pop))
display('BLOCK ROTOR TEST')
vbp=vb/sqrt(3)
pbp=pb/3
re=pbp/(ib*ib)
xe=(sqrt((vbp*vbp*ib*ib)-(pbp*pbp)))/ib
```

C.30 MATLAB CODE FOR EX.7.9

```
clc;clear;
r1=input('Enter the value of R1');
x1=input('Enter the value of X1');
r2=input('Enter the value of R2');
x2=input('Enter the value of X2');
xm=input('Enter the value of Xm');
f=input('Enter the frequency');
p=input('Enter the number of poles');
hp=input('Enter the horse power');
v=input('Enter the voltage');
s=input('Enter the slip');
vp=v/sqrt(3);
```

```
zeq=((r1+(x1*j))*(xm*j))/((r1+(x1*j))+(xm*j))
req=real(zeq)
xeq=imag(zeq)
i2=vp/sqrt((req+(r2/s))^2+(xeq+x2)^2)
(3*i2*i2*(r2/s))
(((2*pi)/60)*((120*f)/p))
pause
t=(3*i2*i2*(r2/s))/(((2*pi)/60)*((120*f)/p))
rr1=3*r1;xx1=3*x1;rr2=3*r2;xx2=3*x2;xxm=3*xm;
zeq=((rr1+(xx1*j))*(xxm*j))/((rr1+(xx1*j))+(xxm*j))
req=real(zeq)
xeq=imag(zeq)
i2=vp/sqrt((req+(rr2/s))^2+(xeq+x2)^2)
(3*i2*i2*(rr2/s))
(((2*pi)/60)*((120*f)/p))
pause
t=(3*i2*i2*(rr2/s))/(((2*pi)/60)*((120*f)/p))
```

C.31 MATLAB CODE FOR EX.8.1

```
clc;clear;
r1=input('Enter the value of R1');
x1=input('Enter the value of X1');
r2=input('Enter the value of R2 ');
x2=input('Enter the value of X2');
xo=input('Enter the value of Xm');
f=input('Enter the frequency');
p=input('Enter the number of poles');
v=input('Enter the voltage');
nr=input('Enter the speed of rotor');
ns=(120*f)/p
s=(ns-nr)/ns
z1=r1+x1*j
num1=r2/(2*s);
num2=(x2/2)*j;
num3=(xo/2)*j;
zf=((num1+num2)*num3)/(num1+num2+num3)
num11=r2/(2*(2-s));
zb=((num11+num2)*num3)/(num1+num2+num3)
zeq=z1+zf+zb
i1=v/zeq
[ti1 mi1]=cart2pol(real(i1),imag(i1))
ef=i1*zf
eb=i1*zb
iff=ef/(num1+num2*j)
ib=eb/(num11+num2*j)
[tff mff]=cart2pol(real(iff),imag(iff))
```

```
[tb mb]=cart2pol(real(ib),imag(ib))
pf=(mff^2)*num1
pb=(mb^2)*num11
pmf=(1-s)*pf
pmb=(1-(2-s))*pb
pnet=pmf+pmb
tf=pf/((2*pi*ns)/60)
tb=pb/((2*pi*ns)/60)
tet=tf-tb
rclf=s*pf
rclb=(2-s)*pb
tcl=rclf+rclb
npo=pnet-tcl
ip=v*mi1*cos(ti1)
eff=(npo/ip)*100
```

C.32 MATLAB CODE FOR EX.8.2

```
clear;clc;
n=input('Enter the speeed of the motor');
v=input('Enter the voltage');
f=input('Enter the frequency');
vm=input('Enter the main winding voltage');
im=input('Enter the main winding current');
wm=input('Enter the main winding power');
va=input('Enter the auxiliary winding voltage');
ia=input('Enter the auxiliary winding current');
wa=input('Enter the auxiliary winding power');
cos_theta_m=wm/(vm*im)
theta_m=acos(cos_theta_m)
cos_theta_a=wa/(va*ia)
theta_a=acos(cos_theta_a)
[rm img]=pol2cart(theta_m, im)
[ra iag]=pol2cart(theta_a,ia)
current_23=(rm+ra)+j*(img+iag);
[th,r]=cart2pol(real(current_23), imag(current_23));
current_115=r*(va/23)
```

C.33 MATLAB CODE FOR EX.8.5

```
clc;clear;
display('Main winding parameter')
vm=input('Enter the value of voltage');
im=input('Enter the value of current');
pm=input('Enter the value of power');
display('Auxiliary winding parameter')
va=input('Enter the value of voltage');
```

```
ia=input('Enter the value of current');
pa=input('Enter the value of power');
display('From Main winding parameter')
zm=vm/im
rm=pm/(im*im)
xm=sqrt(zm*zm-rm*rm)
x1=xm/2
x2d=x1
r1=rm/2
r2d=r1
tm=atan(x1/r1)
zeq=va/ia
req=pa/(ia*ia)
xeq=sqrt((zeq*zeq)-(req*req))
ra=req-r1
xa=xeq-x1
xc=xa-(tan(1.57-tm))*ra
```

C.34 MATLAB CODE FOR EX.10.3

```
clc;clear;
v=input('Enter the voltage value');
f=input('Enter the frequency');
po=input('Enter the power');
pf=input('Enter the power factor');
p=input('Enter the number of poles');
z=input('Enter the synchronous reactance');
ns=(120*f)/p
il=po/((sqrt(3))*v*pf)
ip=il/sqrt(3)
th=acos(pf)
[re,im]=pol2cart(th,ip)
ipc=complex(re,im)
emf=v-((z*j)*ipc)
t=po/((ns*2*pi)/60)
```

C.35 MATLAB CODE FOR V-CURVE PLOT

```
i_f = (38:1:58) / 10;
i_a = zeros(1,21);
x_s = 2.5;
v_phase = 208;
delta1 = -17.5 * pi/180;
e_a1 = 182 * (cos(delta1) + j * sin(delta1));
for ii = 1:21
  % Calculate magnitude of e_a2
  e_a2 = 45.5 * i_f(ii);
```

```
    delta2 = asin ( abs(e_a1) / abs(e_a2) * sin(delta1) );
    e_a2 = e_a2 * (cos(delta2) + j * sin(delta2));
    i_a(ii) = ( v_phase - e_a2 ) / ( j * x_s);
end
plot(i_f,abs(i_a),'Color','k','Linewidth',2.0);
xlabel('Field Current (A)','Fontweight','Bold');
ylabel('Armature Current (A)','Fontweight','Bold');
title ('Synchronous Motor V-Curve','Fontweight','Bold');
grid on;
```

C.36 MATLAB CODE FOR OCC & SCC PLOT OF SYNCHRONOUS MACHINE

```
clear;clc;
vc=[0 17.39 19.3 21.12 22.78 24.39 25.84 27.13 28.45 29.66 30.78 31.65 32.52 33.66 34.08 34.43 34.78
34.95]*1000;
iff=[0 350 400 450 500 550 600 650 700 750 800 850 900 1000 1050 1100 1150 1200];
isc=8960; ifc=500; isc=isc/ifc*iff;
  for i=1:length(vc)
    vag=vc(2)/iff(2)*iff(i);
    if vag>max(vc); break;end
  end
  plot(iff, vc, iff, isc,'-',[0 iff(i)],[0 vag],'--'); grid;
  title('OC and SC Characteristic');
  xlabel('Field current, A');
  ylabel('Line voc, V, V, Isc, A');
  legend ('voc','isc', 'air gap line',2);
```

C.37 MATLAB CODE FOR FIELD CURRENT VERSUS SYNCHRONOUS REACTANCE PLOT OF SYNCHRONOUS MACHINE

```
% plot of synchronous reactance versus field current of synchronous machine
clear;clc;
vc=[0 17.39 19.3 21.12 22.78 24.39 25.84 27.13 28.45 29.66 30.78 31.65 32.52 33.66 34.08 34.43 34.78
34.95]*1000;
iff=[0 350 400 450 500 550 600 650 700 750 800 850 900 1000 1050 1100 1150 1200];
isc=8960; ifc=500; isc=isc/ifc*iff;
xs=[];
for i = 2:length(iff)
  xs=[xs;vc(i)/sqrt(3)/isc(i)];
end
xs=[xs(1);xs];
plot(0,0,iff,xs);
grid;
title('Synchronous reactance');
xlabel('Field current in A');
```

```
ylabel(‘xs in ohm’);
MA
```

C.38 MATLAB CODE TO PLOT THE TORQUE-SPEED CURVE OF THE INDUCTION MOTOR

```
r1 = input(‘Enter the stator resistance’);
x1 = input(‘Enter the stator reactance’);
r2 = input(‘Enter the rotor resistance’);
x2 = input(‘Enter the rotor reactance’);
xm = input(‘Enter the Magnetisation branch reactance’);
v_l = input(‘Enter the line voltage’);
v_phase = v_l / sqrt(3);
n_sync = input(‘Enter the synchronous speed in rpm’);
w_sync = input(‘Enter the synchronous speed in rad/s’);
v_th = v_phase * ( xm / sqrt(r1^2 + (x1 + xm)^2) );
z_th = ((j*xm) * (r1 + j*x1)) / (r1 + j*(x1 + xm));
r_th = real(z_th);
x_th = imag(z_th);
s = (0:1:50) / 50; % Slip
s(1) = 0.001;
nm = (1 - s) * n_sync;
for ii = 1:51
t_ind(ii) = (3 * v_th^2 * r2 / s(ii)) / ...
(w_sync * ((r_th + r2/s(ii))^2 + (x_th + x2)^2) );
end
figure(1);
plot(nm,t_ind,’b-”,LineWidth’,2.0);
xlabel(‘\bf\itn_{m}’);
ylabel(‘\bf\tau_{ind}’);
title (‘\bfInduction Motor Torque-Speed Characteristic’);
grid on;
```

C. 39 MATLAB CODE TO PLOT OF THE INDUCED TORQUE, POWER CONVERTED, POWER OUT, AND EFFICIENCY OF THE INDUCTION MOTOR

```
r1 = input(‘Enter the stator resistance’);
x1 = input(‘Enter the stator reactance’);
r2 = input(‘Enter the rotor resistance’);
x2 = input(‘Enter the rotor reactance’);
xm = input(‘Enter the Magnetisation branch reactance’);
v_l=input(‘Enter the line voltage’);
v_phase = 440 / sqrt(3);
n_sync = input(‘Enter the synchronous speed in rpm’);
w_sync = input(‘Enter the synchronous speed in rad/s’);
```

```
p_mech = input(‘Enter the mechanical losses’);
p_core = input(‘Enter the core losses’);
p_misc = input(‘Enter the miscellaneous losses’);
v_th = v_phase * ( xm / sqrt(r1^2 + (x1 + xm)^2) );
z_th = ((j*xm) * (r1 + j*x1)) / (r1 + j*(x1 + xm));
r_th = real(z_th);
x_th = imag(z_th);
s = (0:0.001:0.1);
s(1) = 0.001;
nm = (1 - s) * n_sync; % Mechanical speed
wm = nm * 2*pi/60; % Mechanical speed
for ii = 1:length(s)
t_ind(ii) = (3 * v_th^2 * r2 / s(ii)) / ...
(w_sync * ((r_th + r2/s(ii))^2 + (x_th + x2)^2) );
p_conv(ii) = t_ind(ii) * wm(ii);
p_out(ii) = p_conv(ii) - p_mech - p_core - p_misc;
zf = 1 / ( 1/(j*xm) + 1/(r2/s(ii)+j*x2) );
ia = v_phase / ( r1 + j*x1 + zf );
p_in(ii) = 3 * v_phase * abs(ia) * cos(atan(imag(ia)/real(ia)));
eff(ii) = p_out(ii) / p_in(ii) * 100;
end
figure(1);
plot(nm,t_ind,’b-’,’LineWidth’,2.0);
xlabel(‘\bf\itn_{m} \rm\bf(r/min)’);
ylabel(‘\bf\tau_{ind} \rm\bf(N-m)’);
title (‘\bfInduced Torque versus Speed’);
grid on;
figure(2);
plot(nm,p_conv/1000,’b-’,’LineWidth’,2.0);
xlabel(‘\bf\itn_{m} \rm\bf(r/min)’);
ylabel(‘\bf\itP\rm\bf_{conv} (kW)’);
title (‘\bfPower Converted versus Speed’);
grid on;
figure(3);
plot(nm,p_out/1000,’b-’,’LineWidth’,2.0);
xlabel(‘\bf\itn_{m} \rm\bf(r/min)’);
ylabel(‘\bf\itP\rm\bf_{out} (kW)’);
title (‘\bfOutput Power versus Speed’);
axis([900 1000 0 160]);
grid on;
figure(4);
plot(nm,eff,’b-’,’LineWidth’,2.0);
xlabel(‘\bf\itn_{m} \rm\bf(r/min)’);
ylabel(‘\bf\eta (%)’);
title (‘\bfEfficiency versus Speed’);
grid on;
```

C. 40 MATLAB CODE TO PLOT OF THE OUTPUT POWER VERSUS SPEED CURVE OF THE INDUCTION MOTOR

```
r1 = input(‘Enter the stator resistance’);
x1 = input(‘Enter the stator reactance’);
r2 = input(‘Enter the rotor resistance’);
x2 = input(‘Enter the rotor reactance’);
xm = input(‘Enter the Magnetisation branch reactance’);
v_l=input(‘Enter the line voltage’);
v_phase = 208 / sqrt(3);
n_sync = input(‘Enter the synchronous speed in rpm’);
w_sync = input(‘Enter the synchronous speed in rad/s’);
v_th = v_phase * ( xm / sqrt(r1^2 + (x1 + xm)^2) );
z_th = ((j*xm) * (r1 + j*x1)) / (r1 + j*(x1 + xm));
r_th = real(z_th);
x_th = imag(z_th);
s = (0:1:50) / 50; % Slip
s(1) = 0.001;
nm = (1 - s) * n_sync; % Mechanical speed (r/min)
wm = (1 - s) * w_sync; % Mechanical speed (rad/s)
for ii = 1:51
t_ind(ii) = (3 * v_th^2 * r2 / s(ii)) / ...
(w_sync * ((r_th + r2/s(ii))^2 + (x_th + x2)^2) );
p_out(ii) = t_ind(ii) * wm(ii);
end
figure(1);
plot(nm,p_out/1000,‘k-’,‘LineWidth’,2.0);
xlabel(‘\bf\itn_{m} \rm\bf(r/min)’);
ylabel(‘\bf\itP_{OUT} \rm\bf(kW)’);
title (‘\bfInduction Motor Ouput Power versus Speed’);
grid on;
```

C. 41 MATLAB CODE TO CALCULATE AND PLOT THE TERMINAL VOLTAGE OF A SYNCHRONOUS GENERATOR AS A FUNCTION OF LOAD FOR POWER FACTORS OF 0.8 LAGGING, 1.0, AND 0.8 LEADING

```
EA = input(‘Enter the generator voltage’);
I = 0:2.406:240.6;
R = input(‘Enter the resistance value’);
X = input(‘Enter the reactance value’);
VP_lag = sqrt( EA^2 - (X.*I.*0.8 - R.*I.*0.6).^2 )- R.*I.*0.8 - X.*I.*0.6;
VT_lag = VP_lag .* sqrt(3);
VP_lead = sqrt( EA^2 - (X.*I.*0.8 + R.*I.*0.6).^2 )- R.*I.*0.8 + X.*I.*0.6;
VT_lead = VP_lead .* sqrt(3);
VP_unity = sqrt( EA^2 - (X.*I).^2 );
```

```
VT_unity = VP_unity .* sqrt(3);
plot(I,abs(VT_lag),'b-','LineWidth',2.0);
hold on;
plot(I,abs(VT_unity),'k--','LineWidth',2.0);
plot(I,abs(VT_lead),'r-.','LineWidth',2.0);
title ('\bfTerminal Voltage Versus Load');
xlabel ('\bfLoad (A)');
ylabel ('\bfTerminal Voltage (V)');
legend('0.8 PF lagging','1.0 PF','0.8 PF leading');
axis([0 260 300 540]);
grid on;
hold off;
```

C. 42 MATLAB CODE TO PLOT THE POWER OUTPUT OF A % SYNCHRONOUS GENERATOR AS A FUNCTION OF THE TORQUE ANGLE

```
delta = (0:1:90);
Pout = 561 .* sin(delta * pi/180);
figure(1)
plot(delta,Pout,'LineWidth',2.0);
title ('\bfOutput power vs torque angle \delta');
xlabel ('\bfTorque angle \delta (deg)');
ylabel ('\bf\itP_{OUT} \rm\bf(kW)');
grid on;
```

C.43 MATLAB CODE FOR EX.11.1

```
clear;clc;
p=input('Enter the number of poles');
t=input('Enter the number of teeths');
rt=input('Enter the number of teeth of the rotor');
ps=p*t
step_angle=((ps-rt)/(ps*rt))*360
```

C.44 MATLAB CODE FOR EX.11.2

```
clear;clc;
sa=input('Enter the step-angle');
pr=360/(3*sa)
display ('IF ps>pr ')
display('\n')
ps=(360*pr)/(360-(pr*sa))
display ('IF pr>ps ')
display('\n')
ps=(360*pr)/(360+(pr*sa))
```

C.45 MATLAB CODE FOR EX.11.3

```
clear;clc;
f=input('Enter the frequency');
p=input('enter the number of poles');
hl=input('Enter the hysteresis loss');
ht=hl/(2*pi);
fprintf('\n Hysteresis torque = %f', ht)
po=((2*pi*(120*f)/p)/60)*ht;
fprintf('\n Power output = %f', po)
el=225*hl;
fprintf('\n Energy losst = %f', el)
pd=el/60;
fprintf ('\n Power dissipated = %f', pd)
```

APPENDIX D

GATE QUESTIONS AND SOLUTIONS

D.1 A single-phase transformer has a turns ratio of 1:2 and is connected to a purely resistive load as shown in Fig.D1. The magnetising current is 1A and the secondary current is 1A. If the core loss and the leakage reactance are neglected, what is the primary current. [GATE–2010]

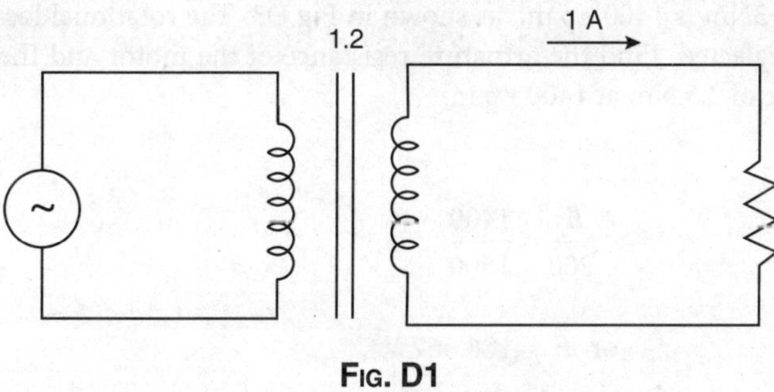

FIG. D1

Solution

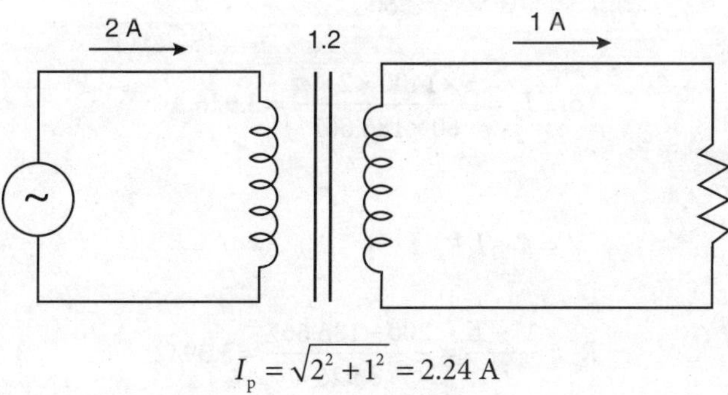

$$I_p = \sqrt{2^2 + 1^2} = 2.24 \text{ A}$$

D.2 A separately excited DC machine is coupled to a 50Hz, three phase, 4-pole induction motor as shown in Fig.D2. The DC machine is energised first and the machine rotates at 1600 r.p.m. Subsequently, the induction motor is also connected to a 50Hz, three-phase source, the phase sequence being consistent with the direction of rotation. What will be the behaviour of the machine in steady state?

[GATE–2010]

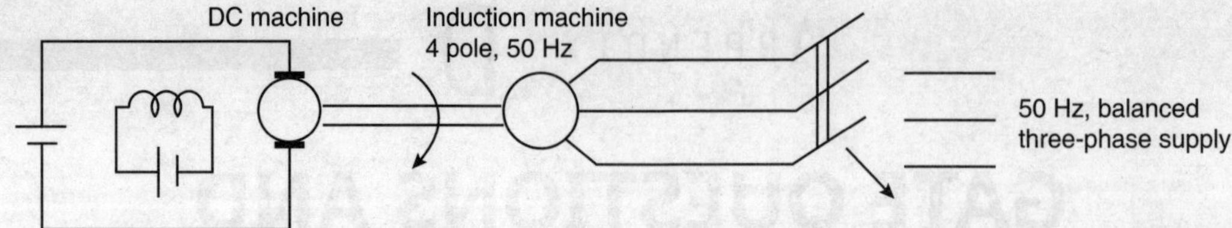

Fig. D2

Solution

The rotor is rotating at 1600 r.p.m.

$$\text{The field is rotating at} \left(\frac{120f}{P}\right) = \left(\frac{120\times 50}{4}\right) = 1500 \text{ r.p.m.}$$

So, the slip will be negative and hence the motor will act as an induction generator.

D.3 A separately excited DC motor runs at 1500 r.p.m. under no-load with 200 V applied to the armature. The field voltage is maintained at its rated value. The speed of the motor when it delivers a torque of 5Nm is 1400 r.p.m., as shown in Fig.D3. The rotational losses and the armature reaction are neglected. Find the armature resistance of the motor and the armature voltage to deliver a torque of 2.5 Nm at 1400 r.p.m. [GATE–2010]

Solution

$$\frac{E_1}{200} = \frac{1400}{1500}$$

$$\text{or } E_1 = 186.667 \text{ V}$$

$$\text{Again, } E_1 I_a = \frac{5\times 1400\times 2\times \pi}{60}$$

$$\text{or, } I_a = \frac{5\times 1400\times 2\times \pi}{60\times 186.667} = 3.926 \text{ A}$$

We know that,

$$V = E + I_a R_a$$

$$R_a = \frac{V-E}{I_a} = \frac{200-186.667}{3.926} = 3.39\ \Omega$$

At a torque of 2.5 Nm

$$E_1 I_a = \frac{2.5\times 1400\times 2\times \pi}{60}$$

$$\text{or, } I_a = \frac{2.5\times 1400\times 2\times \pi}{60\times 186.667} = 1.963 \text{ A}$$

$$V = E + I_a R_a$$

$$= 186.667 + 1.963 \times 3.39 = 193.3 \text{ V}$$

D.4 A single-phase air-core transformer, fed from a rated sinusoidal supply, is operating at no-load. What will be the waveform of a steady-state magnetising current drawn by the transformer from the supply? [GATE–2011]

Solution

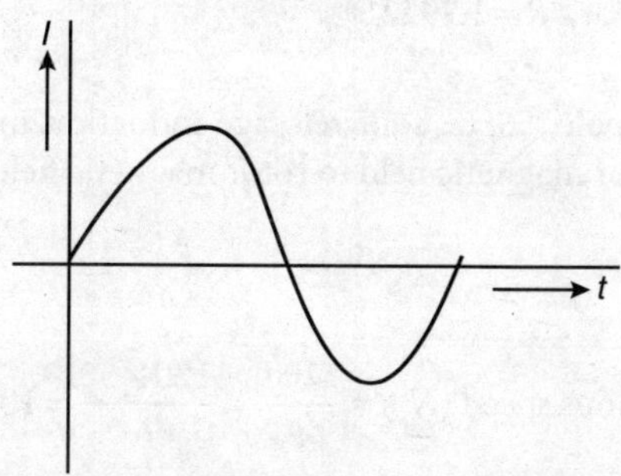

There will be no saturation effect for an air-core transformer.

D.5 A 220 V, DC shunt motor is operating at a speed of 1440 rpm. The armature resistance is 1.0 Ω and armature current is 10A. If the excitation of the machine is reduced by 10%, what is the extra resistance to be put in the armature circuit to maintain the same speed and torque? [GATE–2011]

Solution

$$I_{a1} = 10\,A \text{ (given)}$$

At I_{a1} let the flux be φ_1
Since the flux is decreased by 10%,

$$\varphi_2 = 0.9\varphi_1$$

Since the torque is constant

$$I_{a1}\varphi_1 = I_{a2}\varphi_2$$

$$\text{or } I_{a2} = I_{a1} \times \frac{\varphi_1}{\varphi_2} = \frac{10}{0.9} = 11.11 \text{ A}$$

Let the value of the extra resistance be R.
Then,

$$E_{b1} = 220 - I_{a1} \times R_a = 220 - 10 \times 1 = 210 \text{ V}$$

$$E_{b2} = 220 - I_{a1} \times (R_a + R)$$

Again,

$$\frac{N_1}{N_2}=\frac{E_{b1}}{E_{b2}}\times\frac{\varphi_2}{\varphi_1}$$

$$\text{or, } 1=\frac{210}{220-11.11\times(1+R)}\times\frac{0.9\varphi_1}{\varphi_1}$$

$$\text{or, } R=1.79\,\Omega$$

D.6 A three-phase, 440V, 6-pole, 50Hz, squirrel-cage induction motor is running at a slip of 5%. What is the speed of stator magnetic field to rotor magnetic field and speed of rotor with respect to stator magnetic field? [GATE–2011]

Solution

$$\text{Synchronous speed } (N_s)=\frac{120f}{P}=\frac{120\times 50}{6}=1000 \text{ r.p.m.}$$

$$\text{Rotor speed } (N_r)=1000-0.05\times 1000=950 \text{ r.p.m.}$$

Stator magnetic field speed = N_s = 1000 r.p.m.
Rotor magnetic field speed = N_s = 1000 r.p.m.
So, the relative speed between stator and rotor magnetic fields is zero
Rotor speed with respect to stator magnetic field is = 950 – 1000 = – 50 r.p.m.

D.7 The direct axis and quadrature axis reactances of a salient-pole alternator are 1.2 p.u and 1.0 p.u respectively. The armature resistance is negligible. If this alternator is delivering rated kVA at unity power factor and at rated voltage, then what will be its power angle? [GATE–2011]

Solution

$$\tan\delta=\frac{1\times(1\times 1+0)}{1+1\times 0}=1$$

$$\delta=1$$

D.8 The slip of an induction motor normally does not depends on
Rotor speed
Synchronous speed
Shaft torque
Core loss component [GATE–2012]

Solution

$$\text{Slip}=\frac{N_s-N_r}{N_s}$$

From the equation, the slip depends on rotor speed and synchronous speed. It also depends on torque. If torque increases, rotor speed decreases. Thus, it does not depend on core loss component.

D.9 A 220V, 15 kW, 1000 r.p.m. DC shunt motor has an armature resistance of 0.25Ω. The rated line current is 68A and the rated field current is 2.2 A. What is the change in the field flux needed to obtain a speed of 1600 r.p.m., while drawing a line current of 52.8 A and a field current of 1.8A. [GATE–2012]

Solution

$$I_{a1} = 68 - 2.2 = 65.8 \text{ A}$$

and

$$I_{a2} = 52.8 - 1.8 = 51 \text{ A}$$

So,

$$E_{b1} = 220 - 65.8 \times 0.25 = 203.55 \text{ V}$$

$$E_{b2} = 220 - 51 \times 0.25 = 207.25 \text{ V}$$

Then,

$$\frac{N_1}{N_2} = \frac{E_{b1}}{E_{b2}} \times \frac{\varphi_2}{\varphi_1}$$

$$\text{or, } \frac{\varphi_2}{\varphi_1} = \frac{N_1}{N_2} \times \frac{E_{b2}}{E_{b1}} = \frac{1000}{1600} \times \frac{207.25}{203.55} = 0.6363$$

$$\text{\% decrease} = \frac{\varphi_1 - \varphi_2}{\varphi_2} \times 100 = \left(1 - \frac{\varphi_2}{\varphi_1}\right) \times 100 = 36.37\ \%$$

D.10 A single-phase 10 KVA, 50 Hz transformer with 1 KV primary winding draws 0.5 A and 55 W at rated voltage and frequency on no-load. A second transformer has a core with all its linear dimensions $\sqrt{2}$ times the corresponding dimensions of the first transformer. The primary winding of both the transformers have the same number of turns. If a rated voltage of 2 KV at 50 Hz is applied to the primary of the second transformer, then what will be the no-load current and power respectively? [GATE–2012]

Solution

$$P_{c1} = \left(\sqrt{2}\right)^3 \times 55 = 155 \text{ W}$$

$$P_{c2} = \left(\sqrt{2}\right)^3$$

Then the core-loss component will be

$$I_{c1} = \frac{55}{1000} = 0.055 \text{ A}$$

$$I_{c2} = 2\sqrt{2} \times I_{C1} = 2 \times \sqrt{2} \times 0.055 = 0.155 \text{ A}$$

From magnetising component,

$$I_{p1} = \sqrt{0.5^2 - 0.055^2} = 0.4969 \text{ A}$$

$$\varphi_{m1} = \frac{1000}{\sqrt{2}\times\pi\times f\times N_1}$$

Similarly,

$$\varphi_{m2} = \frac{2000}{\sqrt{2}\times\pi\times f\times N_2}$$

So we can write,

$$\varphi_{m2} = 2\times\varphi_{m1}$$

Again,

$$\varphi_{m1} = \frac{I_{p1}\times N_1}{(\text{Reluctance})_1}$$

Similarly,

$$\varphi_{m2} = \frac{I_{p2}\times N_2}{(\text{Reluctance})_2}$$

But, $\varphi_{m2} = 2\times\varphi_{m1}$
So,

$$\varphi_{m2} = 2\times\frac{I_{p1}\times N_1}{(\text{Reluctance})_1}$$

So,

$$I_{p2} = \frac{(\text{Reluctance})_2}{(\text{Reluctance})_1}\times 2\times I_{p1} = \sqrt{2}\times 0.4969 = 0.702\text{ A}$$

D.11 The locked rotor current in a three-phase, star-connected 15 kW, 4-pole, 230V, 50 Hz, induction motor at rated condition is 50A. Neglecting losses and magnetising current, what will be the locked rotor line current drawn when the motor is connected to a 236V, 57 Hz supply. [GATE–2012]

Solution

$$I_1 \propto \frac{E_1}{f_1} \text{ and } I_2 \propto \frac{E_2}{f_2}$$

So,

$$\frac{I_2}{I_1} = \frac{E_2}{f_2}\times\frac{f_1}{E_1} = \frac{236}{57}\times\frac{50}{230}$$

$$I_2 = 50\times\frac{236}{57}\times\frac{50}{230} = 45\text{ A}$$

D.12 The slip of an induction motor does not normally depend on

i) Rotor speed ii) Synchronous speed
iii) Shaft torque iv) Core loss component [GATE–2012]

Solution

The answer is (iv)

D.13 Leakage flux in an induction motor is

i) flux that leaks through the machine
ii) flux that leaks both stator and rotor winding
iii) flux that leaks none of the winding
iv) flux that leaks the stator winding or the rotor winding but not both. [GATE–2013]

Solution

The answer is (iv)

D.14 Assume an ideal transformer. What will be the Thevenin`s equivalent voltage and impedance as seen from the terminal x and y for the circuit in Fig.D.14? [GATE–2014]

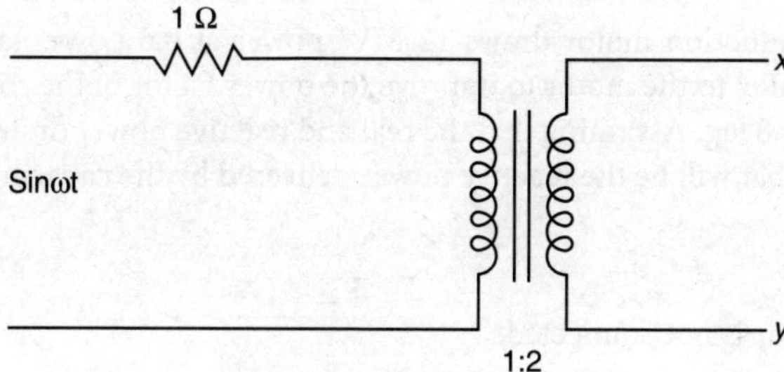

FIG. D.14

Solution

$$\frac{V_{in}}{1} = \frac{V_{xy}}{2}$$

$$\text{or, } V_{xy} = 2V_{in} = 2\sin\omega t$$

$$R_{xy} = R_{th} = 1\times\left(\frac{2}{1}\right)^2 = 4\,\Omega$$

D.15 A single-phase, 50 KVA, 1000/100 V, two-winding transformer is connected as an autotransformer, as shown in Fig.D15. What will be the KVA rating of the autotransformer? [GATE–2014]

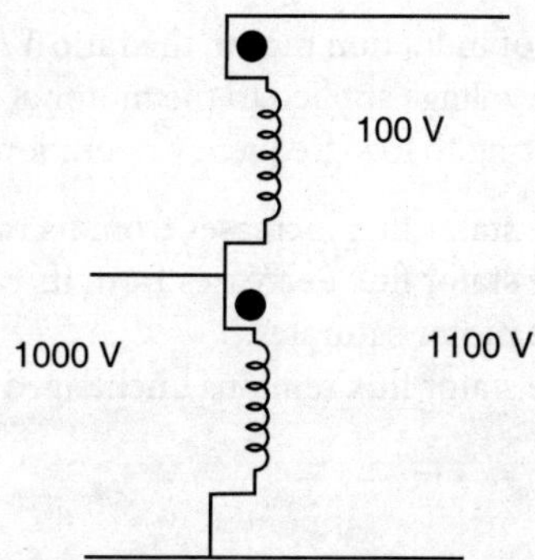

Solution

$$I_2 = \frac{50 \times 10^3}{100} = 500 \text{ A}$$

$$\text{Thus, the KVA rating} = 1100 \times 500 = 550 \text{ KVA}$$

D.16 A three-phase, 4-pole self-excited induction generator is feeding power to a load at a frequency of f_1. If the load is partially removed, the frequency becomes f_2. If the speed of the generator is maintained at 1500 r.p.m. in both the cases, what is the relation between f_1 and f_2? [GATE–2014]

Solution
If the load is removed partially, the speed will increase, as also will the frequency f_2. So, $f_2 > f_1$ and $f_1 f_2 < 50$ Hz.

D.17 A single-phase induction motor draws 12 MW power at 0.6 power factor lag. A capacitor is connected in parallel to the motor to improve the power factor of the combination of the motor and capacitor to 0.8 lag. Assuming that the real and reactive power drawn by the motor remains same as before, what will be the reactive power delivered by the capacitor in MVAR? [GATE–2014]

Solution
When the capacitor is not connected,

$$S_1 = \frac{12 \times 10^6}{0.6} = 20 \text{ MVA}$$

$$\text{Reactive power } (Q_1) = \sqrt{20^2 - 12^2} = 16 \text{ MVAR}$$

With the capacitor,

$$S_2 = \frac{12 \times 10^6}{0.8} = 15 \text{ MVA}$$

$$\text{Reactive power } (Q_2) = \sqrt{15^2 - 12^2} = 9 \text{ MVAR}$$

Since the motor should draw the same reactive power, the MVAR of the capacitor is 16 – 9 = 7 MVAR.

D.18 In a constant *V*/*f* control of induction motor, the ratio *V*/*f* is maintained constant from 0 to base frequency, where *V* is the voltage applied to the motor at fundamental frequency *f*. Which of the following statements relating to low-frequency operation of the motor is true? [GATE–2014]

i) At low frequency, the stator flux increases from its rated value.
ii) At low frequency, the stator flux decreases from its rated value.
iii) At low frequency, the motor saturates.
v) At low frequency, the stator flux remains unchanged at its rated value.

Solution

During constant *V*/*f* control, at low-frequency, the voltage applied to the induction motor is low and hence the stator flux also decreases from its rated value.

D.19 A 250V DC shunt machine has armature circuit resistance of 0.6Ω and field circuit resistance of 125Ω. The machine is connected to 250V supply main. The machine is operated as a generator and then as a motor separately. The line current of the machine in both the cases is 50 A. What is the ratio of the speed as a generator to the speed as a motor? [GATE–2014]

Solution

On operating as a motor,

$$I_f = \frac{250}{125} = 2\text{ A}$$

$$I_a = 50 - 2 = 48\text{ A}$$

$$E_m = V - I_a R_a$$

$$= 250 - 48 \times 0.6 = 221.2\text{ V}$$

On operating as generator,

$$I_f = \frac{250}{125} = 2\text{ A}$$

$$I_a = 50 + 2 = 52\text{ A}$$

$$E_g = V + I_a R_a$$

$$= 250 + 52 \times 0.6 = 281.2\text{ V}$$

Let N_m and N_g be the speed of the machine under motoring action and generating action respectively.

$$\frac{N_g}{N_m} = \frac{E_g}{E_m} = \frac{281.2}{221.2} = 1.27$$

D.20 A 220V, three-phase, 4-pole, 50 Hz induction motor of wound rotor type is supplied at rated voltage and frequency. The resistance, magnetising reactance, and core loss are negligible. The maximum torque produced by the rotor is 225% of full-load torque and it occurs at 15% slip. The actual rotor resistance is 0.03Ω/phase. What is the value of the external resistance (in Ohm) which must be inserted in a rotor phase if the maximum torque is to occur at start? [GATE–2015]

Solution

$$S_m = \frac{r_2}{x_2}$$

$$0.15 = \frac{0.03}{x_2}$$

$$x_2 = 0.2\,\Omega$$

$$\text{Again, } \frac{T_{st}}{T_m} = \frac{2}{\frac{1}{S_m} + S_m}$$

Since, $T_{st} = T_m$

$$\frac{2}{\frac{1}{S_m} + S_m} = 1$$

Solving, we get $S_m = 1$
Thus,

$$r_2 = x_2 = 0.2\,\Omega$$

So, the external resistance = 0.2 – 0.03 = 0.17 Ω/phase.

D.21 Two three-phase transformers are realised using single-phase transformers, as shown in Fig. D21. What is the phase difference in degrees between the voltage V_1 and V_2? [GATE–2015]

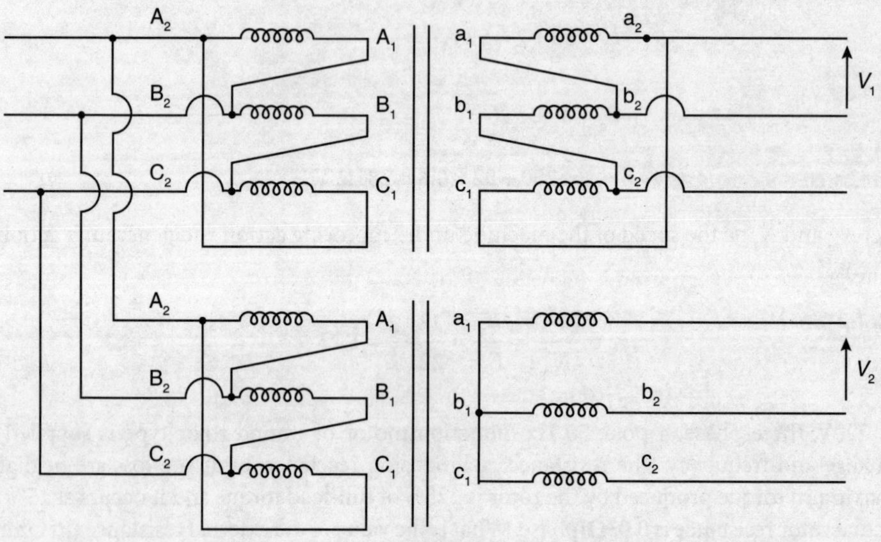

Fig. D21

Solution

As seen from Fig. D21, the secondary of the upper transformer is delta-connected and the secondary of the lower transformer is star-connected. So the phase angle between the delta voltage and star voltage is 30^0.

D.22 A balanced (positive sequence) three-phase AC voltage source is connected to a balanced, star-connected load through a star–delta transformer as shown in Fig.D22. The line-to-line voltage rating is 230V on the star side and 115 V on the delta side. If the magnetising current is neglected and I_s = 100 A, then what is the value of I_p? [GATE–2015]

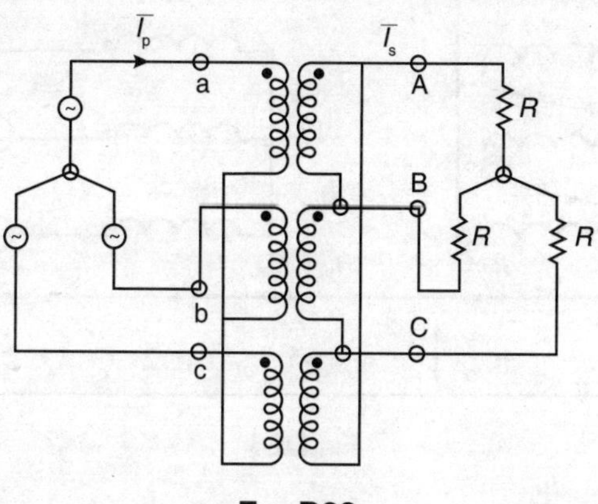

Fig. D22

Solution

$$\left(\frac{230}{\sqrt{3}}\angle -30\right) I_p = 115 \times \frac{110}{\sqrt{3}}$$

$$(115 - j66.39) I_p = 7303.48$$

$$I_p = 47.63 + j27.49 = 54.99\angle 30 \text{ A}$$

D.23 A three-phase, 11kV, 50 Hz, two-pole star-connected, cylindrical rotor synchronous motor is connected to an 11 kV, 50 Hz source. Its synchronous reactance is 50Ω per phase and its stator resistance is negligible. The motor has a constant field excitation. At a particular load torque, its stator current is 100A at unity power factor. If the load torque is increased so that the stator current is 120A, then what is the value of the load angle? [GATE–2015]

Solution

$$E = \frac{11}{\sqrt{3}} \times 10^3 - j100 \times 50 = 6350 - j5000 \text{ V} = 8082.23\angle -38.22 \text{ V}$$

Again,

$$(120 \times 50)^2 = 8082.23^2 + \left(\frac{11}{\sqrt{3}} \times 10^3\right)^2 - 2 \times 8082.23 \times \frac{11}{\sqrt{3}} \times 10^3 \times \cos\delta$$

$$\delta = \cos^{-1}\left[\frac{8082.23^2 + \left(\frac{11}{\sqrt{3}} \times 10^3\right)^2 - (120 \times 50)^2}{2 \times 8082.23 \times \frac{11}{\sqrt{3}} \times 10^3}\right] = 47.27$$

D.24 A three-phase transformer rated for 33kV/11kV is connected in delta–star as shown in Fig. D.24. The current transformers (CTs) on low- and high-voltage sides have a ratio of 500/5. Find the current i_1 and i_2 if the fault current is 300 A as indicated in the Fig.D24. [GATE–2015]

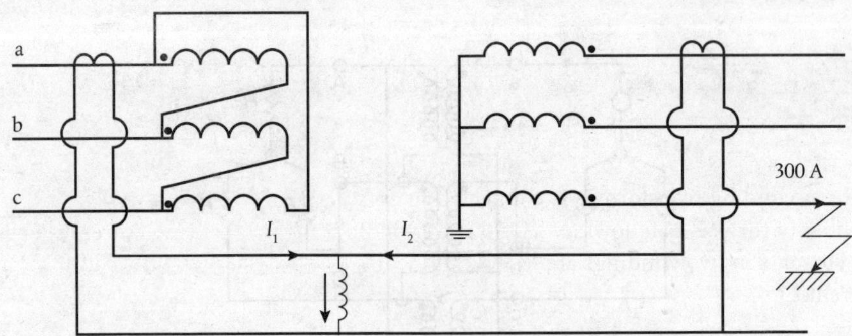

FIG.D24

Solution

Since the entire current will flow through the fault, i_2 will be zero

In a transformer, $(KVA)_{primary} = (KVA)_{secondary}$

$$\sqrt{3} \times 3300 \times I_L = \sqrt{3} \times 1100 \times \left(300 \times \frac{5}{500}\right)$$

$$\text{or, } i_L = 1\text{A}$$

$$\text{Again, } i_L = \sqrt{3}\, I_p = \sqrt{3}\, i_1$$

$$\text{So, } i_1 = \frac{1}{\sqrt{3}}\text{ A}$$

D.25 With an armature voltage of 100V and rated field winding voltage, the speed of a separately excited DC motor driving a fan is 1000 r.p.m. and its armature current is 10 A. The armature resistance is 1Ω. The load torque of the fan is proportional to the square of the rotor speed. Neglecting the rotational losses, what is the value of the armature voltage which will reduce the rotor speed to 500 r.p.m.? [GATE–2015]

Solution

$$E_1 = 100 - 10 \times 1 = 90\text{ V}$$

$$\text{Again, } \frac{E_2}{E_1} = \frac{N_2}{N_1}$$

$$\text{or, } E_2 = \frac{N_2}{N_1} \times E_1 = \frac{500}{1000} \times 90 = 45\text{ V}$$

$$\text{Again, } \frac{I_{a2}}{I_{a1}} = \left(\frac{N_2}{N_1}\right)^2$$

$$\text{or, } I_{a2} = \left(\frac{500}{1000}\right)^2 \times 10 = 2.5 \text{ A}$$

$$\text{So, } V = 45 + 2.5 \times 1 = 47.5 \text{ V}$$

D.26 A three-winding transformer is connected to an AC voltage source as shown in Fig.D26. The number of turns is as follows: N_1 = 100 and N_2 = 50. If the magnetising current is neglected and the currents in two windings are $I_2 = 2\angle 30$ A and $I_3 = 2\angle 150$ A, then what is the value of the current I_1? [GATE–2015]

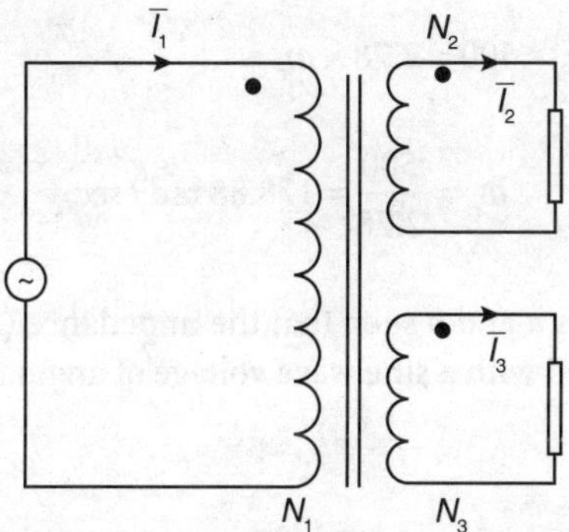

FIG.D26

Solution

For a three-winding transformer,

$$N_1 I_1 = N_2 I_2 + N_3 I_3$$

$$100 \times I_1 = 50 \times 2\angle 30 + 50 \times 2\angle 150$$

$$100 \times I_1 = 50 \times (1.73 + j1) + 50 \times (-1.73 + j1)$$

$$1000 \times I_1 = 0 + j100$$

$$I_1 = j1 \text{ A} = 1\angle 90 \text{ A}$$

D.27 A four-pole separately excited wave-wound DC machine with negligible armature resistance is rated for 230V and 5kW at a speed of 1200 r.p.m. If the same armature coils are reconnected to form a lap winding, what is the rated voltage (in volts) and power (in kW) of the reconnected machine at 1200 r.p.m., if the field circuit is left unchanged? [GATE–2015]

Solution

For wave winding, $A = 2$ and for lap winding $A = P = 4$.

$$\text{So the rated voltage} = \frac{230 \times 2}{4} = 115\ \text{V}$$

Thus, the voltage is reduced to half since the number of parallel paths is increased from 2 to 4 while the power remains the same.

D.28 A shunt-connected DC motor operates at its rated terminal voltage. Its no-load speed is 200 rad/sec. At its rated torque of 500 Nm, its speed is 200 rad/sec. The motor is used to directly drive a load whose load torque, T_L, depends on its rotational speed (in rad/sec) such that

$T_L = 2.78 \times \omega_r$

What will be the steady-state speed of the motor if the losses are neglected? [GATE–2015]

Solution

Since at steady-state, load torque becomes equal to the motor torque

$$500 = 2.78 \times \omega_r$$

$$\omega_r = \frac{500}{2.78} = 178.88\ \text{rad / sec}$$

D.29 Find the transformer ratios a and b such that the impedance (Z_{in}) is resistive and equal to 2.5 Ω when the network is excited with a sine wave voltage of angular frequency 500 rad/sec. Fig.D29. [GATE–2015]

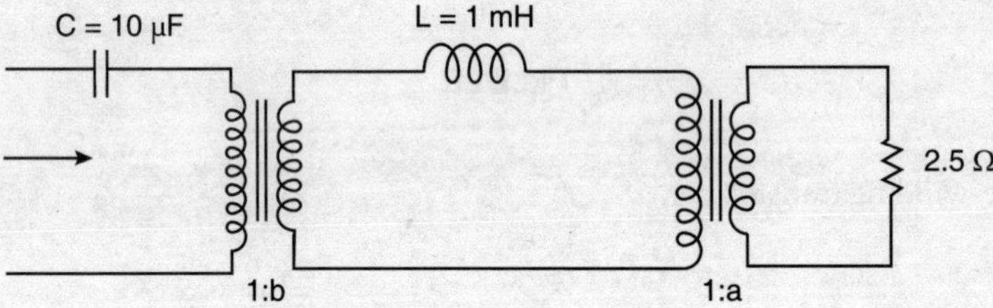

Solution

$$X_c = j20\ \Omega \ \text{ and } \ X_L = j5\ \Omega$$

So, the transformer circuit

$$-20 + \frac{5}{b^2} = 0$$

$$\text{or, } b = 0.5$$

Again,

$$Z_{in} = \frac{1}{b^2}\left[\frac{2.5}{a^2} + j5\right] - j20$$

$$\text{or,}\ \frac{2.5}{a^2b^2} = 2.5$$

Substituting the value of b and solving we get,
$a = 2.$

D.30 A three-phase squirrel-cage induction motor has a starting torque of 150% and a maximum torque of 300% with respect to rated torque at rated voltage and rated frequency. Neglecting the stator resistance and losses, find the value of slip for maximum torque. [GATE–2007]

Solution

$$\frac{T_{st}}{T_{fl}} = 1.5$$

$$\frac{T_m}{T_{fl}} = 3$$

$$\frac{T_{st}}{T_m} = \frac{1.5}{3} = 0.5$$

Again,

$$\frac{T_{st}}{T_m} = \frac{2}{s_m + \dfrac{1}{s_m}} = 0.5$$

Solving we get,

$$s_m = 0.2679 = 26.79\ \%$$

INDEX

S

T

U

V

W

Y

Z